Inorganic Thin Films
and
Coatings

Edited by
Cheng Zhang
Nanchun Chen
Jin Hu

Inorganic Thin Films and Coatings

Selected, peer reviewed papers from the
2012 Workshop on
Inorganic Thin Films and Coatings,
July 16-18, 2012, Guilin, China

Edited by

Cheng Zhang, Nanchun Chen and Jin Hu

Trans Tech Publications Ltd
Kreuzstrasse 10
CH-8635 Durnten-Zurich
Switzerland
http://www.ttp.net

Volume 537 of
Key Engineering Materials
ISSN print 1013-9826
ISSN cd 1662-9809
ISSN web 1662-9795

Full text available online at *http://www.scientific.net*

Distributed *worldwide by*

Trans Tech Publications Ltd
Kreuzstrasse 10
CH-8635 Durnten-Zurich
Switzerland

Fax: +41 (44) 922 10 33
e-mail: sales@ttp.net

and in the Americas by

Trans Tech Publications Inc.
PO Box 699, May Street
Enfield, NH 03748
USA

Phone: +1 (603) 632-7377
Fax: +1 (603) 632-5611
e-mail: sales-usa@ttp.net

printed in Germany

Preface

Aiming to promote the interactions and cooperations among the scientists, engineers, technicians and students engaged in inorganic powders, thin films and coatings, 2012 Workshop on Inorganic Powders, Thin Films and Coatings was held from July 16 to 18, 2012, at Guilin, China. This Workshop was co-sponsored by Guilin University of Technology, Kunming University of Science and Technology and Shanghai Insitute of Technology. 162 participants from universities, institutes and enterprises in China attended the Workshop to give more then 170 presentations.

The organizers received 152 manuscripts from the participants. After review process, 140 papers were finally accepted for publication in the Proceedings. The Proceedings is published as two parts and this one, entitled "Inorganic Thin Films and Coatings", is the first part, which includes 67 papers.

The Editors would like to thank the members of the Organizing Committee and the Academic Committee for their efforts and successful preparation of this event.

We would like to thank all colleagues who have devoted much time and effort to organize this meeting.

The efforts of the authors of the manuscripts prepared for the proceedings are also gratefully acknowledged.

Hopefully this proceeding would be beneficial to all the participantsof the worshop and and the readers.

Prof. Cheng Zhang
Shanghai Institute of Technology

Prof. Nanchun Chen
Guilin University of Technology

Prof. Jin Hu
Kunming University of Science and Technology

Table of Contents

Key Engineering Materials Vol. 537 (2013) pp 1-6
© (2013) Trans Tech Publications, Switzerland
doi:10.4028/www.scientific.net/KEM.537.1

Factors Effecting the Fracture Strength of Freestanding Diamond Films

J.H. Song, F.X. Lu*, L.F. Hei, C.M. Li, W.Z. Tang, G.C. Chen

University of Science and Technology Beijing, Beijing 100083, China

songjianh@263.net, fxlu@mater.ustb.edu.cn, lifu_Hei@163.com, chnegmingli@163.com,
wztang@mater.ustb.edu.cn, guangchaochen@mater.ustb.edu.cn

Keywords: Fracture strength; freestanding diamond film; influencing factors; high power dc arcjet

Abstract. As an emerging brand new type of engineering material for a variety of important high technology applications, the deep understanding of the mechanical behavior of freestanding diamond films has become an emergent task of vital importance. Unfortunately the mechanical behavior of this brand new material is not fully understood. Effects that affect the fracture strength are still not very clear except that the fracture strength of freestanding diamond films is only depended on the grain size (film thickness), and is insensitive to the microscopic defects, and the strength is considerably higher when the nucleation side is in tension than that when the growth side is in tension. However, this is not the full story. Based on the experimental date accumulated in USTB (University of Science and Technology Beijing) for high power dc arcjet diamond films, other factors that may affect the fracture strength of freestanding diamond films are discussed in detail. Effects of the quality of the diamond films, impurities (nitrogen, hydrogen, non-diamond carbon etc.), film morphology and texture, on fracture strength are discussed. Effects of the deposition parameters are explained. Advantages for small amount of nitrogen addition, and for the use of higher substrate temperatures in the increase in fracture strength are demonstrated, which have already been applied in the mass production of tool grade freestanding diamond films. It is hoped that the present paper will be helpful for those who wish to understand and use this brand new type of engineering material.

Introduction

Due to the remarkable progresses achieved as a result of the world wide efforts in the past 20 years, CVD diamond film has been accepted as an emerging engineering material to be used in multidiscipline high technology fields, including mechanical, optical, thermal, electronic, and biomedical applications [1-4]. As an engineering material, the mechanical properties of CVD diamond films have become one of the key issues and a serious concern for researchers and designers and end-users. Presently large area high quality freestanding diamond films up to 120mm in diameter and more than 2mm in thickness can be prepared by high power dc arcjet with arc root rotation and gas recycling with materials properties suitable for mechanical, thermal and optical applications [5-7]. Millions of cubic millimeters of freestanding diamond film products are manufactured yearly by this technique even in only one factory in China, and most of the products are for mechanical applications, such as dressers, cutting tool blanks and wire drawing dies and extraordinary wear resistant components. However, up to now, no systematic data have ever been published in concerning the mechanical properties. The existing data in the literature is rather limited and scattered. The mechanical behavior of this brand new engineering material is still not fully understood. The factors that may affect the fracture strength of freesanding diamond films is not very clear except that both the fracture strength and fracture toughness are depended on the film thickness (the grain size), and the loading conditions, both of which show a Hall-Petch type relationship with the film thickess (the grain size), and the measured strength (toughness) is considerably higher when the load is applied from the growth surface (the nucleation surface is under tension) than that when the load is applied from the nucleation surface (the growth surface is under tension) [8, 9]. Therefore it is the main purpose of the present paper to present a general discussion on the factors that may affect the fracture strength of freestanding diamond films, based mainly on the works at USTB by high power dc arc plasma jet with arc root rotation and gas recycling. It is hoped that the present paper may be of help for those who wish to make use of this brand new type of engineering material.

Effect of the diamond film quality on the fracture strength

Field et al. [8] suggested that the only factor that affects the fracture strength of freestanding diamond films is the grain size (the film thickness, due to the columnar grain feature), and is immune from the defects which might be as large as the grain size. This is in general agreement with Lu et al. [9, 10], and is widely accepted by the diamond film community. However, this may not be completely correct. First of all, the published data of Field et al. and Lu et al. were those from the good quality diamond films (or in other words, those prepared at optimized conditions). Now it is a common sense to commercially grade (optimize) the freestanding diamond films in according to their mechanical or physical properties, e.g., as mechanical grade (tool grade), thermal grade, optical grade, and detector (electronic) grade diamond films. The difference in these different grades diamond films in not only the grain size; instead, the major difference is the quality of the diamond films. The optical grade and the detector grade diamond films are of very high quality, with very narrow FWHM (2-3cm^{-1}) of the diamond characteristic Raman peak and a very low level of impurities (just a few ppm or even down to ppb level). Whilst the FWHM of the mechanical grade diamond films may be larger than 4-6cm^{-1}, and the impurity level may be as high as many tens of ppm, or even higher. However, there were no systematic studies in concerning the effect of the film quality on the mechanical properties of the freestanding diamond films. The only reports from Guo et al. [11] suggested that the fracture strength, thermal conductivity, as well as the IR transmission increased with the increase in the quality (as depicted by the weighted ratio of the area of the diamond characteristic Raman peak and the sum of diamond and non-diamond carbon peak) of the diamond films (see Fig.1). Nevertheless the quality factor used by Guo et al. [11] is only a reflection of the crystallinity as revealed by the Raman spectra. There was no indication about the impurities. It is well known that the fracture strength of the lower quality mechanical grade diamond films is considerably higher than the high quality optical grade diamond films. The reason is simply because the average grain size is much smaller [8, 10]. However, the diamond films shown in fig.1 were prepared at similar conditions with similar thickness; therefore their grain size was also similar.

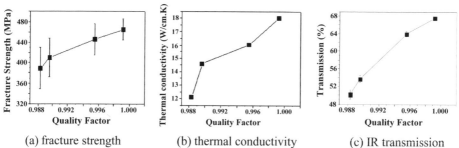

(a) fracture strength   (b) thermal conductivity   (c) IR transmission

Figure 1 Effect of the film quality (the Quality Factor) on the physical properties of the freestanding diamond films prepared by high power dc arcjet [11]

Effect of the impurities

Effect of nitrogen impurity Nitrogen is the most common impurity which could be easily incorporated into the diamond film from the deposition environment due to the use of lower purity source gases, and the leakage through the whole CVD system (e.g. the deposition chamber, the pipe lines, and the vacuum system). Huge amount of publications can be found in the literature; however, most of them were related to the effect of nitrogen in promoting the nucleation and growth of textured diamond films [13-15] and the preparation of nano-crystalline diamond (NCD) or ultra nano-crystalline diamond (UNCD) films [16-18]. No systematic work could be found in regarding the effect of nitrogen on the fracture strength of freestanding diamond films. However, the early works carried out in USTB (University of Science and Technology Beijing) by Lu et al. [19] and Yang et al.

[20] have demonstrated that the fracture strength as well as the dielectric, thermal and optical properties all decreased with the increase in nitrogen added to the H_2-Ar-CH_4 gas stream. Recently, Guo et al. [13] found that the effect of nitrogen on fracture strength of diamond films is complicated. As shown in Fig.2(a), with a relatively large amount of nitrogen, the fracture strength of the freestanding diamond films decreased monotonically with the increase in nitrogen concentration. Whilst, if the nitrogen concentration is small (less than 10ppm), there is a distinct trend that the fracture strength increased with the increase in nitrogen concentration (see Fig.2(b). The reason for this kind of behavior is not clear. However, it become a common sense for the Chinese diamond film companies to specially introduce a small amount of air (75% nitrogen) into the deposition chamber for mass production of tool grade freestanding diamond film products.

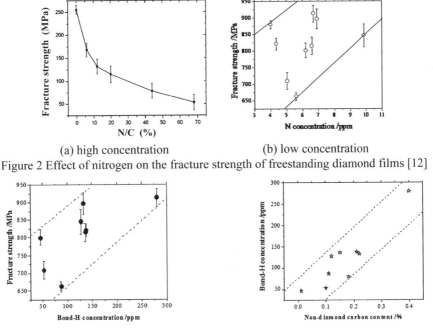

(a) high concentration   (b) low concentration

Figure 2 Effect of nitrogen on the fracture strength of freestanding diamond films [12]

Figure 3 Effect of bonded hydrogen on the fracture strength of freesatnding diamond films [12]

Figure 4 Correlation relationship between non Diamond carbon and the bonded hydrogen

Effect of hydrogen impurity Hydrogen is also the most common impurity in CVD diamond films which is existed at the surface as surface absorbent (for effective termination of the carbon tangling bonds) as well as in the bulk in the diamond lattice which is always associated (bonded) to the non-diamond carbon in the sp2 hybrid bonding configurations and can be experimentally measured from the UV or IR absorption peaks due to the vibration of the C-H bond configurations. The surface hydrogen can be obtained by extracting the bonded hydrogen from the total hydrogen concentration meassured by the nuclear reaction (NRA) methode [12]. The effect of the surface absorbed hydrogen is not clear. However, it was found that the bulk hydrogen did have a distinct influence on the fracture strength of the freestanding diamond films [12]. As shown in fig.3, the fracture strength increased with the amount of bonded hydrogen. The scattering of the data in fig.3 is due to the uncertainties in the measurement of the bonded hydrogen and the statistic nature of the fracture strength data which was the mean of five tested freestanding diamond film samples.

Effect of the non-diamond carbon Non-diamond carbon can also be regarded as impurity, which is closely related (chemically bonded) with the hydrogen atoms. Therefore, there is an approximately linear correlation relationship between the non-diamond carbon content (calculated from the Raman spectra) with that of the bonded hydrogen (see fig.4). As a result, it was found that the fracture strength also increased with the increase in the content of the non-diamond carbon (fig.5) [12].

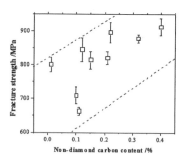

Figure 5 Effect of non-diamond carbon on the fracture strength of freestanding diamond films

Effect of microstructures

Effect of the morphological feature Due to the nature of the columnar grain growth, it is well known that the fracture strength of the freestanding diamond films is considerably higher when nucleation surface is in tension than that when growth surface is in tension [8-10]. Another striking fact is that, as a consequence of the columnar grain growth, the well developed columnar grain structure will result in very low fracture strength [21], simply because the intergranular fracture will become the dominant failure mode (the perfect vertical grain boundaries of the well-developed columnar grains will become the weakest path for the propagation of the cracks), see fig.6.

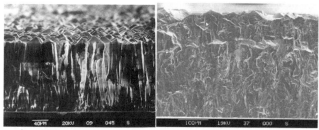

Figure 6 Cross section SEM photographs for low strength (intergranular fracture, left) and high strength (transgranular, right) freestanding diamond films prepared by high power dc arcjet [9, 21]

Effect of diamond film orientations (texture) Based on the existing data obtained at USTB, most of the high strength (tool grade) freestanding diamond films showed a [111]/[110] mixed orientation. With the decrease in the ratio of the intensities of the (111) and (220) X-Ray diffraction peaks $I_{(111)/(110)}$, the fracture strength increases. With the texture evolution from [110] to [111]/[110] and [100], the fracture strength decreases, with the [100] textured diamond film the lowest [21, 22].

Effect of the microscopic defects Because of the high fracture toughness, the fracture strength of freestanding diamond films is generally insensitive to the microscopic defects with dimensions less than the grain size [8, 9]. However, as explained earlier, the impurities (N_2, H_2, non-diamond carbon etc.) do have a pronounced effect on the fracture strength of the freestanding diamond films.

Effect of the deposition parameters

Generally the morphology and texture (orientation), impurities and defects can all be effectively controlled by the properly control of the deposition parameters and the proper design and construction of the CVD system, and the use of the right source gases. In due course, it is possible in certain degree to control the fracture strength of the freestanding diamond films. However, in according to the practice in USTB, it is suggested that, among all the deposition parameters, the substrate temperature may be the most important factor. As shown in fig.7 [9, 12], when the substrate temperature raised from 850°C to 1100°C an increase of 200–300MPa in fracture strength could be realized. The reason for this pronounced increase in fracture strength is possibly due to the increased surface mobility of the carbon atoms at the elevated temperatures, which may lead to the increase in the rigidity (density) of the diamond films. This also has become a common practice in the Chinese diamond film companies in the production of the tool grade freestanding diamond film products. Another important operation is the pretreatment (seeding) of the substrate surface, by which the nucleation density could be effectively controlled, and together with the corresponding optimized deposition parameters, fine grained (high strength) or course grained (high transmission) diamond films could be produced.

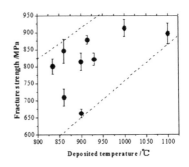

Figure 7 Effect of substrate temperature on the fracture strength of freestanding diamond films

Summary

Based on the accumulated data at USTB for the high power dc arcjet diamond films, the factors that affect the fracture strength of the freestanding diamond films are discussed. Due to the nature of the columnar grain growth, and the relatively high fracture toughness, fracture strength of the freestanding diamond films is generally insensitive to the microscopic defects, and seems to be depended only on the film thickness (grain size). And the fracture strength is considerably higher when the nucleation surface is in tension than that when the growth surface is in tension. However, this is not the complete story. Beside of this, for similar grade diamond films, high quality (by Raman signature) diamond films have higher fracture strength. The effect of impurities is complicated. Nitrogen incorporation in diamond film will generally lower the fracture strength, however, if the concentration is very low (less than 10ppm), there is a distinct trend that the fracture strength increases with the increase in nitrogen concentration. Effect of the surface absorbed hydrogen on the fracture strength is not clear; however the effect of the bonded hydrogen (bulk hydrogen) is obvious. An approximate linear correlation relationship exists between the non-diamond carbon and the bonded hydrogen, and thus the effects of the non-diamond carbon and the bonded hydrogen are similar. Low strength freestanding diamond films generally show a well-developed columnar grain structure and an intergranular dominant fracture surface, whilst in the case of the high strength diamond films, the columnar grain structure is not so obvious, and the fracture surface is transgranular dominant. High strength freestanding diamond films are usually in the mixed [111]/[110] preferred orientation. With the orientation evolution from [110] to [111]/[110] and [100], the fracture strength decreases. Substrate temperature is the most important deposition parameter in affecting the fracture strength of the freestanding diamond films. Substrate pretreatment (seeding) is also very important, together with the corresponding optimized deposition parameters, fine grained (high strength) or course grained (high transmission) diamond films could be produced.

References

[1] N.V. Novikov, Synthesis of superhard materials, Journal of Materials Processing Technology. 161(1-2) (2005) 169-172

[2] A. Gicquel, K. Hassouni, F. Silva, J. Achard, CVD diamond films: from growth to applications, Current Applied Physics. 1(6) (2001) 479-496

[3] Katsuyuki Okada, Plasma-enhanced chemical vapor deposition of nanocrystalline diamond, Science and Technology of Advanced Materials. 8(7-8) (2007) 624-634

[4] X. W. Zhu, D. M. Aslam, CVD diamond thin film technology for MEMS packaging, Diamond & Related Materials, 15 (2006) 254–258

[5] F.X.Lu, G.F. Zhong, J.G. Sun et al. A New Type of DC Arc Plasma Torch for Low Cost Large Area Diamond Deposition. Diamond and related materials, **7/6** (1998) 737-741

[6] F.X. Lu, W. Z. Tang, G.F. Zhong et al. Economical Deposition of Large Area High Quality Diamond Film by High Power DC Arc Plasma Jet Operating in Gas Recycling Mode, Diamond and Related Materials, *9 (9–10)*, (2000) 1655

[7] F.X. Lu, Large Area Optical Grade Freestanding Diamond Films: Deposition, Characterization, and High Technology Application Prosperous. Chinese Surface Engineering, 23 (3), (2010) 1-13

[8] J.E. Field, C.S.J. Pickles, Strength, fracture and friction properties of diamond. Diamond and Related. Materials. 5 (1996) 625-634

[9] L.F. Hei, F.X. Lu, C.M. Li et al., A Review on Mechanical Properties of Freestanding Diamond Films, Advanced materials Rsearch, 490-495 (2012) 3059-3064

[10] F.X. Lu, Z. Jiang, W.Z. Tang, T.B. Huang and J.M. Liu, Accurate measurement of strength and fracture toughness for miniature size thick diamond film samples by three point bending at constant loading rate. Diamond and Related Materials, 10 (2001) 770-774

[11] S.B. Guo, Y.H. Wang, F. Chen, F.X. Lu, W.Z. Tang, Y.M. Tong, Effect of the quality of diamond film deposited by dc arc plasma jet CVD on physical properties, Gongneng Cailiao, 38 (2007) 3802 – 3805

[12] S.B. Guo, Study of the effect of minute impurities on physical properties of diamond films, Ph.D. Thesis, University of Science and Technology Beijing, 2009

[13] H. Chatei, J. Bougdira, M. R emy, et al. Effect of nitrogen concentration on plasma reactivity and diamond growth in a H_2-CH_4-N_2 microwave discharge, Diamond and Related Materials, 6(1) (1997) 107-119

[14] J. Asmussen, J. Mossbrucher, S. Khatami, et al. The effect of nitrogen on the growth, morphology, and crystalline quality of MPACVD diamond films, Diamond and Related Materials, 8(2-5) (1999) 220-225

[15] Y. Avigal, O. Glozman, I. Etsion, et al. [100]-textured diamond films for tribological applications, Diamond and Related Materials, 6(2-4) (1997) 381-385

[16] C.J. Tang, A.J. Neves, Jose´ Gra´cio, A.J.S. Fernandes, M.C. Carmo, A new chemical path for fabrication of nanocrystalline diamond films, Journal of Crystal Growth 310 (2008) 261–265

[17] V. Ralchenko, S. Pimenov, V. Konov et al., Nitrogenated nanocrystalline diamond films: Thermal and optical properties, Diamond & Related Materials 16 (2007) 2067–2073

[18] O.A. Williams, Nanocrystalline diamond, Diamond and Related Materials, 20 (2011) 621-640

[19] F.X. Lu, H.D. Zhang, W.Z. Tang et al., Dielectric property of thick freestanding diamond films by high power arcjet operating at gas recycling mode, Diamond and Related Materials, 13(9), (2004) 1714-1718

[20] J.X. Yang, H.D. Zhang, F.X. Lu et al., Effects of nitrogen addition on morphology and mechanical property of DC arc plasma jet CVD diamond films, Diamond and Related Materials, 13 (2004) 139–144

[21] J.X. Yang, C.M. Li, F.X. Lu, G.C. Chen, W.Z. Tang, Y.M. Tong, Microstructure and fracture strength of different grades of freestanding diamond films deposited by a DC Arc Plasma Jet process, Surface & Coatings Technology 192 (2005) 171– 176

[22] H.D. Zhang, J.Z. Tian, W.Z. Tang, F.X Lu, Correlation Between the Fracture Strength and the Crystal Orientation of thick Freestanding Diamond Films, *Carbon*, 41(3) (2003) 603-606

Key Engineering Materials Vol. 537 (2013) pp 7-11
© (2013) Trans Tech Publications, Switzerland
doi:10.4028/www.scientific.net/KEM.537.7

Tribological Performance of Ceramic Composite Coatings Obtained Through Microarc Oxidation on Ly12 Aluminum Alloy

Shuhua Li[1, a], Yujun Yin[1], Dawei Shen[2], Yuanyuan Zu[3], Changzheng Qu[1]

[1] Shijiazhuang Mechanical Engineering College, Shijiazhuang 050003, China

[2] PLA 62025 Unit, Beijing 102300, China

[3] The Quartermaster Equipment Institute of the General Logistics Department of the PLA, Beijing 100082.China

[a]klddsgz@126.com

Keywords: Aluminium alloy; Micoarc oxidation; Ceramic coating; Tribological performance

Abstract. A dense ceramic oxide coating approximately 30μm thick was prepared on a Ly12 Al alloy by microarc oxidation in an alkali-silicate electrolytic solution. The morphology and microstructure were analyzed by scanning electron microscope and X-ray diffraction. Coating thickness and surface roughness (Ra) were measured after the coating had been synthesized. The tribological performance of the coatings was evaluated using a dry sand abrasion test and a solid particle erosion test. The results show that microarc oxidation coatings consist of the loose superficial layer and the inner dense layer. Both inner layer and out layer are composed of α-Al_2O_3 and γ-Al_2O_3, While the $Al_6Si_2O_3$ phase is observed only in out loose layer. The average of the microhardness of the coating is 2096Hv.

Introduction

Aluminum alloys, which are structurally important materials due to their high specific strength, suffer from moderately poor tribological properties. Surface-modified wear resistant aluminum alloys have been increasingly finding numerous applications in modern times industries[1-2]. A relatively new process called microarc oxidation(MAO) has emerged as a unique technique to produce hard, thick ceramic oxide coatings on different aluminum alloys and some valve metal. Essentially, the MAO process combines electrochemical oxidation with a high-voltage spark treatment. Several studies have been devoted to the synthesis and tribological performance evaluation of ceramic oxide coatings deposited on a large number of Al alloys obtained through the MAO process[3-5].Although several investigations have been carried out, the complete tribological characteristics of MAO coatings high stress levels have not been studied. In the present work, a 30μm thick coating was synthesized on Ly12 Al alloy substrate by means of the MAO process. The evolution of coating surface roughness and thickness after the course of coating deposition was also investigated. Simultaneously, the tribological performance of the MAO coatings has been evaluated. In particular, dry sand abrasive wear tests and solid particle erosive wear tests (with SiO as erodent) have been carried out. These results were utilized to understand the mechanism of coating formation during the MAO process.

Experimental

Materials and coating procedure. Ly12 Al alloy having a nominal composition of 3.9-4.9% Cu, 1.2-1.8% Mg, 0.30-0.90 % Mn was used as the substrate material for all the coatings. Ly12 discs with a diameter of 90 mm and thickness of 10 mm were used as the sample substrate. The surface of the discs was polished with 800#,1000# and 1200# abrasive papers to a roughness of Ra 0.4μm, and then ultrasonically cleaned in distilled water followed in acetone. The discs were used as anodes, while stainless steel plates were used as cathodes in the electrolytic bath. The samples carefully pretreated as above were oxidized in accordance with process conditions detailed in Table 1. An aqueous electrolyte was used, containing mainly Na_2SiO_3, Na_2EDTA, NaF and NaOH to adjust its alkalinity and increase conductivity. A 240 kW AC microarc oxidation device provided constant current which can be set variably depending on the requirements, and the voltage varied with the duration of oxidizing time. Voltage magnitude of the negative pulse equals to that of the 1/9 positive pulse. From our practice, it is the positive pulse rather than the negative pulse plays a decisive role in the microstructure formation of the coatings. The negative pulse is only interspersed within the positive pulses as a mean to interrupt the spark discharges, permitting the surface to cool and inducing the re-conversion of soluble components into metal oxide. Therefore, in order to obtain coatings with optimum surface morphology, a stepped adjustment regime in positive pulse current was applied. The electrolyte was cooled by a cooling system to keep its temperature below $20°C \pm 1°C$.

Table 1 Experimental parameter matrix employed for microarc oxidation

Current Density (A/cm^2) (min)	Voltage (V)	Duty cycles anodes /cathodes	Electrolyte temperature	Per step processing time (°C)	Per step Voltage (min)	Total treatment (V)
0.05~0.2	240-615	20/6	20±1	0.5+1+2.5+5 +5+5+5 +5	240+300+350+400 +450+500+550+615	29

Characterization of coatings. The coating thickness and surface roughness were measured at 10 different randomly selected locations using a coating thickness gauge and surface roughness tester (TT260 and TR210, Time Group Co.). After removing maximum and minimum readings, the rest were averaged to obtain the value of the coating. The surface morphology of the coatings was observed using scanning electron microscopy(SEM, S-570, Hitachi Co., Japan). At a number of randomly selected locations, the scanned regions were photographed and these micrographs were subjected to analyze the coating characteristics. The phase composition of the coatings was identified using an X-ray diffractometer(D/max2200pc,Japan). The tribological performance of the MAO coating was evaluated by conducting a dry-sand rubber-wheel abrasive wear test and a solid-particle erosive wear test. In order to determine the steady-state wear loss, these tests were conducted in discrete steps, with the weight loss for each step measured and converted to volume loss. The abrasive wear test was performed at a normal load of 70 N and with a step size of 250 wheel rotations each. In the case of the erosion wear test, SiO_2 was used as the erodent with a particle velocity of 80 m/s and at 45° and 90° impact angles with a test duration of 5 min each.

Results and Discussion

Coating thickness. Fig. 1 illustrates the influence of MAO treatment time on the thickness of the coatings deposited. It is clear that the kinetics of the coating formation is linear, this aspect is further discussed in Section below of this paper. The linear rate constant is approximately 1.034μm/min, near equal to the average coating thickness deposited per minute of the treatment time.

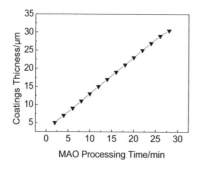

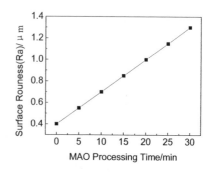

Fig. 1 Influence of MAO treatment time   Fig. 2 Influence of MAO treatment time
       on coatings thickness   on surface roughness

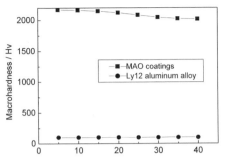

Fig.3. Variation in microhardness of the MAO coating across radius from the edge towards the center of the sample

Coating surface roughness. The variation in surface roughness (Ra) as a function of coating time is illustrated in Fig. 2. It is clear that the surface roughness increased linearly with the increased MAO treatment time. In other words, thinner coatings exhibited lower surface roughness and the thicker coatings exhibited the higher surface roughness.

Microhardness. The average microhardness exhibited by ceramic coatings is 2096 HV and that of Ly12 aluminum alloy substrate material is 98 HV. In the case of MAO coating, a hardness gradient exists across the radius of the sample. Fig. 3 illustrates the variation in microhardness measured from the edge towards the center of the sample-coated. It is evident that the peak microhardness show itself at a distance of 5 mm from the brink of the discs and then exhibits a decreasing trend towards the center of the coated sample.

Microstructure and phase content. SEM micrographs of the MAO coating are presented in Fig.4a(surface) and Fig.4b(cross-sections). The surface micrograph exhibits a dense ceramic coating, witch contains many grains and few ostium with various sizes. The cross sections coatings exhibit a uniform coating thickness with a clean and wavy interface, and the coating appears to be well bonded with the substrate. Simultaneity, the cross sections coatings exhibits the presence of two distinct regions, viz. an inner dense region and an outer granular region. More importantly, the thickness of the outer granular region was less than 4% of the total coating thickness.

XRD spectra of the coating are shown in Fig.5 a and b. Both inner layer and out layer are composed of α-Al_2O_3 and γ-Al_2O_3, While the $Al_6Si_2O_3$ phase is observed only in out loose layer.

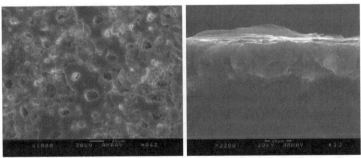

Fig.4 Micrographs of the MAO coating:(a)Surface micrograph; (b)Cross section micrograph

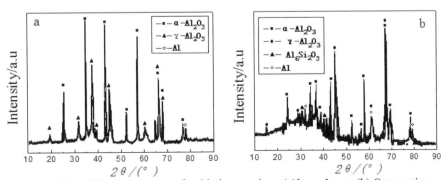

Fig.5 XRD diffraction spectra of oxidation coatings:(a)Inner layer; (b) Out coating

Tribological performance. Fig. 6 illustrates the abrasive wear performance of MAO coatings. It is clearly evident that the Ly12 aluminum alloy exhibits move up abrasive volume-loss right from the first 100 disc rotations, while the MAO coatings reach a near steady-state only after completion of an initial 600 disc rotations. However, in the case of the MAO coating, even after 1500 disc rotations, the coating volume loss still exhibits a decreasing trend at a slower rate. Erosive wear volume-loss per kg of erodent at two different impact angles (45° and 90 °) is shown in Fig.7. Distinctly, Ly12 aluminum alloy exhibits higher volume loss than the microarc oxidation coatings at both impacting angles. Nevertheless, at both the angles of the MAO coatings exhibit almost identical steady-state volume loss rate, while the Ly12 aluminum alloy exhibits higher volume loss at 45° than at 90 ° impacting angles.

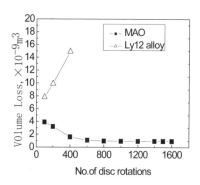 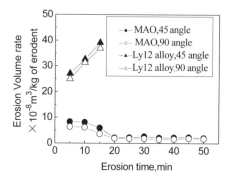

Fig.6. Abrasive wear performance of MAO and Ly12 aluminum alloy

Fig.7. Erosive wear performance of MAO and Ly12 aluminum alloy at 45 ° and 90 ° impact angles.

Results and Discussion

Mechanism of MAO coating formation. Based on the experimental observations reported in this study, a phenomenological mechanism for MAO coating formation can now be proposed. In the MAO process, the observed linear deposition kinetics can be simply and approximately related to the Faraday's laws of electrolysis. However, the micrographs exhibited in Fig. 4 clearly indicate that the coating deposition occurs via the discharge channels. It is also clear, as indicated by the visual observation of continuous movement of the sparks on the coating surface and also by the SEM micrographs, that the discharge channels have moment life. Thus, the discharge channels are continuously formed and closed all through the oxidation process. In our experiments, the experimental results and process indicate that the coating thickness increases linearly with oxidation time and this in turn implies a constant volume rate of coating deposition since oxidation coating area is a constant and deposits material on the coating surface equivalent to the volume of the melting particles surrounding it. As discussed above, the melting particles diameter and fusibility film thickness increases with increasing oxidation time. Since the coating is formed with melting particles as the basic building unit, it should be obvious that the surface roughness of the coating should be of the order of channel diameter as well. Thus, the linear increase in coating roughness is directly related to the fact that the discharge channel diameter also increases linearly with oxidation time.

Structure–property correlations. The rapid solidification of alumina promotes the formation of meta-stable γ-Al_2O_3 phase. Since the surface layers of the MAO coatings are always in contact with the surrounding low temperature electrolyte medium, the surface layers contain γ-Al_2O_3 phase. According to this explanation, the phase contents of the surface layers should comprise predominantly γ-Al_2O_3 phase irrespective of the coating thickness. The identical XRD patterns with predominantly γ-Al_2O_3 phase obtained for the coatings wherein the X-ray penetration is totally within the coating as shown in Fig. 5a and b are in clear support of the above explanation. However, the low thermal conductivity of alumina, causes the underlying layers of the coatings to remain hotter and the temperature is sufficiently high to cause the transformation of originally formed γ-Al_2O_3 to α-Al_2O_3. Thus, it is expected that the proportion of α-Al_2O_3 will continuously increase towards the coating–substrate interface.

Increased α-Al_2O_3 content with increasing distance from the center to brink of sample is the prime reason for the high hardness (Fig. 3) exhibited by the MAO coatings. The surface layer of the MAO coating(Fig. 4a) is granular and irregular, while the inner regions are dense and defect-free. This is the reason for the higher abrasion and erosion wear volume loss of MAO coatings during the initial stages of wear, followed by a dramatic reduction in wear rate. However, under abrasion and erosion conditions, the Ly12 aluminum alloy volume loss is surprisingly high. Apparently , These results connected with the microstructure and surface hardness of the materials.

Summary

On your experiment, the kinetics of MAO coatings is interface-controlled and largely dependent on the applied current density and electrolyte composition. The surface roughness and microhardness of MAO coatings are a linear function of the distance from the edge to center of the coatings, and increases with increasing oxidation time and coatings thickness. Both inner layer and out layer are composed of α-Al_2O_3 and γ-Al_2O_3, While the $Al_6Si_2O_3$ phase is observed only in out loose layer. The microhardness of the coatings increased with the α-Al_2O_3 phase increased in the ceramic coatings. The microhardness characterized by the peak hardness always remaining at the extension of the brim of the sample and then exhibited a decreasing trend towards the center of the sample. The reason for the higher abrasion and erosion wear volume loss of MAO coatings during the initial stages of wear are associated with the surface layer of the MAO coating is granular and irregular, while the inner regions are dense and defect-free.

References

[1] Z.H. Sun, M. Liu, D.P. Guo et al. Formation process of ceramic films fabricated by micro-arc oxidation on 2A12 aluminum alloy. Rare Metal Materials and Engineering, J.36 Suppl.1(2010) 64-69.

[2] W. Xue, W. Chao, Y.l. Li et al. Effect of microarc discharge surface treatment on the tensile properties of Al–Cu–Mg alloy. Materials Letters, J. 56(2002)737-743.

[3] M.H. Zhu, Z.B. Cai, J.Tan et al. Behaviors of fretting wear of microarc oxidation coating, tribology,J.26(2006)306-309.

[4] T.B.Wei, B.G. Guo, J. Liang et al. Tribological properties of micro-arc oxidation coatings on Al Alloy,J.22(2004)564-569.

[5] L. Rama Krishna, K.R.C. Somaraju, G. Sundararajan. The tribological performance of ultra-hard ceramic composite coatings obtained through microarc oxidation. Surface and Coatings Technology,J.163(2003)484–490.

Key Engineering Materials Vol. 537 (2013) pp 12-15
© (2013) Trans Tech Publications, Switzerland
doi:10.4028/www.scientific.net/KEM.537.12

Effects of Substrate Temperature on the Properties of Silicon Nitride Films by PECVD

Chunya Li[a], Xifeng Li[b], Longlong Chen[c] , Jifeng Shi[d], Jianhua Zhang[e]

Key Laboratory of Advanced Display and System Applications, Ministry of Education, Shanghai University, Shanghai 200072, China

[a]lichunya2006@163.com, [b]xifeng.li@hotmail.com, [c]llchen@shu.edu.cn, [d]shijf@shu.edu.cn, [e]jhzhang@staff.shu.edu.cn

Keywords: Silicon nitride films, PECVD

Abstract. Under different growth conditions, silicon nitride (SiNx) thin films were deposited successfully on Si(100) substrates and glass substrates by plasma enhanced chemical vapor deposition (PECVD). The thickness, refractive index and growth rate of the thin films were tested by ellipsometer. The surface morphologies of the thin films were investigated using atomic force microscope (AFM). The average transmittance in the visible region was over 90%.

Introduction

Silicon nitride thin film is a kind of very important electric dielectrics with excellent optoelectronic, insulator, mechanical and passivation properties [1]. It will be widely used in optoelectronics, microelectronics and so on. Silicon nitride thin film is mainly used for inner insulator film, the gate insulator layer of field effect transistor (or thin film transistor), passivation layer, ion stopped layer and encapsulation materials etc [2-3]. Because plasma enhanced chemical vapor deposition (PECVD) succeeded in depositing excellent thin film under lower temperature (<700K), PECVD has been one of the most important methods to prepare silicon nitride thin film in industry [4].

A large number of experimental studies have shown that the composition, structure of the silicon nitride thin films was closely related to the process parameters [5-6].

In this paper, silicon nitride (SiNx) thin films were prepared by plasma enhanced chemical vapor deposition (PECVD).The structural, electrical and optical properties of the SiNx films were investigated.

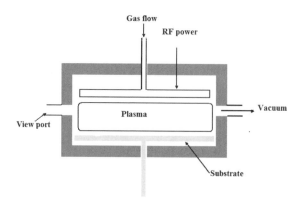

Fig.1. SiNx film fabrication using PECVD system

Experimental methods

The deposition of the SiNx films were conducted by a CME-200E PECVD facility with a 27.12 MHz RF power source. SiNx films were deposited using SiH4 and NH3 as precursors. Fig.1 showed the schematic diagram of the PECVD reactor used in this research.

The temperature of the substrate was vary from 150 0C to 350 0C. After a number of preliminary tests, the optimized deposition conditions of the SiNx films are shown in Table 1.

In this research, SiNx films were deposited respectively on the silicon substrates and 0.7cm-thick glass substrates.The silicon substrates were firstly cleaned with HF acid for 2 minutes. Then they were rinsed in de-ionized water and dried in the oven finally.

The thickness and refractive index was measured using ellipsometer. The surface profile of the film was observed with Nanomavi SPA-400 SPM atomic force microscopy (AFM) system. The optical transmittance measurement was performed with U-3900 spectrophotometer.

Table 1 Deposition conditions of SiNx films

	SiNx
SiH4 flow rate	20 sccm
NH3 flow rate	150 sccm
N2 flow rate	270sccm
Deposition temperature	150°C, 200°C, 250°C, 300°C, 350°C
RF power	150 W
Deposition pressure	100 Pa
Deposition time	6min

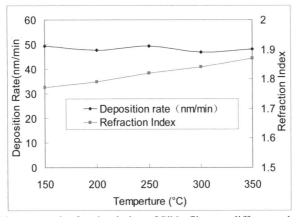

Fig.2 The deposition rate and refraction index of SiNx films on different substrate temperature

Result and discussion

Deposition Rate and Refraction Index. Fig.2 shows the deposition rate and refraction index of SiNx films on the different deposition temperature. The other deposition parameters remained unchanged within the experimental range, the deposition rate showed a downward trend with increasing temperature, refraction index showed an upward trend with increasing temperature.

   Surface roughness. Substrate temperature have a certain influence, as well as the rapid thermal annealing on the surface of the film will also affect the surface morphology of the samples [7-9].

   The experiment studied the surface morphology of the SiNx films by AFM (Fig.3), the scanning range of 4μm × 4μm. Seen from the figure, the SiNx film surface is relatively smooth, dense, uniform particle size distribution. The root-mean-square (Rms) surface roughness of the SiNx films was 1.4 nm. As the temperature rises on a flat film surface began to appear scattered larger particles. The surface defects decreased and high density at the appropriate temperature.

Optical properties. The transmittances of the SiNx films on the glass substrate are shown in Fig. 4. The average transmittance is over 90% at the visual region. With the substrate temperature increasing, the transmittance of the films does not change very much.

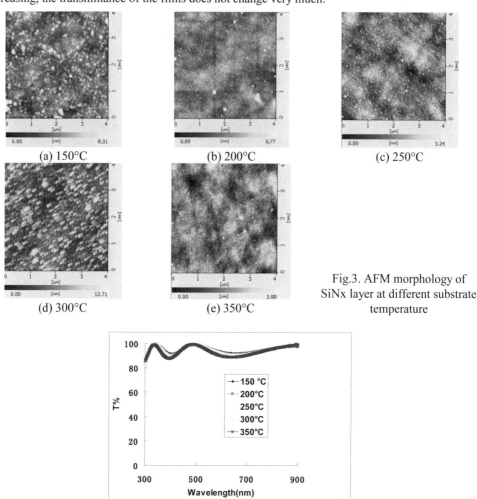

(a) 150°C   (b) 200°C   (c) 250°C

(d) 300°C   (e) 350°C

Fig.3. AFM morphology of SiNx layer at different substrate temperature

Fig. 4 Transmittance of the SiNx films

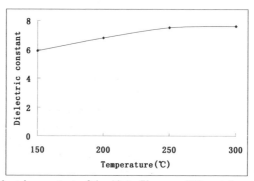

Fig.5 Relative dielectric constant of the SiNx films on different deposition temperature

Dielectric constant. From Fig.5, the dielectric constant of the SiNx films on different deposition temperature is shown. With the deposition temperature rise, the other deposition parameters remained unchanged, there is an upward trend for the dielectric constant of the SiNx films.

Conclusions

Under different growth conditions, silicon nitride (SiNx) thin films were deposited successfully on Si (100) substrates and glass substrates by plasma enhanced chemical vapor deposition (PECVD). The thickness, refractive index and growth rate of the thin films were tested by ellipsometer. The surface morphologies of the thin films were investigated using atomic force microscope (AFM). The average transmittance in the visible region was over 90%.

Acknowledgement

This work was financially supported by the Science and Technology Commission of Shanghai Municipality (Grant No. 10DZ1100102), by national natural science foundation of china (Grant No. 61006005) and by excellent youth project funding of Shanghai Municipal Education Commission.

References

[1] Y. J. Zhao, M. J. Wang, Y. J. Yang, et al., Study on Stress of PECVD SiNx Film, Chin. J. Semicond. 20 (1999) 183-187.

[2] K. Niu, X.Y. Ye, Z.Y. Zhou, et al., Research on mechanical properties of MEMS cantilever actuator based on the reduced order modeling, Piezoelectrics &Acoustooptics. 27 (2007) 507-513.

[3] H. F. W. Dekkers, S. D. Wolf, G. Agostinelli, et al., Requirements of PECVD SiNx :H layers for bulk passivation of mc-Si, Sol. Energy Mater.& Sol. Cells. 90 (2006) 3244-3250.

[4] Z.Y. Xie, C.P. Long, C.Y. Deng, et al., Interfacial structures and properties of SiN layer in a-Si thin film transistors, Chin. J. Vacuum Science and Technology, 27(2007) 341-345.

[5] X.Q. Wang, L. Wang, Z.Q. Xi , et al., Properties of PECVD SiN_x film for solar cells, Acta Energiae Solaris Sinica. 25 (2004) 341-345.

[6] G.W. Zhang, F. Long, Research on technology for Si_3N_4 thin film grown by PECVD, Semicond. Optoelectronics. 22(2001) 201-204.

[7] Y. Li, X.D. Li, Y. Xu, et al., Characterization of As-deposited and annealing SiN films deposited by plasma enhanced chamical vapour deposition , Acta Scientiarum Naturalium Universitatis Sunyatseni. 42 (2003) 74-77.

[8] C.F. Gong, Z.Q. Xi, X.Q. Wang, et al., Optical and electrical properties of annealing SiNx:H thin film, Acta Energiae Solaris Sinica.27 (2006) 300-303.

[9] J.H. Mo, L.M. Liu, Y.J.Tai, et al., Preparation and properties of Silicon Nitride supporting membranes grown by plasma-enhanced chemical vapor deposition, Semicond. Optoelectronic. 28 (2007)804-807.

Key Engineering Materials Vol. 537 (2013) pp 16-19
© (2013) Trans Tech Publications, Switzerland
doi:10.4028/www.scientific.net/KEM.537.16

Study on Inorganic Coating of Porous Nitride Radome

Rong Liao [1], Chuanbing Cheng[2], Chonghai Wang [1], Hongsheng Wang [1], Qihong Wei [1]

[1] Shandong Research and Design Academy of Industrial Ceramics Co., Ltd, Zibo, 255031, China;

[2] University of Jinan, Jinan, 250399, China

Keywords: Inorganic Coating; porous nitride radome

Abstract. The surface of porous nitride radome became very compact by $AlPO_4$-Si_3N_4-SiO_2(n) coating treatment. Coating could immerse into the matrix about 20μm deep. This could also make material be moisture resistance for long time placement. Coating healed the micro cracks on the matrix effectively and combined with it by crossed occlusion, so the strength and the fracture toughness of material increased.

Introduction

Radome materials usually should have lower dielectric constant and loss, so they can be widely applied in the rocket, airship, missile and satellite areas [1]. The dielectric constant of materials relate to the density. The lower is the density, and the lower is the dielectric constant, so effective way of obtaining low dielectric silicon nitride materials is to preparation low density porous material.

Because of the low density, high temperature resistance, corrosion resistance, low dielectric constant and loss, porous nitride ceramics has become one of the main material of broadband electromagnetic transmission missile radome [2]. Because of the capillary force, the pore surface is easily with a molecule of water in the air and condenses into hydration layer. Dielectric loss of porous nitride ceramics extremely sensitive to water. The quality score of the water less than 1% can make the loss tangent increases nearly six times. The absorption of moisture makes dielectric loss increases, seriously influences electromagnetic transmission property.

High performance dense coating in the porous nitride material surface can be solve the above problem. Through closing surface porosity, coating can prevent porous material absorption of moisture. The dielectric properties of the coating must be excellent, so porous materials by coating treatment have smaller dielectric constant and loss in high temperature and high frequency, and meet radome requirements. This research using $AlPO_4$-Si_3N_4-SiO_2 (n) developed low dielectric inorganic coating. The surface of porous nitride radome treated with the inorganic coating. It could make matrix be moisture resistance for long time placement.

Experiment

Raw material. *Si_3N_4 powder:* The Si_3N_4 powder used in this experiment was made by ganister sand nitriding. The components were 97% α-Si_3N_4, 2.28% β-Si_3N_4 and 0.56% fsi. Particle size was less than 1μm. Details can be seen in tables 1 and 2.

Table 1 Chemical analysis of silicon nitride powder (wt.%)

Components	Total Si	N	Al_2O_3	Fe_2O_3	CaO	MgO	Na_2O	K_2O	Si
Content(wt.%)	59.40	36.88	2.26	0.55	0.03	0.005	0.05	0.01	0.56

Table 2 Distribution of particle size of silicon nitride powder

Dia. of Particle(μm)	<0.5	0.5~0.6	0.6~0.8	0.8~1.0	1.0~2.0	2.0~3.0	>3.0
Distribution (%)	19.9	10.3	16.9	11.6	23.6	10.2	7.5

AlPO₄ solution: Colloidal transparent liquid was prepared by hydrothermal method. Analytical reagent H_3PO_4 and $Al(OH)_3$ was used as the starting material.

Nano quartz: Amorphous nano quartz was produced with advanced microwave induction plasma method, average particle size range from 30nm to 50nm, loose density range from 0.1 g/cm³ to 0.15 g/cm³.

Sample preparation. The matrix for coating was 95% Si_3N_4 together with 5% Y_2O_3 and organic pore-forming agent, mixed evenly by ball-milling machine. Then shaped by static pressure. After firing by air pressure (firing temperature is 1700℃, remains 1 hour, nitride pressure is 6MPa), machined into the shape that the experiment requested, then cleaned by regular treatment.

Added Si_3N_4 which particle size less than 1μm and nano quartz to $AlPO_4$ solution. Beat them up for 2 hours under room temperature to make $AlPO_4$-Si_3N_4-SiO_2(n) admixture. Coefficient of expansion of Coating was slightly less than the coefficient of expansion of matrix by adjusting Si_3N_4 and nano quartz proportion. Coated nitride ceramic samples by dipping-lifting into $AlPO_4$-Si_3N_4-SiO_2(n) admixture and sprayed coating. After dried, dealt with heat treatment for 1 hour by 700℃, and then got the ceramic samples changed in nature.

Performance Check. Determined the density and the apparent porosity of the samples by Archimedes method. Processed bending resist intensity test on heat treatment sample and $AlPO_4$-Si_3N_4-SiO_2(n) coated sample separately by WID-10 electronic formula all-purpose testing machine. Testing method is three-point bending. Sample measurement as 36mm×3mm×4mm, span 30mm, loading rate 0.5mm/min. the breaking tenacity testing method is SENB, measurement as 2.5mm×5mm×30mm, cutting depth 2.5mm, span 20mm, loading rate 0.05 mm/min. Used NET-ZSCH-402/EP type inflation instrument to determine thermal expansion coefficient, sample measurement 50mm×5mm×5mm. Adopted SEM method to observe the pattern and the combined status of the sample and the $AlPO_4$-Si_3N_4-SiO_2(n) coating.

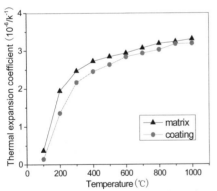

Fig.1 Coefficient of expansion of matrix and coating

Fig.2   The material by coating treatment

Results and Discussion

Analysis on thermal expansion of matrix and coating. Due to matrix and coating closely together, if coefficient of expansion of matrix and coating were different, stress appeared in the coating during cooling or curing, influenced adhesiveness of matrix and coating. Fig. 1 was coefficient of expansion of matrix and coating under different temperature. Because coefficient of expansion of the coating was slightly less than coefficient of expansion of the matrix, coating got compression function of matrix in cooling process, so not big compressive stress produced in the coating. Compressive stress could offset part of tensile stress because of the thermal stress or mechanical force exerted on the surface coating. It could improve tensile strength and thermal stability of coating. Crack wouldn't be between matrix and coating. Fig. 2 was the appearance of the material by coating treatment. As shown by Fig. 2, the surface of the sample was complete. The phenomena of cracking, peel, layering didn't appear. The experiment proved that thickness and quality of the sample didn't change.

Analysis on microscopic structure of the material by coating treatment. Fig. 3 was the microscopic structure of matrix. As shown by Fig. 3, big hole left by decomposition of porogen evenly distributed. Aperture range was 5 ~ 8 um. Fig. 4 was the SEM figure of the material section by coating treatment. From the combination between the coating and the surface of matrix, we could find that the coating immersed into the matrix about 20μm deep, formed the crossed occlusion, larger the faying surface between the coating and the matrix was fine and close.

Fig.3 Microscopic structure of matrix   Fig.4 SEM of the material by coating treatment

Table 3 The material performance compare before and after coating

	Density (g/cm^3)	Porosity (%)	Intensity (MPa)	Breaking Tenacity (MPam$^{1/2}$)
Matrix	1.7	42	55	1.3
Material by Coating Treatment	2.0	3	76	2.4

When the coating immersed into the matrix with some depth and formed the compact film, on one hand, it closed more of the defects on surface; on the other hand, it became the compression stress layer on surface to enhance the intensity and the tenacity of the matrix.

Effect on density and mechanical properties of the material. Table 3 showed the material performance compare before and after coating. From the average data of all tested samples, density of the material after coating treatment increased, porosity decreased obviously; bending resistance intensity of the material before coating treatment was not strong enough but it had been raised up much after coating treatment. The fracture toughness has also improved.

Griffith [3] had pointed out in the ceramic material research in 1920s that there was a close relationship between the surface status and the intensity of the material. There would be slight polishing scratch on the surface after sample be machined by grinding machine, most probably the micro cracks there as well. By coating treatment, because of the smooth of the coating, there could be no obviously macroscopic defects, the quality of the film was good. Compare with the situation before the coating, the polishing scratch on the surface of the sample was lighter and the micro cracks were less also. According to these, we got the conclusion that it would be effective to close the micro cracks on matrix by coating and this method could be better than the effects of only precisely machined material. From the data in table 3, we could see the bending resistance intensity of the material raised from 55MPa to 76MPa after being treated by coating.

Effect on electrical properties of the material. The surface of material by coating treatment was dense, porosity was reduced, so the dielectric constant increased. Fig. 5 was the dielectric constant of material before or after coating treatment at room temperature. As shown by Fig. 5, the dielectric constant of material after coating treatment have improved in different frequency band. Because the dielectric constant of $AlPO_4$-Si_3N_4-SiO_2 (n) system coating was low, the dielectric constant increase of material after coating treatment was not big.

Fig. 6 was the dielectric loss of material before or after coating treatment at room temperature. As shown by Fig. 6, dielectric loss before coating treatment changed a lot in different frequencies. This is because water adsorption existed in the surface of porous nitrides. Dielectric loss for the water was very sensitive. Water polarization not only made dielectric loss angle significantly increased, and produced great fluctuation, it went against application of material. The surface of porous Si_3N_4 by coating treatment didn't form a hydration layer, so dielectric loss curve didn't exist fluctuations caused by water. Stability of dielectric loss improved significantly.

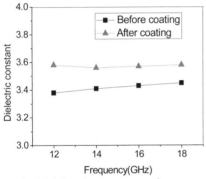

Fig.5 Dielectric constant on frequency

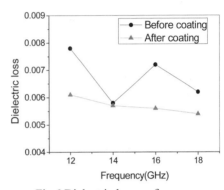

Fig.6 Dielectric loss on frequency

Summary

1. The surface of material became very compact by $AlPO_4$-Si_3N_4-SiO_2(n) coating treatment. And this could also make the matrix be moisture resistance for long time placement.

2. $AlPO_4$-Si_3N_4-SiO_2(n) coating immersed into the matrix about 20μm deep formed the crossed occlusion. Coating could close micro cracks on the surface of matrix effectively and enhance the whole performance.

3. Dielectric constant of $AlPO_4$-Si_3N_4-SiO_2(n) coating was not big, so dielectric constant of material by $AlPO_4$-Si_3N_4-SiO_2(n) coating treatment had little increased. and the stability of the dielectric loss improved effectively.

References

[1] China V A, Copel R L: submitted to Interim Engineering Report 3(1963).

[2] Weipin Ge, Kunyu Zhao, Porous silicon nitride, Yunnan Metallurgy. 2(2004)184-186.

[3] Qingchun Zhang, Mechanics Performance of Ceramics, Science Press, Bei Jing, 1987.

Key Engineering Materials Vol. 537 (2013) pp 20-23
© (2013) Trans Tech Publications, Switzerland
doi:10.4028/www.scientific.net/KEM.537.20

Sduty on Hole-sealing Coating of Porous Silicon Nitride Surface

Qihong Wei[1,2], Chonghai Wang[1,2], Ling Li[1,2], Hongsheng Wang[1,2], Jian Liu[1,2] and Rong Liao[1,2]

[1]Shandong Research and Design Institute of Industrial Ceramic, Zibo 255031,China

[2]Sinoma Advanced Materials Co.,Ltd,, Zibo 255031, China

[a]Wqh81@163.com

Keywords: Porous silicon nitride; Hole sealing; Coating

Abstract. Hole-sealing coating was fabricated on the porous silicon nitridesubstrate by sol-gel method, with Li_2O-Al_2O_3-SiO_2 as the basic materials and other additives were added. The morphology of the coating was tested by scanning electron microscope (SEM). The density, water absorption and porosity of porous silicon nitride before and after coating were tested by Archimedes method. The bending strength was tested by universal testing machine. The results indicated that the water absorption of porous silicon nitride was decreased by 93.73% to 96.74%, the bending strength of porous silicon nitride was increased by 8.21% to 15.56%.

Introduction

Silicon nitride ceramics have the characteristics of high temperature,resistant corrosion, thermal shockresistance, hardness high, good toughness and little thermal expansion coefficient etc.,which are ceramic materials with combined good performance. Porous silicon nitride can be regarded as the complex of solicon nitride and stomata,which has the advandages of both solicon nitride and porous ceramics. Porous silicon nitride meets the requirements of lightweight,heat resistance and broadband wave-transparent,which has broad application prospect in missile antenna mask and spacecraft communication window.But porous ceramic material whose dielectric loss is extremely sensitive to water can easily absobe moisture,which influence the properties of wave-transparent materials seriously. Furthermore, due to the rough surface of porous ceramics,and a large number of micro-cracks exit, the bending strength drop dramatically because of stress concentration[1].So sduty on sealing coating of porous silicon nitride surface has important application value.Sealing coating can prevent the decreace of the dielectric properties because of absorption of moisture,at the same time it also can improve the mechanical properties and resistance ability.So the preparation of dence and high temperature resistance coating on porous silicon nitride surface,which form the composite structure of low density core body and high density surface,has become an important research direction.The composite structure not only has excellent dielectric performance,but also has enough mechanical properties,heat resistance and rain erosion performance,which is considered the most promising wave-transparent material.In this paper,hole sealing coating on porous silicon nitride surface were studied with sol-gel method using Li_2O-Al_2O_3-SiO_2[2-4] as the basic materials.This paper studied mainly changes of water absorption, bending strength and wave-transparent property of porous silicon nitride before and after sealing coating.

Experiment

Preparation of surface coating. Clear LAS sol was prepared according to N(Li:Al:Si:La) = 1:1:2:0.5, with lithium carbonate, nine hydrated aluminum nitrate,lanthanum chloride,etraethyl orthosilicate and nitric acid as main raw material. The sol was dried in drying oven at 70℃, and heat-treated at high temperature. The obtained powder was LAS powder. Stable LAS pulp was obtained with LAS powder, additive and water grinded for 24 hours. LAS pulp was sprayed on the surface of porous silicon nitride, then dried at the temperature 50~80°C for 24h.The dense damp-proof coating on the surface of porous silicon nitride was obtained by sintering in common pressure and nitrogen protection atmosphere.

Testing methods The density, water absorption and porosity of samples were tested by Archimedes method. The bending strength was tested by three point bending method using universal testing machine AG-IC. Netherland FEI Company Sirion200 hot field emission scanning electron microscope was used to observe the surface and section micromorphology of samples.

Results and Discussion

Effects of different sintering temperature on coating surface morphology. Fig.1(a)(b)(c) showed the surface micromorphologies of hole sealing coating on the surface of porous Si_3N_4 samples sintered respectively at 1200°C,1250°C,1300°C. When the temerature was 1200°C, coating was still not fully vitrified,a fewl pores still existed among Si_3N_4 particles.When the temperature was 1250°C, the formed eutectic filled among Si_3N_4 particles,and the dense coating was obtained.So pores of porous Si_3N_4 surface was fully filled, and surface roughness decreased significantly, and no crack existed.And columnar crystal separated out on the coating surface at 1250°C.When the temperature was 1300°C, columnar crystal grains were fused to the glass phase,and glass phase content rised.The result was that dielectric constant rised,and heat resistance decreased. Therefore, the suitable sintering temperature was 1250°C.

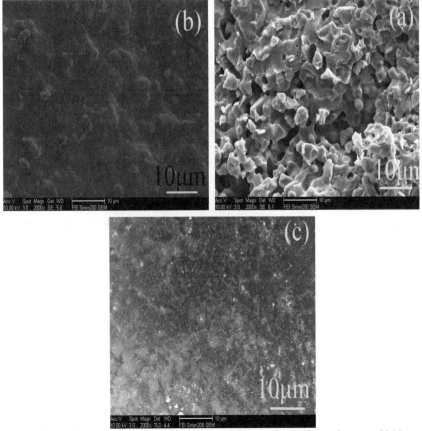

Figure 1.Surface micromorphologies of sealing coating on the surface of porous Si_3N_4 samples sintered at different temperatures

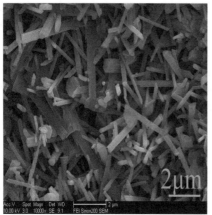

Figure 2 .Surface micromorphologies of porous
Si₃N₄ substrate

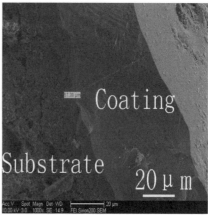

Figure 3 .Section micromorphology of porous
Si₃N₄ substrate after sealing sintered at 1250 ℃

Table 1 Changes of water absorption, apparent porosity and density before and after sealing

Serial Number	Water Absorption/% before/after Sealing	Apparent Porosity/% before/after Sealing	Density/ g/cm³ before/after Sealing	Variation rate of Water Absorption/%	Variation Rate of Apparent Porosity/%
1	45.05/1.56	58.57/2.06	1.30/1.32	-96.54	-96.48
2	37.16/1.29	52.77/1.84	1.42/1.43	-96.53	-96.51
3	45.06/2.65	58.13/3.47	1.29/1.31	-94.12	-94.03
4	41.58/2.60	56.13/3.56	1.35/1.37	-93.73	-93.66
5	36.76/1.20	52.58/1.74	1.43/1.45	-96.74	-96.69

Surface micromorphologies of porous Si₃N₄ substrate before and after sealing. Fig.2 showed surface micromorphologies of porous Si₃N₄ substrate. The porous structure was builded by columnar crystal β- Si₃N₄,and the porosity was more than 50%,which resulted to good thermal shock resistance, low thermal conductivity ,low expansion coefficient and low dielectric constant.But at the same time the existence of the porosity resulted that mechanical property decreased and water absorption increased, the dielectric properties was seriously affected. Compared to Figure 1(b), it showed that surface micromorphologies of porous Si₃N₄ substrate before and after sealing changed obviously. Fig.3 showed the section micromorphologies of porous Si₃N₄ substrate after sealing. It investigated that thickness of coating was about 60μm,and the surface was smooth. Part coating infiltered into the substrate at the bond between coating and porous, and formed nail pierced interface,which made coating and porous combined closely.

Changes of water absorption, apparent porosity and density before and after sealing. It was tested for that changes of water absorption, apparent porosity and density before and after sealing by choosing samples of different porosity. The greater the surface area was, the higher water absorption was, in order to ensure that each sample had similar surface area, every sample was processed into the same size cuboid. The sealing coating was sintered at 1250℃, at which was held for 2h.The testing results were showed in table 1.It investigated that water absorption and apparent porosity decreased obviously after sealing, the biggest change rate was 97.74% and 96.69%, respectively. Density changed little before and after sealing, which was due to little changes of the quality and volume after sealing.

Effect of sealing on the bending strength of porous Si₃N₄ substrate. Select five groups of samples of different porosity, and test bending strength of average. The sample size was 80 mm × 10 mm ×10 mm, and the span was 60 mm. The sealing coating was sintered at 1250℃, at which was held for 2h.The testing results were showed in table 2.It investigated that the bending strength increased obviously after sealing, the biggest change rate was 15.56%. The lower the substrate strength was, the more obvious enhancement effect was [5],which was due to the lower porosity, density and the more crack defects.

Table 2 Changes of bending strength befor and after sealing

Serial Number	substrate Porosity/%	bending strength/MPa befor and after sealing	Variation/%
1	44.34	98.5/106.59	8.21
2	48.13	85.8/93.63	9.13
3	55.46	60.9/70.38	15.56
4	58.57	57.8/64.93	12.34
5	62.77	52.3/60.10	14.92

Conclusion

Damp-proof and enhanced sealing coating was fabricated on the porous Si_3N_4 substrate by sol-gel method, with Li_2O-Al_2O_3-SiO_2 as the basic materials. After sintering at 1250°C,the coating surface was smooth,and coating combined closely with porous Si_3N_4 substrate.After sealing,the water absorption of porous Si_3N_4 ceramic was decreased by 93.73~96.74%, the bending strength was improved by 8.21 ~ 15.56%.

References

[1] S.B. Wang, S.J. Liu, Y. Zhang, Study on damp-proof and enhanced coating of porous Si_3N_4, J. Inorganic Meter., 23 (2008) 769~773.

[2] H.G. Qiu, J.S. Cheng, L.Y. Tang, et al, The influence of heat treatment on the density and coefficient of thermal expansion of Li_2O-Al_2O_3-SiO_2 system glass-ceramics. Glass & Enamel, 35 (2007) 1~6

[3] Y.Zhao, L.Lu, L.J. Zhang, Effects of heat-treatment on the crystallization and properties of Li_2O-Al_2O_3-SiO_2 system glass-ceramics. Bulletn of Chinese Ceramic Society, 26 (2007) 561~566.

[4] Z.X.Hou, Crystallization behavior study on the Li_2O-Al_2O_3-SiO_2 system glass-ceramics. Ordnance Mater. Sci. Engi., 30 (2007) 42~45

[5] D. Leguillon, R. Piat, .Eng. Fract. Mech, 75 (2008) 1840~1853.

Key Engineering Materials Vol. 537 (2013) pp 24-27
© (2013) Trans Tech Publications, Switzerland
doi:10.4028/www.scientific.net/KEM.537.24

The Effect of SiC Coating on the Joint of C/C and LAS

Huanxia Zhu [a], Kezhi Li [b], Jinhua Lu, Hejun Li

Carbon/Carbon composites Research Center, Northwesten Polytechnical University, Xi'an, 710072, China

[a] zyxia777@163.com, [b] likezhi@nwpu.edu.cn

Keywords: C/C composites, SiC coating, joint

Abstract: The SiC coating on C/C composites was prepared by CVD and pack-cementation method in this paper, then SiC coated C/C composites was jointed with LAS glass ceramic at 1423K for 30min and the pressure of 15 MPa. The structure and morphology of the SiC coating and as-received joints were analyzed by SEM and XRD; shear strength of joint was examined through the Instron universal testing tensile machine. The results show that the thickness of SiC coating prepared by CVD is about 30μm, while the thickness of SiC layer by pack-cementation is nearly 100μm, and the layer is easier to connect with LAS glass ceramic owing to the rough surface and wedge-shaped structure on the SiC coating; the shear strength of joint between SiC coating coated C/C and LAS by pack-cementation is greater than the joint of C/C- LAS with SiC coating prepared by CVD.

Introduction

C/C composites have low coefficient of thermal expansion, high modulus and strength, light weight, excellent thermal shock resistance and so on [1-2]. C/C composites have been widely applied for mechanical componets (brakes and driving shafts), for the aeronautic industry (missle and rockets parts), for engines (aerospatial engines and nozzles), for biomedical engineering (bone prosthesis and tendons) and for other fields. LAS glass ceramic which has low thermal expansion coefficient, high thermal stability and high temperature resistant, mainly is used in the field of liquid crystal displays, cooking utensils and high-temperature observation windows [3-6]. Taking into the shortcomings of C/C composites likes the complex shape of weaving preformed difficultly and brittleness of carbon fiber, the connection between C/C composites and LAS glass ceramic is necessary in order to form the structure-functional integration material. Other works [7-12] show that sevral jointing processes have been used to joint C/C and other matrix composites, such as brazing, diffusion bonding, and reaction jointing process. The reaction jointing process is the appropriate one as a low cost jointing method.

However, due to the mismatch of the thermal expansion coefficient and wettability between C/C composites and LAS [7], the joint can not be connected in high temperatures. It is known that glass has a good wetting and spread ability on SiC coating surface [8]. In order to alleviate the mismatch of thermal expansion coefficient and improve shear strength of joint between C/C composites and LAS glass ceramic, SiC coating as a transition layer is introduced. In the present work, SiC coating is prepared by two different methods (pack-cementation and CVD), and the crystal structure and microstructure of SiC coating are analyzed. Microstructure and strength of as-received joints are investigated.

Experimental

Preparation of C/C composites specimens and SiC coating. The specimens (30mm×20mm×3mm) were cut from 2-D C/C composites with a density of $1.68g/cm^3$, C/C composites specimens were cleaned with ethanol, and then dried at 373K for 2h.

SiC coating was prepared by two different methods including pack-cementation technology and chemical vapor deposition (CVD). The powders for preparing the SiC coating by pack-cementation technology were composed of 50–80wt.% Si, 10–20wt.% graphite and 10–25 wt.% Al_2O_3. The specimens and the pack powders were put in a graphite crucible, and then heat-treated at 2173-2473 K for 1-2 h in argon to form the SiC coating.

The CVD method to prepare SiC coating. Used trichloromethyl silane as precursor material and high-purity H_2 as the carrier gas, diluent gas was introduced into the reaction chamber. The specimens were deposited at 1773-1973K for 1-3h with the gas flowing at 10-30sccm.

Preparation of the joint between C/C–LAS. The process preparing the joint was described as follows: firstly, MAS glass powder as a middle layer was dissolved in ethanol and spread on the surface of SiC coated C/C composites, then specimen and LAS powder were sequenced into a graphite mold, finally the graphite mold was put into vaccum sintering furnace to form the "sandwich-like" joint according the certain operation [13].

Performance test. To analyze the microstructure and crystal structure of SiC coating and joint of C/C-LAS, JSM-6460 scanning electric microscopy (SEM) and X-ray diffraction (XRD) measurements were performanced. The shear strength of the joints was tested by the Instron universal testing tensile machine with the specimens of 10mm × 9mm × 8mm.

Results and discussion

Crystal structure and microstructure of SiC coating. Fig. 1 shows the XRD patterns of SiC coating prepared by pack cementation and chemical vapor deposition. Fig. 1(a) shows that the diffraction peaks of SiC coating by pack-cementation are mainly SiC and Si. The strong diffraction peak of SiC indicates that the coating is mostly composed of SiC phase. Fig. 1(b) shows the main phase of SiC coating by the CVD technology is SiC phase.

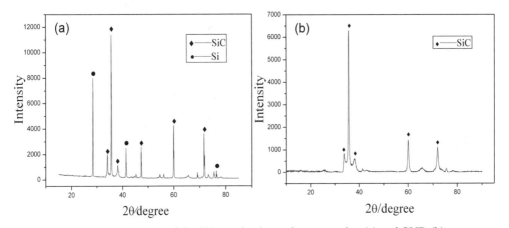

Fig. 1 XRD patterns of the SiC coating by pack-cementation (a) and CVD (b)

Fig. 2 shows the cross-section SEM images of SiC coated C/C composites. From the cross-section SEM images, the thickness of SiC coating by pack-cementation (Fig. 2a) is about 100µm, while the thickness of layer by CVD (Fig. 2b) is nearly 30µm. Fig. 3(a) shows the micrograph with SiC coating by pack-cementation. It can be seen that no defects are on the surface of SiC coating; the coating has rough surface and the big particles. The surface SEM image in Fig. 3(b) shows that SiC coating by CVD is relatively densification, a small amount of micro-cracks exists on the SiC layer and SiC particles stack together with a hill-shaped. In the pack-cementation process, Si element diffuses into the C/C composites through the hole of C/C composites surface to form SiC, while Si element of the coating can be filled in the holes and micro-cracks on the SiC coating by diffusion to form a "pinning" effect, which can alleviate the mismatch of thermal expansion between the SiC coating and C/C composites.

Mechanical property of joint. Table 1 shows the shear strength of joints between C/C composites and LAS glass ceramic. It can be seen that the average shear strength of joint with SiC coating by pack-cementation is nearly 14.93 MPa and the high shear strength ups to 17.38 MPa. The joint of

C/C-LAS with SiC coating by pack-cementation has higher shear strength comparing to the joint of C/C-LAS with SiC coating by CVD. Due to the existence of protruding SiC particles on the SiC coating by pack-cementation in Fig. 3(a), it forms a wedge structure between intermediate layer and the SiC coating which improve the shear strength of joints.

(a)　　(b)

Fig. 2 The cross section SEM images of SiC coating by different methods: (a) the coating by pack-cementation and (b) the coating by CVD

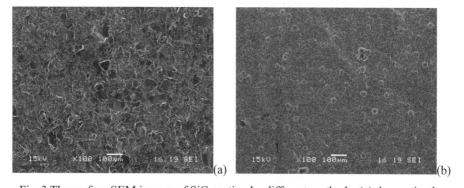

(a)　　(b)

Fig. 3 The surface SEM images of SiC coating by different methods: (a) the coating by pack-cementation and (b) the coating by CVD

Table 1 The shear strength of joints between SiC coated C/C and LAS

Symbol Number	SiC coating by Pack-cementation / MPa	SiC coating by CVD / MPa
1	15.79	9.48
2	11.61	9.05
3	17.38	9.88
Average shear strength	14.93	9.47

Conclusion

SiC coating on the C/C composites was prepared by the pack-cementation and CVD. The thickness of SiC coating prepared by CVD is about 30μm, while the thickness of SiC coating by pack-cementation is nearly 100μm. It has coarse SiC particles and micro-cracks on the SiC coating by pack-cementation comparing with SiC coating by CVD. The highest shear strength of as-received joints by CVD is 9.88 MPa, however the highest joint strength of SiC coated C/C-LAS by pack-cementation is 17.38 MPa; the average strength of joints by pack-cementation is 36.6% higher than the average of joints by CVD.

References

[1]   J.F. Huang, H.J. Li, X.R. Zeng, K.Z. Li, A new SiC/yttrium silicate/glass multi-layer oxidation protective coating for carbon/carbon composites. Carbon. 42 (2004) 2329.

[2]   S.Farhan, K.Z.Li, L.J.Guo, Novel thermal gradient chemical vapor infiltration process for carbon/carbon composites. New Carbon Materials. 22 (2007) 247-252.

[3]   A. Ananthanarayanan, G.P. Kothiyal, L. Montagne, et al., MAS-NMR studies of lithium aluminum silicate (LAS) glasses and glass-ceramics having different $Li2O/Al_2O_3$ ratio. Journal of Solid State Chemistry. 183 (2010) 120-127.

[4]   Y.M. Sung, S.A. Dunn, J.A. Koutsky, The effect of boria and titania addition on the crystallization and sintering behavior of $Li_2O-Al_2O_3-4SiO_2$ glass. Journal of the European Ceramic Society. 14 (1994) 455-462.

[5]   X.Y. Zhai, W.C. Zhou, F. Luo, et al., Microstructure and Dielectric Properties of Chopping SiC_f/LAS Composites. Rare Metal Materials and Engineering. 38 (2009) 2089-2092.

[6]   P.F.Jame, Glass ceramics: new compositions and uses. Journal of Non-Crystal Solids. 181 (1995) 1-15.

[7]   F.T. Lan, K.Z. Li, H.J. Li, Q.G. Fu, et al., Vitreous jointing of SiC-coated carbon/carbon composites. Materials Letters. 62 (2008) 2347-2350.

[8]   M. Salvo, P. Lemoine, M. Terraris, J. Am. Ceram. Joining of carbon-carbon composites for thermonuclear fusion applications. Journal of the American Ceramic Society. Soc. 80 (1997) 206.

[9]   C. Xi, H.J. Li, K.Z. Li, Research of the joint of tungsten phenolic resin and carbon/carbon composite. Rare Met. Mater. Eng. 34 (2005) 367.

[10] J. L. Li, J. T. Xiong, F. S. Zhang.Transient liquid-phase diffusion bonding of two-dimensional carbon/carbon composites to niobium alloy. Mater Sci Eng A. 483-484(2008) 698~702

[11] Dadras P, Mehrotra G M. S. Joining of carbon/carbon composites using boron and titanium disilicide interlayers. Journal of the American Ceramic Society. 76 (1993) 1274-1280.

[12] Pan Y, Baptjista J L. Low-Temperature Sintering of Silicon Carbide with $LiO_2-Al_2O_3-SiO_2$ Melts as Sintering Aids. Eur. Ceram. Soc. 16 (1996) 1221-1230.

[13] X.Q. Lin, K.Z. Li, H.J. Li, et al., oining of Surface Modified Carbon/Carbon Composites and LAS Glass Ceramic. Acta Aeronauticaet Astronautica Sinica. Acta Aeronauticaet Astronautica Sinica. 30 (2009) 380-384.

Key Engineering Materials Vol. 537 (2013) pp 28-31
© (2013) Trans Tech Publications, Switzerland
doi:10.4028/www.scientific.net/KEM.537.28

Crack Formation and Distribution of SiC Coating Prepared by CVD on Carbon/carbon Composites

Junliang Zhu[a], Xiaohong Shi[b], Hejun Li and Zibo He

C/C Composites Technology Research Center, Key Laboratory of Superhigh Temperature
Structural Composite Materials, Northwestern Polytechnical University, Xi'an 710072, P. R. China

[a]zhujunliang@mail.nwpu.edu.cn, [b]npusxh@nwpu.edu.cn

Keywords: C/C composites; SiC coating; Cracks

Abstract. In order to further study the cracks initiation and distribution mechanism of SiC coating, then optimize the coating structure and composition on surface of C/C composites, CVD method is used to deposit SiC coating on C/C composites and pyrolytic carbon. Through the analysis on the reasons of cracks initiation via SEM, the influence of different kinds of substrates and the toughening of SiC whisker on the cracks initiation and distribution are researched. The results show that the cracks easily form in the area near the defects and sample edge, and the cracks also mainly distribute in the direction which are perpendicular to the first fiber layer. Compared with pyrolytic carbon matrix, cracks on C/C composites matrix are width and numerous, and SiC whisker toughening coating can significantly reduce the number of cracks.

Introduction

The research of carbon/carbon (C/C) composites has been intensively improved since this material was explored in 1958. Because the fiber and the matrix have the same element, carbon, C/C composites possess the common characters of carbon material, such as high strength, high modulus, and high thermal and chemical stability. Additionally, the composite structures can be tailored to meet different physical and thermal requirements by weaving architecture design. C/C composites had been considered as the potential materials for aerospace, navigate, medical and other civil fields. The extremely low thermal expansion of C/C and the high-strength retention of C/C at elevated temperature make this material greatly endure over 3000K temperature [1, 2].

The C/C materials are easily oxidized at as low as about 773K temperature in oxygen-containing environment [3]. Coating method is used to overcome this problem. The ceramic materials are usually produced on the surface of C/C composites due to their physical and chemical compatibility with carbon substrates. But because of the structural difference between coating and substrate, cracks would be generated that could induce the failure of coating. Thus, analysis on the crack formation and distribution is very important to promote the coating research [4, 5].

In this paper, SiC coatings and SiC whiskers toughened SiC coating were produced on C/C composites and pyrolytic carbon by CVD method, crack morphology and distribution were investigated to show the crack formation mechanism.

Experiment

The samples, with a dimension of $5 \times 5 \times 5 mm^3$, were cut from C/C composites and pyrolytic carbon, polished with 400 and 1500 grit SiC papers and cleaned ultrasonically with ethanol then dried at 373K for 2h.

The SiC whiskers were formed by the chemical vapor deposition (CVD), using a mixture of high-purity SiO_2, Si and C powders as the raw materials. The chemical composition of these powders was given as follows: 55-75 wt. % SiO_2 (purity: 99.7%) and 10-20 wt. % Si (purity: 99.0%), 15-30 wt. % graphite (purity: 99.5%). The samples were placed over the mixture powder in a graphite crucible, and then the crucible was placed into the center of a vertical resistance heating furnace and heated to

1773 K for 2 h in argon protection atmosphere to obtain the porous SiC whisker layer. The SiC coating was also formed by CVD, using CH_3SiCl_3, and H_2 gases. The coating was deposited at 1273K under a total gas pressure of 0.01Mpa.

The crack morphology and distribution on the coating were investigated through scanning electron microscopy.

Result and discussion

Cracks of SiC coating on surface of pyrolytic carbon and C/C composites. Fig.1 represents the surface morphology of pyrolytic carbon and C/C composites. As shown in Fig. 1(a), the pyrolytic carbon has a smooth surface with no cracks. The diameter of the pores is about 10μm. Different with pyrolytic carbon, the surface morphology of C/C composites is rough and the pores are large. The diameter of the pores is close to 100μm.The fiber bundles also can be seen on it (Fig. 1(b)).

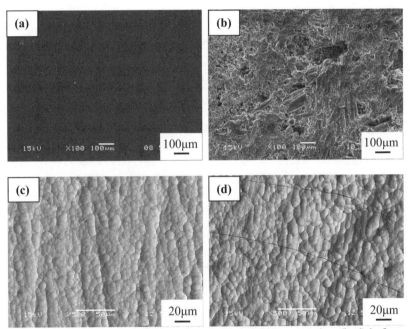

Fig.1 Surface morphology of pyrolytic carbon (a, c) and C/C composites (b, d) before and after deposition SiC coating

The surface morphology of SiC coating after the same deposition processes on the two substrates is shown in Fig.1(c, d). It can be easily distinguished that no cracks in Fig.1(c), the substrate of which is pyrolytic carbon. The cracks on the coating of C/C composites are evident in Fig.1(d), the crack width and length of which is respectively 3-5μm and 300-500μm through SEM photogram.

It's well known that thermal stress could result in coating cracks [6]. Therefore, the thermal stress distribution has a major influence on crack initiation and growth. The pyrolytic carbon substrate has fewer defects and smooth surface, so the thermal stress uniformly distributed, result in the cracks difficultly generated. The C/C composites substrate has a rough surface with more large pores and cracks, the thermal stress easily focus on the defects, causing the generation and growth of the cracks.

Cracks of SiC coatings on C/C composites with and without SiC whisker toughen. Fig.2(a) shows the pre-produced SiC whisker on C/C composite. The SiC whisker porous structure consist of a large amount of random-oriented and distribution-uniform SiC whisker, and the diameters are ranges from 0.5μm to 3μm. Besides, there are also large amounts of SiC particles with a diagram of 0.5μm in the pre-produced SiC whisker porous structure.

During the deposition processes, SiC begin to grow on the particles and whiskers, then fill the porous and finally form a dense SiC coating, as shown in Fig.2 (b). To contrast with Fig.2 (b), Fig.2 (c) is SiC coating produced with no whiskers. It can be seen that less crack exist in the SiC whisker toughen coating, the surface of coating is rough, and some columnar SiC covered whiskers appear on the coating, while the normal coating have apparent cracks which are intersected to form a reticular structure. In addition, the pores are not sealed by the coating, as shown in the inferior part of Fig.2 (c).

The fact that no obvious defects on the surface of the coating suggest the fracture toughness and elastic modulus are increased owing to the introduction of SiC whiskers, since the excellent Young's modulus of SiC whiskers and the uniform dispersion of SiC whiskers in the SiC coating allow elastic recovery and improve the hardness of SiC coating. The SiC whiskers improve the fracture toughness and elastic modulus via three modes: pullout, bridging and deflect. The pullout whiskers can absorb a vast amount of fracture energy. The whisker bridging between a microcrack retards the further widening of the crack. And the propagation path of the microcrack is prolonged upon deflecting, when more energy is needed [7].

Fig.2 SiC whisker porous structure on the C/C composites(a); Surface morphology of SiC whisker toughened coating (b) and non-toughened coating (c)

Formation of cracks of SiC coating on C/C composites. The thermal stress result in coating cracks formation and expand, while the thermal stress generated when the coating cools from the production temperature to the ambient temperature due to the mismatch of the thermal expansion coefficient between the SiC coating(4.3×10^{-6}/K) and C/C composites(1.0×10^{-6}/K). However, the reinforcing phase of C/C composites is a carbon fiber weave structure with a large amount of porous and microcracks, which will cause the production of more cracks.

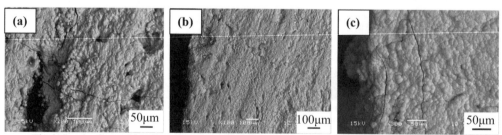

Fig.3 Cracks formed by the substrate defects(a); Cracks direction on SiC coating(b); Cracks initiated by the sample edge(c)

Fig.3(a) shows clearly the cracks formed by the substrate defects. Due to the short deposition time, the defects are not filled with SiC coating completely, the pores still exist on the surface of the coating. The coating defects result in concentration of stress, so the cracks initiation by the pores and connect to other pores and the crack width is comparatively large by the pores. With the cracks propagated, the width become narrow because of the stress release.

Fig.3(b) shows the relation between crack distribution and the first fiber layer. In the direction of perpendicular to the first fiber layer, the cracks are in a large number and parallel to each other. The crack width is narrow. This is because the thermal expansion coefficient of the C/C substrate in the out-of-plane directions is much larger than that of SiC.

Fig.3(c) shows the cracks initiated by the edge of the sample, and then propagated parallel to the arris direction of the sample. It has been verified that the sharp edge of the sample will lead to the concentration of stress. It can be noticed that the crack width become narrow and no cracks in the inner area because of the truth that the stress level decrease gradually with the larger distance to the boundary.

Summary

SiC coatings were deposited on C/C composites and pyrolytic carbon via CVD method to show the cracks formation and propagate mechanism during the coatings cooled down from the production temperature to the ambient temperature. The cracks width and length on the C/C composites sample were respectively 3-5μm and 300-500μm, and rarely cracks on the pyrolytic carbon sample due to more defects on C/C composites than pyrolytic carbon. Cracks in the SiC whisker toughened coating could be neglected while the cracks in the non-toughened coating were apparent because the whiskers could impede the cracks' formation and propagate. The cracks distribution on C/C composites sample show that cracks are easily formed in the thermal stress focus zone, such as the area near the defects and sample edge. The cracks are also easily formed in the direction of perpendicular to the first fiber layer.

Acknowledgments

This work has been supported by the National Natural Science Foundation of China under Grant Nos. 50832004 and 50902111, the "111" Project (Grant No. B08040).

References

[1] A. V. K. Westwood, B. Rand, S. Lu, Oxidation resistant carbon materials derived from boronated carbon–silicon alloys, Carbon. 42 (2004) 3071–3080.

[2] C. A. A. Cairo, M. Florian, M. L. A. Graca, Kinetic study by TGA of the effect of oxidation inhibitors for carbon–carbon composite, Mater Sci Eng A. 358 (2003) 298–303.

[3] C. Friedrich, R. Gadow, M. Speicher, Protective multilayer coatings for carbon–carbon composites, Surf. Coat. Technol. 151–152 (2002) 405–411.

[4] J.F. Huang, X.R. Zeng, H.J. Li, X.B. Xiong, Y.W. Fu. Influence of the preparation temperature on the phase microstructure and anti-oxidation property of a SiC coating for C/C composites, Carbon. 42 (2004) 1517–1521.

[5] Q.G. Fu, H.J. Li, X.H. Shi, Microstructure and anti-oxidation property of $CrSi_2$–SiC coating for carbon/carbon composites, Appl. Surf. Sci. 252 (2006) 3475–3480.

[6] R. Z. Wang, H. B. Wen, F. Z. Cui, H. B. Zhang, and H. D. Li, Observations of damage morphologies in nacre during deformation and fracture, J. Mater. Sci. 30 (1995) 2299-304.

[7] Y. H. Chu, Q. G. Fu, H. J. Li, SiC coating toughened by SiC nanowires to protect C/C composites against oxidation, Ceramics International. 38 (2012) 189-194.

Key Engineering Materials Vol. 537 (2013) pp 32-35
© (2013) Trans Tech Publications, Switzerland
doi:10.4028/www.scientific.net/KEM.537.32

A C/SiC Graded Coating on C/C Composites by LPCVD in the Methyltrichlorosilane/Hydrogen System

He Zibo[a], Li Hejun[b], Shi Xiaohong[c]

State Key Laboratory of Solidification Processing, Northwestern Polytechnical University, Xi'an 710072, PR China

[a]zibohe@yahoo.com.cn, [b]lihejun@nwpu.edu.cn, [c]npusxh@nwpu.edu.cn

Keywords: LPCVD; C/SiC; graded coating; oxidation resistance;

Abstract. In order to improve the anti-oxidation of C/C composites, a C/SiC functionally graded coating for C/C composites was prepared by low pressure chemical vapor deposition (LPCVD) using methyltrichlorosilane (MTS, CH_3SiCl_3) and H_2 as precursors. The relative amount of C and SiC in coatings was varied by controlling the input ratio of H_2 to MTS. The phase composition and morphology were examined by scanning electron microscope (SEM) and X-ray diffraction (XRD), and the content distribution of C and SiC phases were investigated by energy dispersive spectroscopy (EDS). The isothermal oxidation test was evaluated at 900°C and 1500°C respectively. The results showed that the as-obtained coatings were possessed of a dense and uniform structure, and the SiC content in coatings increased with an increment of the molar ratio of H_2 to MTS. The C/SiC coating had a good oxidation resistance for C/C composites.

Introduction

C/C composites are distinguished for the excellent high temperature mechanical properties that are suitable for the high-temperature structural components. However, they are restricted to an inert or vacuum environment due to the oxidation of C/C composites above 400°C in an oxidizing atmosphere [1-4].

The coating technique is considered as the best method to improve the anti-oxidation of C/C composites at high temperature, and in particular SiC ceramic coating is the most promising candidate for C/C composites thanks to a high decomposition temperature, good thermal shock resistance and good compatibility with C/C composites [5][6]. Nevertheless, the thermal stress resulting from the CTE (thermal expansion coefficient) mismatch of SiC ($\alpha_{SiC}=4.5\times10^{-6}$/°C) and C/C composites ($\alpha_{C/C}=1\times10^{-6}$/°C) may cause cracks in SiC coatings during preparation procedure or service period. In order to reduce the thermal stress and formation of cracks, a C/SiC graded coating is introduced on C/C composites by chemical vapor deposition (CVD) method. The gaseous precursors involved in the deposition of C/SiC graded coatings included CH_4-$SiCl_4$-H_2 system [7][8], C_3H_8-$SiCl_4$-Ar-H_2 system [9][10] and CH_3SiCl_3-C_2H_2-H_2 system [11][12], whereas the only precursor of CH_3SiCl_3/H_2 system without a separate carbon source was not reported for the deposition of C/SiC graded coatings.

In this paper, coatings with various C/Si ratios were separately deposited on C/C composites by the LPCVD method at 1250°C using CH_3SiCl_3/H_2 as precursors, and then the C/SiC graded coating was obtained in one step by controlling the input ratio of CH_3SiCl_3 to H_2. The composition and microstructure of each separate coating and the C/SiC graded coating were evaluated. The oxidation resistance of the C/SiC graded coating was tested at 900 °C and 1500 °C.

Experimental

Experimental details. Small specimens with a dimension of 10 mm×10 mm×4 mm used as substrates were cut from a bulk 2D-C/C composite (prepared in Xi'an, China) with a density of 1.65 g/cm^3. They were abraded with a sequence of 100, 400 and 800 grit SiC paper, cleaned ultrasonically with ethanol and dried at 90 °C for 24 h.

The CVD process was carried out within an alumina tube in a vertical furnace, demonstrated schematically in Fig.1. The as-dried specimens were suspended with molybdenum wires and carbon fibers in the alumina tube. The liquid CH_3SiCl_3 precursor was transported into the reaction chamber of the furnace using H_2 as a carrier gas which was bubbled through the liquid CH_3SiCl_3. The flow rate of CH_3SiCl_3 was kept constant by putting the liquid CH_3SiCl_3 in a 20 ± 1 °C bath and fixing the inflow of H_2. The flow rate of the dilute H_2 was manipulated from 0 to 1200 ml/min, respectively named as H0, H250, H500, H750, H1200. The deposition temperature, time and pressure were maintained 1250 °C, 10 h and 5 ~ 10 KPa, respectively.

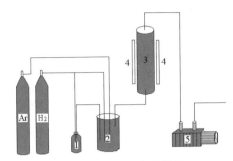

Fig.1 Schematic diagram of LPCVD system: 1. CH_3SiCl_3 bubbling bottle; 2. Mixing tank; 3. Alumina tube; 4. Heating element; 5. Vacuum pump

The compositions and morphologies of coatings were analyzed by X-ray diffraction (XRD, X'Pert MPD PRO) and scanning electron microscopy (SEM, Tescan VEGA TS5136XM). The distribution of C and SiC phases in C/SiC layer was characterized by EDS, attached to SEM. The hardness of the coatings was tested with a microidentation. The isothermal oxidation tests were performed at 900 °C and 1500 °C in an electric furnace in air. After oxidized for 2 h, the specimens were put out of the furnace to air and cool down to room temperature quickly, and then the specimens were weighted by an electronic balance with a sensitivity of ±0.1 mg. The process was repeated for 5 times, and the weight loss of the specimens, as a function of the oxidation time, was calculated with the following equation:

$$\Delta W\% = \frac{m_0 - m}{m_0} \times 100\% \tag{1}$$

where m_0 and m are the weight of the specimens before and after oxidation.

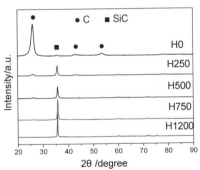

Fig.2 XRD spectra of the separate coatings with varied ratio of H_2 to CH_3SiCl_3

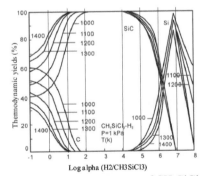

Fig.3 The CVD phase diagram of CH_3SiCl_3/H_2 system[14]

Results and discussion

Fig.2 shows the XRD spectra of the prepared coatings with the increase of the flow rate of the dilute H_2. From Fig.2, the prepared coatings consist of C phase and SiC phase. As the flow rate of dilute H_2 increases, the relative intensity of the C peak to SiC peak decreases rapidly. For the H0 specimen, the major phase is C and the minor phase is SiC, while it consists of pure SiC phase for the H1200 specimen. In the CVD process of CH_3SiCl_3/H_2 system [13-15], the major reaction of CH_3SiCl_3 was to break the Si-C bond to originate CH_3 and $SiCl_3$ radicals which were decomposed and reacted further to form the deposits of SiC. During course of reaction H_2 would suppress the formation of C and stimulate that of Si in the deposits, which could result for the varied content of C and Si in the obtained coating and prepare the functionally graded coating. Fig.3 clarified the relationship between the deposits of the CH_3SiCl_3/H_2 system and the input ratio of H_2 to CH_3SiCl_3.

Fig.4 exhibits the backscattered electron (BSE) image and the corresponding element line scanning EDS analyses of the cross-section of the C/SiC coating prepared in one step. In Fig.4 (a), it reveals that the as-obtained coating is composed of five layers which have a distinct interface with each other, and they are well bonded without any delamination. With the increasing of the flow rate of dilute H_2, the deposition rate of coatings increases. For Fig.4 (b), it can be found that the intensity of C element decreases and that of Si element increases slowly from the substrate surface to the outer layer. As mentioned above in Fig.2, the as-obtained coatings are just composed with C phase and SiC phase, and it can be concluded that the graded C/SiC coating was deposited by manipulating the input ratio of H_2 to CH_3SiCl_3 in the CH_3SiCl_3/H_2 system.

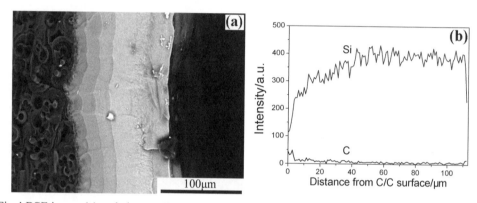

Fig.4 BSE image (a) and element line scanning analyses (b) of the cross-section of the C/SiC graded coating

The oxidation resistance of the specimens with C/SiC graded coating is demonstrated in Fig.5. In the isothermal oxidation tests, the number of the thermal shock between the oxidation temperature and room temperature was 5, and then the oxidation time was 10 h. From Fig.5, it implies that the coating can protect C/C composites with a little weight loss of 0.5% at 1500 °C. Moreover, the weight loss of the specimen is 4.5% at 900 °C, which is well improved compared to the pure SiC coating on C/C composites reported in Ref [11]. During the oxidation process, the oxygen would diffuse to C/C composites through microcracks on the coating surface, and react with C/C composites to bring on the weight loss. At the same time, SiC coating could react with oxygen to generate SiO_2 glass, which would seal the microcracks because of the fluidity of the SiO_2 glass at 1500 °C. While the formed SiO_2 would not spread the coating surface to seal the microcracks at 900 °C, resulting from the viscosity of it.

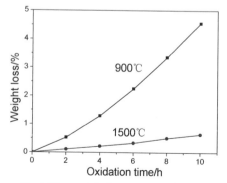

Fig.5 The weight loss curves of coated samples at 900 °C and 1500 °C

Conclusion

C/SiC graded coating was deposited on C/C composites by LPCVD method using CH_3SiCl_3/H_2 as precursors, and the as-obtained coating had a dense and uniform structure. The composition of C/SiC coating could be easily controlled by manipulating the input ration of CH_3SiCl_3 to H_2. The C/SiC graded coating on C/C composites exhibited the good oxidation resistance and thermal shock resistance both at high temperature (1500 °C) and at moderate temperature (900 °C).

References

[1] J.D. Buckley, Carbon-carbon, an overview, American Ceramic Society Bulletin 67 (2) (1988) 364-368.

[2] H.J. Li, Carbon/carbon composites, New Carbon Materials 16 (2) (2001) 79-80.

[3] T. Feng, H.J. LI, Q.G. Fu, et al., High-temperature erosion resistance and aerodynamic oxidation mechanism of multi-layer MoSi2-CrSi2-Si/SiC coated carbon/carbon composites in a wind tunnel at 1873K, Carbon 50 (2012) 2171-2178.

[4] Y.L. Zhang, H.J. Li, X.F. Qiang, et al., C/SiC/MoSi2-Si multilayer coatings for carbon/carbon composites for protection against oxidation, Corrosion Science 53 (2011) 3840-3844.

[5] M.E. Westwood, J.D. Webster, R.J. Day, et al., Oxidation protection for carbon fiber composites, Journal of Materials Science 31 (1996) 1389-1397.

[6] M.H. Van De Voorde, M.G. Hocking, V. Vasantasree, A comprehensive review on the development of ceramic coatings for thermo mechanical applications, High Temperature Materials and Processes 7 (2&3) (1986) 107-121.

[7] W. Kowbel, J.C. Witilers, CVD and CVR silicon-based functionally gradient coatings on C-C composites, Carbon 33 (4) (1995) 415-426.

[8] Yoshio Wakamatsu, Tetsuo Shoji, Isao Hino, et al. Oxidation damage process of C/C composites with functionally graded C/SiC coating. 9th Int. Space Planes and Hypersonic Systems and Technologies Conference, American Institute of Aeronautics &Astronautics, AIAA-99-4913, 1-4 November 1999, Norfolk, Virginia

[9] Yootaek: Kim, Jun-Tae Choi, Jong Koen Choi, et al., Effect of source gas composition on the synthesis of SiC/C functionally gradient materials by CVD, Materials Letters 26 (1996) 249-257.

[10] Y. Wang, M. Sasaki, T. Hiral, Density and microstructure of CVD SiC-C nanocomposites, Journal of Materials Science 26 (1991) 6618-6624.

[11] J.I. Kim, W.-J. Kim, D.J. Choi, et al., Design of a C/SiC functionally graded coating for the oxidation protection of C/C composites, Carbon 43 (2005) 1749-1757.

[12] Joung Il Kim, Weon-Ju Kim, Doo Jin Choi, et al., Deposition of compositionally graded SiC/C layers on C-C composites by low pressure chemical vapor deposition, Journal of Nuclear Materials 307-311 (2002) 1084-1087.

[13] Y.D. Xu, X.T. Yan, Chemical Vapor Deposition-An Integrated Engineering Design Approach, first ed., Springer-Verlag, London, 2010.

[14] R. Naslain, F. Langlais, R. Fedou, The CVI-processing of ceramic matrix composites, Journal De Physique 50 (5) (1989) 191-207.

[15] W.G. Zhang, K.J. Hüttinger, CVD of SiC from Methyltrichlorosilane. Part II: Composition of the Gas Phase and the Deposit, Chemical Vapor Deposition 7 (4) (2001) 173-181.

[16] M.D. Allendor, C.F. Melius, Theoretical study of the thermlchemistry of molecules in the Si-C-Cl-H system, Journal of Physical Chemistry 97 (1993) 720-728.

Key Engineering Materials Vol. 537 (2013) pp 36-41
© (2013) Trans Tech Publications, Switzerland
doi:10.4028/www.scientific.net/KEM.537.36

Effect of Yb$_2$SiO$_5$ Ceramic Layer Thickness on the Thermal Cycling Life of Yb$_2$SiO$_5$/LaMgAl$_{11}$O$_{19}$ Coating Deposited on C/SiC Composites

Ying Wang[1, 2], Binglin Zou[1, a], Xizhi Fan[1, 2] and Xueqiang Cao[1, b]

[1]State Key Laboratory of Rare Earth Resources Utilization, Changchun Institute of Applied Chemistry, Chinese Academy of Sciences, Changchun 130022, Jilin, China

[2]Graduate School of the Chinese Academy of Sciences, Beijing, 100049, China

[a] zoubinglin@ciac.jl.cn, [b] xcao@ciac.jl.cn

Keywords: C/SiC composites; Coating thickness; thermal cycling; Failure mechanism

Abstract: C/SiC composites were plasma sprayed with Yb$_2$SiO$_5$/LaMgAl$_{11}$O$_{19}$ (LMA) coatings with varying Yb$_2$SiO$_5$ layer thickness. The effect of Yb$_2$SiO$_5$ layer thickness on the thermal cycling life of the Yb$_2$SiO$_5$/LMA coatings was investigated. The results showed that the thermal cycling life is significantly dependent on the Yb$_2$SiO$_5$ layer thickness. It decreased from 130 cycles to 35 cycles as Yb$_2$SiO$_5$ layer thickness increased from 50 μm to 100 μm. Further increasing Yb$_2$SiO$_5$ layer thickness to 200 μm made it decrease to 2 cycles. The influencing mechanism of Yb$_2$SiO$_5$ layer thickness for the thermal cycling life was clarified based on the thermal expansion behavior, the chemical stability at high temperature and the microstructure analysis.

Introduction

Carbon fiber reinforced silicon carbide ceramic matrix composites (C/SiC) are the promising candidate materials for advanced structural components in propulsion and aerospace applications [1]. Unfortunately, C/SiC composites are apt to oxidation in oxidizing environment, which significantly reduce the mechanical strength. Coating protection is a better approach to improve the oxidation resistance of C/SiC composites at high temperature [2].

Recently, the silicon-containing ceramic coatings were prepared on C/SiC composites, which can well protect C/SiC composites against oxidation [2-9]. However, the oxidation protection of these coatings seems to be only satisfied at relatively low temperatures below 1873 K. High temperature up to 2273 K or higher will be created in new generation of rocket propulsion and hypersonic aircraft. The C/SiC composites may fail even though they are protected by the developed coatings.

In order to develop oxidation protection coatings for C/SiC composites at higher temperature, our group raised a new idea in which the coating consists of inner layer of rare-earth element silicate and top layer of thermal barrier ceramic. Due to the higher melting point, lower thermal conductivity, lower oxygen permeability and superior oxidation resistance, the top thermal barrier ceramic layer will provide good heat insulation and thus enhance the heat resisting and thermal stability of the inner layer. For this reason, the rare-earth element silicate layer may protect C/SiC composites against oxidation at higher temperature. In a previous paper, authors designed an Yb$_2$SiO$_5$/LaMgAl$_{11}$O$_{19}$ (LMA) oxidation protection coating for C/SiC composites [10]. The coated samples were subjected to dynamic thermal cycling (DTC) tests with gas flame temperature about 2273 K. It was found that the weight loss for the sample coated on one-side was 4.1% after 11 cycles of heating for 85 min, for the uncoated sample it was as high as 20.6% [10].

The coating thickness is one of the crucial factors that can manipulate the thermal cycling life. In this work, Yb$_2$SiO$_5$/LMA coatings with varying Yb$_2$SiO$_5$ layer thickness were fabricated on C/SiC composites by atmospheric plasma spraying (APS). The effect of Yb$_2$SiO$_5$ layer thickness on the thermal cycling life was investigated and the influencing mechanism was proposed.

Experimental procedure

Yb_2SiO_5 and LMA powders were synthesized by solid-state reactions using commercial powders of Yb_2O_3, SiO_2, MgO, Al_2O_3 and La_2O_3 (all purities >99%). The synthesized powders with proper additions of de-ionized water, Gum Arabic and Tri-Ammonium Citrate were ball milled for 72 h. The formed suspensions were spray-dried (GZ-5, Yangguang Ganzao), and the obtained powders were then sieved to collect those with particle size of 32-100 μm for APS.

Two-dimensional C/SiC composites provided by Institute of Metal Research, Chinese Academy of Sciences were used. Small substrates with the dimensions of 35 mm × 45 mm × 4 mm were cut from the bulk C/SiC composites and hand-grounded using 600 grit SiC papers, followed by cleaning with ethanol and drying at 373 K in an oven. The substrates were properly grit blasted with alumina sand prior to APS. The feed powders were plasma sprayed onto C/SiC substrates by a Sulzer Metco plasma spraying unit with the F4-MB gun. Deposition efficiency of the feed powders were predicted prior to produce the coatings, from which the layer thickness in the coating can be well controlled. All samples were coated on one-side with the dimensions of 35 mm × 45 mm.

DTC tests were performed for the coated samples on a burner-rig setting. The temperature of the gas flame at the position of the coating surface was monitored as about 2273 K by using a pair of W-5%Re/W-26%Re thermocouples via controlling the flowing rates of natural gas and oxygen. For each cycle, the sample was heated for 6 min and then cooled to room temperature within 2 min by compressed air jet. The samples were thermal cycled until an area of about 5% of the coating was peeled off or macro-cracks were observed using optical microscopy.

Phase composition for powders and coatings was identified by X-ray diffraction (XRD, Bruker D8 Advance diffractometer, Cu Kα radiation). Microstructure of the coated samples before and after DTC tests was examined using scanning electron microscopy (SEM, XL-30 FEG, Philips). Thermal expansion behavior of the coupons of C/SiC composites, Yb_2SiO_5 powder and LMA powder was investigated by using a Netzsch DIL 402C Dilatometer.

Fig. 1 Typical SEM images on the (a) surface and (b) cross-section of Yb_2SiO_5/LMA coating deposited on C/SiC composites.

Results and discussion

The phase compostion of LMA layer in the Yb_2SiO_5/LMA coating consists of LMA and a small amount of $LaAlO_3$, and that of Yb_2SiO_5 layer is composed of Yb_2SiO_5 [10]. Figs. 1(a-b) show the typical surface and cross-section microstructures of Yb_2SiO_5/LMA coatings deposited on C/SiC composites. The surface feature of Yb_2SiO_5/LMA coatings can be described by the presence of fully melted and partly melted regions [see Fig. 1(a)]. The deposits in the fully melted region exhibit the amorphous structure due to the rapid cooling of the droplets in APS deposition (>10^6 K/S). As shown in Fig. 2(b), the Yb_2SiO_5 layer is bright and the LMA layer is grey, which are both evenly distributed on the substrate.

The coated samples with varying Yb_2SiO_5 layer thickness and LMA layer thickness about 100 μm were subjected to the DTC test. The thermal cycling lives for the Yb_2SiO_5/LMA coatings with Yb_2SiO_5 layer thickness about 50 μm, 100 μm and 200 μm were found to be about 130 cycles, 35 cycles and 2 cycles, respectively. Clearly, an increase in Yb_2SiO_5 layer thickness progressively decreased the thermal cycling life. Fig. 2 shows the macrographs of the samples after DTC test. As shown in Fig. 2(a), the Yb_2SiO_5/LMA coating with 50 μm Yb_2SiO_5 layer remained intact after 50 cycles, but a small piece of the coating was peeled off after 130 cycles [see Fig. 2(b)]. From Fig. 2(c), the Yb_2SiO_5/LMA coating with 100 μm Yb_2SiO_5 layer showed macrocracks and spallation after 35 cycles. From Fig. 2(d), the macrocracks almost propagated through the Yb_2SiO_5/LMA coating with 200 μm Yb_2SiO_5 layer only after 2 cycles. It was interestingly found that the single-layer Yb_2SiO_5 coating did not fail even after 180 cycles [see Fig. 2(e)]. Note that the oxidation protection of single-layer Yb_2SiO_5 coating for C/SiC composites was not so good as that of the Yb_2SiO_5/LMA coating with 50 μm Yb_2SiO_5 layer.

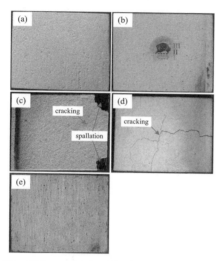

Fig. 2 Macrographs of (a-d) the Yb_2SiO_5/LMA coatings with varying Yb_2SiO_5 layer thickness deposited on C/SiC composites: (a-b) Yb_2SiO_5 layer thickness about 50 μm after (a) 50 cycles and (b) 130 cycles, (c) Yb_2SiO_5 layer thickness about 100 μm after 35 cycles, (d) Yb_2SiO_5 layer thickness about 200 μm after 2 cycles, and (d) the single-layer Yb_2SiO_5 coating with thickness about 50 μm deposited on C/SiC composites after 180 cycles.

Fig. 3 (a) SEM image on the single-layer Yb_2SiO_5 coating after 180 cycles, (b-d) SEM images for regions I, II and III in Fig. 3(b) for the Yb_2SiO_5/LMA coating with 50 μm Yb_2SiO_5 layer after 130 cycles, respectively, (e) SEM image on the Yb_2SiO_5/LMA coating with 100 μm Yb_2SiO_5 layer after 35 cycles and (f) SEM image on the Yb_2SiO_5/LMA coating with 200 μm Yb_2SiO_5 layer after 2 cycles.

Fig. 3 shows the surface microstructures of the thermal cycled samples. From Fig. 3(a), network cracks appeared on the single-layer Yb_2SiO_5 coating. The recrystallization of the amorphous phase led to the formation of platelet-like grains and thus the porous structure [see the insert in Fig. 3(a)]. Figs. 3(b-d) show the microstructures for regions I, II and III in Fig. 2(b), respectively. From Fig. 3(b), a dense microstructure was developed, especially liquid sintering occurred among the grains [see the insert in Fig. 3(b)]. From Fig. 3(c), bumps and cracks are observed on the periphery of the failed coating. Moreover, many small bubbles are present on the substrate closing to the periphery of the failed coating. The gases of CO and/or CO_2 produced by the oxidation of substrate may not promptly escape from the sintered coating due to the dense microstructure; as a result, the bubbles were formed. From Fig. 3(d), the bubbles grew up, which accelerated the formation of bumps and cracks and therefore the delamination of the coating from the substrate, leading to the peeling off of the coating.

Figs. 3(e-f) show the microstructures of the cycled Yb_2SiO_5/LMA coatings with Yb_2SiO_5 layer thickness 100 μm and 200 μm, respectively. As shown, cracks appear. The amorphous phase in the as-sprayed coating [see Fig. 1(a)] recrystallizatied into the platelet-like grains, but liquid sintering among the grains did not occurred after the failure of the coatings [see the inserts in Figs. 3(e-f)].

In order to explore the influencing mechanism of Yb_2SiO_5 layer thickness for the thermal cycling life, the thermal expansion behavior of Yb_2SiO_5 and LMA as well as the chemical stability between Yb_2SiO_5 and LMA at high temperatures were investigated.

Fig. 4 shows the variations in $\Delta L/L_0$ (L_0 is the coupon length at room temperature and ΔL is the length change) with temperature for LMA, Yb_2SiO_5 and C/SiC coupons. As indicated, the values of $\Delta L/L_0$ for LMA and C/SiC coupons nearly exhibit a linear increase with temperature increasing. A similar expansion behavior is observed for the Yb_2SiO_5 coupon before 1415 °C. But after 1415 °C, a parabolic decrease in $\Delta L/L_0$ is clearly observed, implying the occurrence of volume contraction in the coupon. Linear fits between $\Delta L/L_0$ and temperature (see the dash lines in Fig. 4) indicate that the coefficients of thermal expansion (CTEs) for C/SiC, Yb_2SiO_5 and LMA coupons are about 2.57, 7.11 and 8.94 ppm/°C, respectively.

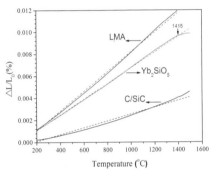

Fig. 4 Variations in $\Delta L/L_0$ with the temperature for the coupons of C/SiC composites, LMA powder and Yb_2SiO_5 powders.

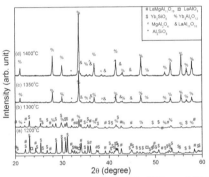

Fig. 5 XRD patterns for the Yb_2SiO_5 and LMA powder mixture heat-treated at different temperatures (a) 1200 °C, (b) 1300 °C, (c) 1350 °C and (d) 1400 °C for 12 h in the furnace.

Fig. 5 shows the XRD patterns for the heat-treated Yb_2SiO_5 and LMA powder mixture at different temperatures. Phase composition of the mixture did not vary after heat-treatment at 1200 °C [see Fig. 5(a)]. However, a small amount of $Yb_3Al_5O_{12}$ and Al_2SiO_5 were detected after heat-treatment at 1300 °C [see Fig. 5(b)], implying the occurence of chemical reactions. The reactants disappeared and the products consisted of $Yb_3Al_5O_{12}$, Al_2SiO_5, $MgAl_2O_4$ and $LaAl_{11}O_{18}$ after heat-treatment at 1350 °C and 1400 °C [see Figs. 5(c-d)]. The diffraction intensities of Al_2SiO_5, $MgAl_2O_4$ and $LaAl_{11}O_{18}$ are weak while that of $Yb_3Al_5O_{12}$ is strong. The Al_2SiO_5 and $MgAl_2O_4$ may form a glass melt and thus gave the weak diffraction intensities. The $LaAl_{11}O_{18}$ grain may dissolve into the glass melt, which made its diffraction intensity also weak. The strong diffraction intensity of $Yb_3Al_5O_{12}$ phase implied that it crystallized well. The increasing temperature made the $Yb_3Al_5O_{12}$ grains grow, and thus its diffraction intensity increased [See Figs. 5(c-d)].

The coupon consisting of Yb_2SiO_5 and LMA powder layers was prepared and heated at 1400 °C for 12 h in the furnace. Fig. 6 shows the cross-section microstructures of the heat-treated coupon. As shown in Fig. 6(a), a transition interface marked by a rectangle was formed between Yb_2SiO_5 and LMA powder layers, which implies the significant interdiffusion of atoms. Figs. 6(b-d) show the high magnification SEM images for Yb_2SiO_5 powder layer, the transition interface and LMA powder layer, respectively. The morphology in the transition interface is quite differrent from those in the powder layers. Dense microstructure was formed in the transition interface, which may result from chemical reactions and the subsequent melt formation possiblly due to eutectic reaction among the products

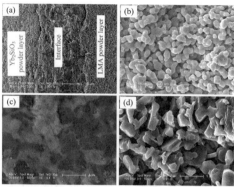

Fig. 6 (a) Cross-section microstructure of the heat-treated coupon consisting of Yb_2SiO_5 and LMA powder layers and (b-d) the corresponding high magnification SEM images for (b) the Yb_2SiO_5 powder layer, (c) the interface and (d) the LMA powder layer, respectively.

and/or the melting of the resultant products. The formed melt was solidified into the grey and black microstructures [see Fig. 6(c)]. Combining Fig. 5 with Fig. 6, it is believed that the chemical reactions between Yb_2SiO_5 and LMA occured at high temperatures and subsequently glass melt was formed, leading to the dense microstrucutre in the final product.

According to above results, the effect of Yb_2SiO_5 layer thickness on the thermal cycling life is not difficult to understand. For thicker Yb_2SiO_5 layer (e.g. ≥100 μm), the CTE mismach between the substrate and Yb_2SiO_5 layer and the volume contraction of Yb_2SiO_5 after 1415 °C are the crucial factors manipulating the thermal cycling life of the coating. More thicker the Yb_2SiO_5 layer, more larger the interface thermal stress between the substrate and the Yb_2SiO_5 layer because of the CTE mismach and the volume contraction of Yb_2SiO_5. As a result, the thermal cycling life decreased from 35 cycles to 2 cycles when the Yb_2SiO_5 layer thickness increased from 100 μm to 200 μm. For thin Yb_2SiO_5 layer (e.g. ≤50 μm), the thermal stress by the CTE mismach and the volume contraction of Yb_2SiO_5 is considered to be very small. Moreover, the developed porous strucutre resulted in high strain tolerrance [11]. As a result, the single-layer Yb_2SiO_5 coating did not peel off even after 180 cycles. The failure of Yb_2SiO_5/LMA coating with 50 μm Yb_2SiO_5 layer after 130 cycles should be contributed to on one hand the liquid sintering of the coating, which resulted from the chemical reactions due to the interdiffusion of atoms between the double layers and the subsequent melt formation, and on the other hand the formation of bubbles between the substrate and the coating.

Summary

The Yb_2SiO_5 layer thickness plays a significant role on manipulating the thermal cycling life of Yb_2SiO_5/LMA coating deposited on C/SiC composites by APS. The thermal cycling life decreased from 130 cycles to 35 cycles when the thickness of Yb_2SiO_5 layer in the Yb_2SiO_5/LMA coating increased from 50 μm to 100 μm. It further decreased to 2 cycles with the Yb_2SiO_5 layer thickness increasing to 200 μm. The influencing mechanism of Yb_2SiO_5 layer thickness for the thermal cycling life is as follows: in the case of thin Yb_2SiO_5 layer (e.g. ≤50 μm), the failure of Yb_2SiO_5/LMA coating should be contributed to on one hand the liquid sintering of the coating, which resulted from the chemical reactions between the double layers and the subsequent melt formation, and on the other hand the formation of bubbles between the substrate and the coating. In the case of thick Yb_2SiO_5 layer (e.g. ≥100 μm), the CTE mismach between the substrate and the Yb_2SiO_5 layer as well as the volume contraction of Yb_2SiO_5 after 1415 °C are the crucial factors manipulating the failure of the coating; more thicker the Yb_2SiO_5 layer, more larger the thermal stress between the substrate and the Yb_2SiO_5 layer, which thus gave rise to a significant decrease in thermal cycling life.

References

[1] P.D. Sarkisov, N.V. Popovich, L.A. Orlova, Yu.E. Anan'eva, Barrier coatings for type C/SiC ceramic-matrix composites (Review), Glass Ceram. 65 (2008) 366-371.

[2] L.F. Cheng, Y.D. Xu, L.T. Zhang, R. Gao, Oxidation behavior of C/SiC composites with a Si-W coating from room temperature to 1500 °C, Carbon 38 (2000) 2133-2138.

[3] S.J. Wu, L.F. Cheng, L.D. Zhang, Y.D. Xu, Oxidation behavior of 2D C/SiC with a multi-layer CVD SiC coating, Surf. Coat. Technol. 200 (2006) 4489-4492.

[4] Z.Q. Yan, X. Xiong, P. Xiao, *et al.*, Si-Mo-SiO$_2$ oxidation protective coatings prepared by slurry painting for C/C–SiC composites, Surf. Coat. Technol. 202 (2008) 4734-4740.

[5] M. Aparicio, A. Durán, Yttrium silicate coatings for oxidation protection of carbon-silicon carbide composites, J. Am. Ceram. Soc. 83 (2000) 1351-1355.

[6] J.D. Webster, M.E. Westwood, F.H. Hayes, *et al.*, Oxidation protection coatings for C/SiC based on yttrium silicate, J. Euro. Ceram. Soc. 18 (1998) 2345-2350.

[7] X.H. Zheng, Y.G. Du, J.Y. Xiao, *et al.*, Double layer oxidation resistant coating for carbon fiber reinforced silicon carbide matrix composites, Appl. Surf. Sci. 255 (2009) 4250-4254.

[8] X.H. Zheng, Y.G. Du, J.Y. Xiao, *et al.*, Celsian/yttrium silicate protective coating prepared by micro- wave sintering for C/SiC composites against oxidation, Mater. Sci. Eng. A 505 (2009) 187-190.

[9] M. Aparicio, A. Durán, Preparation and characterization of 50SiO$_2$-50Y$_2$O$_3$ sol-gel coatings on glass and SiC(C/SiC) composites, Ceram. Int. 31 (2005) 631–634.

[10] B.L. Zou, Z.S. Khan, L.J. Gu, *et al.*, Microstructure, oxidation protection and failure mechanism of Yb$_2$SiO$_5$/LaMgAl$_{11}$O$_{19}$ coating deposited on C/SiC composites by atmospheric plasma spraying, Corros. Sci. 62 (2012) 192-200.

[11] X.Q. Cao, Y.F. Zhang, J.F. Zhang, *et al.*, Failure of the plasma-sprayed coating of lanthanum hexaluminate, J. Euro. Ceram. Soc. 28 (2008) 1979-1986.

Key Engineering Materials Vol. 537 (2013) pp 42-45
© (2013) Trans Tech Publications, Switzerland
doi:10.4028/www.scientific.net/KEM.537.42

High Temperature *in-situ* Antioxidation Coating Fabricated by AAC Method

Zhenting Wang, Gang Liang and Guogang Zhao

College of Materials Science & Engineering, Heilongjiang Institute of Science & Technology, Harbin 150027, P. R. China

wangzt2002@163.com

Keywords: AAC(argon arc cladding); Antioxidation; Ceramic Particle

Abstract. In the surface of graphite electrode, the in-situ synthesized high temperature antioxidation composite coating is prepared, depending on argon arc cladding and the raw materials of Si and B_4C powder. The coating consists of SiC and $B_{13}C_2$ ceramic particles. The results show that: the reaction generates B_2O_3 and SiO_2 with high temperature play a role in healing crack and preventing oxygen diffusion; a kind of continuous interface is present between the cladding layer and the graphite substrate, no obvious flaws; burning at 1573 K and 10 h, oxidation weightlessness rate is 0.912%.

Introduction

Graphite electrode plays an important role in conductive and structure materials, used in high temperature metallurgy, which is easy to processing, light weight, high temperature resistant, thermal stability. But, its oxidation consumption is serious under high temperature. With the rise of temperature, the oxidation consumption increases sharply [1-3]. The consumption of graphite electrode is abnormal consumption (electrode broken, surface flaking) and chemical reaction oxidation loss (electrode end volatile, electrode wall oxidation). Through manual operation, abnormal consumption can be avoided, so the study of chemical reaction oxidation loss of graphite electrode is particularly significant [4-8]. In graphite electrode surface, covered antioxidation coating is an effective way to solve the problem of oxidation, which requires a higher adhesive strength between the coating and substrate, within the coating. At higher temperatures another type of ceramics, commonly known as ultra high temperature ceramics (UHTCs), is under study. These include the transition metal and group IV of compounds. This kind of material has good mechanical properties and oxidation resistance.

This paper uses the heat source of AAC and clad Si and B_4C powder in graphite electrode surface in-situ synthesized composite coating, and studies oxidation resistance of the coating.

Experimental

In this experiment, we choose the substrate material of graphite electrode, the average grain size of 10μm Si powder and 50μm B_4C powder as raw material [Figure 1]. First the solid charge is dry mixed for 30 min by ball milling in a sealed container, then mixing powers glue to the substrate by sodium silicate, thickness in 1.0~1.2 mm. Using MW3000 Digital Tungsten Inert Gas welding machine, the optimal process parameters for AAC are current 100 A, voltage 9 V, argon gas flow 8 ml/min, deposition rates 10 mm/s.

Analyze organization structure, micro-area component and phase composition of the cladding layer by MX-2600 FE scanning electron microscope (SEM), OXFORD energy dispersive spectrometer (EDS) and XD-2 X-ray diffraction (XRD). In the dry air, the sample is oxidized for 10h by SK-G05123K open-system atmosphere tubular furnace, cooling it in drying basin and weighing with electronic balance of which the precision is 0.01 mg. Though these, oxidation weightlessness rate can be calculated, and we test the coating performances of oxidation resistance in high temperature. In air from 1573 K to 298 K and vice versa, thus determine thermal shock resistance of the composite coating.

Fig.1 SEM images (a) silicon power (b) boron carbide power

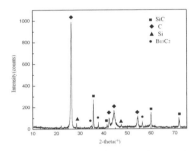

Fig. 2 XRD pattern of composite coating

Result and Discussion

Fig. 2 displays XRD pattern of composite coating. XRD analysis results show that the coating consists of C, Si, SiC and $B_{12}C_2$ ceramic particles. No B_4C raw powder diffraction peak in the pattern, so it explains the combination bond of all B_4C compounds has opened in high temperature. B, C and Si are restructured into SiC and $B_{12}C_2$ particles, with C and Si elementary substance.

Fig. 3 shows the line scanning of element of cladding of composite coating. Combined with EDS and XRD, black particles are $B_{13}C_2$ and the rest are SiC. Due to the atom radius of B and C are similar, so a mass of B and a small amount of Si seep into graphite substrate under the high temperature effect. Graphite substrate surface about 200 μm zones are SiC transition layer.

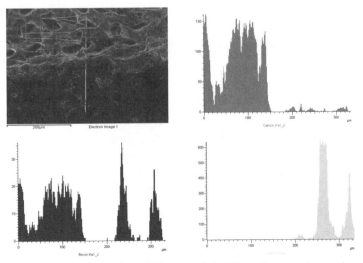

Fig.3 The line scanning of element of cladding of composite coating

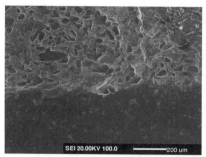

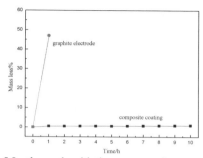

Fig.4 Microstructure of cross-section of
composite coating

Fig.5 Isothermal oxidation curves of composite
coating in air at 1573K

Fig. 4 shows microstructure of cross-section of composite coating. There is a continuous and equable interface between cladding layer and substrate, without impurity, porosity, aliquation. Because in the cladding layers bottom temperature gradient is large and fast growth, organizations of the transition layer and the substrate occur to distortion, forming machine rivets binding mode. With increasing the distance of the melting from the bottom of the cladding layer, the organization varies significantly. Cladding layer appears in a lot of in-situ particles, so there is the particle edge effect in the Figure 4.

Weightlessness rate of the sample is 0.912% after oxidation at 1573 K and 10 h, far higher than the green graphite electrode[Figure 5]. The generation of ceramic particles by AAC have excellent high temperature stability and good structure compact. Oxidizing temperature is higher than the deposition temperature of SiC (1273 K) and linear expansion coefficient is similar to graphite electrode, in order to don't exist serious mismatch and cracks. According to Rizzo [9], research shows that: increasing oxidation temperature, B and $B_{13}C_2$ generate a lot of glass phase B_2O_3. B_2O_3 has lower melting point and good liquidity, filling and healing the crack of coating, preventing oxygen spread, so coating have been slow oxidation and prevent role. The main reason of oxidation weightlessness: the tiny cracks of the surface coating become to the channel of oxygen. In the oxidation process, coating generates CO and CO_2 gaseous materials such as to escape will make many holes on the cladding layer. Under high temperature B_2O_3 and SiO_2 have a good healing ability, but theirs stability will decline with increasing temperature, gradually volatile. This causes the cracks and holes aren't promptly effective healed, then these cracks and holes provide short channel for oxygen diffusion. Above factors led to oxidation weightlessness.

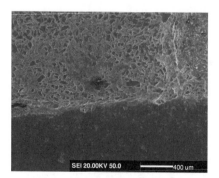

Fig. 6 The cross-section SEM of sample after psychro-thermal cycles

In air ten times from 1573 K to 298K and vice versa, when heated sample produce thermal stress, then putting into cold water, cooling contraction, that substrate produce tensile stress for the coating, but coating and substrate don't make a delaminated phenomenon. The figure 6 shows a continuous combine is between coating and substrate, and this interface is twists and turns.

Conclusion

High temperature antioxidation coating is directly in-situ synthesized on the surface of graphite electrode by AAC. The cladding is composed of SiC and $B_{13}C_2$ ceramic particles, and the cladding layer is dense, no cracks and holes defects. B_2O_3 and SiO_2 can heal crack and prevent oxygen diffusion. Weightlessness rate of the sample is 0.912% after oxidation at 1573 K and 10 h. In air ten times from 1573 K to 298K and vice versa, coating and substrate don't make a delaminated phenomenon. A continuous combine is between coating and substrate, and this surface is twists and turns.

Acknowledgements

This work was supported by Scientific Research Fund of Heilongjiang Provincial Education Department (No: 12511469).

References

[1]　H. Behzad, H. Hassan, G. Lo, Electrochemical behavior and application of Prussian blue nanoparticle modified graphite electrode, Sensors and Actuators B: Chemical. 147 (2010) 270–276.

[2]　S. Jitendra, K.S. Ashok, A.K. Jain, Fabrication of novel coated graphite electrodes for the selective nano-level determination of Cd2+ ions in biological and environmental samples, Electrochimica Acta. 56 (2011) 9095-9104.

[3]　Y. Kong, J. Yuan, Z.L. Wang, S.P. Yao, Z.D. Chen, Application of expanded graphite/attapulgite composite materials as electrode for treatment of textile waste water, Applied Clay Science. 46 (2009) 358–362.

[4]　W. Wang, A.L. Fan, Investigation on property of oxidation resistance of Al2O3 coating for common graphite, Carbon. 138 (2009) 23-25.

[5]　Y. Kong, X.H. Chen, J.H. Ni, S.P. Yao, W.C. Wang, Z.Y. Luo, Z.D. Chen, Palygorskite–expanded graphite electrodes for catalytic electro-oxidation of phenol, Applied Clay Science, 49 (2010) 64-68.

[6]　S. Zhang, W.E. Lee, Improving the water-wettability and oxidation resistance of graphite using Al_2O_3/SiO_2 sol-gel coatings, Journal of the European Ceramic Society. 23 (2003) 1215-1221.

[7]　G.S. Zeng, D.J. Yang, G. Xie, Current status of high temperature oxidation resistant for graphite electrode, Southern Metals. 132 (2003) 17-20.

[8]　G.S. Jiao, H.J. Li, K.Z. Li, Y.L. Zhang, Multi-composition oxidation resistant coating for SiC-coated carbon/carbon composites at high temperature, Materials Science and Engineering: A. 486 (2008) 556-561.

[9]　H.F. Rizzo, J.A. Kohn, W.F. Nye, et al, Boron Synthesis Structure and Properties, New York: plenum press. (1960) 175.

Key Engineering Materials Vol. 537 (2013) pp 46-51
© (2013) Trans Tech Publications, Switzerland
doi:10.4028/www.scientific.net/KEM.537.46

TiC-TiB$_2$ Coating on Ti-6Al-4V alloy Substrate in Graded Composition Achieved by Fusion Bonding and Atomic Interdiffusion under Combustion Synthesis in High-gravity Field

Xuegang Huang[a], Zhongmin Zhao[b], Long Zhang, Liqing Shi

Mechanical Engineering College, Shijiazhuang 050003, China

[a]emei127@126.com, [b]zhaozm2007@yahoo.cn

Keywords: ceramic coating; fusion bonding; atomic interdiffusion; surface alloying; graded joint

Abstract. Based on fusion bonding of liquid TiC-TiB$_2$ ceramic film and molten Ti-6Al-4V substrate to induce the surface alloying of Ti-6Al-4V substrate, the TiC-TiB$_2$ ceramic coating on Ti-6Al-4V substrate was achieved in graded composition by combustion synthesis in high-gravity field. XRD, FESEM and EDS results showed that TiC-TiB$_2$ ceramic coating was composed of fine TiB$_2$ platelets, TiC irregular grains, Cr metallic phase and a few Al$_2$O$_3$ inclusions. The mechanical properties showed that the relative density, microhardness and fracture toughness of TiC-TiB$_2$ ceramic coating reached 98.5%, 25.8 ± 3.5 GPa and 16.5 ± 2.5 MPa · m$^{0.5}$, respectively. The fusion-bonding area and heat-affected zone of Ti alloy were obviously observed between the ceramic coating and Ti alloy substrate, and within the fusion bonding area a number of ultralfine TiB platelets and fine Ti-enriched carbides were embedded in the matrix of α'-Ti martensite, whereas at the heat-affected zone of Ti-6A1-4V substrate some C, B atoms were also detected. As a coupled result of fusion bonding and inter-diffusion between liquid ceramic film and molten Ti alloy substrate, followed by TiB$_2$-Ti peritectic reaction and subsequent eutectic reaction in TiC-TiB-Ti ternary system, the ceramic coating of TiC-TiB$_2$ on Ti-6Al4V substrate was achieved in continuously-graded composition, presenting the transitional change in microhardness from the ceramic coating to Ti alloy substrate with high shear strength of 450 ± 35 MPa between the coating and the substrate.

Introduction

The Ti-6Al-4V alloys have been widely used in structural and engine parts of ultrasonic airplanes, materials for petrochemical plants, and surgical implants due to their attractive properties such as high specific strength, stiffness, and excellent corrosion resistance [1,2,3]. However, their poor resistance to wear at room temperature and oxidation at high temperatures limited the applications under extreme conditions, it is necessary to do some surface treatments [4,5,6]. Hence, in order to improve the surface properties of Ti-6Al-4V alloys, many methods have been developed to enhancing surface hardness and wear resistance in recent years.

Surface alloying is a method for forming an surface-alloyed layer on a substrate to improve the resistance to corrosion, wear, and heat. Recently, this has been developed by direct irradiation using high-energy heat sources such as pulsed laser or electron beam, the high thermal energy from irradiation can easily melt the ceramics. When the metal substrate, on which ceramic powders are evenly deposited, is irradiated with electron beam, both ceramic powders and the substrate surface are melted. In these methods, metal/ceramic surface-alloyed materials can be fabricated as ceramic elements are penetrated into the substrate and re-precipitated while ceramic powders and the substrate are melted [7]. However, these high-energy irradiation methods were difficult to obtain thicker surface-alloyed layer of several millimeters in thickness due to penetration depth is totally depended on the energy of the irradiation.

Recently, TiC-TiB$_2$ coating on Ti-6A1-4V substrate in continuously-graded composition was achieved by combustion synthesis in high-gravity field, as a result, microstructures, properties and formation mechanism of the ceramic coating on Ti alloy substrate by the non-conventional and innovative process were discussed.

Experimental

Raw materials 30-63μm in particle size were prepared from high purity (>99.9%) CrO_3 and Al powders, while purity (>97%) B_4C powder with particle size < 3 μm and high purity (>99%) Ti powder with particle size < 34 μm. The molar ratio (Ti to B_4C) of 3:1 was chosen as the starting composition, so the composition of the solidified TiC-TiB_2 composite was determined by Eq. 1. In order to ensure that all reaction products are full-liquid in combustion reaction, the adiabatic temperature of the whole combustion system was designed as 3200°C, and the (CrO_3+Al) subsystem was added as the activators for increasing the adiabatic temperature according to Eq. 2. In order to protect the ceramic coating of TiC-TiB_2 liquid from the oxidation, CaF_2 powders (density of 3.18 g · cm^{-3}) and a melting temperature of 1423 °C) were used as flux and were mixed with the thermit of (CrO_3+Al) by designing the adiabatic temperature of the mixed reactive system as same as 3200°C, and mass fraction ratio of the primary system (Ti+B_4C+CrO_3+Al) to the flux- mixed system (CaF_2+CrO_3+Al) was determined as 1:1.

$$3Ti+B_4C \rightarrow TiC+2TiB_2 \tag{1}$$

$$CrO_3+2Al \rightarrow Al_2O_3+Cr+1094kJ \tag{2}$$

After mixing the above powders, a Ti-6Al-4V alloy plates with 10 mm in thickness and 150 mm in diameter were putted at the bottom of the crucibles, then, the crucibles were filled with the reactive blends of the above powders at constant pressure and inserted into two combustion chambers at the end of the rotating arms of the centrifugal machine. The whole combustion system were firstly ignited with the W wire (diameter of 0.5 mm) at the top of the reactive blends after the centrifugal machine had provided a high-gravity acceleration in the combustion chambers greater than 300g (g=9.8m · s^{-2}). As the combustion chambers were cooled to ambient temperature, the centrifugal machine was stopped and the crucibles were taken out of the combustion chambers. Finally, TiC-TiB_2 coating of 4~5 mm in thickness on Ti-6A1-4V alloy substrate were obtained after the samples were taken out of the crucibles and Al_2O_3-CaF_2 flux layer at the top of the sample was eliminated by grinding.

Results and discussion

XRD, FESEM and EDS results showed that the ceramic coating was comprised of fine TiB_2 platelets, irregular TiC grains and a few of Cr metallic binder and black particles of Al_2O_3 inclusions, as shown in Fig. 1 and Fig. 2. The density, relative density, microharness and fracture toughness of the ceramic coating measured 4.20 g · cm^{-3}, 98.5%, 25.8 ± 3.5 GPa and 16.5 ± 2.5 MPa · $m^{0.5}$, respectively. FESEM images of crack propagation paths showed that fracture mode of the ceramic coating presented one mixed mode of intergranular fracture along TiB_2 platelets with transgranular fracture in TiC grains and Al_2O_3 inclusions, as shown in Fig. 3, promising that fine TiB_2 platelets as the reinforcements played a predominant role in toughening ceramic coating by inducing a coupled toughening mechanism of crack deflection, crack bridging and pull-out.

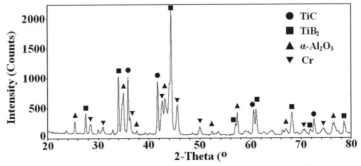

Fig. 1 The XRD patterns of the TiC-TiB_2 ceramic coating.

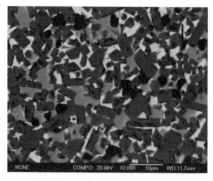 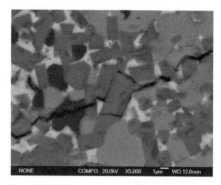

Fig. 2 FESEM microstructure of TiC-TiB$_2$ Fig. 3 FESEM image of crack propagation
            ceramic coating   path at polished ceramic coating

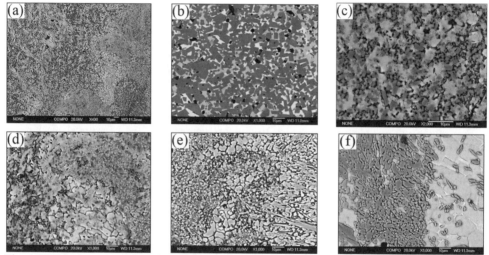

Fig. 4 FESEM images of the microstructures from TiC-TiB$_2$ ceramic coating to Ti alloy substrate
(a) joint between the ceramic coating and Ti alloy substrate (b) the ceramic coating far away from
the joint (c) the ceramic coating nearby the joint (d) joint area (e) Ti alloy substrate nearby the joint
(f) Ti alloy substrate far away from the joint

According to the literature [8, 9], the procedure of preparing bulk solidified TiC-TiB$_2$ compsoites
by combustion synthesis in high-gravity field is described in brief as follows: by selecting
(B$_4$C+Ti+C) as the primary system and the thermit of (CrO$_3$+Al) as the subsystem, and designing the
adiabatic temperature above the melting point of all products, combustion synthesis was conducted in
the presence of a high-gravity acceleration larger than 300g, finally, near-full-density TiC-TiB$_2$
composites with fine even ultrafine microstructures were achieved through rapid solidification of
liquid TiC- TiB$_2$ ceramics. Moreover, it can be concluded that the new processing for preparing the
solidified TiC-TiB$_2$ composites involves three stages: formation of the immiscible liquids consisting
of Ti-Cr-C-B liquid and oxide liquid following thermit reaction and the decomposition of B$_4$C during
the first stage, formation of the layered melt in high-gravity field due to density difference between the
immiscible liquids of Ti-Cr-C-B liquid and oxide liquid during the second stage, and formation of
TiC-TiB$_2$ solidified fine even ultrafine microstructures due to rapid coupled growth of TiC and TiB$_2$
phases from Ti-Cr-C-B liquid during third stage. Hence, silimar to the preparation of bulk solidified
TiC-TiB$_2$ composite, the formation of the ceramic coating on Ti alloy substrate is a result of the rapid

deposition of TiC-TiB$_2$-Cr liquid film, followed by rapid solidification of the liquid film at the top of Ti alloy substrate; however, because of adding a large amount of CaF$_2$ flux into combusion system, TiC-TiB$_2$-Cr liquid is very less by comparing to the one for preparing bulk solidified TiC-TiB$_2$ composite, resulting in the achievement of the ceramic coating of 4~5 mm in thickness on Ti alloy.

XRD, FESEM and EDS results at the joint of ceramic coating and Ti alloy substrate showed that there was only the microstructure transformation rather than a clear joint interface between the ceramic coating and Ti alloy substrate, as shown in Fig. 4. Meanwhile, compared to the ceramic coating far away from the joint, the microstructure of the ceramic coating nearby the joint was obviously refined so that the ultrafine-grained microstructure characterized by the average thickness of fine platelets smaller than 1 μm was achieved, respectively, as shown in Fig. 4(b) and Fig. 4(c). In addition, a few isolated Al$_2$O$_3$ inclusions were observed only at the area of the ceramic coating, as shown in Fig. 3(b).

The XRD patterns and FESEM images show that the joint area was composed of martensitic Ti, TiB and Ti-enriched carbides, as shown in Fig. 5 and Fig. 6. A considerable amount of utralfine hexagonal-pillar-shaped primary TiB grains (diameter of 0.5 to 1.5 μm and length of 2 to 5 μm) and fine spherical Ti-enriched carbides are distributed in the α'-Ti martensitic matrix, as shown by the arrow in Fig. 6.

Because of instant release of high chemical enegy of combustion system in high-gravity field, the surface of Ti alloy substrate at the bottom of the liquid TiC-TiB$_2$ is rapidly molten, while the high-gravity filed also brings about thermal vacuum circumstance around the full-liquid products to protect the molten substrate of Ti alloy from the oxidation. Subsequently, C and B atoms in liquid TiC-TiB$_2$ ceramic rapidly diffuse into the molten Ti alloy substrate, while Ti atoms in the molten substrate of Ti alloy simultaneously diffuse into the nearby liquid TiC-TiB$_2$ ceramic. As a result, at initial stage of solidification a number of TiB$_2$ solids as the primary phases precipitate from TiC-TiB$_2$ liquid, follwed by peritectic reaction (TiB$_2$ + Ti → TiB) due to the presence of liquid Ti alloy, finally, a number of ultrafine TiB platelets (diameter of 0.5 to 1.5 μm and length of 2 to 5 μm) come to existence around the joint of ceramic to Ti alloy, resulting in the achievement of the ultralfine-grained microstructure in joint region and the ceramic coating nearby the joint, as shown in Fig. 4(c) and Fig. 6. Meanwhile, because of high diffusion rate of C relative to B in molten Ti alloy and the isotropy of TiC in crystallography, a number of Ti-enriched carbides develop at the joint area due to rapid growth of TiC crystals, moreover, some fine Ti-enriched carbides are also found at the area nearby unmolten Ti solid alloy due to rapid atomic diffusion of C toward solid substrate of Ti alloy.

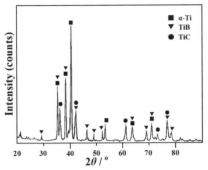

Fig. 5 The XRD patterns of the fused joint area

Fig. 6 The FESEM image of fusion-induced joint between TiC-TiB$_2$ ceramic and Ti alloy

Fig. 7 is an high-magnification FESEM micrograph of the substrate and heat-affecting zone (HAZ) of Ti-6Al-4V alloy. The Ti-6Al-4V substrate consists of equiaxed α and intergranular β phases; however, the heat-affected zone of Ti-6Al-4V substrate nearby the joint area shows the presaence of a number of needle-shaped grains, as shown in Fig. 4(c) and Fig. 7. XRD and EDS results further determined that those needle-shaped grians embedded in the Ti-6Al-4V substrate were in fact α'-Ti martensites. By considering the high heat conductivity of Ti alloy substrate, a part of α-Ti phases at Ti alloy substrate awary from the joint area transform into β-Ti phases once the temperature at the area is somewhat higher than transformation temperature of $\alpha \rightarrow \beta$, as a result, some needle-shaped α'-Ti martensites inevitably come to existence in α-Ti matrix through martensitic transformation of $\beta \rightarrow \alpha'$ due to rapid cooling of solid Ti alloy substrate, as shown in Fig. 7.

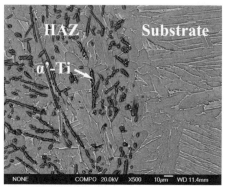

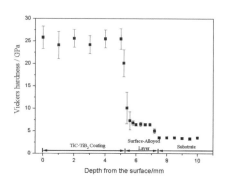

Fig. 7 FESEM images of the HAZ and the substrate of Ti-6Al-4V alloy

Fig.8 The dependence of the microhardness on the distance from the coating to the substrate

The microhardness was measured from the ceramic coating to the substrate of Ti alloy, and the change in microhardness with the distance form the coating to the substrate was shown in Fig. 8. The maximum microhardness measured 25.8 ± 3.5 GPa at TiC-TiB$_2$ coating, and high hardness of TiC-TiB$_2$ coating is considered to benefit from the achievement of fine-grained microstructure in near-full-density ceramic coating. The microhardness of the surface-alloyed substrate of Ti alloy measured 6.4 to 7.3 GPa, which was much lower than that of TiC-TiB$_2$ coating due to the presence of martensitic Ti, but presented a near-double increase over that that (about 3.2 to 3.5 GPa) of Ti alloy substrate, and the phenomanon is considered to be a results of the achievement of the surface-alloyed substrate of Ti alloy containing a number of hard TiC and TiB phases.

Meanwhile, the transitional change in microhardness was observed clearly from the ceramic coating to Ti alloy substrate, whereas shear strength of TiC-TiB$_2$ ceramic coating to Ti-6Al-4V substrate measured 450 ± 35 MPa. As dicussed above, the ceramic coating on Ti alloy substrate is achieved by fusion bonding of liquid TiC-TiB$_2$ ceramic film and molten Ti-6Al-4V substrate to induce the surface alloying of Ti-6Al-4V substrate, thus, the joint of the cermic coating to Ti alloy substrate is also achieved in continouslly-graded composition by atomic interdiffusion between liquid film of TiC-TiB$_2$ and molten substrate of Ti alloy, resulting in the transition change in microhardness and high shear strength between the ceramic coating and the substrate of Ti alloy.

Summary

Based on taking combustion synthesis in high-gravity field to prepare high-performance solidified TiC-TiB$_2$ composite, a novle available process for preparing TiC-TiB$_2$ ceramic coating on the substrate of Ti-6Al-4V alloy in continuously-graded composition has been achieved. The TiC-TiB$_2$ ceramic coating was composed of a number of fine TiB$_2$ platelets, irregular TiC grains and a few of Cr metallic binder and Al$_2$O$_3$ inclusions, and physical, mechanical properites showed the density, relative density, microhardness and fracture toughness of TiC-TiB$_2$ ceramic coating reached 4.20 g $\cdot$ cm^{-3}, 98.5%, 25.8 ± 3.5 GPa and 16.5 ± 2.5 MPa $\cdot$ m$^{0.5}$, respectively, and high fracture toughness of the ceramic coating benefited mainly from a coupled toughening mechanism of crack deflection, crack bridging and pull-out by fine TiB_2 platelets. XRD, FESEM and EDS results showed that there was not a clear joint interface but the microstructure transformation between the ceramic coating and Ti alloy substrate, i.e. TiC-TiB_2 ceramic coating, surface-alloyed zone of Ti alloy, HAZ of Ti alloy and raw substrate of Ti-6Al-4V, and within surface-alloyed zone of Ti alloy a number of ultrafine TiB platelets and fine Ti-enriched carbides were embedded in the matrix of α'-Ti martensite, whereas in ceramic coating nearby surface-alloyed zone of Ti alloy the ultralfine microstructure characterized by a average thickness of TiB platelets samller than 1 μm was also observed, so the ceramic coating on Ti alloy substrate was actually achieved by fusion bonding of liquid TiC-TiB_2 ceramic film and molten Ti-6Al-4V substrate to induce the surface alloying of Ti-6Al-4V substrate, and the achievement of a number of ultrafine TiB platelets is considered a result of peritectic reaction of TiB_2 and liquid Ti alloy following the nucleation of TiB_2 as the primary phase in Ti-enriched melt of liquid TiC-TiB_2. Moreover, the joint of the cermic coating to Ti alloy substrate is achieved in continuouslly-graded composition by atomic interdiffusion between liquid film of TiC-TiB_2 and molten substrate of Ti alloy, resulting in the transition change in microhardness and high shear strength between the ceramic coating and the substrate of Ti alloy, so combustion synthesis in high-gravity field is considered a non-conventional, innovative and economical processes to achieve low cost and high-performance ceramic coating of TiC-TiB_2 on the substrate of Ti-6Al-4V alloy in continuously-graded composition.

Acknowledgements

This work is supported by National Natural Science Foundation of China (Grant No. 51072229).

References

[1] R. Boyer, G. Welsch and E. W. Collings: *Material Properties Handbook* (ASM INTERNATIONAL, New york 1994).

[2] A. Molinari, G. Straffelini and B. Tesi: Wear. Vol. 208 (1997), p. 105

[3] G. Straffelini and A. Molinari: Wear. Vol. 236 (1999), p. 328

[4] B. Courant, J.J. Hantzpergue and S. Benayoun: Wear. Vol. 236 (1999), p. 39

[5] H. Xin, C. Hu and T.N. Baker: J. Mater. Sci. Vol. 35 (2000), p. 3373

[6] J. D. Majumdar, B. L. Mordike and I. Manna: Wear. Vol. 242 (2000), p. 18

[7] J. C. Oh, K. Euh and S. Lee: Scripta Mater. Vol. 39 (1998), p. 1389

[8] Z. M. Zhao, L. Zhang, Y.G. Song and W. G. Wang: Scripta Mater. Vol. 61 (2009), p. 281

[9] X. G. Huang, L. Zhang, Z. M. Zhao and C. Yin: Mater. Sci. Eng., A. (2012), 10.1016/j.msea.2012.05.099

Key Engineering Materials Vol. 537 (2013) pp 52-57
© (2013) Trans Tech Publications, Switzerland
doi:10.4028/www.scientific.net/KEM.537.52

Ceramic Coating of TiC-TiB$_2$ on 1Cr18Ni9Ti Substrate Achieved by Combustion Synthesis in High-gravity field

Zhongmin Zhao[a], Long Zhang, Yigang Song and Minquan Wang

Mechanical Engineering College, Shijiazhuang 050003, China

[a]zhaozm2007@yahho.cn

Keywords: Ceramic coating; fusion bonding; atomic diffusion; intermediate; ceramic-metal bonding

Abstract. Based on fusion bonding of liquid TiC-TiB$_2$ ceramic film to stainless steel substrate of 1Cr18Ni9Ti, the ceramic coating on stainless steel substrate was achieved in graded composition by combusion synthesis in high-gravity field. XRD, FESEM and EDS results showed the ceramic coating was composed of a number of fine TiB$_2$ platelets, irregular TiC grains and a few of Cr metallic binder and the inclusions of Al$_2$O$_3$ partilces, and physical and mechanical properties showed density, relative density, microhardness and fracture toughness of the ceramic coating were 4.25 g · cm^{-3}, 98.2%, 24.6 GPa and 14.5 ± 3.5 MPa · m$^{0.5}$, respectively. The joint of TiC-TiB$_2$ ceramic coating on stainless steel substrate consisted of 3-layer structures, i.e. the ceramic coating, the intermediate and stainless steel substrate, and within the intermediate of Fe-Cr-Ni-Ti metallic matrix Ti-Fe enriched carbides decreased gradually both in volume fraction and particle size from the ceramic coating to stainless steel substrate in the presence of atomic unidirectional diffusion of Ti and C, while some isolated α-Al$_2$O$_3$ fine particles were also observed in the intermediate besides the presence of both some coarsened α-Al$_2$O$_3$ inclusions nearby the ceramic coating and the barrier layers of Al$_2$O$_3$ between the ceramic and the intermediate, so it is considered that the ceramic coating on stainless steel substrate is achieved partially by fusion bonding due to the poor incompatibilities between Al$_2$O$_3$ inclusions and the intermediate, resulting in the moderate shear strength of 125 ± 35 MPa between the ceramic coating and stainless steel substrate.

Introduction

Recently, TiB$_2$-TiC composites attract more and more interests because they represent the promising materials for use as wear-resistant parts and high-temperature structural components [1]. Besides, cermets and metal–matrix composites (MMCs) reinforced by TiB$_2$-TiC also exhibit improved mechanical properties compared with the composites reinforced using conventional reinforcements [2] . The basic principle of the advantageous properties of such MMCs is that TiB$_2$ and TiC possess many desirable properties, such as high melting temperature (respectively as 3325°C and 3067°C of TiB$_2$ and TiC), high hardness (respectively as 29.4 GPa and 28.0 GPa of TiB$_2$ and TiC) [3], good wetting behavior with metal matrix, high electrical conductivity and chemical stability [1]. Additionally, unlike in situ MMCs reinforced with a monolithic ceramic which often show particle agglomeration or abnormal growth (coarsing of reinforcements), TiB$_2$ and TiC particles even with high volume fraction can exist in fine size in the composites due to the separate nucleation and growth of these particles. This synergetic effect is particularly useful for producing wear-resistant composites in which the size and distribution of reinforcements are major concerns [4, 5] . Although bulk materials of TiB$_2$–TiC reinforced composites are desirable in commercial applications, there is also a need to employ these composites as coating materials, especially for wear and corrosion industry where destruction of materials usually begins from the surface.

A number of surface modification techniques has been attempted to fabricate this kind of coating. These include but are not limited to sintering reaction [6], low oxygen partial pressure fusing technique [7], and utilizing the self-propagating high-temperature synthesis (SHS) reactions of Ni-Ti-B4 and Ni-Ti-B$_4$C-C systems during casting [8]. Nevertheless, it has been well established that highly confined and controlled local heat generated by the laser beam can yield the formation of a composite coating with metallurgical joint to the substrate. This led to the rapid development of so-called laser surface engineering (LSE) in recent decades [9]. Recently, large-bulk fined-grained TiC-TiB$_2$ composites are achieved by a novel process called as combustion synthesis in high-gravity field [10], and the experiment is renewed by adjusting the composition of combustion system in order to achieve ceramic coating of TiC-TiB$_2$ on 1Cr18Ni9Ti substrate by fusion bonding.

Experimental

Raw materials 30-63μm in particle size were prepared from high purity (>99.9%) CrO$_3$ and Al powder, while purity (>97%) B$_4$C powder with particle size < 3 μm and high purity (>99%) Ti powder with particle size <34 μm were used. The molar ratio (Ti to B$_4$C) of 3:1 was chosen as the starting composition, then, the composition of the solidified TiC-TiB$_2$ composite was determined as TiC-66.7 mol % TiB$_2$, as shown in Eq. 1. In order to ensure full-liquid products after combustion reaction, the adiabatic temperature of the whole combustion system was designed as 3200°C, and the (CrO$_3$+Al) subsystem was added as the activators for increasing the adiabatic temperature according to Eq. 2. In order to protect the ceramic coating of TiC-TiB$_2$ liquid from the oxidation, CaF$_2$ powders (density of 3.18 g · cm^{-3}) and a melting temperature of 1423 °C) were used as flux and were mixed with the thermit of (CrO$_3$+Al) by designing the adiabatic temperature of the mixed reactive system as same as 3200°C, and mass fraction ratio of the primary system (Ti+B$_4$C+CrO$_3$+Al) to the flux-mixed system (CaF$_2$+CrO$_3$+Al) was determined as 1:1.

$$3Ti+B_4C \rightarrow TiC+2TiB_2 \tag{1}$$

$$CrO_3+2Al \rightarrow Al_2O_3+Cr+1094kJ \tag{2}$$

After mixing the above powder, and followed by putting the plates of stainless steel at the bottom of the crucible, the crucible were filled with the reactive blends at constant pressure and inserted into two combustion chambers at the end of the rotating arms of the centrifugal machine. The combustion systems were ignited with the W wire (diameter of 0.5 mm) after the centrifugal machine had provided a high-gravity acceleration in the combustion chambers greater than 300g. As the combustion chambers were cooled to ambient temperature, the centrifugal machine was stopped and the crucibles were taken out of the combustion chambers. Finally, TiC-TiB$_2$ ceramic coating of 4~5 mm in thickness on the substrate of stainless steel was achieved after the samples were taken out of the crucibles and the flux layer at the top of the sample was eliminated by grinding.

Results and discussion

Microstructures and properties of TiC-TiB$_2$ ceramic coating. XRD patterns of the ceramic coating showed that the ceramic coating was composed of TiB$_2$, TiC and a few of Al$_2$O$_3$ inclusions and Cr metallic phases, as shown in Fig. 1. According to standard XRD patterns of stoichiometric TiC and TiB$_2$, the diffraction angle and crystal plane interval at (111), (200) and (220) crystal planes of stoichiometric TiC were (35.960°, 2.4954Å), (41.760°, 2.1612 Å) and (60.521°, 1.5285 Å) in succession, while the diffraction angle and crystal plane interval at (001), (100) and (101) crystal planes of stoichiometric TiB$_2$ were respectively (27.561°, 3.2338 Å), (34.119°, 2.6256 Å) and (44.420°, 2.0378 Å); correspondingly, the diffraction angle and crystal plane interval at same crystal planes of TiC in ceramic coating measured (36.019°, 2.4914 Å), (41.860°, 2.1563 Å) and (60.70°, 1.5245 Å) in succession, while the diffraction angle and crystal plane interval at same crystal planes of TiB$_2$ in ceramic coating measured (27.560°, 3.2339 Å), (34.079°, 2.6286 Å), and (44.401°, 2.0386 Å) respectively. Hence, TiC and TiB$_2$ in ceramic coating prepared in current experiment are both close to their respective stoichiometry, which promises the achievement of stoichiometric TiC-TiB$_2$ coating by combustion synthesis in high-gravity field.

FESEM images and EDS results showed a large number of randomly-orientated, fine TiB_2 platelets (presented by the dark areas in Fig. 2) were uniformly embedded in the irregular TiC grains (presented by the grey areas in Fig. 2) and Cr metallic phases (presented by the white areas in Fig. 2), and a few inclusions of α-Al_2O_3 black particles were also observed, as shown by the arrow in Fig. 2.

For three equilibrium products of TiC, TiB_2 and Al_2O_3 from combustion synthesis, in terms of the thermodynamics of chemical reaction among the elements of Al, O, Ti, C and B, their respective Gibbs free energy is calculated to decreases linearly along with the temperature below 3000K, as shown in Fig. 3, whereas at the same temperature Al_2O_3 relative to TiC and TiB_2 presents the strongest chemical stability due to its very low Gibbs free energy, so it is demonstrated that Al_2O_3 will come to existence independently in the products of the combustion reaction even if it is at liquid state. According to the equilibrium diagram of the TiC-TiB_2 system [1], the melting temperature of TiC-66.7% TiB_2 ceramic is around 2900 °C, which is smaller than the adiabatic temperature of 3200 °C involved in current experiments, thus, the full-liquid products of TiC, TiB_2, Cr, Al_2O_3 and CaF_2 can be achieved in thermodynamics of the materials. Because of the immiscibility and density different between liquid TiC-TiB_2-Cr and Al_2O_3-CaF_2, the Stocks immigration characterized by float-up of liquid Al_2O_3-CaF_2 and settle-down of liquid TiC-TiB_2, rapidly takes place in high-gravity field with gravity coefficient larger than 300, resulting in the presence of the layered melt in the crucible, finally, the layered sample is formed to present the solidified TiC-TiB_2 ceramic with Cr metallic binder covered by the flux layer of Al_2O_3-CaF_2, and a few inclusions of fine Al_2O_3 black particles also remain in TiC-TiB_2 ceramic along with the irregular coarsened Al_2O_3 inclusions at the region between the ceramic and the flux layer.

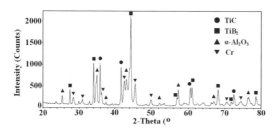

Fig. 1 XRD patterns of the solidified TiC-TiB_2 ceramic coating

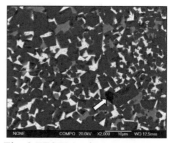

Fig. 2 FESEM microstructures of solidified TiC-TiB_2 ceramic coating

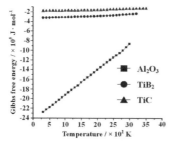

Fig. 3 Dependence of Gibbs free energy of Al_2O_3, TiB_2 and TiC ceramics on the temperature

Fig.4 FESEM image of crack propagation path at polished ceramic coating

Physical and Mechanical properties showed that density, relative density, microhardness and fracture toughness of TiC-TiB_2 ceramic coating were 4.25 g · cm^{-3}, 98.2%, 24.6 GPa and 14.5 ± 3.5 MP · $m^{-0.5}$. Meanwhile, FESEM images of crack propagation paths showed that crack deflection and crack bridging are considered the main interaction mechanisms between the crack and TiB_2 platelets, as shown in Fig. 4, resulting in the achievement of pull-out toughening by TiB$_2$ platelets. Thus, greater crack-opening displacement within the bridging zone is achieved, whereas the stress concentration around crack tip is relieved due to the presence of the closure stress behind the crack tip, resulting in greater resistance to crack propagation in the ceramic. Hence, high fracture toughness of TiC-TiB$_2$ ceramic coating results from the coupled toughening mechanisms of crack deflection, crack bridging and pull-out by TiB$_2$ platelets along with plastic deformation by Cr metallic binder.

Joint of TiC-TiB$_2$ ceramic coating on the substrate of stainless steel. FESEM images showed the joint of TiC-TiB$_2$ ceramic coating on the substrate on stainless steel consisted of three-layer structures, i.e. ceramic, the intermediate and stainless steel, as shown in Fig. 5. XRD patterns detected in the intermediate between the ceramic and stainless steel varied greatly, as shown in Fig. 6, in other words, in the intermediate nearby the ceramic coating α-Al$_2$O$_3$ inclusions presented the highest diffraction peak in XRD pattern in contrast with ones of TiB$_2$ and TiC phases, which indicated there were a large number of α-Al$_2$O$_3$ inclusions in the intermediate nearby the ceramic coating, as shown in Fig. 5 and Fig. 6(a), respectively. Meanwhile, in the intermediate nearby stainless steel a number of non-stoichiometric TiC phases were also determined in Fe-Cr-Ni-Ti alloy with Ti mass content higher than one in stainless steel of 1Cr18Ni9Ti, as shown in Fig. 6(b), and the non-stoichiometric TiC phase was confirmed as the Ti-enriched carbide according to its deviation from stoichiometric TiC in diffraction angle and d value of XRD pattern.

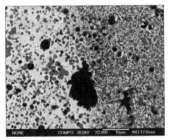

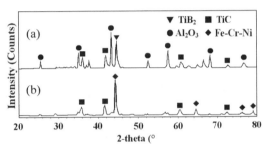

Fig. 5 FESEM images of the intermediate between ceramic coating and stainless steel

Fig. 6 XRD patterns of the intermediate (a) nearby the ceramic coating (b) nearby 1Cr18NiTi substrate

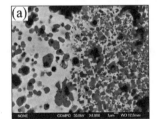
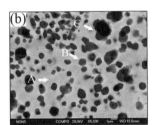
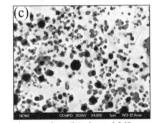

Fig. 7 FESEM images of the intermediate (a) the area nearby the ceramic coating (b) the middle area at the intermediate (c) the area nearby stainless steel substrate

Table 1 Chemical compositions at different area of the intermediate [weight %]

Location	C	O	Mg	Al	Ti	Cr	Fe	Ni	Total
A				2.68	3.42	12.32	79.69	1.89	100.00
B	14.60			0.80	31.31	3.67	48.63	1.00	100.00
C		40.07	2.57	23.83	16.01	1.94	15.57		100.00

FESEM images showed that the volume fraction and the average particle size of Ti-enriched carbides in the intermediate both gradually decreased from the area nearby the ceramic coating to the one nearby stainless steel substrate, as shown in Fig. 7. In addition, chemical composition at different area of the intermediate detected by EDS were shown in Fig. 7 (b) and Table 1, which not only promised there were both diffusion of C and Ti atoms from the liquid TiC-TiB_2 ceramic coating to stainless steel substrate during combustion synthesis in high-gravity field, but also demonstrated a large number of Fe atoms dissolved into Ti-enriched carbides in the intermediate.

By combining with current achievements in the solidified TiC-TiB_2 composites by combustion synthesis in high-gravity field [6], it is considered that the introduction of high-energy thermit not only increases the release of high chemical energy from combustion system, but also brings about the increment of high-density metal liquid in the products out of combustion reaction, and high-gravity field also makes the metallic liquid consisted of Ti-C-B-Me (Me stands for the metal reduced from thermit reaction) rapidly settle down, thus, Ti powder in unreacted blends is accelerated to melt or dissolve in metallic liquid, while B_4C powder in unreacted blends is also promoted to decompose in metallic liquid, making combustion velocity of the whole combustion system so much increased that combustion reaction in high-gravity field is similar to one of thermal explosion. Because of instant release of high chemical energy in combustion system, full-liquid products are rapidly achieved, and the stainless steel substrate at the bottom of liquid products is also molten, subsequently, Ti and C atoms in liquid TiC-TiB_2 both rapidly diffuse into the molten stainless steel due to high diffusion rate of C relative to B in metallic liquid [1]. Finally, the intermediate that the Fe-Ti enriched carbides are embedded with graded composition in Fe-Cr-Ni-Ti alloy is achieved between the ceramic coating and stainless steel substrate.

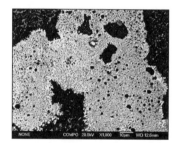 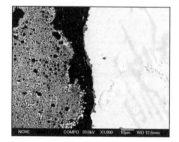

Fig. 8 Coarsened Al_2O_3 inclusions at the intermediate nearby the ceramic   Fig. 9 Barrier layers of Al_2O_3 between the ceramic and the intermediate

However, because of rapid solidification of liquid TiC-TiB_2 ceramic coating at the top of the intermediate, a large number of liquid Al_2O_3 hardly float up from the liquid intermediate alloy and have to reside in liquid intermediate, so that the coarsened Al_2O_3 inclusions and even the barrier layers of Al_2O_3 are formed between the ceramic coating and the intermediate, respectively, as shown in Fig. 8 and Fig. 9. Meanwhile, just because of float-up of liquid Al_2O_3 from the liquid intermediate in high-gravity field, volume fraction and average size of isolated Al_2O_3 particles both decrease from stainless steel substrate to the ceramic coating, as shown Fig. 7. Hence, it is just the presence of the coarsened Al_2O_3 inclusions and the barrier layers of Al_2O_3 between the ceramic coating and the intermediate that the joint by fusion-induced diffusion between TiC-TiB_2 ceramic coating and stainless steel substrate is achieved partially, resulting in shear strength of 125 ± 35 MPa of ceramic coating on stainless steel substrate. Hence, it is considered in future works that the improvement in bonding strength of the ceramic coating on stainless steel substrate can only be achieved by increasing the high-gravity acceleration to eliminate the inclusions Al_2O_3 between the intermediate and ceramic coating, followed by achieving surface alloying on stainless steel substrate of 1Cr18Ni9Ti to promote atomic diffusion.

Summary

Based on fusion bonding of liquid TiC-TiB_2 ceramic film to stainless steel substrate of 1Cr18Ni9Ti to induce atomic diffusion of Ti and C toward liquid and solid substrate of stainless steel, the ceramic coating of TiC-TiB_2 on stainless steel substrate of 1Cr18Ni9Ti was achieved in graded composition by combusion synthesis in high-gravity field. XRD, FESEM and EDS results showed the ceramic coating was composed of a number of fine TiB_2 platelets, irregular TiC grains and a few of Cr metallic binder and the inclusions of Al_2O_3 partilces, and physical and mechanical properties showed density, relative density, microhardness and fracture toughness of the ceramic coating were 4.25 g · cm^{-3}, 98.2%, 24.6 GPa and 14.5 ± 3.5 MPa · m$^{0.5}$, respectively. Meanwhile, the joint of TiC-TiB_2 ceramic coating on stainless steel substrate of 1Cr18Ni9Ti consisted of 3-layer structures, i.e. the ceramic coating, the intermediate and stainless steel substrate, and within the intermediate of Fe-Cr-Ni-Ti metallic matrix the particles of Ti-Fe enriched carbides decreased gradually both in volume fraction and particle size from the ceramic coating to stainless steel, while some isolated α-Al_2O_3 fine particles were also observed in the intermediate besides the presence of both some coarsened α-Al_2O_3 inclusions nearby the ceramic coating and the barrier layers of Al_2O_3 between the ceramic and the intermediate. The formation of the joint is considered a result of the procedure that instant release of high chemical energy induced both by the thermit in combustion system and Stocks flow in high-gravity field not only achieve the full-liquid layered products, but also make the surface of stainless steel substrate under liquid TiC-TiB_2 to be molten rapidly, subsequently, Ti and C atoms come to diffuse into the molten stainless steel from liquid TiC-TiB_2, finally, Ti-Fe enriched carbides are embedded with the graded composition in the intermediate of Fe-Cr-Ni-Ti; however, because of rapid solidification of liquid TiC-TiB_2 ceramic film above the liquid intermediate, liquid Al_2O_3 inclusions hardly float up from the liquid intermediate, resulting in the presence of both the coarsened α-Al_2O_3 inclusions nearby the ceramic and the barrier layers of Al_2O_3 between the ceramic and the intermediate, so the joint of TiC-TiB_2 ceramic to stainless steel is achieved partially by fusion bonding due to the poor incompatibilities between Al_2O_3 inclusions and the intermediate, which presents the moderate shear strength of 125 ± 35 MPa between the ceramic coating and stainless steel substrate. Hence, it is considered that the improvement in bonding strength of the ceramic coating on stainless steel substrate can only be achieved by increasing the high-gravity acceleration to eliminate the inclusions Al_2O_3 between the intermediate and ceramic coating, followed by achieving surface alloying on stainless steel substrate of 1Cr18Ni9Ti to promote atomic diffusion.

Acknowledgements

This work is supported by National Natural Science Foundation of China (Grant No. 51072229).

References

[1] D. Vallauri, I.C. Atlas and A. Adrian: J. Eur. Ceram. Soc. Vol. 28 (2008), p. 1697

[2] F. Akhtar: J. Alloys Compd. Vol. 459 (2008), p. 491

[3] B. Bhushan and B. Gupta: *Handbook of Tribology Materials* (McCraw Hill, New York 1991).

[4] K. Komvopoulos and K. Nagarathnam: J. Eng. Mater. Technol. Vol. 112 (1990), p. 131

[5] K. Komvopoulos and K. Nagarathnam: Metall. Trans. Vol. A 24 (1993), p. 1621

[6] F. Akhtar: Surf. Coat. Technol. Vol. 201 (2007), p. 9603

[7] W. Zhou, Y. Zhao, W. Li, X. Mei and Q. Jiang: Surf. Coat. Technol. Vol. 202 (2008), p. 1652

[8] Y. Yang, H. Wang, Y. Liang, R. Zhao and Q. Jiang: Mater. Sci. Eng. Vol. A445-446 (2007), p. 398.

[9] B. Du, S. R. Paital and N. B. Dahotre: Scripta Mater. Vol. 59 (2008), p. 1147

[10] Z. Zhao, L. Zhang, Y. Song and W. Wang: Scripta Mater. Vol. 61 (2009), p. 281

Key Engineering Materials Vol. 537 (2013) pp 58-62
© (2013) Trans Tech Publications, Switzerland
doi:10.4028/www.scientific.net/KEM.537.58

Investigation of Boron Nitride Prepared by Low Pressure Chemical Vapor Deposition at 650~1200 °C

Fang Ye [a], Litong Zhang, Yongsheng Liu [b], Meng Su, Laifei Cheng
and Xiaowei Yin

Science and Technology on Thermostructural Composite Materials Laboratory, Northwestern
Polytechnical University, Xi'an, Shaanxi 710072, P. R. China

[a] plusring@163.com, [b] yongshengliu@nwpu.edu.cn

Key words: Boron nitride; Chemical vapor deposition; Deposition mechanism; Thermodynamics; Kinetics

Abstract. BN coatings were deposited on carbon substrates by low-pressure chemical vapor depositionin a large temperature range of 650~1200 °C, employing BCl_3-NH_3-H_2 reaction system. The effects of depositing temperature on the yield, control step of deposition progress (deposition mechanism), microstructure, and crystallization degree of BN coating were investigated. Results show that BN deposition rate first increases and then decreases as the rising temperature and the maximum deposition rate occurs at 900~1000°C. By the determination of the Arrenius relationship, there are three temperature regions with different active energies and controlled by different deposition mechanisms, i.e. chemical reaction, mass transport and depletion of reactants. Through the surface morphology observation by scanning electron microscopy, chemical composition analyses by energy dispersion spectroscopy and crystallization degree and grain size comparison by Raman spectroscopy, it can be drawn that interphase-used BN is suitable to be deposited at 1000 °C.

Introduction

Boron nitride (BN) has been used as the interphase material of continuous fiber reinforced ceramic matrix composites (CMCs) owing to the layered structure and the excellent oxidation resistance [1,2]. Interphase-used BN has generally been prepared by chemical vapor deposition (CVD) from BCl_3-NH_3-H_2 [3]. Although the early work has demonstrated that the deposition temperature of BN can be lowered to 500 °C [4], too low deposition temperature always causes a poor crystallization degree of as-deposited BN, leading to the poor oxidation resistance of BN interphase [5,6]. Many researches [3,7,8] and our previous work [9] have shown that the crystallization degree of BN can be improved using heat treatment. However, heat treatment (especially above 1300 °C) is not advisable for CMCs because high-temperature can result in the dramatic decrease of the strength of fiber reinforcements [10]. On the other hand, below 1300 °C, although a higher depositing temperature often corresponds to a higher crystallization degree of CVD BN [8], no detailed work has demonstrated the essential relationship between them and a lot of problems and challenges exist during applying CVD BN as the practical interphase, e.g. how the deposition temperature to influence BN depositing process such as deposition rate and depositing mechanism, BN product's properties such as yield, structure and crystallization degree and so on. Focusing on these issues, here, the BN preparations in a large temperature range of 650~1200°C were studied systematically, employing BCl_3–NH_3–H_2 reaction system. The effects of depositing temperature on the yield, control step of deposition progress (deposition mechanism), microstructure, and crystallization degree of BN coating were investigated.

Experimental

Materials and deposition conditions. BCl_3 (purity 99.99%) and NH_3 (purity 99.99%) were used as boron and nitrogen source for BN preparation. H_2 (purity 99.999%) was used as carrier gas. Carbon cloth (woven with T300 carbon fibers) or high-purity graphite (20 × 10 × 2 mm) was used as deposition substrates. To avoid the effect of surface roughness on BN deposition rate, graphite substrates were grinded and polished before experiments. BN coatings were deposited at a temperature range of 650~1200 °C, adopting a total pressure of 400 Pa and 1:3 mole ration of B and N source which are determined by thermodynamic analysis. After deposition, all samples were kept in vacuum drying apparatus.

Microstructure analysis. Microstructures of BN products were examined by a scanning electron microscopy (SEM, 6700S, Hitachi) equipped with an energy dispersive X-ray spectrometer (EDS). Further microstructure analyses were resorted to Raman spectroscopy (InVia, Renishaw).

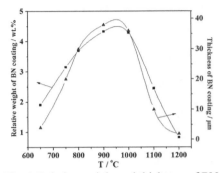 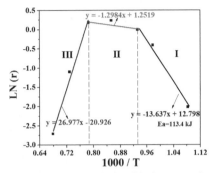

Fig. 1 Relative weight and thickness of BN coating on the graphite as a function of deposition temperature (t=30 h)

Fig. 2 Arrhenius relationship between the deposition rate of BN coating and temperature

Results and discussion

Fig. 1 shows the BN deposition rate dependence on the temperature varied from 650 to 1200 °C. It indicates that BN deposition rate increases with the temperature up to 950 °C and then decreases gradually. It can be inferred that the BN deposition at 650~1200 °C involves different controlling mechanisms, which is also identified by the greatly different apparent reaction activation energies (E_a) (Fig. 2). Here, E_a was calculated by Arrhenius equation:

$$r = A\exp(-\frac{E_a}{RT}) \tag{1}$$

where r is the deposition rate, E_a the apparent activation energy, R the gas constant, and T the temperature.

Three regions can be divided. At low temperature (region I, 650~800 °C), the growth rate is limited by chemical kinetics according to a high Ea of ~113.4 kJ/mol, which is temperature dependent and increases exponentially with temperature. Since the deposition rate is limited by chemical kinetics, uniform coating thickness can be obtained by minimizing temperature variation [11]. For region II (800~1000 °C), the growth rate is nearly independent of temperature, where well known mass-transport limited growth regime as typically observed in CVD controls the rate [12]. The flux of reactants to the deposition substrate is proportional to the concentration of the limiting precursor, which has the further advantage of allowing simple control of the magnitude of the growth rate [13]. At higher temperatures (region III, 1000~1200 °C), the growth rate decreases because of an increased desorption rate, depletion of reactants on reactor walls, and especially the enhanced gas phase nucleation, which can drastically reduce the permeability and uniformity of BN deposited in fiber preform [14]. Besides, the intensive decomposition of NH_3 at over high temperatures can also lead to a decline in the growth rate. Fig. 3 (SEM observation on cross section) provides some intuitive messages on BN products under 1000 °C after 30 h CVD. It is shown that the grain size decreases drastically from 650 °C to 900 °C, and then increases again from 900 °C to 1000 °C. On the other hand, it is also found that the orientation degree of BN (as labeled by white lines in Fig. 6f) increases dramatically as temperature rising, which is further proved by the following Raman analyses.

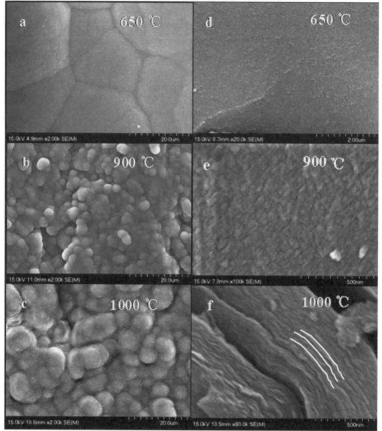

Fig. 3 SEM images of surface morphology (a, b, c) and cross section (d, e, f) of the BN coatings
prepared at different temperatures (t=30 h)

Tab. 1 EDS analysis of BN coatings obtained at different temperatures (t=30 h)

T /°C	At % of elements contained in BN coating			
	B	N	C	O
650	19.65	17.43	58.92	04.00
900	49.08	47.7	0	02.50
1000	48.28	51.14	0	00.58

Microstructures of BN coatings obtained at different temperatures were investigated by Raman
spectra (Fig. 4). All spectra of BN coatings deposited at different temperatures have a characteristic
bond at about 1366 cm^{-1}, which reveals the vibration of B-N bonding. With the increase of
temperature, the bond becomes narrow and the peak intensity becomes high, indicating both the
crystallization degree and the layer-structure characteristic of BN coatings increasing [15]. It should
be noted that the graphite sheet has the substrate effect on the spectra of 650 °C, 1100 °C and 1200 °C
(uncharted) because the coatings prepared at these temperatures (especially for 1100 °C and 1200 °C)
are very thin. Therefore, the spectra have a small bond at 1580 cm^{-1} which is referred as the G band of
carbon [16]. In addition, the variation of element content in the coatings was also determined by EDS
as shown in Tab. 1. The analysis results show that the higher deposition temperature, the lower
oxygen content, which is conducive to the stability of BN coatings

Conclusions

BN deposition rate first increases and then decreases as the rising temperature and the maximum deposition rate occurs at 900~1000 °C. By the determination of the Arrenius relationship, there are three temperature regions controlled respectively by chemical reaction, mass transport and depletion of reactants. Based on the comprehensive consideration of the orientation, the oxygen content, the yield and the deposition controllability of the BN products, 1000 °C can be a suitable deposition temperature for preparing BN as the interphase of CMCs.

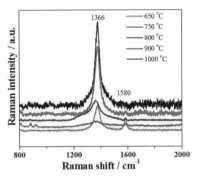

Fig. 4 Raman spectra of the BN coatings obtained at different temperatures (t=30 h)

Acknowledgments

This work has been supported by natural science foundation of China (No. 51002120, 51032006, 50820145202) and the research fund of the state key laboratory of solidification processing (NWPU), China (Grant No. 45-QP-2010).

References

[1] R. Naslain, A. Guette, F. Rebillat, *et al.*, Boron-bearing species in ceramic matrix composites for long-term aerospace applications, J. Solid. State. Chem. 177 (2004) 449-456.

[2] J.S. Li, C.R. Zhang, B. Li, F. Cao, S.Q. Wang, Boron nitride coatings by chemical vapor deposition from borazine, Surf. Coat. Technol. 205 (2011) 3736-3741.

[3] S. Gallet, G. Chollon, F. Rebillat, *et al.*, Microstructural and microtextural investigations of boron nitride deposited from BCl_3–NH_3–H_2 gas mixtures, J. Eur. Ceram. Soc. 24 (2004) 33-44.

[4] V. Cholet, L. Vandenbulcke, J.P. Rouan, Characterization of boron nitride films deposited from BCl_3-NH_3-H_2 mixtures in chemical vapour infiltration conditions, J. Mater. Sci. 29 (1994) 1417-1435.

[5] J. Minet, F. Langlais, R. Naslain, On the chemical vapour deposition of zirconia from $ZrCl_4$-H_2-CO_2-Ar gas mixture: II. An experimental approach, J. Less-Common. Met. 95 (1987) 273-287.

[6] T. Matsuda, Stability to moisture for chemically vapour-deposited boron nitride, J. Mater. Sci. 24 (1989) 2353-2357.

[7] C.G. Cofer, J. Economy, Oxidative and hydrolytic stability of boron nitride — A new approach to improving the oxidation resistance of carbonaceous structures, Carbon. 33 (1995) 389-395.

[8] M. Leparoux, L. Vandelbulcke, C. Clinard, Influence of isothermal chemical vapour deposition and chemical vapour infiltration conditions on the deposition kinetics and structure of boron nitride, J. Am. Chem. Soc. 82 (1999) 1187-1195.

[9] Y. Cheng, X.W. Yin, Y.S. Liu, *et al.*, BN coatings prepared by low pressure chemical vapor deposition using boron trichloride–ammonia–hydrogen–argon mixture gases, Surf. Coat. Technol. 204 (2010) 2797-2802.

[10] A. Udayakumar, A. Ganesh, S. Raja, *et al.*, Effect of intermediate heat treatment on mechanical properties of SiCf/SiC composites with BN interphase prepared by ICVI, J. Eur. Ceram. Soc. 31 (2011) 1145-1153.

[11] J.E. Crowell, Chemical methods of thin film deposition: Chemical vapor deposition, atomic layer deposition, and related technologies, J. Vac. Sci. Technol. 21 (2003) 88-95.

[12] J.M. Jasinski, B.S. Meyerson, B.A. Scott, Mechanistic Studies of Chemical Vapor Deposition, Annu. Rev. Phys. Chem. 38 (1987) 109-140.

[13] C.Y. Lu, L.F. Cheng, C.N. Zhao, L.T. Zhang, Y.D. Xu, Kinetics of chemical vapor deposition of SiC from methyltrichlorosilane and hydrogen, Appl. Surf. Sci. 255 (2009) 7495-7499.

[14] E.P. Foryst, K. Fitzner, Deposition of boron nitride on graphite at 1300 K, Mikrochim. Acta 125 (1997) 73-77.

[15] R.J. Nemanich, S.A. Solin, R.M. Martin, Light scattering study of boron nitride microcrystals, Phys. Rev. B 23 (1981) 6348-6356.

[16] A. Sadezky, H. Muckenhuber, H. Grothe, *et al.* Raman microspectroscopy of soot and related carbonaceous materials: Spectral analysis and structural information, Carbon 43 (2005) 1731-1742.

Key Engineering Materials Vol. 537 (2013) pp 63-66
© (2013) Trans Tech Publications, Switzerland
doi:10.4028/www.scientific.net/KEM.537.63

Hardness and Wear Resistance of Ti-B-C-N Nanocomposite Coatings Synthesized by Multi-Target Magnetron Co-Sputtering

C.K. Gao[1], M.Y. Liu[1], D.J. Li[1, a], L. Dong[1], H.Q. Gu[2,3], R.X. Wan[2,3]

[1]College of Physics and Electronic Information Science, Tianjin Normal University, Tianjin, 300387, P.R. China

[2]Tianjin Institute of Urological Surgery, Tianjin Medical University, Tianjin, 300211, P.R. China

[3]Ninth People's Hospital, Shanghai Jiao Tong University, School of Medicine, Shanghai 200011, P.R. China

[a]dejunli@mail.tjnu.edu.cn (corresponding author)

Keywords: multi-target magnetron co-sputtering, nanocomposite, Ti-B-C-N coating, deposition parameter, hardness.

Abstract. Ti-B-C-N nanocomposite coatings were synthesized on Si(100) and stainless steel substrate by multi-target magnetron co-sputtering. XRD and XPS were employed to measure the structure and chemical states of the coatings. The measurements of nanoindenter and multi-functional tester indicated that the maximum hardness and elastic modulus were 42.8 GPa and 424 GPa when the work pressure was 0.5 Pa, the powers of Ti target and B_4C were 60 and 150 W, the flows of N_2 and Ar were 4 and 36 sccm, and the substrate bias was -100 V.

Introduction

Multicomponent and multiphase coatings are of increasing interest as protective coating systems because of their excellent combination of physical, chemical and mechanical properties [1]. These coatings have high hardness, good wear resistance, low friction coefficient, and high oxidation resistance from room temperature to above 700°C [2-4]. Nanostructure multiphase composites and multicomponent materials in which the dimensions of the individual phases or components are in the range from 3 to 20 nm represent a new class of engineering materials [5,6], These materials consists of nanometer sized crystals embedded in a preferable matrix which may be either amorphous or nanocrystalline [7]. Recently, some new studies were focused on introduction of a third element such as Al and C into transition metal nitride coatings for improving hardness and/or oxidation resistance.

In this work, we introduced B and C into TiN to synthesize quaternary Ti-B-C-N coatings using magnetron co-sputtering from Ti target and B_4C target respectively. The structure and mechanical properties of the coatings were investigated. Our purpose is to understand the growth of Ti–B–C–N coatings and develop a non-sticking, oxidation resistant, and wear resistant coating system for forming tools [8].

Experimental

The coating depositions were carried out in a computer controllable multi-target magnetron co-sputtering system [9]. Si (100) wafer and stainless steel (for friction coefficient testing) substrates were ultrasonically cleaned consecutively in acetone and ethanol for 20 and 15 min, respectively, before being placed into the chamber. The chamber was evacuated to a residual pressure lower than 4.0×10^{-4} Pa prior to deposition. Before deposition, the chamber was filled with argon to 5.0 Pa and the substrate bias was set to -600 V, producing a glow discharge etching for cleaning the substrate. The Ti and B_4C targets were sputtered to synthesize Ti-B-C-N coatings using DC power and RF power in an argon–nitrogen sputtering atmosphere of 10% nitrogen. A pulsed (frequency of 40 kHz and duty ratio of 40%) bias of −100 V was applied to the substrate holder rotated at a speed of 6 rpm. During deposition, the work pressures kept 0.5Pa and deposition temperature was at room temperature.

The crystal structure of the coatings was characterized using monochromatic Cu-Kα radiation on a Rigaku D/MAX 2500 X-ray diffractometer (XRD). The coating elemental concentrations and chemical states were investigated by X-ray Photoelectron Spectroscopy (XPS) using an Al Kα X-Ray source (1486.6 eV). The Ambios XP-2 surface profilometer was used to measure thickness and substrate curvature, from which the residual stress of coatings was calculated with [10].

$$\sigma_f = - \frac{E_s t_s^2}{6t_f(1-v_s)R}$$
(1)

with σ_f the in-plane stress component in the coating, t_f the thickness of the coating, E_s Young's modulus of the substrate, v_s Poisson's ratio for the substrate, t_s the thickness of the substrate and R the radius of curvature of the initial flat substrate after deposition of the coating. Hardness and Young modulus of the coatings were obtained using a Nano Indenter XP™ with a Berkovich indenter. The properties were evaluated using the Oliver and Pharr's analysis technique [11]. The instrument was operated in the continuous stiffness mode (CSM) and the indentations were made using a constant nominal strain rate of 0.05 s^{-1} and a frequency of 45 Hz. The harmonic displacement was 2 nm. Poisson's ratio of 0.25 was used to calculate the elastic modulus. Ten indents were taken and the results presented here were the average of 10 values. The friction coefficient was measured using a multi-functional tester. In this test, the sliding Si$_3$N$_4$ ball, 2 mm in diameter, held in a pen-like ball holder moved 200 circles with a radius of 5 mm at a speed of 104 mms^{-1} on the surface of the coatings in ambient air. The applied load was 2 N.

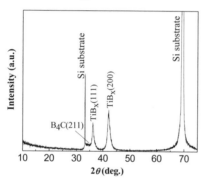

Fig.1. XRD pattern of Ti–B–C–N coating

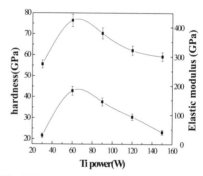

Fig.2. Hardness and elastic modulus of Ti-B-C-N coatings at different power of Ti.

Result and discussion

Fig.1 shows the XRD pattern of the Ti-B-C-N coating prepared at 60 W of Ti power and 150 W of B$_4$C power in an argon–nitrogen sputtering atmosphere of 10% nitrogen. The B$_4$C(104), TiB$_X$(111) and TiB$_X$(200) peaks appear in pattern. The weak B$_4$C(104) peak indicates the coating is crystallization. The sputtered Ti and B atoms act in an argon–nitrogen atmosphere to form TiB$_X$ phase by Ti+B4C=TiB$_X$+C. The broad lower intensity reflections, TiB$_X$(111) and TiB$_X$(200) peaks, stand for smaller (<10 nm) grains, which corresponds to amorphous or nanocrystalline phase [12]. The formation of nanocomposite structures is connected with a segregation of the one-phase to the grain boundaries of the second phase, which is responsible for stopping of the grain growth. This may contribute to the hardness enhancement according to the literature [13].

Fig.2 shows the change in hardness and elastic modulus of the Ti-B-C-N coatings at different powers of Ti. The hardness and modulus increase initially and then decrease with Ti power increasing. Their maximum values reach 42.8 GPa and 424 GPa at the 60 W of Ti power. The hardness enhance- ment may be attributed to suppression of dislocation movement in small grains and the narrow space between the nanocrystalline particles and amorphous phases [14]. The strong interfaces between nanocrystalline and amorphous phases can increase the cohesive energy of the interface boundaries, restraining the grain boundary sliding. It is well known that 5 ~ 20 nm small nanocrystalline grains are free of any defect and dislocation which are easy to divert from inner grains to the grain boundaries due to relatively high total free energy and short diffusion length in the small grains [15]. Then, the hardness and modulus decrease as Ti power increase due to excess soft phase of Ti.

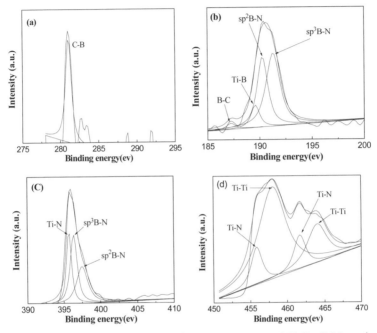

Fig.3. (a) C_{1s} (b) B_{1s} (c) N_{1s} and (d) Ti_{2p} XPS spectrum of Ti–B–C–N coating

Fig.3 shows the C_{1s}, B_{1s}, N_{1s} and Ti_{2p} XPS spectrum of Ti–B–C–N coating. The C_{1s} spectrum shows that the B–C (281 eV) bond exists in the coating. The B_{1s} spectrum is composed by four peaks centered at 191, 190.1, 189.4 and 187 eV, which can be attributed to the sp^3B-N, sp^2B-N, Ti-B and B-C bonds. The N_{1s} spectrum consists of three components centered at 395.8, 396.5 and 397.8 eV corresponding to the Ti–N, sp^3B–N and sp^2B–N bonds, respectively. The Ti_{2p} spectrum is composed of spin doublets, and the $Ti_{2p1/2}$ peak can be fitted into peaks stated at 456.1 and 458.2 eV, which are attributed to the Ti–N and Ti–Ti bonds. The $Ti_{2p3/2}$ peak can be fitted into two peaks at 462 and 464 eV, which corresponds to Ti–N and Ti–Ti, respectively [16].

Fig.4 shows the friction coefficient of the Ti-B-C-N coatings at different powers of B_4C when Ti keeps 60 W. The lower friction coefficient of 0.16 appears at the 180 W. Then its value increases gradually from 0.16 to 0.4 at 200 cycles. But the coating is not wear-out after 200 cycles. The lower friction coefficient and higher wear resistance than others can be attributed to self-lubricant of carbon. The coating becomes easier to wear-out and higher friction coefficient at lower B_4C power. When the power of B_4C is 150 W, the coating appears wear-out after 120 cycles. As the power continues to decrease, the coating wears out in 100

Fig.4. Friction coefficient of Ti-B-C-N coatings at different power of B_4C

cycles. These results clearly show that when the C content increases in the coating, the friction coefficient decreases. This result demonstrates that the multicomponent Ti–B–C–N coating can be designed to combine high hardness along with low friction and optimized tribological behavior.

Conclusion

The effects of the different powers of Ti and B_4C on the hardness and wear resistance of the Ti-B-C-N nanocomposite coatings deposited by multi-target magnetron co-sputtering were investigated. 60 and 150 W were optimum powers of Ti and B_4C targets for significant improved hardness and wear resistance. The maximum hardness and elastic modulus of 42.8 GPa and 424 GPa were reached at these powers. The coating also revealed the lower friction coefficient and higher wear resistance than other coatings synthesized at other powers.

Acknowledgements

This work was supported by National Basic Research Program of China (973 Program, 012CB933604), National Natural Science Foundation of China (51272176, 11075116), and the Key Laboratory of Beam Technology and Material Modification of the Ministry of Education, Beijing Normal University, China.

References

[1] D. Zhong, E. Sutter, J.J. Moore, *et al.*, Composition and oxidation resistance of Ti–B–C and Ti–B–C–N coatings deposited by magnetron sputtering, Surf Coat Technol. 163-164 (2003) 50.

[2] G. Tang, X. Ma, Magnetron sputtering deposition Ti-B-C-N films by Ti/B_4C compound target, Surf Coat Technol. 203 (2009) 1288-1291.

[3] Y.G.Shen, Y.W. Mai, Residual stress, microstructure, and structure of tungsten thin films deposited by magnetron, Journal of Applied Physics 87 (2000) 177.

[4] D.Zhong, E. Mateeva, I.Dahan, *et al.*, Wettability of NiAl, Ni-Al-N, Ti-B-C, and Ti-B-C-N films by glass at high temperatures, Surf Coat Technol. 133-134 (2000) 8

[5] H.Gleiter, Nanostructured materials: basic concepts and microstructure, Acta Mater 48 (2000) 1-29

[6] P. Eh. Hovsepian, D.B. Lewis and W. D. Münz, Recent progress in large scale manufacturing of multilayer/superlattice hard coatings, Surf Coat Technol. 133 (2000) 166-175.

[7] A. Vyas, Y.H. Lu, Y.G. Shen, Mechanical and tribological properties of multicomponent Ti–B–C–N thin films with varied C contents, Surf Coat Technol. 204 (2010) 1528-1534.

[8] D. Zhong, E. Sutter, J.J. Moore, *et al.*, Mechanical properties of Ti–B–C–N coatings deposited by magnetron sputtering, Thin Solid Films 398-399 (2001) 320-325.

[9] M.Y. Liu, J.Y. Yan, S. Zhang, *et al.*, The Effect of deposition temperature and work pressure on Ti-B-C nanocomposite coating prepared by multitarget cosputtering, Plasma Science 39 (2011) 3115-3119.

[10] G.C.A.M. Janssen, M.M. Abdalla, F. van Keulen, B.R. Pujada and B. van Venrooy, Celebrating the 100th anniversary of the Stoney equation for film stress: Developments from polycrystalline steel strips to single crystal silicon wafers, Thin Solid Films 517 (2009) 1858–1867.

[11] W.C. Oliver and G.M. Pharr, An improved technique for determining hardness and elastic modulus using load and displacement sensing indentation experiments, J. Mater. Res. 7 (1992) 1564-1583.

[12] J. Musil and J. V.ceka, Magnetron sputtering of hard nanocomposite coatings and their Properties, Surf Coat Technol. 142 (2001) 557-566.

[13] J. Musil, P. Zeman, H. Hrubý and P. H. Mayrhofer, Hard and superhard nanocomposite coatings, Surf Coat Technol. 125 (2000) 322-330.

[14] S. Veprek, New development in superhard coatings: the superhard nanocrystalline amorphous composites, Thin Solid Films 317 (1998) 449-454.

[15] S. Veprek, R.F. Zhang, M.G.J. Veprek-Heijmana, *et al.*, Superhard nanocomposites: Origin of hardness enhancement properties and applications, Surf Coat Technol. 204 (2010) 1898–1906.

[16] X.Y. Chen, Z.H. Wang, *et al.*, Microstructure, mechanical and tribological properties of Ti–B–C–N films prepared by reactive magnetron sputtering, Diam. Relat. Mater. 19 (2010) 1336-1340.

Key Engineering Materials Vol. 537 (2013) pp 67-70
© (2013) Trans Tech Publications, Switzerland
doi:10.4028/www.scientific.net/KEM.537.67

Preparation and Salt Spray Corrosion Resistance of Ni/BN and NiCrAl/BN Coatings

Feng Zhang[1,2], Chuanbing Huang[1], Wei liu[1], Kui Zhou[1], Wenting Zhang[1], Lingzhong Du[1], Weigang Zhang[1,*]

[1] State Key Laboratory of Multi-phase Complex Systems, Institute of Process Engineering, Chinese Academy of Sciences, Beijing 100190, China

[2] Graduate University of Chinese Academy of Sciences, Beijing 100049, China

Keywords: salt spray corrosion, plasma spray technology, abradable sealing coating, turbo engines

Abstract. Ni/BN and NiCrAl/BN abradable sealing coatings used in turbo engines were prepared by plasma spray technology. The phases and the microstructures of the coatings were characterized with X-ray diffraction (XRD) and scanning electron microscopy (SEM). Corrosion behaviors of these coatings were investigated with open-circuit potential (OCP) and salt spray corrosion test. The results showed that the NiCrAl/BN possess better corrosion resistance as compared with Ni/BN.

Introduction

Naval aircrafts serving in the high salt spray atmosphere suffer serious corrosion, especially on the turbo engines and the abradable sealing coatings inside [1]. In order to increase efficiency and reduce fuel consumption, abradable sealing coatings are always necessary to reduce the blade-tip clearances in a compressor as well as on the tip of turbo blades [2-4]. Abradable sealing coatings are typical multiphase materials which compose of metal phase, non-metal phase and a defined level of porosity. The role of the metal phase in the composite is of providing the coating strength and resistance to its environment (includes erosion, corrosion, and high temperature oxidation). The non-metal phase and the porosity offer a good abradability and lubrication. However, the pores in the coatings provide penetrating paths for corrosion medium and formed a large numbers of galvanic cells [5-7]. Furthermore, galvanic corrosion between the top coating and bond coating might occur when the corrosion medium wetted the interface between the surface coatings and the bonding layer, which might significantly reduce their bonding strength and lead to failure [8, 9]. Therefore, it is necessary to investigate the corrosion behavior of these functional coatings, as the coatings with a porous multi-phase structures are prone to be corroded.

In this study, two composite coatings of Ni/BN and NiCrAl/BN were prepared by plasma spraying. The corrosion behaviors of these two coating systems were studied via electrochemical tests and salt spray tests.

Experiments

The Ni/BN and NiCrAl/BN composite powders were prepared and used for producing the coatings, the nominal compositions and particle sizes are listed in Table1. The cross sectional micrographs of these powders are shown in Fig.1. A continuous metal shell provides protection to BN core, which provides the abradabilty of the coating. TC4 titanium alloy plates with dimensions of φ25mm×5mm were used as substrates. The Ni/BN and NiCrAl/BN coatings were prepared using APS-2000K plasma spray equipment. Details of the spraying parameters are given in Table 2.

Table1 Composition and particle size of the powders

Powder	Chemical composition [wt.%]	Particle size[μm]
Ni/BN	Ni(70) BN(30)	45-150
NiCrAl/BN	Ni(balance) Cr(14~17) Al(1~4) BN(20~25)	45-150

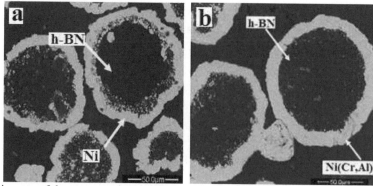

Fig.1 SEM images of the cross sections of the used composite powders: (a) Ni/BN; (b) NiCrAl/BN

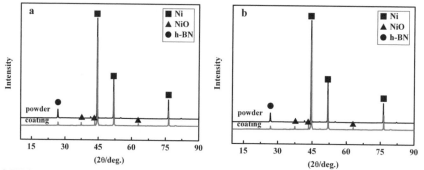

Fig.2 XRD patterns of the used powders and the derived coatings: (a)Ni/BN; (b) NiCrAl/BN

Table2 Plasma spray process parameters

Ni/BN		NiCrAl/BN	
Arc current	500A	Arc current	400A
Voltage	64V	Voltage	60V
Spray distance	100mm	Spray distance	100mm
Scan times	20	Scan times	20
Plasma gas(Ar)flow rate	40l/min	Plasma gas(Ar)flow rate	40l/min
Coating thickness	2.0mm	Coating thickness	1.8mm

Phases of the used powders and the derived coatings were investigated by X-ray diffraction (XRD). The microstructures of the samples were examined by a QUANTA 200 FEG scanning electron microscope (SEM). The OCP was measured by a CHI660D electrochemical measurement system, using a three-electrode cell with the coating working electrode platinum counter electrode and a saturated calomel electrode (SCE) reference electrode. The OCP was monitored for 180min.

Salt spray corrosion test was carried out in a salt spray chamber under a 5wt.% NaCl aqueous spray at a rate of 1~2 ml/h·cm^2. The specimens subjected to 480h corrosion at a temperature of 35 ºC. The salt spray corrosion(SSC)tests following the national standard GB/T10125-1997(ISO9227) procedure.

Results and discussions

The XRD patterns of the used powders and the derived coatings are shown in Fig.2. The main phases of the Ni/BN powder are Ni and BN, and the Ni/BN coating mainly consists of Ni and BN with the presence of NiO as a minor phase, caused by weak oxidizing of Ni during spraying. The main phases of the Ni(Cr, Al) powder are Ni(Cr, Al) and BN. While, for NiCrAl/BN coating, there was no phase transformation occurred during spraying. The peaks of Ni(Cr,Al) exhibited a shift to lower angle after sprayed. The result can be attributed to the additional macroscopic stress which was induced by plastic deformation during spraying [10].

Fig.3 shows the cross sectional morphologies of Ni/BN coating and NiCrAl/BN coating. It can be obviously observed that the structures of the as sprayed coatings are typical lamella microstructure. The metal is highly homogeneous with interconnected pores and the BN particles uniformly distributed in the metal matrix.

Fig.3 Cross sectional SEM morphologies: (a) Ni/BN coating; (b) NiCrAl/BN coating

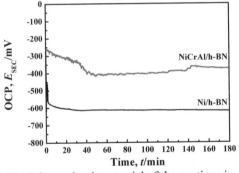

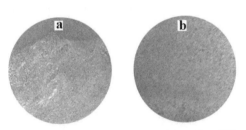

Fig.4 Open-circuit potential of the coatings in 5wt.%NaCl solution

Fig.5 Surface images of the coatings after 480h salt spray corrosion: (a) Ni/BN; (b) NiCrAl/BN

Fig.4 indicates the change of OCP of the Ni/BN and NiCrAl/BN coatings. The OCP was measured as a function of time in 5wt.% NaCl solution. The OCP of the Ni/BN coating is negative to that of NiCrAl/BN coating. The curve of the NiCrAl/BN coating was characterized by frequent potential fluctuations. This indicates that the localized corrosion such as crevice corrosion or pitting was carried out [11]. The OCP of Ni/BN coating was found to drift with time during the first 10min of electrode immersion in 5wt.% NaCl solution. The main reason of the initial potential decrease can be associated to the changing in the surface of the coating. As the electrolyte penetrates the coating, the oxide layer has been dissolved [9].

As shown in Fig.5, after tested in the salt spray chamber for 480h, green corrosion products were accumulated on the surface of the Ni/BN coating. On the contrast, no corrosion phenomenon could be found on the surface of NiCrAl/BN coating. This indicated that NiCrAl/BN coating possess the corrosion resistance better than Ni/BN coating in salt spray environment. The result is in agreement with that obtained by open-circuit potential measurement.

Conclusions

Two kinds of composite coatings of Ni/BN and NiCrAl/BN were prepared by plasma spray technology, which characterized by the difference matrix of pure nickel and nickel based alloy, respectively. The open-circuit potential and salt spray corrosion behavior of the coatings were studied. It was found that:

(1) Both of the prepared Ni/BN and NiCrAl/BN abradable sealing coatings exhibit typical lamella microstructures.

(2) The results of OCP and salt spray corrosion tests indicate that NiCrAl/BN coating possess better corrosion resistance than Ni/BN coating in salt spray environment.

Acknowledgements

The authors are grateful for the financial support of the National Natural Science Foundation of China (grant no. 50901071 and 51001093), and the Natural Science Foundation of Jiangsu Province, China (No. BK2011452).

References

[1] C.G. Xu, L.Z. Du, B. Yang, W.G. Zhang, Study on saltspray corrosion of Ni–graphite abradable coating with 80Ni20Al and 96NiCr–4Al as bonding layers, Surf. Coat. Technol. 205 (2011) 4154-4161.

[2] H.I. Faraoun, T. Grosdidier, J.L. Seichepine, et al., Improvement of thermally sprayed abradable coating by microstructure control, Surf.Coat.Technol. 201 (2006) 2303-2312.

[3] H.I. Faraoun, J.L. Seichepine, C. Coddet, H. Aourag, J. Zwick, N. Hopkins, D. Sporer, M. Hertter, Modeling route for abradable coatings, Surf.Coat.Technol. 200 (2006) 6578-6582.

[4] J. Matejicek, B. Kolman, J. Dubsky, K. Neufuss, N. Hopkins, J. Zwick, Alternative methods for determination of composition and porosity in abradable materials, Mater.Charact. 57 (2006) 17-29.

[5] F.C. Walsh, C. Ponce de León, C. Kerr, S. Court, B.D. Barker, Electrochemical characterisation of the porosity and corrosion resistance of electrochemically deposited metalcoatings, Surf. Coat. Technol. 202 (2008) 5092-5102.

[6] A. Pardo, M.C. Merino, M. Mohedano, P. Casajús, A.E. Coy, Corrosion behaviour of Mg/Alalloys with composite coatings, Surf.Coat.Technol.203 (2009) 1252-1263.

[7] M. Campo, M. Carboneras, M.D. López, et al., Corrosion resistance of thermally sprayed Aland Al/SiC coatings on Mg, Surf.Coat.Technol.203 (2009) 3224-3230.

[8] A. Lekatou, D. Zois, A.E.Karantzalis, D. Grimanelis, Electrochemical behaviour of cermet coatings with a bond coaton Al7075:Pseudopassivity, localized corrosion and galvanic effect considerations in a saline environment, Corros.Sci.52 (2010) 2616-2635.

[9] C.G. Xu, L.Z. Du, B.Yang, W.G. Zhang, The effect of Al content on the galvanic corrosion behavior of coupled Ni/graphite and Ni–Al coatings, Corros. Sci. 53 (2011) 2066-2074.

[10] X.T. Luo, G.J. Yang, C.J. Li, Multiple strengthening mechanisms of cold-sprayed cBNp/NiCrAl composite coating, Surf. Coat.Technol. 205 (2011) 4808-4813.

[11] K. Alvarez, S.K. Hyuu, H. Tsuchiya, S. Fujimoto, H. Nakajima, Corrosion behaviour of Lotus-type porous high nitrogen nickel-free stainless steels, Corros. Sci. 50 (2008) 183-193.

Key Engineering Materials Vol. 537 (2013) pp 71-75
© (2013) Trans Tech Publications, Switzerland
doi:10.4028/www.scientific.net/KEM.537.71

Microstructure and Properties of CrN Multilayer Coatings

Ting Guo[1, a], Chen Wang[1,b] and Bing Yang[2,c]

[1] Division of Fine Ceramics, Institute of Nuclear and New Energy Technology, Tsinghua University, Beijing Key Lab of Fine Ceramics, Beijing, 100084, China

[2]School of Power and Mechanical Engineering, Wuhan University, Wuhan, 430072, China

[a]guot10@mails.tsinghua.edu.cn, [b]wangchen@mail.tsinghua.edu.cn, [c]toyangbing@163.com

Keywords: arc ion plating; Cr/CrN coatings; multilayer; microstructure; properties

Abstract:The CrN and Cr/CrN multilayer coatings with different modulation ratios were deposited onto stainless steel and cemented carbide substrates by arc ion plating technology. The phase composition, surface and cross-sectional morphology were analyzed by using X-ray diffraction (XRD), atomic force microscope (AFM) and scanning electron microscope (SEM). The micro Vickers was used to test the hardness. The results indicated that the coatings were mainly composed of Cr, CrN and Cr_2N phase. No columnar grain growth was observed in Cr/CrN multilayer coatings, and the crystalline orientation of the Cr/CrN multilayer was different from CrN film. There exists an optimal modulation ratio of 10:10 to the coating deposition, under which the surface was compact and smooth with a relatively lower roughness. The coatings hardness increased with decreasing modulation ratio and the hardness of $CrN+Cr_2N$ was lower than that of single CrN phase.

Introduction

In the 21st century, the automobile industry is faced with many challenges. Energy saving, environmental protection and safety are the most important and difficult issues. With the increasingly stringent requirements of human on the ecological environment, the voice, emphasis on low emission and low fuel consumption of gasoline engine or diesel engine, is also more and more intense. Piston ring-cylinder liner is an important influence factor of the engine ideal combustion. Their tribological properties exhibit a direct impact on environmental protection and energy saving of the entire engine. A variety of surface treatment techniques such as electroplating Cr, sprayed molybdenum, laser strengthening, nitriding and coating technology have been widely used in piston rings, but they can't meet the requirements of piston ring on wear resistance and environmental protection. In recent years, multilayer films have been concerned gradually. Usually, the multilayer films formed of two materials A and B, the thickness of adjacent two layers is known as modulation period, while the ratio of thickness of A and B as modulation ratio. The crystal structures of layers can exist in various types, such as single crystal, polycrystalline or amorphous, and thus a very complex interface structure will be formed. The interface of A and B can prevent the growth of columnar grains and coarse grains, refine grains and improve the ability of plastic deformation; then hinder dislocation glide, inhibit the formation and expansion of cracks, so as to improve the strength and impact resistance of films[1-3].

In this paper, the nanocrystallization and multilayer composite film technology were used to significantly reduce the size of CrN grain with the diameter below 100 nm. Besides, the strengthening effect of nanocrystalline and multilayer composite structure caused a substantial increase of wear resistant and lubricating properties of Cr / CrN coatings.

Materials and Methods

The domestic arc ion plating machine was used as the deposition system and the chromium target with a purity of 99.99% as the cathode target. 1Cr18Ni9Ti Stainless steel and YT15 cemented carbide were selected as substrates, with dimension of 40mm × 15mm × 1mm and 15mm × 15mm × 5mm, respectively. Substrates were grinded by metallographic sandpaper progressively to 1400 #, polished with 0.1 μm diamond plaster, and then cleaned by ultrasonic.

The distance from the axis of the substrates holder to targets was approximately 250 mm. A temperature controller was applied to control the heater by which the temperature of the substrate was fixed at about 300 °C. Prior to the deposition, the substrates were cleaned by glow discharge at -800V bias voltage, with 80% duty cycle for 30min. The bombarding voltage was fixed at 800 V by using Cr^+ for 10 min to improve the adhesion of the coatings. Then the bias voltage was down to-150V to deposite Cr transition layer for 5min with Cr target current of 80A at 0.5Pa. Bias voltage was maintained at -150V, work pressure of 2.3Pa and cathodic arc current of 80A. The deposition pressure of pure CrN was fixed at 2.3Pa with 150 V bias voltage and 80A Cr target current. By controlling the nitrogen flow meter, four kinds of Cr/CrN multilayer films with modulation ratio of 10:4,10:10,4:10 and 1:10 and pure CrN coatings were prepared. The deposition time was 60min.

The cross-sectional morphologies of the composite coatings were analyzed by using Nova 400 Nano scanning electron microscope. The surface roughness was measured by SPM-9500J3. The crystal structures and preferred orientation were analyzed by D8 advanced X-ray diffractormeter with a Cu Kα radiation (0.15418 nm). The micro Vickers was used to test hardness by using diamond indenter with 25g applied load. The hardness value was the average of five measurements.

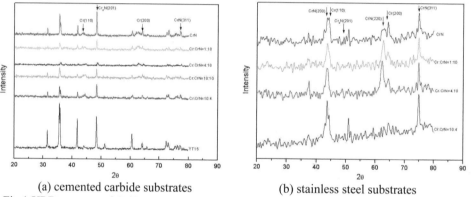

(a) cemented carbide substrates   (b) stainless steel substrates
Fig.1 XRD patterns of Cr/CrN coatings on different substrates with different modulation ratios

Results and Discussion

Analysis of Crystal Structures. Figure 1(a) shows the XRD patterns of Cr/CrN coatings on cemented carbide substrates with different modulation ratios.

As shown in figure 1(a), the diffraction peaks appeared at 31 °, 36 °, 44 °, 48 °, 65 °and 76 °. The diffraction peaks of YT15 had been removed and the calibration results were shown in figure 1(a). It can be seen that modulation ratios exhibit a significant effect on the structure of Cr/CrN multilayer films. Cr_2N (201), Cr (110), Cr (200) and CrN (311) were presented while the ration of Cr:CrN was 10:4, and Cr_2N (201) and Cr (200) as preferred orientation. With the reducing ratio of Cr/CrN, the diffraction peak intensity of Cr (110) and Cr (200) decreased significantly. At the same time, more Cr_2N appeared. This maybe relate with the decreasing ratio of Cr/N, which promote the combination of Cr^+ and N. As a result, more Cr_2N phase were easily formed in the films instead of CrN. Furthermore, due to relatively few crystallization, some diffraction peaks were not obvious.

It can be found from the XRD diffraction pattern that either peaks of CrN or Cr_2N were wide. The main reason is because CrN is fcc structure and N atoms are in the octagonal clearance while Cr_2N is hcp structure. Due to ion bombardment, which produced a cascade of collisions, radiation damage and displacement spike effect, CrN_x cannot strictly meet the stoichiometry with N atoms absence. Lattice constants of CrN films have a certain degree of uncertainty, so that the CrN diffraction peaks broaden.

Fig1(b) shows XRD patterns of Cr/CrN coatings on stainless steel substrates with different modulation ratios.

It can be seen that substrates materials exhibited a great influence on coating's crystal structures. Comparing figure 1(a) with figure 1(b), it can be found that there was a great difference of the diffraction intensity and preferred orientation of as-deposited coating with the same modulation ratio on different substrates. As modulation ratio was 10:4, preferred orientation of coating on cemented carbide substrate were Cr_2N (201) and Cr (200), while preferred orientation of coating on stainless steel were CrN (200) and Cr (110).

Analysis of Surface Morphology and Roughness. Figure 2 is AFM photos of Cr/CrN coatings on cemented carbide with different modulation ratios.

From figure 2, it can be seen that there is a great influence of modulation ratio on the surface morphology of films. As increasing modulation ratio, the surface exhibit a big fluctuation and big particle sizes. With the decreasing ratio of Cr/CrN, the particle sizes decreased and the surface became dense. The film surface became loose and porous again when the modulation ratio was further reduced. The Root Mean Square roughness obtained from AFM images and modulation ratios were plotted in Figure 4. The same trend was obtained. It can be concluded that there is an optimal modulation ratio of 10:10 to the coating deposition. Under the ratio, the surface is compact and smooth with a lower roughness.

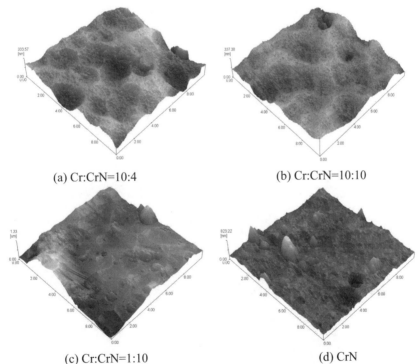

(a) Cr:CrN=10:4   (b) Cr:CrN=10:10

(c) Cr:CrN=1:10   (d) CrN

Fig.2 AFM photos of Cr/CrN coatings on cemented carbide with different modulation ratios

With the high modulation ratio of Cr/CrN, Cr_2N phase was easily formed. Metal particles and N ion not fully reacted, leaving more particles on the film surfaces. Large particles peeled off under stress, and pits were formed on the surfaces. These particles freely accumulated on the substrates, resulting in a loose structure and rough surfaces. However, with the modulation ratio further reduced, until to the pure CrN coating, the columnar crystals of chromium nitride formed. Due to the presence of columnar crystals, surface roughness increased.

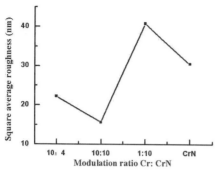

Fig.3 Relationship between the roughness
and modulation ratios of coatings

It also can be seen from figure 3 that there exist uneven particles on coating surfaces. This is because that the large droplet particles attached to the substrate surfaces at a faster rate. Subsequent fine particles took large particles as nuclei and began to deposit. Some particles were buried and some reached the film surfaces, which were easy to fall off. The white dotted areas, as shown in figure 2 (a), were Cr droplet buried in layers during the deposition process. The presences of these large droplets not only affect the surface morphology, but also reduce other properties of the films, such as corrosion resistance and mechanical properties.

Cross-sectional Morphology of Composite Coatings. Figure 4 shows the cross-sectional morphology of Cr/CrN on stainless steel substrate with different modulation ratios.

As shown in figure 4, multilayer structures of Cr/CrN composite coatings were successfully prepared. The coatings were nearly 2 ~ 3 μm and consist of large quantity Cr and CrN layers, It was discontinuous between layers and connections were not very dense in figure 4(a). This may be due to surface cleaning was not enough, so that the adhesion weakened. A clear and dense layered structure can be seen in figure 4(b). The coating were combined by 5 layers. The minimum thickness of single-layer was 457 nm, while the maxmum was 593 nm. The entire coating thickness was 2.23μm. The percentage of N in the coating was 32.3at% by EDS analysis, which was greater than that in CrN. It was indicated that there was a rich N region in the coating. There was no layered structure occurred in the coating from figure 4(c), which mainly due to the unclean surface was not clean and the layered structure was covered by stains. Figure 4(d) showed the pure CrN coating. The thickness of as-deposition coating was 2.51μm. The multilayered structure was clear and the coating was compact in figure 4(e). The entire coating thickness was more than 8 microns, with layer thickness of 700nm.

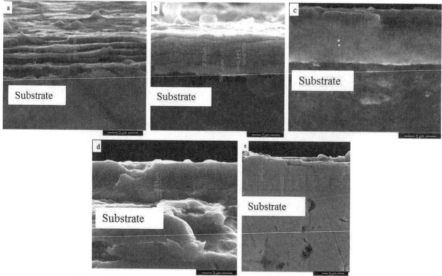

Fig.4 Cross-sectional morphologies of Cr/CrN composite coatings with different modulation ratios and pure CrN coating. (a) Cr:CrN=10:4, (b) Cr:CrN=4:10, (c) Cr:CrN=1:10, (d) CrN and (e) Cr:CrN=1:15

Comparing five images in figure 4, it can be found that there was not a columnar crystal in multilayer films. This is because that Cr is body-centered cubic, while CrN is a face-centered cubic. Their alternating deposition process blocked the growth of columnar crystals, which make the films more dense.

Influence of Modulation Ratios on Coating Hardness. Figure 5 shows the relationship of coating hardness and modulation ratios. The theoretical hardness of Cr_2N and CrN were 2175 $HV_{0.05}$ and 2740 $HV_{0.05}$, respectively[4]. As shown in figure 5, the coating hardness increased with decreasing ratio of Cr and CrN. That is mainly due to the hardness of Cr is lower and the increasing content of CrN is conducive to the increase of coating hardness. It also was found that two-phase film hardness was lower than that of single-phase film. The literature suggests [5] the reason for lower hardness of two-phase film is that the film composition deviates from stoichiometry and that will change the Fermi level of the compounds. Thus the energy of bonding and anti-bonding electrons of d energy band are changed, and then the bonding strength of the materials were changed, too.

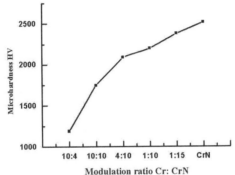

Fig.5 Relationship between the hardness and modulation ratios of coating

Conclusion

1. Films were mainly composed of Cr, CrN and Cr_2N, and Cr_2N (201) as preferred orientation; no columnar crystal structure in multilayer films were observed; there exist a significant difference of the crystalline orientation between the multilayer films and single-layer CrN; there was an optimal modulation ratio of 10:10 to the coating deposition, under which the surface was compact and smooth with a low roughness.

2. Thick coatings can be prepared by arc ion plating technology. The thicknesses of coatings were several microns, with single-layer of a few hundred nanometers. The hardness of coatings increased with decreasing ratio of Cr and CrN, and the hardness of two-phase film of $CrN + Cr_2N$ is lower than that of CrN single-phase film.

References

[1] B.D. Narendra, S. Nayak, Nanocoatings for engine application, Surface and Coatings Technology. 194 (2005) 58-67.

[2] J. A. Picas, A. Fom, G.M. atthaus, HVOF coatings as an alternative to hard chrame for pistons and valves, Wear. 261(2006) 477-484.

[3] B. Warcholinski, A. Gilew icz, Z. Kuklinski, et. al, Arc-evaporated CrN, CrN and CrCN coatings, Vacuum. 83 (2009) 715-718.

[4] B. Zhong, W. Gou, G.Q. Li, Y.R. Hu, Effect of nitrogen ratio on phase structure and tribological properties of CrN_x coating, Transactions of materials and heat treatment. 28(2007) 134-137,144.

[5] M.B. Tian, D.L. Liu, Thin film science and technology manual, Mechanical industry press, Beijing, 1991.

Key Engineering Materials Vol. 537 (2013) pp 76-81
© (2013) Trans Tech Publications, Switzerland
doi:10.4028/www.scientific.net/KEM.537.76

Influence of Power Supply Frequency on Microstructure and Properties of Micro-arc Oxidation Coating on Aluminum Alloy

XIN ShiGang [a], LE Jun [b], SONG LiXin [c]

The key library of Inorganic coatings, Shanghai Institute of Ceramic, Chinese Academy of Sciences, Shanghai 200050, People's Republic of China

[a] sgxin@mail.sic.ac.cn, [b] lejun@mail.sic.ac.cn, [c] lxsong@sunm.shcnc.ac.cn

Keywords: Micro-arc oxidation; Frequency; Coating; Aluminum alloy

Abstract. This paper studies the influence of current frequency on the micro-arc oxidation (MAO) process, microstructrue and the properties of the formed alumina coatings. The amount of discharge sparks increases and the spark moves more quickly with increasing the frequchlo of current pulse. SEM results show that the size of the discharge products decreases. The anodic cell voltage decreases with increasing the frequency of power supply, however, the cathodic cell voltage increases. The thickness and roughness of the coating produced using frequency of power supply 50 Hz are high. Increasing the frequency, the thickness of the coating decreases, but coating surface becomes even, and the coating possesses better corrosion protective property.

Introduction

The spark discharge phenomena were discovered on anodic surface at high polarization by Güntherschlze and Betz in the 1930s, however, and then the spark on anodic surface was considered to be bed for the anodizing coating [1]. Up till the end of 1960s, its practical benefits were exploited [2]. It was noticed that a ceramic coating could be formed during the process of spark discharge. The coating possesses good anti-wear, corrosion resistance and electric insulating protective properties [3-5]. This process is named micro-arc oxidation (MAO).

MAO treatment of aluminum alloys is typically conducted in an alkaline silicate or aluminate bathes with a power supply capable of voltage output of 300-1000[6, 7]. Direct current (DC) power supply with a high output voltage is used widely, but the produced coating is thin and porous and typically do not possess high adhesion to the substrate [2, 5, 8]. The application of pulsed DC allows for controlling interruption of the process and the arc duration. However, a pulsed current with a high peak current can give rise to additional polarization of the electrode surface by a creation of a charged double layer, and the produced coating is porous still [9].

It is noticed that an oxide-hydroxide can be formed on aluminum alloy using cathodic polarization. The additional polarization of the electrode can be avoided by combining anodic and cathodic polarizations. The application of unbalanced alternative current power supply with different positive and negative voltages provides a wide possibility to controlling the coating composition and microstructure. The coating possesses lower porosity, and high adhesion to the substrate, and is thoroughly different from the coating produced by DC and pulsed DC power supplies [2, 9, 10].

The effect of voltage and current on the composition and properties is investigated widely [11-12], but the influence of the frequency does not study. In this work, the alumina coatings were produced using different frequency of power supply on aluminum alloy, the influence of the frequency on the MAO process, microstructure and properties is studied.

Experimental procedure

Aluminum alloy 2024 is used as substrates in the experimental work. The samples are a disk 40mm in diameter and 2.5mm thick. The surface finish is improved by mechanical polishing to Ra~0.22±0.005µm prior to MAO coating deposition. The samples are connected to the power supply output through an aluminum alloy holder. The holder with a dimension ($100 \times 10 \times 2$ mm^3) is insulated with the electrolyte by PVC sheath.

The MAO alumina coating is fabricated using a 10kW alternative current (AC) MAO unit, which consists of an insulated electrolyte bath, a thermometer, a stirrer and a high voltage DC power supply with approximate 1000V of the maximum voltage amplitude. A stainless steel tank with a gas agitation containing 10 l of electrolyte is used. The electrolyte is prepared from a solution of 10g/l NaAlO$_2$ and 2g/l KOH in distilled water. One output of the power supply is connected to the electrolyte bath and another output to a specimen immersing into the electrolyte. A constant current density (4A/dm^2) at the coating surface is maintained by controlling the voltage amplitude. Stirring and cooling systems maintain the temperature of electrolyte at 20~30°C during the process. More details about the MAO equipment can be found elsewhere [10].

The phase composition of the coating is investigated using D/max-rB X-ray diffractometer (Cu kα radiation). The X-ray generator settings are 50kV and 50mA, and the scanning range is acquired from 20° to 90° (in 2θ).

The surface morphologies of the MAO coating produced for 1hour treatment time using frequencies of power supply 30, 50 and 100Hz are observed using Hitachi S-570 scanning electron microscope (SEM).

Surface roughness of the coatings is measured by a profilometer (Hommelwerke T8000-C, Germany). An eddy current thickness meter (CTG-10) is used to determine the thickness of the coating formed using different frequency of power supply.

The corrosion resistance of aluminum alloy with ceramic coating is evaluated by means of potentiodynamic polarization. The aluminum alloy is treated for 1 hour at density of 4A/dm2 using different frequencies of power supply. A 3.5% NaCl solution is selected for the test that is conducted in a three-electrode experimental corrosion cell. The counter electrode is platinum and a saturated calomel electrode (SCE) is used as reference. The dimension of working electrode is 1cm2 in contact with the electrolyte. The potentiodynamic polarization voltage swept from 250mV below to 500mV above the open circuit potential at a sweep 10mV/s.

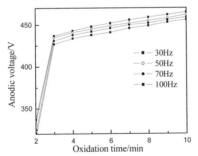 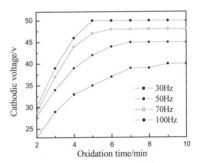

Fig.1 Anodic (a, left) and Cathodic (b, right) voltage change with power supply frequency during MAO

Results and Discussion

Voltage Evolution and surface observation. The frequency of AC power supply gives a influence on the spark lifetime and polarization of electrode in MAO process. The discharge lifetime should not be longer than the duration of positive biasing in the AC waveform employed. Increasing the frequency of power supply, the lifetime of every discharge spark decreases and the amount of formed spark increases during MAO process, so it is observed in the experiment that the spark moves more quickly.

The discharge phenomena occurring during MAO using frequency 50, 70 and 100 Hz is similar, but using 10 and 30 Hz, the discharge sparks appear on substrate surface is very discrete flashes of light. The size of the discharge spark decreases obviously with increasing the frequency. Because the

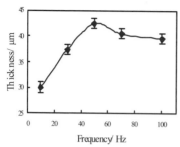

Fig.2 The thickness of MAO coating formed by different power supply frequency

amount of spark formed using high frequency (more than 50Hz) is more than that using low frequency, the sound produced by breakdown of oxidation film increases. The anodic and cathodic voltages of MAO process vary when the frequency of power supply changes, the results are shown in Fig.1.

It is observed that both anodic and cathdic cell voltages increases with oxide time. According to the curves, both the anodic and cathodic processes are divided into two stages. The anodic cell voltage reaches quickly 420-430V during initial 3min treatment time (Stage 1), then the voltages begin to increase slowly with a rate of about 4V/min (Stage 2). The anodic cell voltage decreases with increasing the frequency of power supply (Fig.1), and the voltage in Stage 1 gives a sufficient influence on final voltage of MAO process. The Stage 1 consists of both the formation of barrier film and its breakdown process. The difference of the anodic cell voltages reveals that increasing the frequency of power supply decreases the voltage of breakdown of barrier film.

The cathodic cell voltage have a steady value after about 5-8 min, and this time period is designated as Stage 1 of cathodic polarization process. It is seen from Fig.1(b) that the time of reaching the steady value decreases from 8min to 5min with increasing the frequency of power supply. However, at Stage 2, the cathodic cell voltages are constant, and the final voltage increases with increasing the frequency of power supply. The cathodic cell voltage is contributed by the electrical conductivities of MAO coating, this indicates that increasing the frequency of power supply decreases the electrical conductivities of MAO coating.

The Coating Thickness and Roughness. The variety of the coating thickness formed during 1hour as a function of the frequency of power supply is shown in Fig.2.

It is noticed that the frequency gives a obviously influence on the thickness of the ceramic coatings. Because the discharge sparks is very discrete using frequency 10 Hz, the formed coating is only 30μm, and the coating is brown with a number of black specks of 1mm diameter. The thickness of the coating formed by frequency 30 Hz increases, and the coating is faint black with a few of black speck. The maximum coating thickness produced using frequency of power supply 50 Hz, and then the thickness of the coating decreases with increasing the frequency. When the frequency 50, 70 and 100 Hz is used, the coatings are black. It is observed visually that the coating formed using high frequency is more compact and even.

The Roughness of surface of the MAO coating produced using different frequencies of power supply during 10min is listed in table 1. It shows that the surface roughness of the coating treated by MAO compared with uncoated metal substrate (0.22±0.005μm) is high. It is well known that the growth of the MAO coating is the plasma discharge event over the substrate surface, and the discharge is a treatment process with high temperature and high pressure, which increases the roughness of the metal surface. On one hand, the random discharge on surface leads increase of the coating roughness partly, on other hand, the porous outer layer produced by the discharge also increases the roughness. The surface roughness of MAO coating formed decreases with increasing the frequency of power supply, but the roughness of the coating formed by frequency 30Hz is low, this is because the coating is thin. The lifetime of plasma discharge decreases using high frequency, and this leads the mass of product of every discharge is less, and it avoids the appearance of the destruct big spark, so the roughness of the coating is low.

Table 1 Influence of the frequency of power supply on the roughness of MAO coating surface

Frequency of power supply/ Hz	30	50	70	100
Roughness of surface/μm	0.46±0.03	0.47±0.11	0.37±0.01	0.35±0.01

The composition and morphologies of the coating. The composition of the coating produced using different frequencies is analogous. Fig.3 illustrates the XRD patterns of the MAO coating fabricated using the frequency of power supply 50Hz for 1 hour treatment time. It shows that the coating consists of α-Al_2O_3 and γ-Al_2O_3.

The surface morphologies of the MAO coating formed using the frequencies of power supply 30, 50 and 100Hz is shown in Fig.4a-c. It can be seen that the surface of the MAO coating using frequency of 50 and 100Hz contains many obvious discharge products, which appear as crater-mouth like traces, but the surface of the coating formed by frequency of 30Hz is porous without obvious

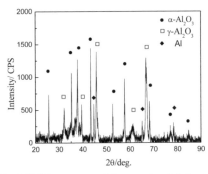

Fig.3 XRD patterns of the coating fabricated by MAO

discharge products. The discharge channels distribute over the coating surface, and the amount of such channels increases with increasing the frequency of power supply while the diameter of the discharge product decreases clearly with increasing treatment time. The diameter of discharge product on the coating fabricated using frequency 50Hz is about 50-70μm, and that for frequency 100Hz is about 20-40μm.

Fig.4 The surface morphologies for MAO coatings on aluminum using different frequency of power supply (a) 30Hz; (b) 50Hz; (c) 100Hz

When a plasma discharge appears, plasma, thermal chemical and electrochemical reactions occur between the aluminum from the substrate and the oxygen and other ions from the electrolyte. Although these reactions are quite complex in nature, but finally high melting point alumina is contributed to the formation of ceramic coating. The alumina should be produced in the channels of plasma discharge with high temperature and pressure, and it extends continually into interior of the substrate and spreads on interface of the coating/electrolyte, then the discharge channel is occluded. The frequency of power supply controls the time of discharge sparks interruption, which influences on the outspread area (diameter) of the discharge products on the coating surface. Moreover, the anodic cell voltage using high frequency of power supply is low, which decreases the dimension of the discharge channel formed by the breakdown of the coating, so the outspread area of the discharge products reduces.

In the MAO process, a large number of discharge sparks are visible and move quickly on the metal surface, it lead to the superposition of the discharge products on the coating surface, therefore, the thickness of the coating and the roughness of the coating increases. The dimension of the discharge products using high frequency of power supply is low, so the roughness of the coating decreases.

Corrosion Resistance of aluminum with MAO coating. Fig.5 shows the potentiodynamic polarization curves of the alumina coated aluminum alloy samples using the frequencies of power supply 30, 50 and 100Hz by MAO and the untreated metal substrate. The corrosion potentials and corrosion current densities of the potentiodynamic corrosion are given in Table 2. The results show that the alumina coated aluminum alloy samples have an excellent corrosion protective property compared with the aluminum alloy substrate. The corrosion potentials of MAO treated aluminum alloy sample increase from –1.31V for the aluminum alloy to –0.8~ –0.65. It is noticed that the corrosion potentials for the samples with the alumina coating decrease with increasing the frequency of the power supply. The corrosion currents of the coated aluminum alloy decrease evidently increasing the frequency of power supply. It is noticed that the corrosion current of the coating formed using frequency of power supply50 Hz reaches $18.2\mu A/cm^2$, which is higher than that for other coatings. The high corrosion current may be due to corresponding increase in coating roughness. Because the coating produced using frequency of power supply 30Hz is thin and uneven with black specks, so according to the corrosion current, the coating formed using frequency of power supply100 Hz possesses a better corrosion protective property.

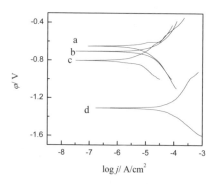

Fig.5 Potentiodynamic polarization curves of untreated metal substrate (d) and coated the alumina coating using frequency 30Hz (a), 50Hz (b) and 100Hz (c)

Table 2 The results of the potentiodynamic corrosion test in 3.5% NaCl solution

Samples coated by different requency	30Hz	50Hz	100Hz	uncoated
E_{corr}/ V	-0.65	-0.70	-0.80	-1.31
j_{corr}/ $\mu A/cm^2$	6.54	18.2	4.13	52.2

Summary

Micro-arc oxidation treatment can be used to form an alumina coating on aluminum alloy, consisting of α-Al_2O_3 and γ-Al_2O_3. The frequency of power supply controls the lifetime of the discharge sparks interruption, so the amount of formed spark increases and moves more quickly with increasing the frequency of power supply. The anodic cell voltage decreases with increasing the frequency of power supply, however, the cathodic cell voltage increases. The thickness of the coating produced using frequency of power supply 50 Hz is maximum, and its roughness is high.

The surface of the MAO coating contains many obvious discharge products, and the diameter of the discharge products decreases with increasing the frequency, so the roughness decreases also correspondingly. The alumina coated aluminum alloy samples have an excellent corrosion protective property compared with the aluminum alloy substrate, the corrosion current of MAO treated aluminum alloy sample using 100Hz decreases from $52.2\mu A/cm^2$ for the aluminum alloy to $4.13\mu A/cm^2$.

References

[1] Güntherschlze and H. Betz. Die Electronenstromung in Isolatoren bei Extremen Feldstarken, Z.Phys. 91(1934)70.

[2] Q. B. Nguyen, M.Gupta, Enhancing compressive response of AZ31B using nano-Al_2O_3 and copper additions, Journal of Alloys and Compounds, 490 (2010) 382.

[3] S. V. Gnednekov, O. A. Khrisanfova, A. G. Zavidnaya, Composition and adhesion of protective coatings on aluminum, Surf. Coat. Technol. 145 (2001) 146.

[4] A.A. Voevodin, A.L.Yerokhin, V.V. Lyubinov, *et al.*, Characterization of wear protective Al-Si-O coatings formed on Ai-based alloy by micro-arc discharge treatment, Surf. Coat. Technol. 86-87 (1996) 516.

[5] V. S. Rudnev, T. P. Yarovaya, D. L. Boguta, L. M. Tyrina, P. M. Nedozorov, P. S. Gordienko, Anodic spark deposition of P, Me(II) or Me(III) containing coatings on aluminium and titanium alloys in electrolytes with polyphosphate complexes, J. Electroanal. Chem. 497 (2001) 150.

[6] B. Rajasekaran, S.G.S. Raman, S.V. Joshi, *et al.*, Effect of microarc oxidised layer thickness on plain fatigue and fretting fatigue behavior of Al–Mg–Si alloy, Int. J. Fatigue, 30 (2008) 1259.

[7] F. Liu, J.L. Xu, D.Z. Yu, *et al.*, Effects of cathodic voltages on the structure and properties of ceramic coatings formed on NiTi alloy by micro-arc oxidation, Mater. Chem. Phys., 121 (2010) 172.

[8] J. M. Li, H. Cai, B.L. Jiang, Growth mechanism of black ceramic layers formed by microarc oxidation, Surf. Coat. Technol. 201 (2007) 8702.

[9] J. Liang , L.T. Hua, J.C. Hao, Improvement of corrosion properties of microarc oxidation coating on magnesium alloy by optimizing current density parameters, Appl. Surf. Sci. 253 (2007) 6939.

[10]S.G.Xin, L.X. Song, R.G. Zhao, X.F.Hu, Properties of aluminum oxide coating on aluminum alloy produced by micro-arc oxidation, Surf. Coat. Technol. 199 (2-3) (2005) 184.

[11]V. Raj, M. M. Ali, Formation of ceramic alumina nanocomposite coatings on aluminium for enhanced corrosion resistance, Journal of Materials Processing Technology, 209 (2009) 5341.

[12]A.L.Yerokhin, X.Nie, A.Leyland, A.Matthews, S. J. Dowey, Plasma Electrolysis for Surface Engineering. Surf. Coat. Technol. 122(1999)73.

Key Engineering Materials Vol. 537 (2013) pp 82-86
© (2013) Trans Tech Publications, Switzerland
doi:10.4028/www.scientific.net/KEM.537.82

Fabrication of Zirconia Coating on the Surface of Al₂O₃-SiC-C Castable

Yucheng Yin[a], Yonghe Liang[b], Shan Ge[c], Jiuchang Lu[d], and Zhiqiang Liu[e]

The Hubei Province Key Laboratory of Refractories and Ceramics, Ministry-Province
jointly-Constructed Cultivation Base for State key Laboratory, Wuhan University of Science and
Technology, Wuhan, China

[a]yucheng.yin@gmail.com, [b]wkdlyh@126.com, [c]geshanss@163.com

Keywords: Al₂O₃-SiC-C castable, zirconia Sol, zirconia coating, Pretreatment

Abstract. Using zirconia Sol as zirconia source, zirconia coating has been fabricated on the surface of Al₂O₃-SiC-C castable by spraying technology after pretreated at 110°C, 300°C, 500°C, 700°C and 900°C. The oxidation resistance of Al₂O₃-SiC-C castable with zirconia coating was compared with that without coating, at the same time using microanalysis technique, effects of pretreatment temperatures and spraying times on the fabricated zirconia coating were investigated. Results suggest that better zirconia coating might be achieved after pretreated at 300°C and spraying one time. At this condition, fabricated zirconia coating has good continuity, closely bonded with substrate castable and even has a little penetration into the castable, and thus improved the oxidation resistance of Al₂O₃-SiC-C castable.

Introduction

Silicon carbide and carbon do not react with slag when they are not oxidized, thus, well improve the corrosion resistance of Al₂O₃-SiC-C castable. While, once they have been oxidized, silicon carbide would transform into SiO_2, which could react with slag components easily, and induce the formation of compounds with low melting points. These compounds would be blow away by molten iron and/or slag, and result in continuous corrosion of refractories materials; meanwhile, the oxidation of carbon would cause pores formation inside, and this would accelerate the corrosion of refractory materials, because Pores are main throughs for oxygen diffusion and slag penetration. So according to the above analysis, the oxidation resistance of Al₂O₃-SiC-C castable has a strong effect on its corrosion resistance. If the oxidation resistance of Al₂O₃-SiC-C castable could be well improved, also its corrosion resistance would be improved.

On the other hand, it is well known for us, much bigger the viscosity of slag, means much smaller depth of slag penetrating into refractories, and this also indicates little corrosion of refractories. So, to increase the viscosity of slag might be an effective measure to reduce the corrosion[1,2]. While, what we can change are refractories themselves, not slag. In fact, not the whole viscosity of slag affect the corrosion of refractories, but the local viscosity of slag which contact with refractories, and this suggests us, it is possible to increase the local viscosity of slag by means of other compounds introduction into refractories, such as Cr_2O_3, TiO_2 or ZrO_2 with higher melting points[3].

Nowadays, coating technology is widely used to improve the oxidation resistance of carbon containing refractories[4,5]. However, for Al₂O₃-SiC-C castable, there are few researches about coating fabrication on it. As mentioned above, specific kind of coatings might not only help to improve the oxidation resistance of Al₂O₃-SiC-C castable, but also its corrosion resistance. In order to improve the oxidation and corrosion resistance of Al₂O₃-SiC-C castable at same time, we have tried to fabricate a zirconia coating on Al₂O₃-SiC-C castable, with commercial zirconia Sol as zirconia source, and using spraying technology. The processes and results are presented in this paper.

Experimental procedure

Al₂O₃-SiC-C castable samples preparation. Raw materials listed in Table.1 were used to prepare Al₂O₃-SiC-C castable samples. Balancing 3kg raw materials according to Table 1, and then mixing for half one minute, successively, 5.5wt% water was added into mixture, keep mixing, the total mixing time is 5 minutes. The mixed mixture was put into stainless steel modules, and cast into Al₂O₃-SiC-C castable samples with dimension of 40mm×40mm×40mm. Samples were cured in modules at room temperature for 24h, and then taken out from modules, dried at 110°C for 24h, finally, samples were pretreated at 300 °C,500°C, 700°C and 900°C, respectively.

Coating fabrication and characterization. The next step is to fabricate zirconia coating on surfaces of Al₂O₃-SiC-C castable samples. Commercial zirconia Sol was used as zirconia source, its composition was shown in Table 2. Zirconia Sol was sprayed on the surfaces of Al₂O₃-SiC-C castable samples for different times, from 1 to 5 times. Here, spraying for one time, means that keep spraying until zirconia Sol flowing down along the surfaces, and then dried at 110°C for 24h, cooling to room temperature. Taking out samples with one time spraying, repeating spraying as the first time, coating fabricated by spraying two times was obtained. Similarly, fabricating coatings with three times, four times and five times spraying. Coated Al₂O₃-SiC-C castable samples then were heated up to 1450°C with a heating rate of 5°C/min in air, and dwell for 3h, samples cooling to room temperature in the furnace, to simulate its real working condition.

Taking out from furnace and cut into two pieces, cross section views were observed and compared with each other, also scanning electronic microscope(SEM, XL30TMP) and energy dispersive spectrometer(EDAX, Phoenix) were employed to make micro-analysis of samples.

Table 1 Compositions of Al₂O₃-SiC-C castable

Raw materials	Purity[wt%]	Particle Size	Content[Wt%]
Fused Corundum	>98.0	1-8mm	63
SiC	>97.0	0-1 mm	15
α-Al₂O₃	>98.5	<3μm	10
Cement	-	<0.045 mm	3
Silica Fume	>93	<1μm	3
Ball Pitch	>46	1-0.5 mm	3
Silicon	>98.5	<0.088 mm	3
Additives			0.25

Table 2 Chemical compositions of the zirconia Sol[wt%]

Composition	ZrO₂	CaO	SiO₂	Al₂O₃	Fe₂O₃	Total water
Content	23.87	0.01	0.054	0.062	0.046	75.95

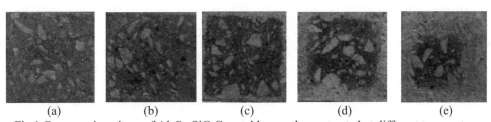

| (a) | (b) | (c) | (d) | (e) |

Fig.1 Cross section views of Al₂O₃-SiC-C castable samples pretreated at different temperatures, (a) 110°C, (b) 300°C, (c) 500°C, (d) 700°C, (e) 900°C

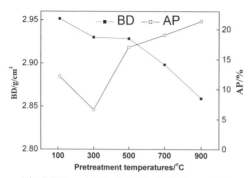

Fig.2 BD and AP of Al$_2$O$_3$-SiC-C castable
samples after pretreated at different temperatures

Results and Discussions

Effect of pretreatment temperatures. Al$_2$O$_3$-SiC-C castable samples pretreated at different temperatures were cut into two pieces, and their cross section views were shown in Fig.1. We can found that, samples pretreated at 110°C and 300°C had no obvious oxidation layer, this indicates that carbon in samples had not been oxidized in this process. However, samples pretreated at temperatures higher than 300°C seemed to have obvious oxidation layers, and the thickness become thicker and thicker with the increasing of pretreatment temperatures from 300°C to 900°C. Bulk density (BD) and apparent porosity (AP) of Al$_2$O$_3$-SiC-C castable samples pretreated at different temperatures were determined and shown in Fig.2. It can be seen from Fig.2 that, BD of pretreated Al$_2$O$_3$-SiC-C castable samples decreased quickly when temperature up to 300°C, this is because the water evaporation during heating at 110°C, which resulted in some weight loss, while the samples' volume keep constant, thus the BD would become smaller. Meanwhile, the AP of samples decreased too. Normally, if the BD decreased, the AP would increase. Why this happened? The carbon source of Al$_2$O$_3$-SiC-C castable was ball pitch, which has a soften point of about 120°C. When pretreated at 300°C, water in sample will evaporate firstly; this must induce some pores formation inside sample. While, with temperature rising, ball pitch will melt and become liquid state, and pores would be filled by it. This is just the reason why although BD decreases, AP decreases too.

When pretreatment temperature rising and higher than 300°C, the BD of Al$_2$O$_3$-SiC-C castable samples had a slight decrease, when pretreatment temperature increased from 300 to 500°C, and then decreased quickly after 500°C, see Fig.2. At the same time, the AP had a contrary change tendency. This may be attribute to the carbon oxidation when pretreated at temperatures higher than 300°C. Much higher the pretreatment temperature, much more carbon had been oxidized, and thus much thicker the decarburized layer, see Fig.2 (c), (d) and (e). It is well known that carbon loss would leave pores, much thicker the decarburized layer, means much more carbon loss and much more pores formed. So the BD decreased, while the AP increased.

Carbon plays an important role to improve the corrosion resistance, and coatings are also employed with the aim to prevent oxidation of carbon, thus to achieve good corrosion resistance. In order to obtain good coatings, the pretreatment technology is used to modify the surface structure, while according to the above experimental results, it is suggested that, carbon in Al$_2$O$_3$-SiC-C castable would be oxidized, if pretreatment temperatures are higher than 300°C, especially sever with temperature rise. Meanwhile, in practical operation, the pretreatment of Al$_2$O$_3$-SiC-C castable main trough has three stages, that are 300°C, 500°C, and 900°C for different periods, respectively. That is, the best pretreatment temperature should be 300°C.

Effect of spraying times on zirconia coating. Based on the above results, Al$_2$O$_3$-SiC-C castable samples pretreated at 300°C were selected to be coated, using zirconia Sol as zirconia source, and by spraying method. Spraying times are 1 to 5 times, respectively. Cross section views of Al$_2$O$_3$-SiC-C castable samples pretreated at 300°C and coated with different spraying times, after heated up to 1450°C and dwelled for 3h, were shown in Fig.3.

Fig.3 Cross section views of Al$_2$O$_3$-SiC-C castable samples pretreated at 300°C and coated with different spraying times, (a) 1 time, (b) 2 times, (c) 3 times, (d) 4 times, (e) 5 times, after heated up to 1450°C and dwelled for 3h

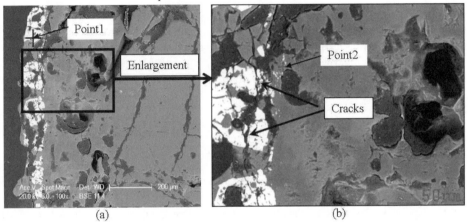

Fig.4 SEM photographs of Al$_2$O$_3$-SiC-C castable samples pretreated at 300°C and coated by spraying one time, after heated up to 1450°C and dwelled for 3h, part of (a) marked with a rectangle was enlarged and shown as (b).

Table 3 Compositions of point1 and point2 [wt%]

Elements	Point1	Point2
C		12.34
O	18.44	22.14
Al		2.38
Si	1.42	16.57
Zr	80.14	46.58

It is shown by Fig. 3 that, Al$_2$O$_3$-SiC-C castable samples pretreated at 300°C and coated by spraying one time have better oxidation resistance than the others. This may be attribute a suitable zirconia coating has formed on the surface of Al$_2$O$_3$-SiC-C castable samples with this process. Microstructure analysis has been done on the Al$_2$O$_3$-SiC-C castable samples pretreated at 300°C and coated by spraying one time, after heated up to 1450°C and dwelled for 3h. SEM photographs were shown in Fig. 4, and compositions of marked point1 and point2 have been done by EDAX, and shown in Table 3.

From Fig.4 (a), we can found the fabricated zirconia coating (see compositions of point1 shown in Table 3) is relatively continuous and its thickness is about 100μm. With the enlarged part in Fig.4 (a), a little penetration of zirconia into the sample has been observed in Fig.4 (b), as the zirconia content decrease, while the other main compositions of Al$_2$O$_3$-SiC-C castable increase, and see compositions of point1 and point2 shown in Table 3.

This zirconia coating might fill some pores on the surface, thus help to prevent the oxygen diffusion into inner part of Al$_2$O$_3$-SiC-C castable, and finally improve its oxidation resistance. However, as we know, zirconia has a phase transformation as follows, and there is a volume change about 7% during this transformation [6].

$$m - ZrO_2 \underset{950^\circ C}{\overset{1170^\circ C}{\rightleftharpoons}} t - ZrO_2 \qquad (1)$$

This volume change during heating or cooling, might result in some cracks, see Fig.4(b), and with the spraying times increase, the quantity of zirconia coated on the surface would increase, accordingly, the cracks induced by this volume effect of zirconia would increase too. Much more cracks is certainly harmful for the oxidation resistance of Al_2O_3-SiC-C castable, this may be the reason why with the spraying times increase, the oxidation resistance become poorer and poorer.

Summary

In order to fabricate zirconia coating on Al_2O_3-SiC-C castables, Pretreatment is needed to modify the surface structure, and the pretreatment temperature should be $300^\circ C$, to insure as little carbon oxidized as possible. With spraying method, and spraying for one time, the fabricated zirconia coating has a better effect on improving the oxidation resistance of Al_2O_3-SiC-C castables, than that fabricated with more than one time.

References

[1] C.X. Wang, J.Hou, L. Zhao, etal, Damage of Refractories and Supression Techniques, Damage of Refractories Induced by Molten Slag, Metallurgy Industry Press, Beijing, 2009, pp.136.

[2] N. Li, Reactions between Refractories and Iron and Steel and Their Effects on the Quality of Iron, Basic Knowledge of Refractions between Refractories and Molten Iron and Steel, Metallurgy Industry Press, Beijing, 2005, pp.18.

[3] A.J. Kessman, K. Ramji, N.J. Morris, etal, Zirconia sol-gel coatings on alumina-silica refractory material for improved corrosion resistance, Surface & Coatings Technology. 204(2009) 477-483.

[4] J. Liu, L.Y. Cao, J.F. Huang, etal, A ZrSiO4/SiC Oxidation proctive coating for carbon/carbon composites. Surface& Coatings Technology. 206(2012) 3270-3274.

[5] O.S. Kwon, S.H. Hong, H. Kim, The improvement in oxidation reistance of carbon by a graded SiC/SiO$_2$ Coating. Journai of the European Ceramic Society. 23(2003) 3119-3124.

[6] R.H.G. Kiminami, The monoclinic-tetragonal phase transformation of zirconia in the system ZrO$_2$-Fe$_2$O$_3$, Journal of materials science letters. 9(1990) 373.

Key Engineering Materials Vol. 537 (2013) pp 87-91
© (2013) Trans Tech Publications, Switzerland
doi:10.4028/www.scientific.net/KEM.537.87

Preparation and Properties of Asymmetric Porous Aluminium-Oxide Ceramic HollowFibre Membranes

Haibo Mu, Guizeng Hao, Xiaowei Li and Bo Meng

School of Chemical Engineering, Shandong University of Technology, Zibo 255049, PR China

muhaibo111@126.com, mb1963@sdut.edu.cn

Keywords: hollow fibre membrane; Al_2O_3; phase inversion; non-solvent

Abstract. Asymmetric porous aluminium-oxide ceramic hollow fibre membranes have been prepared by the phase inversion / sintering technique. The effect of non-solvent such as ethanol, isopropanol and ethylene glycol monomethylether(2-methoxyethanol) on the geometry and performance of hollow fibres was investigated. Morphologies of Al_2O_3 ceramic hollow fibre membranes were characterized using a scanning electron microscope (SEM). The effective porosity and the mechanical strength were determined by Archimedes method, and three point method, respectively. The prepared Al_2O_3 hollow fibre membranes show the asymmetric structure with a finger-like layer and a sponge-like layer. The effective porosity of the prepared hollow fibre membranes exceeds 47%, and the bending strength of the hollow membranes exceeds 63 MPa. The Al_2O_3 hollow fibre membranes with moderate permeation characteristics for gas and pure water are prepared by the introduction of nonsolvent in membrane casting solution. The separation factors of H_2 to N_2 or CO_2 of the hollow fibers with nonsolvent are over 2.0.

Introduction

Due to the excellent mechanical, thermal and chemical stabilities of Al_2O_3 membranes, Al_2O_3 membranes were extensively applied in micropore filtration, ultrafiltration, high temperature separation and purification of gases and organic solvents, and membrane catalysis, et al areas[1-9]. Porous Al_2O_3 membranes are usually used carrier of catalyst and support of dense composite membrane in membrane catalysis. The widely used porous Al_2O_3 supports are usually in the form of discs and tubes. Solid state particles sintering technique is usually adopted for preparing these porous Al_2O_3 support membranes. In order to reduce the sintering temperature, to enhance the mechanical strength and the permeability of the membrane, during the prepared process, it is necessary introducing pore-forming agent, binding agent, sintering assistant, or pre-coating the alumina powders. However, some drawbacks of the Al_2O_3 support membrane prepared by the method are heterogeneous pore size distribution, low permeability, low packing density and thick tube wall [3,10-12]. The ceramic hollow fibres prepared by phase inversion and sintering method possess the advantages of thin fiber wall and high packing density. Moreover, the structure of the hollow fibers can be controlled by modifying the process parameters [13,14]. The macrostructure of the sintered fiber is largely determined by the macrostructure of the precursor fiber. The precursor structure can be controlled through the choice of composition of the spinning mixture and the conditions of phase separation. Both symmetric and asymmetric structures with different porosity and tortuosity can be prepared [13,15]. In this work, the effects of non-solvent such as ethanol, isopropanol and ethylene glycol monomethylether on the Al_2O_3 hollow fiber membranes and their performances have been carefully studied.

Experimental

The α-Al_2O_3 hollow fibre precursor was fabricated by the dry–wet spinning method as described elsewhere [13]. 0.8-3.1 wt% of ethanol, isopropanol and ethylene glycol monomethylether was used as non-solvent additives. Tap water was used as bore liquid and outer coagulant. The hollow fibre precursors were sintered at 1450–1650°C for 4 h to form the α-Al$_2$O$_3$ hollow fibre membranes. The α-Al$_2$O$_3$ hollow fibre membranes prepared with ethanol, isopropanol and ethylene glycol monomethylether non-solvent as additives are defined as MEA, MIPA, MEME, and the hollow fibers of no-additive named ME.

Microstructure and morphology of the α-Al$_2$O$_3$ hollow fibres were observed by scanning electron microscopy(SEM) (FEI Sirion-200, the Netherlands).

The values of three point bending strength measurements were performed with Instron Model 5544. A hollow fiber sample was fixed on the sample holder at a distance of 32.0 mm. The bending strength, σ_F, was calculated from the following equation[13].

$$\sigma_F = \frac{8FLD}{\pi(D^4 - d^4)} \tag{1}$$

Here, F is the measured force at which fracture take place; L, D and d are the length, the outside diameter and the inner side diameter, respectively.

The effective porosity was determined by testing the volume of water absorbed by hollow fiber sample on the Archimedes law. The volume of water is the same with the opened pores. This volume was then expressed as a percentage as an effective porosity percentage.

$$\varepsilon_e = \frac{W_2 - W_1}{\rho_{H_2O} \cdot V_h} \times 100\% \tag{2}$$

Here, ε_e is the effective porosity; W_1, W_2 are the weight of the dry sample and wetted sample, respectively; ρ_{H2O} is the density of the water under the test temperature; V_h is the apparent volume of the sample.

The clean water flux and the gas permeability of the sintered hollow fiber membranes were determined on the homebuilt test set under the room temperature. The fiber membranes were sealed at one end with epoxy resin, and the other end of the fiber membranes was left open. The clean water flux was calculated by the following equation:

$$Q = \frac{V_{H_2O}}{S \cdot t \cdot \triangle P} \tag{3}$$

Here, Q is the clean water flux; V_{H2O} is the volume of the water through the fiber membranes; S is the area of fiber membranes; t and $\triangle P$ is the time of the test and the pressure difference across the fiber membranes, respectively. The gas permeability was calculated by the following equation: $J = Q/S \cdot \triangle P$. Here, J is the gas permeability; Q is the total gas permeation rate; S and $\triangle P$ is idem. The selectivity α is defined as the ratio of the permeances of pure H$_2$ to pure N$_2$ or CO$_2$.

Results and Discussion

Morphologies of α-Al$_2$O$_3$ hollow fibre membranes. Fig. 1 shows the SEM micrographs of the α-Al$_2$O$_3$ hollow fiber membranes sintered at 1550 °C, spun from four different dopes containing 2.3 wt% nonsolve. Comparing the fiber dimensions Fig 1A reveals that the fibers prepared from the casting solution containing non-solvent have the asymmetric structure with a finger-like layer and a closely packed sponge layer, the fiber prepared from the casting solution without non-solvent has the sandwich structure with two finger-like layers near the outer and inner walls of the fiber and a sponge-like layer at centre of the hollow fiber. The difference in the fiber structures can be attributed to the difference of the coagulation rate. By introducing a nonsolvent in the dope, the viscosity of the casting solution is increased, and the viscosity of the casting solution in the outer region increases more rapidly than that in the inter region. An increase in casting solution viscosity results in the delay of the phase separation and the deceleration of the coagulation rate, and then inhibits the formation of the finger-like void. It can also be seen from Fig 1A that the sponge layer is thicker than the finger-like layer. Fig.1B and C are the inner surface and the outer surface of the hollow fiber membranes, respectively. Comparing the SEM photos between the inner and outer surface, reveals that the porosity is higher and the pore size is bigger in the inner surface than that in outer surface for the membranes made from the spinning dope with the nonsolvent existence. The porosity of the surface for the membrane prepared from the casting solution without non-solvent is lower. This is probably attributed to the rapid precipitation occurred for the casting solution without non-solvent.

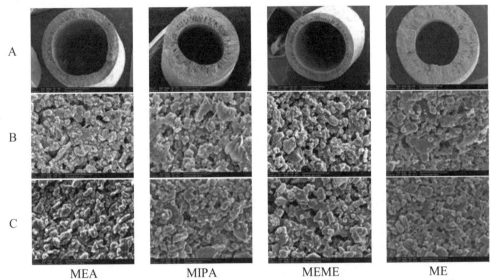

| MEA | MIPA | MEME | ME |

Fig. 1 The effect of non- solvent in the dope on the Morphologies of Al_2O_3 hollow fibre
A Cross-section; B Inner surface; C Outer surface

Table 1 Three point bending strength and effective porosity of the hollow fiber membranes

Membranes	Porosity / (%)	Bending strength / (MPa)	Dope viscosity / mPa s
MEA	52.63	77.95	86400
MIPA	47.84	84.26	117400
MEME	55.52	69.66	53000
ME	57.89	63.00	47200

Effective porosity and mechanical strength. The effective porosity and the mechanical strength of the α-Al_2O_3 hollow fiber membranes sintered at 1550 °C, spun from four different dopes containing 2.3 wt% nonsolve, and the viscosity of the spinning dope are given in table 1.

The result given in Table 1 indicate that the porosity of the fiber membranes from the spinning dope containing nonsolve is lower than that from the spinning dope without nonsolve, however the three point bending strength is higher; the three-point values of all the membranes exceed 63.0 MPa, and are increased as the increasing of the viscosity values. This is probably due to the thinker sponge layer. Comparing with the SEM photos, the values of the porosity and the bending strength are correspond with the SEM graphs showed. The results in the table 1 also show that the viscosity of the spinning dope can be controlled by adding a nonsolve in the dope.

Gas permeation properties and clean water flux. The experimental results of the gas permeation permeability and the clean water flux are shown in table 2. The clean water flux was determined at a water pressure of 4.0 bars. The clean water flux of the hollow fiber prepared from the spinning dope without nonsolve is low relative to its porosity. This is probably owed to its denser surfaces of inner and outer. Similarly, due to the denser inner and outer surfaces and the thicker sponge layer, the clean water flux of the hollow fiber with isopropanol as additive is the lowest.

The gas permeation permeability of the hollow fibers with nonsolve is higher than that without nonsolve. The selectivity α of the H_2 to N_2 or CO_2 is over 2.0, except the hollow fiber without nonsolve. Combination of the results of the SEM, the porosity, the bending strength and the dope viscosity analysis, we can conclude that the nonsolve has the effect on the viscosity of the spinning dope, further, effect the structure and the performance of the hollow fiber membranes.

Table 2 Pure water flux and gas permeation permeability of the α-Al_2O_3 hollow fiber membranes

Membrane	MEA	MIPA	MEME	ME
Clean water flux / ($m^3 \cdot m^{-2} \cdot h^{-1} \cdot bar^{-1}$)	5.12	3.82	7.18	4.68
Gas permeation permeability / ($mol.m_{-2} \cdot Pa^{-1} \cdot s^{-1} \times 10^5$)	21.76(H_2) 9.41(CO_2) 9.76 (N_2)	18.69(H_2) 7.87(CO_2) 8.86 (N_2)	26.88(H_2) 12.57(CO_2) 13.2166(N_2)	22.68(H_2) 11.81(CO_2) 11.93(N_2)
Selectivity α	2.31(H_2/CO_2) 2.23(H_2/N_2)	2.37(H_2/CO_2) 2.11(H_2/N_2)	2.14(H_2/CO_2) 2.03(H_2/N_2)	1.92(H_2/CO_2) 1.90(H_2/N_2)

Conclusions

Aluminium-oxide ceramic hollow fibre membranes with the asymmetric porous have been successfully prepared by introducing a nonsolve into the casting solution. Due to the sponge-like structures existence, the Al_2O_3 hollow fibre membranes have high mechanical strength; further the finger-like structures located at inner walls of the fibre also result in the appreciable gas and clean water permeation characteristics

Acknowledgements

The authors gratefully acknowledge the National Natural Science Foundation of China (No. 20976098 and No. 21176146)

References

[1] G Q Lu, J C Diniz da Costa, M Duke, et al. Inorganic membranes for hydrogen production and purification: a critical review and perspective, J Colloid Interf Sci 2007; 314: 589–603.

[2] I M Atadashi, M K Aroua, A R Abdul Aziz, et al. Refining technologies for the purification of crude biodiesel, Applied Energy, 2011, 88: 4239–4251.

[3] M D Irfan Hatim, X Y Tan, Z T Wu, K Li. Pd/Al_2O_3 composite hollow fibre membranes: Effect of substrate resistances on H2 permeation properties, Chem. Eng. Sci., 2011, 66: 1150-1158.

[4] E Gbenedio, Z T Wu, I Hatim, B F K Kingsbury, K Li. A multifunctional Pd/alumina hollow fibre membrane reactor for propane dehydrogenation, Catalysis Today 2010,156: 93–99.

[5] L J Shan, J Shao, Z B Wang, Y S Yan. Preparation of zeolite MFI membranes on alumina hollow fibers with high flux for pervaporation, J of Membr Sci, 2011, 378: 319-329.

[6] G B Sun, K Hidajat, S Kawi. Ultra thin Pd membrane on α-Al_2O_3 hollow fiber by electroless plating: High permeance and selectivity, J of Membr Sci, 2006, 284: 110–119.

[7] C Yang, P Huang, N P, Xu, et al. Preparation of zirconia-alumina composite microfiltration membrane by sintering, Membr Sci Technology, 1998, 18(1): 18-22.

[8] L Shi, G F Zeng, H Y Xu. Characterization and performance of high-flux PdAu/ceramic composite membranes, Chinese Journal of Catalysis, 2010, 31(6): 711-715

[9] G L Chen, H Qi, W B Peng,et al. Static corrosion of porous alum ina support of ceramic membrane in HNO3 solutions, J of Nanjing University of Technology, 2006, 28(5): 1-5.

[10] J M Hu, H Qi, Y Q Fan, et al. Porous ceramic support of coated alumina prepared by low-temperature sintering, Journal of the Chinese Ceramic Society, 2009,37(11): 1818-1822.

[11] Y Z Liu, L Gu, H Y Shen, et al. Effect of pore-forming agent on properties of Al2O3 supports, Membr Sci Technology, 2008,28(6): 24-38.

[12] P Wang, W H Xing, N P Xu, et al. Preparation of alumina MF membranes with high permeability, membrane science and technology, 1998, 18(4): 22-33.

[13] S M Liu, K Li, R Hughes. Preparation of porous aluminium oxide (Al_2O_3) hollow fibre membranes by a combined phase-inversion and sintering method, Ceramics International 2003, 29: 875–881.

[14] I H Choi, I C Kim, B R Min, K H Lee. Preparation and characterization of ultrathin alumina hollow fiber microfiltration membrane, Desalination, 2006, 193: 256–259.

[15] J D Jong, N E Benes, G H Koops, M Wessling. Towards single step production of multi-layer inorganic hollow fibers, Journal of Membrane Science, 2004, 239: 265–269.

Key Engineering Materials Vol. 537 (2013) pp 92-96
© (2013) Trans Tech Publications, Switzerland
doi:10.4028/www.scientific.net/KEM.537.92

Effects of Al₂O₃ Nanoparticles on Microstructure and Performance of Ceramic Coatings by Micro-arc Oxidation

Yujun Yin[1], Shuhua Li[1, a], Dawei Shen[2], Yuanyuan Zu[3] and Changzheng Qu [1]

[1]Shijiazhuang Mechanical Engineering College, Shijiazhuang 050003, China

[2]PLA 62025 Unit, Beijing 102300, China

[3]The Quartermaster Equipment Institute of the General Logistics Department of the PLA, Beijing 100082. China

[a]weihuyanghua@163.com

Keywords: Micro-arc oxidation; Ceramic coatings; Al₂O₃ nanoparticles ; Surface properties

Abstract. A dense ceramic oxide coating approximately 45 μm thick was prepared on a Ly12 aluminum alloy by micro-arc oxidation in an alkali-phosphate electrolytic solution. Coating thickness and surface roughness (Ra) were measured after the coating had been synthesized. The effects of Al₂O₃ nanoparticles in electrolyte on phase composition, microstructure and microhardness of the micro-arc oxidation ceramic coatings on Ly12 aluminum alloy were investigated by means of XRD, SEM and hardness experimentation. The results show that the ceramic coatings become more dense and its microhardness increased by adding Al₂O₃ nanoparticles in electrolyte. In addition, the roughness of the micro-arc oxidation ceramic coatings is obviously improved by addition of Al₂O₃ nanoparticles.

Introduction

The method of micro-arc oxidation (MAO) in the electrolytic solution has been widely investigated recently [1]. Ceramic coatings obtained using the above technique possess a number of valuable properties and can be used. In addition, this technique is economic efficiency, ecological friendly and characterized by high productivity [2]. MAO is based on the conventional anodic oxidation of processing metals and alloys in aqueous electrolyte solutions under the additional condition of plasma discharge at exceeding the critical values of the polarization potential, and the fact that discharge leading to localized substrate oxides but of more complex oxides containing compounds which involve the components presented in the electrolyte [3]. Both intrinsic factors and extrinsic factors affect the formation and microstructure of micro-arc oxidation coatings. Therein, composition and concentration of electrolyte and electrical parameters during the process play a crucial role in obtaining the desired coatings of special phase component and microstructure [4]. Among them, it is assumed that the effect of electrolytes compositions may be summarized as follows: firstly, as the medium of current conduct, transmitting the essential energy needed for micro-arc oxidizing occurring in the interface of metal/electrolyte. Secondly, electrolyte compositions providing the oxygen source in the form of oxysalt needed for oxidation. Thirdly, components presenting in the electrolyte incorporated into the coatings can further improve the properties of oxidation coatings. Various specially selected electrolytes and their combinations have been successfully developed in order to provide protective coatings of corrosion and wear resistance and diverse functional coatings [5,6]. The earlier works are focused on optimizing electrolytes composition and concentration in order to obtain excellent micro-arc oxidation coatings with special phase composition [7]. However, few data at present are available concerning the nanoparticles, microstructure and mechanical properties of such coatings. The primary objectives of this paper are: (1) a new aqueous electrolyte containing (NaPO₃)₆ was developed by adding the Al₂O₃ nanoparticles , then the growth process and corresponding microstructure evolution of the coatings were investigated; (2) the microhardness and roughness, as well as the phase of the micro-arc oxidation coatings between without Al₂O₃ nanoparticles and with Al₂O₃ nanoparticles in electrolyte were evaluated.

Experimental

In the present study, Ly12 aluminum alloy having a nominal composition of 3.8%-4.9%Cu, 1.2%-1.8%Mg, 0.3%-0.9%Mn, balance Al was used as the substrate material for all the coatings. The coatings were accomplished with an MAO coating unit designed and built by the authors. The unit consists of an asymmetric pulse bipolar current power (f=50 Hz) supply in which both cathodic and anodic potentials can be adjusted by oneself up to 800 V, a stainless steel container used as an electrolyte cell, a stirring system was used to mix round electrolyte in order to keeping the solution uniformity. A chiller was provided to maintain the temperature of the electrolyte during the process at any times ±1 °C variation. The Ly12 specimen was used as an anode, while the stainless steel container was used as a cathode. A essential aqueous solution containing 20 g/L $(NaPO_3)_6$, 2 g/L Na_2SiO_3, 5 g/L NaOH and variety concentrations Al_2O_3 nanoparticles of analytically pure was used as an electrolyte. Micro-arc oxidation processes were carried out under the conditions of anodic and cathodic voltages in the range of 230–600 V and 0–90 V, respectively. Deposition time was 40min, the bath being water-cooled and its temperature being maintained 20°C±1 °C. During coating deposition, at frequent intervals the power supply was switched off and samples were taken out of the electrolytic bath, washed in running water, ultrasonically cleaned with acetone and dried at room temperature for microstructure analysis and phase composition testing. The thickness of the oxide coating was measured on a TT260 coating thickness gauge (Time group Co.) based on the eddy current technique. The surface roughness (Ra) was measured using a TR210 surface roughness tester (Time Group, Co) and the average value was calculated. The morphology and fractography of the oxide coatings were observed on a JSM-5600LV scanning electron microscope (SEM). The phases present in MAO were identified using an X-ray diffractometer (D8-Advance, Brooker, Germany).

Results and Discussion

Surface roughness and coating thickness. The average surface roughness (Ra) measured on the patterns of ceramic coatings prepared in the electrolyte without Al_2O_3 nanoparticles s was 0.937±0.02µm, while with Al_2O_3 nanoparticles 2g/L and 4g/L exhibited 0.851±0.02µm and 0.831±0.02µm Ra, respectively. The average coatings thickness was 43µm, 45µm and 46µm for without Al_2O_3 nanoparticles, with 2g/L Al_2O_3 nanoparticles and with 4g/L Al_2O_3 nanoparticles in electrolyte, respectively.

Microstructure and phase content. SEM surface micrographs of the MAO coating without Al_2O_3 nanoparticles and with 2g/L Al_2O_3 nanoparticles in electrolyte are presented in Fig.1. The above micrographs clearly indicate the presence of discharge channels appearing as dark circular spots distributed all over the surface of the coating (Fig. 1a). It is also apparent that the number of such channels are decreasing and diminishing with adding Al_2O_3 nanoparticles in the electrolyte. Moreover, the micro-arc oxidation ceramic coatings become denser by adding Al_2O_3 nanoparticles in electrolyte (Fig. 1b).

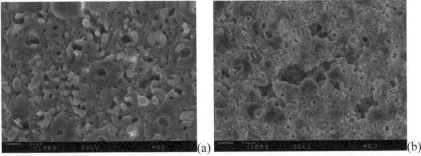

Fig.1 Surface micrographs of the MAO coating: (a) without Al_2O_3 nanoparticles; (b) with 2%Al_2O_3 nanoparticles

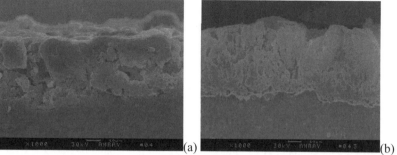

Fig.2 Across section micrographs of the MAO coating: (a) without Al$_2$O$_3$ nanoparticles; (b)with 2%Al$_2$O$_3$ nanoparticles

Mechanism of MAO coating formation. The Fig.1 evidence suggests that MAO coating formation is essentially an interface controlled process. The mechanism of electrical breakdown of the coating followed by spark discharge was explained by co-workers [2]. For further investigation, the as-coated surface of the MAO coating was examined under SEM. Fig.1 illustrate SEM images of the surface layers synthesized in different electrolyte solution. Fig. 1a indicates that the presence of circular spots surrounded by molten alumina rings are rapidly solidified due to sudden a quenching effect of the surrounding electrolyte. Fig.1b is also evident that some of the region on the surface of the coating was covered with some of Al$_2$O$_3$ nanoparticles existed in the electrolyte. Because of the presence of well-distributed discharge channel during the initial stage of the process, the coating containing Al$_2$O$_3$ nanoparticles exhibits lower surface roughness. As the Al$_2$O$_3$ nanoparticles increases in the electrolyte, the number of such discharge channels gradual decreased, leading to the coatings come into uniformity, which caused the surface roughness gradual improved. As observed in Fig.2a and b, the presence of micro-cavity across the diameter of the solidified circular regions indicate that these opening and spots are actually the cylindrical discharge channels through which the molten oxide flows through and onto the surface. These micro-cavity are due to sudden volumetric shrinkage associated with rapid quenching of the molten oxide.

The section micrographs are presented in Fig.2. Both the MAO coatings without Al$_2$O$_3$ nanoparticles and with 2% nanoparticles in electrolyte exhibit a uniform coating thickness with a clean and wavy interface, and the coating appears to be well bonded with the substrate. The layer with Al$_2$O$_3$ nanoparticles exhibits dense region and the without Al$_2$O$_3$ nanoparticles coating appears granular and cavity region.

XRD study of the MAO coating at different Al$_2$O$_3$ nanoparticles concentrations in electrolyte are shown in Fig.3. The substrate material consist of γ-Al$_2$O$_3$ and Al phase (see Fig.3a), the inner layer consist of γ-Al$_2$O$_3$, α-Al$_2$O$_3$ and Al phase (see Fig.3b), while the Fig.3 c without Al$_2$O$_3$ nanoparticles and Fig.3 d with 2% Al$_2$O$_3$ nanoparticle reveal that the coatings mainly consist of the γ-Al$_2$O$_3$, α-Al$_2$O$_3$, and Al phase, respectively. The content of Al is little, and γ- Al$_2$O$_3$ is soft so that we can simply use the content of the α-Al$_2$O$_3$ phase to represent the quality of the coating. At the same time, it is found that the content of the α-Al$_2$O$_3$ phase increases as the concentration of Al$_2$O$_3$ nanoparticles increased in electrolyte, because the high concentration of Al$_2$O$_3$ nanoparticale leads to the low working voltage and the transformation rate of a- Al$_2$O$_3$ phase became strong.

From the Fig.3, it is found that among these layers with Al$_2$O$_3$ nanoparticles and without Al$_2$O$_3$ nanoparticles in electrolyte are mainly composed of α-Al$_2$O$_3$ and γ- Al$_2$O$_3$. The α-Al$_2$O$_3$ content increased in with Al$_2$O$_3$ nanoparticles increased in electrolyte. This is mainly caused by a variation in the current power and cooling rate of the molten oxide in the micro-arc zone. In other words, the hardness of the coating directly connected with the solution composition and parameter of the process. The basic electrolyte has a higher cooling rate to favor the formation of γ-Al$_2$O$_3$ phase during the solidification of alumina; while the contained Al$_2$O$_3$ nanoparticles in electrolyte layer retains higher temperature which is high enough to transform the newly formed γ-Al$_2$O$_3$ to α-Al$_2$O$_3$, owing to low thermal conductivity of the alumina oxide, which could be attributed to the more severe thermal energy accumulation in a thicker coating. Therefore it can be inferred that the deposition process of a relatively coating is dominated by the growth of γ-Al$_2$O$_3$, while that of a thicker coating is dominated by the growth of α-Al$_2$O$_3$. Such differences in the compositions of the micro-arc oxidation coatings are also confirmed by the corresponding XRD analytical results.

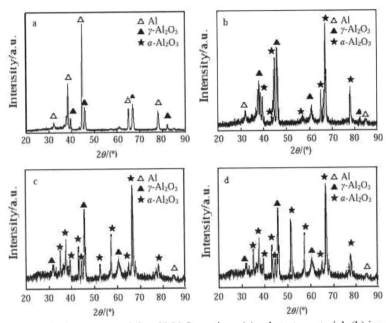

Fig.3. XRD spectra of substrate material and MAO coatings:(a) substrate material; (b) inner layer; (c) without Al$_2$O$_3$ nanoparticles ; (d) with 2% Al$_2$O$_3$ nanoparticle

Micro hardness. The average micro hardness of the coating exhibited by without Al$_2$O$_3$ nanoparticles in electrolyte is 1779 HV and that coatings with Al$_2$O$_3$ nanoparticles 2g/L and 4g/L in electrolyte are 1927 and 1882 HV, respectively. In the case of MAO coatings, a micro hardness gradient exists from edge to center of the coatings. Fig.4 illustrates the variation in micro hardness measured from the brim of coating towards the center of the coating. It is evident that the microhardness is maximum (peak hardness) at a distance of 10-20μm from the brim of sample and then exhibits a decreasing trend towards the center of the sample.

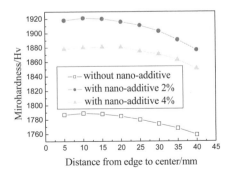

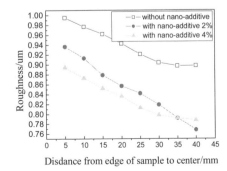

Fig. 4 Microhardness of the coating for without nanoparticles and with nanoparticles

Fig. 5 Roughness of the coating for without nanoparticles and with nanoparticles

Evolution of surface roughness. Surface roughness (Ra) measured after the coatings had been synthesized is presented in Fig.5. It is clearly evident that the surface roughness decreased along with the Al_2O_3 nanoparticles increased in electrolyte. Among these results, the average surface roughness of coating with 2% and 4% Al_2O_3 nanoparticles in electrolyte had been improved. In particular, Surface roughness of the coating with 2% Al_2O_3 nanoparticles in electrolyte exhibits a better smoothness layers than the coating with 4% Al_2O_3 nanoparticles in electrolyte. On second thoughts, above results illustrate that the performance of the coatings can been improved by adding Al_2O_3 nanoparticles , while at the same time, the additive quantity of the Al_2O_3 nanoparticles in electrolyte should be matching with the composition.

Summary

In conclusion, the kinetics of MAO coatings is interface-controlled and largely dependent on the applied current density and electrolyte composition. The surface roughness and microhardness of MAO coatings are a near linear function of the distance from the edge to center of the sample, and increases with increasing coatings thickness. The MAO coatings at different Al_2O_3 nanoparticale concentrations in electrolyte mainly consist of the γ-Al_2O_3, α-Al_2O_3, and little Al phase, and found that the content of the α-Al_2O_3 phase increases as the concentration of Al_2O_3 nanoparticles increased. The micro hardness of the coatings increased with the α-Al_2O_3 phase increased in the ceramic coatings and the peak hardness always remaining at the extension of the brim of 10-20μm on the coating and then exhibited a decreasing trend towards the center of the sample.

References

[1] X.B. Suo, J. Qiu, J.Y. Liu. Effects of SiO_2 Nanoparticles in Electrolytes on Growth Process and Surface Properties of Alumina Coatings Formed on 7A52 Aluminium Alloy by Micro-arc Oxidation, China Surface Engineering, 23(2010)42-44.

[2] G.Sundararajan, L.Rama Krishna. Mechanisms underlying the formation of thick alumina coatings through the MAO coating technology, Surface and Coatings Technology,176(2003) 269–277.

[3] J. Zhao, R.G. Song, H.X. Li et al. Effects of Nanoparticles on microstructure and properties of micro-arc oxidation coatings on 6063 aluminum alloy, Trans.Mater. Heat Treat. 31 (2010) 125-128.

[4] X.W. Zhang, B.J. Ma. Effect of additive in phosphate system electrolyte on properties of Mg alloy oxidation film prepared by micro-oxidation, Hot Working Technology, 1(2008)82-84.

[5] Y.J. Miao, C.J. Shen, D.K. Wang. Effect of additional agent on the wear ability of micro-arc oxidation ceramic coating on aluminum alloy, J. Heat Treatment, 22(2007)34-36.

[6] H.H. Luo, Q.Z. Cai, B.K. Wei et al. Effect of additive concentration on microstructure and corrosion resistance of ceramic coatings formed by micro-arc oxidation on AZ91D Mg alloy, The Chinese Journal of Nonferrous Metals,18(2008)1082-1088.

[7] W.C. Gong, G.H. Liu. Characterization of ceramic coatings produced by plasma electrolytic oxidation of aluminium alloy, Sci. Eng. A Struct. Mater. 5(2007)158-162.

[8] S.G. Xin, L.X. Sun, R.G. Zhou et al. Influence of cathodic current on composition, structure and properties of Al_2O_3 coatings on aluminium alloy prepared by micro-arc oxidation process, Thin Solid Films, 9(2006)326-332.

Key Engineering Materials Vol. 537 (2013) pp 97-100
© (2013) Trans Tech Publications, Switzerland
doi:10.4028/www.scientific.net/KEM.537.97

Influence of Single Layer-Thickness on Residual Thermal Stresses in $Sm_2Zr_2O_7$/YSZ Thermal Barrier Coatings

Hu Renxi[1, a], Chen Xiaoge[2,b] and Li Gang[3,c]

[1]Mechanic Design Teaching and Research Section, Ordnance Engineering College, Shijiazhuang 050003, China

[2] Department of Construction Engineering, Henan Institute of Engineering, Zhengzhou 451191, China

[3]Department of Mechanical Engineering, Henan Institute of Engineering, Zhengzhou 451191, China

[a]hurenxi2000@163.com(corresponding author), [b]chenxg831025@163.com, [c]ligang@163.com

Keywords: Thermal barrier coatings; Residual thermal stresses; Layer thickness; Finite element

Abstract. In this paper, influence of single-layer thickness on residual stresses in $Sm_2Zr_2O_7$/YSZ thermal barrier coating was analyzed by finite element method. Results show that the radial stress remains stable in x range 0-12mm, and it decreases abruptly at edge of the sample. The distribution of axial stress resembles that of radial stress, the shear stress increase abruptly at edge of sample. In three typical residual stresses, radial stress has the highest value, axial stress and shear stress can be ignored. The best thickness combination of $Sm_2Zr_2O_7$/YSZ TBCs should be 0.1mm-NiCoCrAlY layer, 0.05-0.1mm -TGO, 0.1mm-YSZ and 0.9mm-$Sm_2Zr_2O_7$

Introduction

Now, searching for new ceramic materials has become an important direction in thermal barrier coating (TBC) field [1]. Selection of new ceramic materials must meet to a few of basic requirements, such as low thermal conductivity, high thermal expansion coefficient, excellent phase stability at high temperature, high melting point, better chemistry inertness and bonding strength between coating and substrate [2]. In these requirements, low thermal conductivity and high thermal expansion coefficient are always regarded as the most important requirements. In recent years, rare earth zirconates of $A_2Zr_2O_7$ type have become the most potential ceramic materials for new thermal barrier coatings because of their lower thermal conductivity and higher thermal expansion coefficient [3-5]. Because their better thermophysical property, research about corresponding TBCs has attracted more and more attentions in recent years, especially the TBCs with double ceramic layer [6,7]. It is well known that thickness of single layer has important on residual stress of the double-ceramic-layer TBCs. However, research about the layer-thickness on residual stresses in $Sm_2Zr_2O_7$/YSZ TBCs was not reported up to now. Therefore, influence of layer-thickness on residual stress in $Sm_2Zr_2O_7$/YSZ TBCs was investigated using finite element technique, and results can provide theory instruction for preparing actual $Sm_2Zr_2O_7$/YSZ TBCs.

Analysis model

The distribution of coating residual stresses was analyzed by finite element code ANSYS (revision 10.0), and the thermal-structure PLAN13 was selected. For the modeling of residual stresses after plasma spraying, the model (shown in Fig.1) represents a cylinder for the nickel substrate of 36mm in diameter and 10mm in thickness. The coating and substrate are assumed to be isotropic for simplicity in this body. The analytical model is a perfect elastic body without plastic deformation. An axisymmetric element mode is chosen to reduce computer calculation and data manipulation time, Y axis, shown in Fig.1, was chosen as the symmetric axis. Four-layer TGO/NiCoCrAlY/YSZ/$Sm_2Zr_2O_7$ coatings with different dimensions of single-layer thickness are computed. A dense mesh is used at the ceramic layer. The material properties of every layer are shown in Table 1. Then the influence of single layer-thickness on the distribution of residual stresses in the coatings is reflected. In order to investigate the effects of single layer-thickness on the residual stress level, four aspects of this investigation are shown as follows. 1. Calculation of residual stresses was owing to a uniform cooling from an assumed uniform stress-free manufacturing temperature of 525°C to room temperature; 2. Calculation of residual stresses was processed using stead state analysis type of FEM.3. In the analysis, the convection coefficients to environment were taken to be 150W/m^{2}k. Because failure usually occurs at interface between ceramic layers, residual stresses at interfaceceramic layers and surface of top-layer were mainly analyzed in this paper[8, 9].

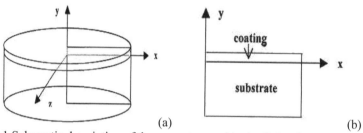

(a)   (b)

Fig.1 Schematic description of the geometry used in the finite element modeling

Table 1 Physical properties of Sm$_2$Zr$_2$O$_7$/YSZ thermal barrier coating

	Thermal conductivity (W/m.K)	Specific heat (J/kg.K)	Coefficient of thermal expansion ($\times 10^{-6}$)	Possion's ratio	Elastic modulus (GPa)	Denstity (Kg/m^3)
Sm$_2$Zr$_2$O$_7$	0.35	436	9.29	0.27	157	6517.4
YSZ	1.1	656	8.2	0.25	48	6037
TGO	25.2	857	5.1	0.27	380	3978
NiCoCrAlY	16.1	501	11.6	0.3	214.5	7320
Ni	73.9	460	16.4	0.312	150	8880

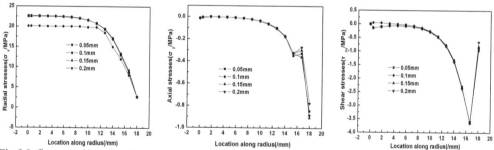

Fig.2 Influence of bonding-layer thickness on residual stresses: (a, left) radial stress; (b, middle) axial stress and (c, right) shear stress

Results and discussion

Effect of bond-layer thickness. In this part, the bonding-layer thickness is named a variable bt. Influence of bt on residual stresses is listed in Fig.2. It can be observed in Fig.2 (a) that radial stress remains stable in x range 0-12mm, and decreases at the edge of sample. Radial stress reaches the lowest value at bt=0.1mm.In Fig.2(b), the distribution of axial stress resembles that of radial stress, however, the shear stress increase abruptly at edge of sample. It can also be seen from Fig.2 that bt have no influence on axial stress and shear stress, the magnitude of axial stress and shear stress are smaller than radial stress in the coating. It is considered that the NiCoCrAlY bondcoat improve effectively the stress distribution in the whole specimen. The finite element analysis result show that axial stress and shear stress can be ignored, which are consistent with most of theoretical calculation [10]. In the following part of the current paper, only the influence of single-layer thickness on radial residual stress was analyzed.

Effect of TGO thickness. In the working process of thermal barrier coatings, the bond-coat temperature in gas-turbine engines typically exceeds 700°C , resulting in bond-coat oxidation and the inevitable formation of a new layer—the thermally grown oxide (TGO)—between the bond-coat and the ceramic top-coat [11]. Fig.3 shows the influence of TGO thickness on radial stress in coating. It can be noted in Fig.3(a) that the 0.05mm TGO can lead to the highest residual stress at interface between two ceramic layers, residual stress in 0.1-TGO coating is close to that in 0.15-TGO coating. However, the radial stress at surface of top-layer decrease with increasing TGO-thickness. In light of the analytical result and the actual thickness of TGO, we think that the appropriate thickness of TGO is in the range 0.05-0.1mm.

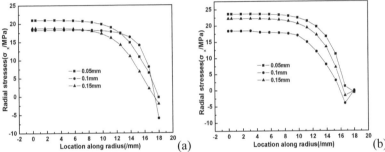

Fig.3 Influence of TGO thickness on residual stresses (a) interface between ceramic layers (b) surface of to-layer

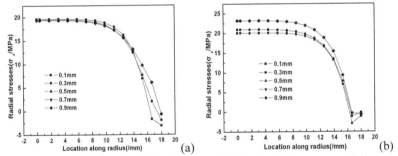

Fig.4 Influence of TGO thickness on residual stresses (a) interface between ceramic layers (b) surface of to-layer

Effect of ceramic-layer thickness. In this section, the total thickness of two ceramic layers is 1mm, the thicknesses of bonding-layer and TGO are 0.1mm and 0.05mm, respectively. Fig.4 depicts the influence of YSZ-layer thickness on radial stress distributions at interface of two ceramic layers and top-layer surface. As can be seen from Fig.4(a), the YSZ-layer thickness has almost no influence on magnitude of radial stress in x range 0-12mm, however, 0.1mm YSZ-thickness results in the lowest value of radial stress at edge of sample. In Fig.4(b), radial stress at surface of the top-layer decreases gradually in x range 0-12mm, 0.1mm YSZ-thickness also has the lowest radial stress. Low residual stress helps to improve the adhesion and working life of thermal barrier coating, we think that 0.1mm is the most appropriate YSZ thickness. Thus, the best thickness combination of $Sm_2Zr_2O_7$/YSZ TBCs should be 0.1mm- NiCoCrAlY layer, 0.05-0.1mm -TGO, 0.1mm-YSZ and 0.9mm-$Sm_2Zr_2O_7$.

Summary

(1) The radial stress remains stable in x range 0-12mm, and it decreases abruptly at edge of the sample. The distribution of axial stress resembles that of radial stress, the shear stress increase abruptly at edge of sample.

(2) In three typical residual stresses, radial stress has the highest value, axial stress and shear stress can be ignored.

(3) The 0.05mm TGO can lead to the highest residual stress at interface between two ceramic layers, the radial stress at surface of top-layer decrease with increasing TGO-thickness. The appropriate thickness of TGO is in the range 0.05-0.1mm.

(4) The YSZ-layer thickness has almost no influence on magnitude of radial stress in x range 0-12mm, however, 0.1mm YSZ-thickness results in the lowest value of radial stress at surface of top-layer.

(5) The best thickness combination of $Sm_2Zr_2O_7$/YSZ TBCs should be 0.1mm-NiCoCrAlY layer, 0.05-0.1mm -TGO, 0.1mm-YSZ and 0.9mm-$Sm_2Zr_2O_7$

References

[1] R. Vaben, M. O. Jarligo, T. Steinke, et al, Overview on advanced thermal barrier coatings, Surf. Coat. Technol, 145 (2010) 938-942.

[2] H. S. Zhang, S. R. Liao, X. D. Dang, et al, Preparation and thermal conductivities of Gd2Ce2O7 and (Gd0.9Ca0.1)2Ce2O6.9 ceramics for thermal barrier coatings, J. Alloys Compds. 509 (2011) 1226 -1230.

[3] L. Ling, X. Qiang, W. F. Chi, Z. H. Song, Thermophysical Properties of Complex Rare-Earth Zirconate Ceramic for Thermal Barrier Coatings, J. Amer. Ceram. Soc. 91 (2008) 2398-2401.

[4] Z. H. Song, X. Qiang, W. F. Chi, et al, Preparation and thermophysical properties of (Sm0.5La0.5)2 Zr2O7 and (Sm0.5La0.5)2(Zr0.8Ce0.2)2O7 ceramics for thermal barrier coatings, J. Alloys. Compds. 475 (2009) 624-628

[5] Z. G. Liu, J. H. Ouyang, Y. Zhou, Preparation and thermophysical properties of (NdxGd1-x)2Zr2O7 ceramics, J. Mater. Sci. 43 (2008) 3596-3603.

[6] G. Moskal, L. Swadzba, M. Hetmanczky, et al, Characterisation of the micro-structure and thermal properties of Nd2Zr2O7 and Nd2Zr2O7/YSZ thermal barrier coatings, J. Eur. Ceram. Soc.32 (2012) 2035-2042.

[7] L. Wang, Y. Wang, X. G. Sun, et al, Finite element simulation of residual stress of double-ceramic-layer La2Zr2O7/8YSZ thermal barrier coatings using birth and death element technique, Comput. Mater. Sci. 53 (2012) 117-127.

[8] Z. H. Xu, S. M. Hei, L. M. He, et al, Novel thermal barrier coatings based on La2(Zr0.7Ce0.3)2O7/ 8YSZ double-ceramic-layer system deposited by electron beam physical vapor deposition, J. Alloys Compd. 509 (2011) 4273-4283.

[9] X. Q. Cao, R. Vassen, F. Tietz, D. Stoever, New double-ceramic-layer thermal barrier coatings based on zirconia-rare earth composite oxides, J. Eur. Ceram. Soc. 26 (2006) 247-251.

[10] Y. Jiang, B. S. Xu, H. D. Wang, et al, Finite element modeling of residual stress around hole in the thermal barrier coatings, Comp. Mater. Sci. 49 (2010) 603-608.

[11] N. P. Padture, M. Gell, E. H. Jordan, Thermal barrier coatings for gas-turbine engine applications, Sci. 4 (2002) 280-284.

Key Engineering Materials Vol. 537 (2013) pp 101-104
© (2013) Trans Tech Publications, Switzerland
doi:10.4028/www.scientific.net/KEM.537.101

Water-Quenching Thermal Shocking Property of Sm₂Zr₂O₇/NiCoCrAlY Function Grade Thermal Barrier Coatings

Wang Xinli[1, a], Bi Jianping[1,b]

¹Department of Mechanical Engineering, Henan Institute of Engineering, Zhengzhou 451191, China

[a]wangxinli01@sina.com, [b]bijianping02@sina.com

Keywords: Thermal barrier coatings; Finite element method; Thermal shocking

Abstract. In the present manuscript, the influence of interface location, coating structure (layer number) and layer-material composition on water-quenching thermal-shocking property of $Sm_2Zr_2O_7$/NiCoCrAlY functional graded thermal barrier coatings (TBCs) were investigated using finite element method. Results show that radial thermal stress at surface of top-layer have the highest value. Radial thermal stress in TBCs decrease gradually with increase of layer number, and function graded TBCs have better thermal-shocking resistance compared with other TBCs. 1357 function graded TBCs have the most excellent thermal shocking resistance in these investigated TBCs in the current manuscript.

Introduction

Thermal barrier coatings (TBCs) have been widely used to protect turbine engines operating in high temperature environment [1, 2]. Typical TBCs consist of a ceramic top-coat and an oxidation resistant bond-coat. Due to low thermal conductivity, high phase stability, high thermal-expansion coefficient and high fracture toughness compared to other ceramics, yttria-stabilized zirconia (YSZ) has been known as the most favorite top-layer material for many years [3,4]. Although such layers have been used for several years, their durability is still a fundamental problem. Because of this issue, intensive research has been carried out that aims to replace the YSZ oxide with a new type of ceramic material that will surpass it in properties and durability. The scope of most of the research has concerned on compositions based on the ZrO_2 phase, modified with one or two rare earth element type oxides (RE_2O_3), but different from Y_2O_3. Rare earth ziaconates with the pyrochlore-type structure and general formula of $RE_2Zr_2O_7$ and other types of materials that have been tested. Results show that rare earth zirconates of $RE_2Zr_2O_7$ have potentials to be used as new ceramic materials for TBCs because of their excellent thermophysical properties [5, 6]. Now, research about $RE_2Zr_2O_7$ -type TBCs has attracted more and more attention, and some works about the $RE_2Zr_2O_7$-type TBCs have been reported [6-8]. However, the current works were focused on the double-ceramic-layer TBCs, research about functional graded TBCs has not been reported up to now. This paper will try to investigate the thermal shocking stress distribution in $Sm_2Zr_2O_7$/ NiCoCrAlY functional graded TBCs using finite element method.

Analysis model

The distribution of coating thermal shocking stresses was analyzed by finite element code ANSYS (revision 10.0), and the thermal-structure PLAN13 was selected. For the modeling of thermal shocking stresses after plasma spraying, the model (shown in Fig.1) represents a cylinder for the nickel substrate of 36mm in diameter and 10mm in thickness. The coating and substrate are assumed to be isotropic for simplicity in this body. The analytical model is a perfect elastic body without plastic deformation. An axisymmetric element mode is chosen to reduce computer calculation and data manipulation time, Y axis, shown in Fig.1, was chosen as the symmetric axis. In this research, thermal shocking stresses of two kinds of graded TBCs were analyzed. Besides metal bonding layer and substrate, these two graded TBCs are three-layer TBCs and five-layers TBCs, respectively. In five-layer TBCs, there are two series volume percentage of $Sm_2Zr_2O_7$. One is 10%, 30%, 50%, 70% and 100% (1357), the other is 20%, 40%, 60%, 80% and 100%(2468). The corresponding three-layer TBCs were named 57 and 68, respectively. The thicknesses of bonding layer and every grade-layer are 0.1mm and 0.2mm, respectively. Thermophysical properties of bonding layer and $Sm_2Zr_2O_7$ layer are measured experimentally, corresponding properties of other graded layers are calculated using mathematic model in ref[11]. The thermophysical parameters used in the current work are listed in Table 1. In this analysis, the thermal shocking stress in these TBCs was considered to be caused by cooling from a uniform temperature of 1000°C to water temperature (20°C). The convection coefficient of water is 3000W/ (m^2 °C).

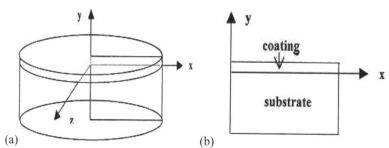

(a)   (b)

Fig.1 Schematic description of the geometry used in the finite element modeling

Table 1 Thermophysical properties of every geaded coatings

$Sm_2Zr_2O_7$ Vol%	0	10	20	30	40	50	60	70	80	100
E(GPa)	214	207.3	200.5	194.1	188	182.2	176.7	171.4	166.4	157
v	0.3	0.299	0.294	0.290	0.28	0.281	0.28	0.279	0.276	0.27
$\rho(kg/m^3)$	7320	7232	7147	7063	6980	6600	6820	6902	6666	6517
$\alpha(10^{-6}°C^{-1})$	11.6	11.5	11.3	11.1	10.9	8.6	10.5	10.2	9.9	9.29
$\lambda(W/m.K)$	16.1	5.2	3.9	3.2	2.6	2.2	1.8	1.4	1.1	0.35
Cp(J/kg.K)	501	493.8	486.8	479.6	473	466.7	460.3	454	447.9	436

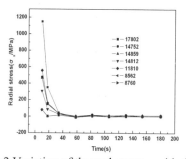

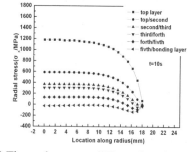

Fig.2 Variation of thermal stresses with time   Fig.3 Thermal stress at different location in 2468

Results and discussion

Thermal stresses in different location Fig.2 lists variation of radial thermal stress of different nodes with time. It can be noted from Fig.2 that thermal stresses of different nodes in 2468 decrease gradually before 40s, and they have the greatest values at 10s. Therefore, thermal stresses at the beginning were mainly analyzed in the following sections. Fig.3 presents the distribution of radial thermal stresses at different interfaces in 2468. As can be seen from Fig.3, radial thermal stress remains stable in x range 0-12mm, and they decrease gradually at edge of sample. From the interface between the fifth layer and bonding layer to top layer, radial thermal stress in 2468 increases gradually, surface of top-layer has the greatest radial stress in the whole TBCs. In the following paragraph, radial thermal stress at surface of top-layer was the only analysis object.

Effect of coating structure In order to investigate the influence of TBC structure on thermal -shocking stress. Radial stress of TBCs with different structure was analyzed, and results are depicted in Fig.4. It can be clearly observed from Fig.4 that distribution of radial thermal stress resemble that in Fig.3, and it decrease with increase of TBC layer number. In Fig.4(a), radial stress in single ceramic-layer TBC decreases from 1230MPa to 87MPa in x range 10-18mm. However, radial stresses in 68 and 2468 decrease from 1090MPa and1050MPa to 76.4MPa and 84MPa in the same x range, respectively. In Fig.4(b), radial stresses in 57 and 1357 decrease from 1060MPa and 100MPa to 82.3MPa and89MPa. It is well known that lower thermal shocking stress always imply better thermal-shocking resistance and longer working life of TBCs. These actual analytical results indicate effectively that increasing in TBC layers help to improve its thermal-shocking resistance, the functional graded TBCs with five-layers have better thermal-shocking property.

Effect of layer material composition Fig.5 displays the influence of layer material composition on radial thermal stress in TBCs. It can be noted clearly from Fig.5(a) that radial stress in 57 is slightly lower than that in 68. The similar results can also be observed in Fig.5(b). According to the analytical

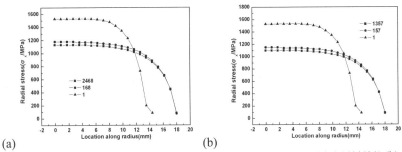

(a)   (b)

Fig.4 Distribution of radial thermal stress in TBCs of different structures (a) 2468/68/1 (b) 1357/57/1

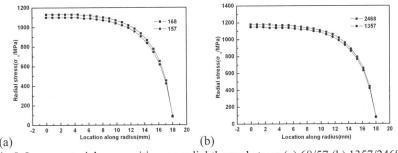

(a)   (b)

Fig.5 Layer material composition on radial thermal stress (a) 68/57 (b) 1357/2468

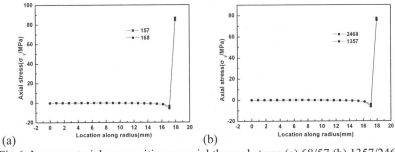

(a)   (b)

Fig.6 Layer material composition on axial thermal stress (a) 68/57 (b) 1357/2468

results displayed in Fig.4 and Fig.5, it can be concluded that the 1357 have the most excellent thermal-shocking resistance in these TBCs in the current paper. In order to investigate further the influence of layer-material-composition on thermal-shocking property of TBCs, axial thermal stress of different TBCs are also analyzed, which are presented in Fig.6. As can be seen in Fig.6, axial stress is transferred to tensile stress at the edge of the sample, which is different from the distribution of radial stress. Fig.6 also indicates that layer-material composition have no effect on axial thermal stress, and the magnitude of axial thermal stress is much lower than that of radial thermal stress, which is consistent with theory calculation result.

Summary

(1) In the functional graded TBCs, radial thermal stress remains stable in x range 0-12mm, and it decreases at the edge of sample. Radial thermal stress at surface of top-layer has the highest value.

(2) Radial stress in these TBCs decrease gradually with the increase of layer number, the five-layer TBC has the lowest radial thermal stress.

(3) In these function graded TBCs, 1357 has the best thermal resistance. The magnitude of axial thermal stress is much lower than that of radial thermal stress. The layer-material composition has no influence on axial thermal stress.

References

[1] N. P. Padture, M. Gell, E. H. Jordan, Thermal barrier coatings for gas-turbine engine applications, Sci. 4(2002) 280-284.

[2] Y. Shinsuke, T. Takanori, M. Takuji, Thermal and mechanical properties of $SrHfO_3$, J. Alloy Compds. 381 (2004)295-299.

[3] G. L. Carlos, Emerging materials and processes for thermal barrier systems, Current Opinion in Solid State Mater Science. 8 (2004) 77-91.

[4] L. Wang, Y. Wang, X. G. Sun, et al, A novel structure design towards extremely low thermal conductivity for thermal barrier coatings-experimental and mathematical study, Mater. Des, 35 (2012) 505-517.

[5] C. Zhu, P. Li, A. Javed, et al, An investigation on the microstructure and oxidation behavior of laser remelted air plasma sprayed thermal barrier coatings, Surf. Coat. Technol. 206 (2012) 3739-3746.

[6] G. Moskal, L. Swadzba, M. Hetmanczyk, et al, Characterization of the microstructure and thermal properties of $Nd_2Zr_2O_7$ and $Nd_2Zr_2O_7$/YSZ thermal barrier coatings, J. Eur. Ceram. Soc. 32 (2012) 2035-2042.

[7] G. Moskal, L. Swadzba, M. Hetmanczyk, et al, Characterization of microstructure and thermal properties of $Gd_2Zr_2O_7$-type thermal barrier coating, J. Eur. Ceram. Soc. 32 (2012) 2025-2034.

[8] M. H. Habibi, L. Wang, S. M. Guo, Evolution of hot corrosion resistance of YSZ, $Gd_2Zr_2O_7$, and $Gd_2Zr_2O_7$+YSZ composited thermal barrier coatings in Na_2SO_4+V_2O_5 at 1050° C, J. Eur. Ceram. Soc. 32 (2012) 1635-1642

[9] L. Wang, Y. Wang, X. G. Sun, et al, Finite element simulation of residual stress of double-ceramic-layer $La_2Zr_2O_7$/8YSZ thermal barrier coatings using birth and death element technique, Compt. Mater. Sci. 53 (2012) 117-127

[10] L. Wang, Y. Wang, W. Q. Zhang, et al, Finite element simulation of stress distribution and development in 8YSZ and double-ceramic-layer $La_2Zr_2O_7$/8YSZ thermal barrier coatings, App. Surf. Technol. 258 (2012) 3540-3551

[11] T. Ishikawa, Thermal deformation & thermal stress of FGM plates under steadily gradient temperature fields. Proc The Frist International symp FGM, Sendai,1990,11-16.

[12] Y. Jiang, B. S. Xu, H. D. Wang, et al, Finite element modeling of residual stress around holes in the thermal barrier coatings, Compt. Mater. Sci. 49 (2010) 603-608.

Key Engineering Materials Vol. 537 (2013) pp 105-108
© (2013) Trans Tech Publications, Switzerland
doi:10.4028/www.scientific.net/KEM.537.105

Investigation on the Representation Method for Microwave Absorbing Coat

Liu Heng[a], Zeng Zhaoyang and Guo Yiming

PLA University of Science and Technology, Nanjing, 210007, China

[a]amma1973@163.com

Keywords: coat; microwave absorption; matching; attenuation

Abstract. Based on the transmission line theory, influence factors of microwave absorbing ability for slab coat were analyzed and calculated. It was found that the parameter absorption rate could not effectively describe the different influence from material character, geometry dimension and microwave. As a solution, the matching property and absorption property needed to be discussed simultaneously. According to the new representation, matching property of coat depends only on the material character, whereas the absorbing ability resulted from the interaction between different interfaces within the sample. The influence of microwave frequency is reflected by interfere between different interfaces. Relatively, in the new representation, the influence from different factors was clear and intuition, which means the applicability of the new representation.

Introduction

The absorption ability of microwave absorption coat depends on two key factors: the matching property and the attenuation ability [1,2]. The matching property describes the energy rate of the entering wave, while the attenuation ability tells the absorption caser within the coat. However, reflection rate is usually used as simple index for microwave absorption coat, the influence of the two key factors are mixed together and it is hard to distinguish the influence from different factors. In this paper, ferrite slab coat were used as an example to illustrate the disadvantage of using reflection rate as single index. As a result, new representation was proposed to comprise reflection rate, matching coefficient and attenuation coefficient. Only in this way, the complicated interaction within coat can be demonstrated clearly.

Theoretical Analysis

According to transmission line theory, the input impedance of the number k interface within a multi-layer slab sample is

$$Z_{in}^k = Z_k \frac{Z_{in}^{k+1} + Z_k th(r_k d_k)}{Z_k + Z_{in}^{k+1} th(r_k d_k)} \tag{1}$$

where the superscript is the sequence of interfaces, Z_{in}^{k+1} is the impedance of the number k +1 interface, Z_k is the intrinsic impedance of the number k interface, r_k and d_k are the transmission factor and thickness of the layer, respectively. The first layer is the backed metal, thus the impedance for this layer is zero, $Z_{in}^1 = 0$. If the total number of layers is n, then, the reflection of the sample is,

$$\Gamma(dB) = 20 \lg \left| \frac{Z_{in}^n - Z_0}{Z_{in}^n + Z_0} \right| \tag{2}$$

where $Z_0 = (\mu_0 / \varepsilon_0)^{1/2}$ is the intrinsic impedance of air. For the convenience of discussion, multi-layer sample can be treated as single layer sample, with the equivalent permittivity ε_r and permeability μ_r [3,4,5]. Then, the Equ.2 can be rewritten as

$$\Gamma = 20 \lg \left| \frac{\sqrt{\dfrac{\mu_r}{\varepsilon_r}} th(rd) - 1}{\sqrt{\dfrac{\mu_r}{\varepsilon_r}} th(rd) + 1} \right| \tag{3}$$

where $r = j2\pi f \left(\mu_r \varepsilon_r \mu_0 \varepsilon_0 \right)^{1/2}$ is the equivalent transmission constant. If the reflection is regarded as a superimposed effect of reflection from upper interface and lower interface of the sample, then Equ.3 is equivalent to the following equation,

$$\Gamma = 20 \lg \left| \frac{\eta - 1}{\eta + 1} + \sum_{n=1}^{\infty} \frac{4\eta}{1 - \eta^2} \left(\frac{\eta - 1}{\eta + 1} e^{-2rd} \right)^n \right| \tag{4}$$

with $\eta = \sqrt{\mu_r / \varepsilon_r}$. The transmission constant r usually is complex, $r = \alpha + j\beta$. The real part α is the attenuation index, which depends on ε_r and μ_r,

$$\alpha = \frac{2\pi f}{c} \sqrt{ \frac{ \left(\varepsilon''\mu'' - \varepsilon'\mu' \right) + \sqrt{ \left(\varepsilon'\mu' - \varepsilon''\mu'' \right)^2 + \left(\varepsilon'\mu'' + \varepsilon''\mu' \right)^2 } }{2} } \tag{5}$$

In this way, the sample reflection can be written as

$$\Gamma = 20 \lg \left| \frac{\eta - 1}{\eta + 1} + \sum_{n=1}^{\infty} \frac{4\eta}{1 - \eta^2} \left(\frac{\eta - 1}{\eta + 1} e^{-2j\beta d} \right)^n \cdot e^{-2\alpha nd} \right| \tag{6}$$

From Equ.6, it is obvious that the sample reflection comprises two parts: the first term in Equ.6 which is resulted from sample matching property, and the second term which illustrates the sample absorption ability. Therefore, we suggest that three parameters should be used to describing the microwave property of coat, these parameters are: sample reflection rate in Equ.6, matching parameter $R1 = 20 \lg \left| (\eta - 1)/(\eta + 1) \right|$, and the attenuation parameter defined in Equ.5.

Ferrite Example

In order to illustrate the rationality of the new representation mentioned above, hexagonal ferrite coat $Ba(Co_2TiZn)_xFe_{12-4x}O_{19}$ is used as an example in this paper, the parameters of which is shown in Table 1.

Table1. Parameters of $Ba(Co_2TiZn)_xFe_{12-4x}O_{19}$

f/GHz	ε'	ε''	$\varepsilon''/\varepsilon'$	μ'	μ''	μ''/μ'
8.5	11.99	0.084	0.007	14.42	1.34	0.092
9.0	9.78	0.073	0.007	11.84	1.61	0.135
9.5	7.32	0.079	0.010	9.54	2.17	0.230
10.0	7.53	0.090	0.010	8.38	2.64	0.320
10.5	7.00	0.474	0.067	4.54	2.25	0.490
11.0	6.42	0.080	0.012	3.86	1.60	0.410
11.5	5.88	0.072	0.010	3.51	1.41	0.410
12.0	5.31	0.054	0.010	3.01	1.03	0.400
12.4	4.85	0.086	0.017	2.78	0.91	0.320

The permittivity of ferrite coat can be adjusted by doping conductor grains, and the effective parameters of the doped coat can be calculated under the theory of effective medium model. Then, the microwave property of pure and doped ferrite coat can be compared. Fig.1 shows the coat reflection data for ferrite coat before and after doping. It is clear that doping improves the absorption ability of the coat. However, the reason for the improvement is unclear from this figure.

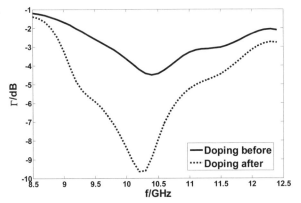

Fig.1   The reflection curve for ferrite coat

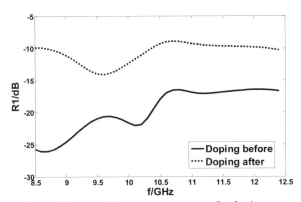

Fig.2   The matching parameter curve for ferrite coat

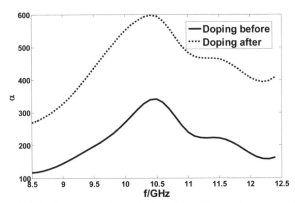

Fig.3   The frequency dependence of the attenuation parameter

From Fig.2 it can be seen that the matching effective is damaged by doping, a natural result if the permittivity is increased much by conductor grain doping. Now, the conclusion is a bit ambivalent, with the decreased reflection and damaged matching effect. As conductor grain doping usually results in much increased imaginary part of permittivity, the attenuation parameter should be improved obviously. This is just the case in Fig.3. From Fig.2 and Fig.3, it is clear that doping results in both beneficial and harmful consequences; the beneficial part is the improved attenuation parameter, while the harmful part is the damaged matching effect. The net effect, of course, is positive for the ferrite coat in this paper, as the data shown in Fig.1. However, the doping process should be controlled carefully, because big attenuation parameter is useless only if the matching effect is bad enough.

Summary

Low reflection rate is the only target for many microwave absorbing coat research, however, low reflection rate depends on several properties of the coat, including matching effect with air, attenuation ability of the material, and the interfere within the sample interfaces. Thus, representation of microwave absorbing coat should take all these properties into account. Only in this way, coat representation is helpful for sample fabrication.

References

[1] V. Korenivski, Magnetic films for GHz applications, J. Appli. Phys. 81(1997) 1024-1026.

[2] T. Giannakopoulou, A. Oikonomou, Double-layer microwave absorbers based on materials with large magnetic and dielectric losses, J. Magnetism and Magnetic Materials. 271(2004) 224-229.

[3] L. Gao, L.P. Gu, Y.Y. Huang, Effective medium approximation for optical bistability in nonlinear metal-dielectric composites, Solid State Communications. 129(2004) 593-598.

[4] L. Gao, J.Z. Gu, Effective dielectric constant of a two-component material with shape distribution, J. Phys. D: Appl. Phys.. 35(2002) 267-271.

[5] S. Giordano, Effective medium theory for dispersions of dielectric ellipsoids, J. Electrostatics, 58 (2003) 59-76.

Key Engineering Materials Vol. 537 (2013) pp 109-113
© (2013) Trans Tech Publications, Switzerland
doi:10.4028/www.scientific.net/KEM.537.109

Preparation and Properties of Dy doped La and Sc solution of BiFeO$_3$ Film

Xiwei Qi [a], Xiaoyan Zhang, Xuan Wang, Haibin Sun and Jianquan Qi

School of Resources and Materials, Northeastern University at QinHuangDao Branch,
QinHuangDao 066004, P. R. China

[a] qxw@mail.neuq.edu.cn

Keywords: Sol-gel; BiFeO$_3$ film; Dy doped; La and Sc solution.

Abstract. A series of Dy doped La and Sc solution of BiFeO$_3$ thin films have been prepared by using spin-coating process on conductive indium tin oxide (ITO)/glass substrates, which a simple sol-gel possess is applied and annealed at 500°C. With the increase of content of Dy, the strongest peak (110) of La and Sc solution BiFeO$_3$ film tends to further broaden. There is no second phase existence within the present Dy doping level. Cross section scanning electron microscope (SEM) pictures revealed that the thickness of BiFeO$_3$ film was about 370 nm. For Dy doping level is 0.05, the maximum double remanent polarization 2P$_r$ of as-prepared BiFeO$_3$ thin film is15.44 µC/cm^2. Image of atomic force microscopy indicated that the root-mean-square surface roughness value of as-prepared BiFeO$_3$ thin film is 2.11 nm. The dielectric constant of as-prepared films tends to firstly increase and then decrease with the increase of Dy content.

Introduction

Multiferroic materials have recently attracted widely attentions [1-5] because of simultaneously exhibiting ferroelectric, ferromagnetic performances and new coupling effects in a certain range of temperature. For example, they have high dielectric permittivity and high magnetic permeability, and could therefore replace the discrete inductor and capacitor in integrated circuits with a single component [6], further miniaturize electronic devices. Multiferroic materials can also demonstrate new coupling effects such as magnetoelectric coupling effect-magnetic (electric) field induction of polarization *P* (magnetization *M*), which can be used in the field of multifunctional devices, solar energy devices, ferroelectric random access memory, spintronics, and other potential application [7-9]. Up to now, BiFeO$_3$ is one of the rare multiferroic materials with both ferroelectric behavior (Curie temperature T$_C$ = 830 °C) and antiferromagnetic ordering (Neel temperature T$_N$ = 380°C) above the room temperature [10]. Several methods have been to prepared BiFeO$_3$ (BFO) films, for example, hydrothermal epitaxy [11], sol-gel [12, 13], ion beam sputtering [14], and magnetron sputtering [15], etc. However, there are a few literatures on using ITO as bottom electrode to prepare Dy doped BFO thin films.

In this paper, the thickness of BiFeO$_3$ film with 370 nm was successfully prepared using sol-gel spin coating process on conductive indium tin oxide (ITO)/glass substrates. The phase structure, microstructures, ferroelectric properties, dielectric properties of as-prepared films were investigated.

Experimental

A series of Bi$_{0.9-y}$Dy$_y$La$_{0.1}$Fe$_{0.9}$Sc$_{0.1}$O$_3$ films (y =0.05, 0.1, 0.15 and 0.2) are prepared by using sol-gel spin coating process on conductive indium tin oxide (ITO)/glass substrates. According to the stoichiometric composition, analytical reagent grade bismuth nitrate (Bi (NO$_3$)$_3$ ·5H$_2$O), iron nitrate (Fe (NO$_3$)$_3$ ·9H$_2$O), lanthanum nitrate (La (NO$_3$)$_3$ ·6H$_2$O), scandium oxide (Sc$_2$O$_3$) and dysprosium oxide (Dy$_2$O$_3$) were used as starting materials, and 2-methoxyethane was used as the solvent. In addition, acetic anhydride was used as dehydrate and mono-ethanolamine was adopted to adjust the viscosity and pH value. First, scandium oxide (Sc$_2$O$_3$) and dysprosium oxide (Dy$_2$O$_3$) were dissolved in concentrated nitric acid stoichimetric by stirring at 80°C, respectively. According to $Bi_{0.9-y}Dy_yLa_{0.1}Fe_{0.9}Sc_{0.1}O_3$ (y =0.05, 0.1, 0.15 and 0.2), the specified amounts of reactive reagent were mixed and dissolved in 2-methoxyethane by stirring for 1 h at 80°C water bath. Thereafter acetic anhydride was dropped to dehydrate by stirring 30 min. Then, the mono-ethanolamine was added to above solutions for viscosity and pH value adjustment. Last, the stable precursor solutions of 0.3 M were obtained. Dy doped BFO thin films were spin-coated on ITO/glass substrates at 800 rpm for 0-6 s and 6000 rpm for 6-20 s. The detail process is reported in the literature [16].

The structure of as-prepared films was performed using X-ray diffractometry (XRD, DX-2500, Dandongfangyuan Co. Ltd., China) with Cu K_α radiation operated at 30 kV and 25 mA in a 2θ range of 10-70°. Field emission scanning microscopy (S-4800, Japan) was carried out for the observation of surface and cross section microstructures. The surface morphology is performed using the atomic force microscope (SPA-300HV, Seiko, Japan). For electric properties measurement, top Pt electrodes were deposited on the surface of the films. Ferroelectric hysteresis loops were carried out on IF analyzer 2000 (aixACCT Co.). The dielectric properties are measured using the impedance analyzer 4284A (Agilent, USA).

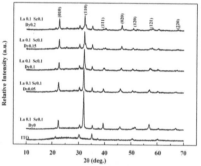

Fig.1 XRD patterns of $Bi_{0.9-y}Dy_yLa_{0.1}Fe_{0.9}Sc_{0.1}O_3$ films doped with different Dy content   Fig.2 XRD of (110) peak for as-prepared films

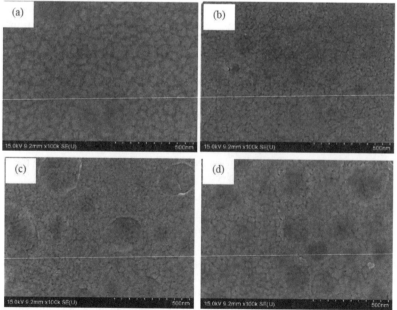

Fig.3 SEM photographs of as-prepared $Bi_{0.9-y}Dy_yLa_{0.1}Fe_{0.9}Sc_{0.1}O_3$ films:
(a) y=0.05, (b) y=0.10, (c) y=0.15 and (d) y=0.20

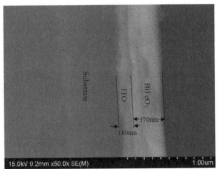

Fig.4 SEM photograph of cross-section for
$Bi_{0.8}Dy_{0.1}La_{0.1}Fe_{0.9}Sc_{0.1}O_3$ film

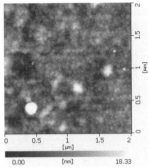

Fig.5 Surface morphology by AFM

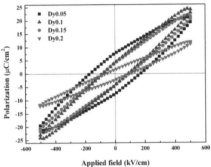

Fig.6 P-E hysteresis loops of as-prepared films

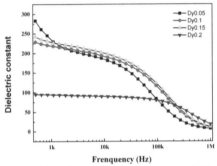

Fig.7 Dielectric constant of as-prepared films

Results and Discussion

Phase identification. Fig.1 shows the XRD patterns of $Bi_{0.9-y}Dy_yLa_{0.1}Fe_{0.9}Sc_{0.1}O_3$ films doped with different Dy level. It can be seen that all diffraction peaks are indexed using JCPDS diffraction files 74-2016. No second phases are detected within the present Dy doping level. Fig.2 shows the refine structure of (110) peak for $Bi_{0.9-y}Dy_yLa_{0.1}Fe_{0.9}Sc_{0.1}O_3$ films. As shown in Fig.2, the (110) peak tends to broaden and shift towards high angel degree with the increase of Dy doping level.

  Microstructures. Fig.3 (a-d) shows the SEM photographs of surface morphologies of as-prepared $Bi_{0.9-y}Dy_yLa_{0.1}Fe_{0.9}Sc_{0.1}O_3$ films with y =0.05, 0.1, 0.15 and 0.2. The average grain size of as-prepared $Bi_{0.9-y}Dy_yLa_{0.1}Fe_{0.9}Sc_{0.1}O_3$ films is less than 100 nm within the Dy doping level. The grain size has the tendency to grow larger when the Dy doping level is excess 0.2. As shown in Fig.3, The films are relatively   dense   and   uniform.   Fig.4   shows   the   SEM   image   of   cross-section   for $Bi_{0.8}Dy_{0.1}La_{0.1}Fe_{0.9}Sc_{0.1}O_3$ film. It can be seen that the thickness of the film is about 370 nm.The surface morphology is also identified by Atomic force microscopy as shown in Fig.5. The roughness of the films can be quantitatively identified by the root-mean-squared (rms) value of the roughness (R_{rms}). The R_{rms} of $Bi_{0.8}Dy_{0.1}La_{0.1}Fe_{0.9}Sc_{0.1}O_3$ film annealed at 500°C is 2.11 nm, indicating the surface of the film is quite flat.

  Ferroelectric properties. The effective method for determining the ferroelectric nature of a material is to experimentally measure the polarization-electric field (P-E) hysteresis loops. P-E hysteresis loops of as-prepared $Bi_{0.9-y}Dy_yLa_{0.1}Fe_{0.9}Sc_{0.1}O_3$ films with different Dy doping level are shown in Fig.6. The films exhibit un-saturated hysteresis characteristics. With increase the Dy doping level, the ferroelectric properties tend to decrease. For Dy doping level is 0.05, the maximum double remanent polarization $2P_r$ of as-prepared $BiFeO_3$ thin film is 15.44 $\mu C/cm^2$.

Dielectric properties. Fig.7 shows the dielectric constant of as-prepared $Bi_{0.9-y}Dy_yLa_{0.1}Fe_{0.9}Sc_{0.1}O_3$ films with different Dy doping level. It can be seen that the dielectric constant firstly increases and then decreases with the increase of Dy doping level. For example, the dielectric constant for the as-prepared film with the y=0.1 is 190 at 10 kHz, whereas the as-prepared film with the y=0.2 is about 90 at 10 kHz. Fig.8 shows the dielectric loss of as-prepared $Bi_{0.9-y}Dy_yLa_{0.1}Fe_{0.9}Sc_{0.1}O_3$ films with different Dy doping level. As shown in Fig.8, the dielectric loss tends to decreases with the increase of Dy content.

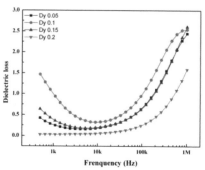

Fig.8 Dielectric loss of as-prepared films

Summary

A series of $Bi_{0.9-y}Dy_yLa_{0.1}Fe_{0.9}Sc_{0.1}O_3$ films (y =0.05, 0.1, 0.15 and 0.2) are prepared by using sol-gel spin coating process on conductive indium tin oxide (ITO)/glass substrates. The thickness of as-prepared thin films is approximate 370 nm. All films annealed at 500°C exhibit polycrystalline rhombohedral structure without the second phase. The root-mean-squared value for as-prepared films is about 2.11 nm. The maximum double remanent polarization $2P_r$ of $Bi_{0.85}Dy_{0.05}La_{0.1}Fe_{0.9}Sc_{0.1}O_3$ film is 15.44 $\mu C/cm^2$. The dielectric constant of as-prepared films tends to firstly increase and then decrease with the increase of Dy content

Acknowledgement

This work was supported by the National Natural Science Foundation of China under Grant No. 50902019, program for New Century Excellent Talents in University (NCET-10-0304), the Special Fund for Basic Scientific Research of Central Colleges, Northeastern University under Grand No. N100123003.

References

[1] T. Ahmed, A. Vorobiev, S. Gevorgian, Growth temperature dependent dielectric of $BiFeO_3$ thin films deposited on silica glass substrates, Thin Solid Films 520 (2012), 4470-4474

[2] J. H. Jo, S. G. Lee, S. H. Lee, Structural and pyroelectric properties of sol-gel derived multiferroic BFO thin films, Mater. Res. Bull. 47 (2012), 409-412

[3] A. R. Damodaran, E. Breckenfeld, A. K. Choquette, et al., Stabilization of mixed-phase structures in highly strained $BiFeO_3$ thin films via chemical-alloying, Appl. Phys. Lett. 100 (2012), 082904-4

[4] H. Deng, H. M. Deng, P. X. Yang, J. H. Chu, Effect of Cr doping on the structure, optical and magnetic properties of multiferroic $BiFeO_3$ thin films, J. Mater. Sci. Mater. El. 23 (2012), 1215-1218

[5] J. G. Wu, J. Wang, D. Q, Xiao, J. G. Zhu, A method to improve electrical properties of $BiFeO_3$ thin films, ACS Appl. Mater. Inter. 4 (2012), 1182-1185

[6] J. Yu, J. H. Chu, Progress and prospect for high temperature single phased magnetic ferroelectrics, Chinese Science Bulletin, 53 (2008), 2097-2112

[7] V. E. Wood, A. E. Austin, Possible applications for magnetoelectric materials, Int. J. Magn. 5(1973), 303-315

[8] Y. Tokura, Multiferroics as quantum electromagnets, Science, 312 (2006), 1481-1482

[9] N. A. Spaldin, M. Fiebig, The Renaissance of magnetoelectric multiferroics, Science, 309 (2005), 391-392

[10] C. Michel, J. M. Moreau, G. D. Achenbach, R. Gerson, W. J. James, The atomic structure of $BiFeO_3$, Solid State Commun. 7(1969), 701-704

[11] S. H. Han, C. I. Cheon, H. G. Lee, H. W. Kang, H. I. Hwang, Low temperature hydrothermal epitaxy of heteroepitaxial BiFeO3 film, Ceram. Int. 38 (2012), S391-395

[12] H. R. Liu, Z. L. Liu, Q. Liu, K. L. Yao, Ferroelectric properties of $BiFeO_3$ films grown by sol-gel process, thin solid films 500(2006), 105-109

[13] Y. Wang, C. W. Nan, Integration of BiFeO3 thin film on Si wafer via a simple sol-gel method, thin solid films 517 (2009), 4484-4487

[14] S. Nakshima, Y. Tsujita, H, Fujisawa, T. Kanashima, M. Okuyama, M. Shimizu, Characterization of epitaxial BiFeO3 thin films prepared by ion beam sputtering, Curr. Appl. Phys. 11 (2011), S 244-246

[15] S. J. Chiu, Y. T. Liu, H. Y. Lee, G. P. Yu, J. H. Huang, Growth of $BiFeO_3$/$SrTiO_3$ artificial superlattice structure by RF sputtering, J. Cryst. Growth. 334 (2011), 90-95

[16] X. W. Qi, X. Y. Zhang, X. Wang, H. B. Sun, J. Q, Qi, Preparation and characterization of $BiFeO_3$ film via sol-gel spin-coating process, Key Eng. Mater. 492 (2012), 202-205

Key Engineering Materials Vol. 537 (2013) pp 114-117
© (2013) Trans Tech Publications, Switzerland
doi:10.4028/www.scientific.net/KEM.537.114

Ferroelectric Property and Microstructures of La-doped $Bi_4Ti_3O_{12}$ Thin Films

X. A. Mei [a], R. F. Liu, C. Q. Huang and J. Liu

School of Physics and Electronics, Hunan Institute of Science and Technology, Yueyang, 414000, China

[a]xamei276@yahoo.com.cn

Keywords: Ferroelectric; Dielectric; Films

Abstract. La-doped bismuth titanate ($Bi_{4-x}La_xTi_3O_{12}$: BLT) and pure $Bi_4Ti_3O_{12}$ (BIT) thin films with random orientation were fabricated on $Pt/Ti/SiO_2/Si$ substrates by rf magnetron sputtering technique. These samples had polycrystalline Bi-layered perovskite structure without preferred orientation, and consisted of well developed rod-like grains with random orientation. For the samples with x=0.25 and 1.0 the current-voltage characteristics exhibited negative differential resistance behaviors and their P-V hysteresis loops were characterized by large leakage current, whereas for the samples with x=0.5 and 0.75 the current-voltage characteristics showed simple ohmic behaviors and their P-V hysteresis loops were the saturated and undistorted hysteresis loops. The remanent polarization (P_r) and coercive field (E_c) of the BLT ceramic with x=0.75 were above $20\mu C/cm^2$ and 85KV/cm, respectively.

Introduction

The most popular ferroelectric material for nonvolatile memories is lead zirconate titanate (PZT). However, the PZT thin films on platinum electrode show serious problems of degradation due to oxygen vacancies created at the interface[1]. As alternative a new class of ferroelectric based on Bi-layered perovskites has been attempted. The electrical conductivity, dielectric and optical properties of many doped (La, Sm, Nd…) bismuth titanate perovskites based on $Bi_4Ti_3O_{12}$ compound have been studied [2-5]. Recentlly, the ferroelectric property of bismuth titanate substituted by some rare-earth ions, such as La or Nd, a Bi-layered perovskite oxide with platinum electrode has received increasing attention on ferroelectric application, such as nonvolatile memory [6, 7]. These rare-earth-doped materials have many attractive properties, such as well-known fatigue-free and large values of remnant polarization. Obviously, the substitution of by rare-earth ions influences on the ferroelectric properties of this material dramatically. It is known that the role of A-site substitution is to displace the volatile Bi with rare-earth ions to suppress the A-site vacancies which are accompanied by oxygen vacancies that act as space charges. It has been reported that ferroelectric properties were improved by ion doping on A- or B-site [8, 9]. In the present study, La-doped BIT ($Bi_xLa_xTi_3O_{12}$, BLT) films were prepared and the effects of the Bi-site substitution by La ion on the microstructure and ferroelectric properties of BIT-based film were investigated.

Experimental

The BLT targets with x = 0.25, 0.5, 0.75, and 1.0 were prepared by solid-state reaction of a mixture of Bi_2O_3 (99.9%), La_2O_3 (99.95%) and TiO_2 (99.9%) powder with 1% mol Bi_2O_3 excess. We grew BLT thin films on $Pt/Ti/SiO_2/Si$ substrates by rf magnetron sputtering technique. The as-grown thin films were annealed for an hour in an oxygen atmosphere at $650^\circ C$. For electrical measurements, the Pt electrodes of 100 nm thickness were deposited onto BLT films at $400^\circ C$ through a shadow mask on an area of 2×10^{-4} cm^2. The phase constitutions were characterized by a Rigaku X-ray diffractometer (D/MAX-RB) with Cu Kα radiation. The surface morphology was investigated by Scanning Electron Microscopy (SEM) (JEOL JSM-6300). The ferroelectric hysteresis loops and the leakage current were measured with the Radiant RT66A Unit. The current-voltage characteristics were measured by means of a d.c. power supply and a digital multimeter controlled by a computer.

Results and Discussions

Fig. 1 shows XRD patterns of the BLT films annealed at temperature 650 °C, which indicates that all the BLT films consist of single phase of a bismuth-layered structure, showing a highly (117) oriented preferential growth with a minor fraction of (00l) orientation. The results show that La-doping did not affect the crystal orientation of the BIT film. According to the results, it is possible that the Bi-layered structure is always maintained in BLT and is rather insensitive to the amount of La. The Bi-layered perovskite structure of BIT can be maintained for such a large amount of La doping because the radii of Bi and La are very close, and some Bi ions are only substituted near Ti-O octahedron layers by La ions.

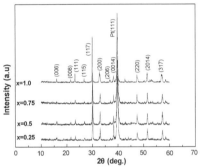

Fig.1 XRD patterns of BLT films annealed at 650°C

The SEM micrographs of the BLT thin films with x = 0.25 and 0.75 are shown in Fig. 2(a) and (b), respectively, which indicate that the BLT films consist of well-developed grains with random orientation. The average grain length of the films with x = 0.25 and 0.75 are approximately 200 nm and 250 nm, respectively. The average grain diameter of the film with x = 0.75 is about 125nm, larger than that of the film with x = 0.25. It implies that the film with x = 0.75 promotes bismuth titanate grain growing greater than the film with x = 0.25.

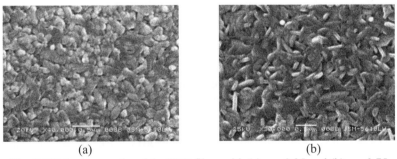

(a)   (b)

Fig. 2 SEM micrographs of the BLT films with (a) x = 0.25 and (b) x = 0.75

The P-E hysteresis loops of BLT thin Films at room temperature are shown in Fig.3. Hysteresis loops of BLT thin Films with x=0.25 and 1.0 are distorted due to leakage current. However, well-saturated loops are measured in BLT Films with x=0.5 and 0.75.The remanent polarization (P_r) and the coercive field (E_c) of the BLT film with x=0.75 are 20 µC/cm^2 and 82 kV/cm, respectively. Without controlling the crystal orientation, the P_r of the polycrystalline BLT film with x=0.75 can exceed 20mC/cm^2, a value larger than that of the BLT film [3] and BST films [8], and comparable to that of the BNTV film [9]. Meanwhile, the E_c is remarkably lower than that of the BNTV film (about 150 kV/cm) [9] and comparable to that of the BLT film [3]. Thus, the BLT film with large P_r and low E_c seems to be more suitable for application as FRAM.

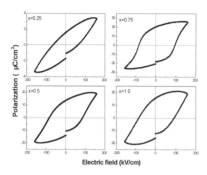   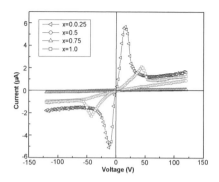

Fig. 3 P-E hysteresis loops of BLT films   Fig. 4 I-V curves of $Bi_{4-x}La_xTi_3O_{12}$ films.

The current-voltage characteristics of BLT samples are shown in Fig. 4. There is an obviously different characteristic among the I-V curves of x=0.25, 0.5, 0.75 and 1.0 samples. In the same region of applied voltage the current values of the La-doped samples with x=0. 5 and 0.75 are about 1~2 order of magnitude lower than that of the samples with x=0.25 and 1.0. The samples with x=0.5, 0.75 exhibit near ohmic behavior, whereas the sample with x=0.25 shows a voltage-controlled negative differential resistance (VNDR) behavior. The I-V curve of x=0.25 sample gives a near ohmic behavior in low voltage region (≤15V). With increasing of voltage beyond 15V the voltage-controlled negative differential resistance appears: the current falls into a nonzero minimum value at a voltage around 55V and then although some very small peaks characterized by VNDR still exist, the current tends to increase slowly as voltage increases. The VNDR has been observed in electroformed metal-insulator-metal (MIM) thin film sandwich structures, amorphous semiconductors and semiconductor quantum devices, and some electrical models have been suggested [10]. Chen Min et al. [10] assumed that the electroforming process gave rise to an array of conducting filaments that span the two electrodes and the filaments might arise from the diffusion of the electrode material or carbon into the insulator. As the voltage applied between the electrodes increased, the filaments were ruptured progressively due to thermal processes, resulting in an eventual decrease in current. Chen Min et al. [10] have shown that the current will fall gradually into zero after a maximum value according to conducting filament model. Obviously, the sandwiched structure of MIM devices with VNDR characteristics must be very thin because of the limit of solid phase diffusion according to conducting filament model. For example, a typical value of filament length used by S. Gravano and R. D. Gould was $4×10^{-8}$m. In our work, however, the samples were as thick as $3×10^{-4}$ m, The distance between two electrodes of the sample was too large to form the active conducting filaments with direct diffusion of electrode material. We assume that the origin of the conducting filaments should be related to small structural change of bismuth titanates and consequent change of electronic bands due to La doping. Therefore the VNDR characteristics of La-doped bismuth titanates can be explained based on the thermal model of filamentary conductivity. However, when voltage is higher than 60V the sample appears an intrinsic ohmic characteristic due to non-filamentary conducting effect since most of conducting filaments were ruptured at so high voltages. On the other hand, these filaments must be revertible because of approximately symmetric shape of I-V curve for the two directions of applied electrical field. Clearly, The I-V curve of the sample x=0.25 is not entirely symmetric when the direction of the applied field is reversed because of polarization response of these materials. The sample with x=1.0 also exhibits a voltage-controlled negative differential resistance (VNDR) behavior, but the current of the sample with x=1.0 is lower than those of the sample with x=0.25. Although all samples appear Bi-layered structure similar to that of $Bi_4Ti_3O_{12}$, some microstructure differences in region of grain boundary due to Bi and O ion vacancies cause different electrical properties.

Conclusions

In summary, La doped Bismuth titanate films with composition $Bi_{4-x}La_xTi_3O_{12}$ were prepared by rf magnetron sputtering technique. XRD pattern revealed that all the films consisted of a single phase of Bi-layered perovskite structure. La ions simultaneous substitution for Bi^{3+} ions are effective to derive enough ferroelectricity by efficient decreasing in space charge density caused by Bi^{3+}-site substitution. Although all samples appear Bi-layered structure similar to that of $Bi_4Ti_3O_{12}$, some microstructure differences in region of grain boundary due to Bi and O ion vacancies cause different electrical properties.

Acknowledgement

This work has supported by the Project of National Natural Science Foundation of China (grant No. 50774034).

References

[1] J. F. Scott, C. A. P. De Araujo, Ferroelectric memories. Science. 246 (1998) 1400-1404.

[2] M. Chen, Z. L. Liu, Y. Wang, Electrical characteristics and microstructures of Sm_2O_3-doped Bi4Ti3O12, Chinese Physics Letters. 21(2004) 1181-1185.

[3] B. H. Park, B. S. Kang, S. D. Bu, Lanthanum-substituted bismuth titanate for use in non-volatile memories, Nature. 401 (1999) 682-686.

[4] T. Kojima, T. Sakai, T, Watanabe, large remanent polarization of $(Bi,Nd)_4Ti_3O_{12}$ epitaxial thin films grown by metalorganic chemical vapor deposition, Appl. Phys. Lett. 80 (2002) 2746-2750.

[5] Y. Noguchi, M. Miyayama, Large remanent of polarization of vanadium-doped $Bi_4Ti_3O_{12}$, Appl Phys Lett. 78(2001) 1903-1907.

[6] Y. C. Chang, D. H. Kuo, The improvement in ferroelectric performance of $(Bi_{3.15}Nd_{0.85})_4Ti_3O_{12}$ films By the addition of hydrogen peroxide in a spin-coating solution, Thin Solid Films. 515(2006) 1683-1687.

[7] T. Watanabe, H. Funakubo, M. Osada, Effect of cosubstitution of La and V in $Bi_4Ti_3O_{12}$ thin films on the low-temperature deposition, Appl. Phys. Lett. 80 (2002) 2229-2233.

[8] U. Chon, K. B. Kim, H. M. Jang, et al: Fatigue-free samarium-modified bismuth titanate $(Bi_{4-x}Sm_xTi_3O_{12})$ film capacitors having large spontaneous polarizations, Appl. Phys. Lett. 79 (2001) 3137-3141.

[9] H. Uchida, H. Yoshikawa, I. Okada, Approach for enhanced polarization of polycrystalline bismuth titanate films by Nd3+/V5+ cosubstitution, Appl. Phys. Lett. 81 (2002) 2229-2233.

[10] H. N. Lee, D. Hesse, N. Zakharov, Ferroelectric $Bi_{3.25}La_{0.75}Ti_3O_{12}$ films of uniform a-axis orientation on silicon substrates, Science. 296 (2002) 2006-2010.

Key Engineering Materials Vol. 537 (2013) pp 118-121
© (2013) Trans Tech Publications, Switzerland
doi:10.4028/www.scientific.net/KEM.537.118

Dielectric and Ferroelectric Properties of Er$_2$O$_3$-doped Bi$_4$Ti$_3$O$_{12}$ Thin Films

X. B. Liu [a], X. A. Mei, C. Q. Huang and J. Liu

School of Physics and Electronics, Hunan Institute of Science and Technology, Yueyang, 414000, China

[a]lxb_hnist@yahoo.com.cn

Keywords: Ferroelectric; Dielectric; Films

Abstract. Er$_2$O$_3$-doped bismuth titanate (Bi$_{4-x}$Er$_x$Ti$_3$O$_{12}$, BET) and pure Bi$_4$Ti$_3$O$_{12}$ (BIT) thin films with random orientation were fabricated on Pt/Ti/SiO$_2$/Si substrates by rf magnetron sputtering technique. These samples had polycrystalline Bi-layered perovskite structure without preferred orientation, and consisted of well developed rod-like grains with random orientation. Er-doping into BIT caused a large shift of the Curie temperature (T$_C$) from 675°C to lower temperature and a improvement in dielectric property. The experimental results indicated that Er doping into BIT also result in a remarkable improvement in ferroelectric property. The P$_r$ and the E$_c$ values of the BET film with x=0.75 were 21 μC/cm^2 and 80 kV/cm, respectively.

Introduction

The most popular ferroelectric material for nonvolatile memories is lead zirconate titanate (PZT). However, the PZT thin films on platinum electrode show serious problems of degradation due to oxygen vacancies created at the interface[1]. As alternative a new class of ferroelectric based on Bi-layered perovskites has been attempted. The electrical conductivity, dielectric and optical properties of many doped (Sm, La, Nd…) bismuth titanate perovskites based on Bi$_4$Ti$_3$O$_{12}$ compound have been studied [2-5]. Recentlly, the ferroelectric property of bismuth titanate substituted by some rare-earth ions, such as La or Nd, a Bi-layered perovskite oxide with platinum electrode has received increasing attention on ferroelectric application, such as nonvolatile memory [6, 7]. These rare-earth-doped materials have many attractive properties, such as well-known fatigue-free and large values of remnant polarization. Obviously, the substitution of by rare-earth ions influences on the ferroelectric properties of this material dramatically. It is known that the role of A-site substitution is to displace the volatile Bi with rare-earth ions to suppress the A-site vacancies which are accompanied by oxygen vacancies that act as space charges. It has been reported that ferroelectric properties were improved by ion doping on A- or B-site [8-10]. In the present study, Er$_2$O$_3$-doped BIT (Bi$_x$Er$_x$Ti$_3$O$_{12}$, BET) films were prepared and the effects of the Bi-site substitution by Er ion on the dielectric and ferroelectric properties of BIT-based film were investigated.

Experimental

The BET targets with x = 0.25, 0.5, 0.75, and 1.0 were prepared by solid-state reaction of a mixture of Bi$_2$O$_3$ (99.9%), Er$_2$O$_3$ (99.95%) and TiO$_2$ (99.9%) powder with 1% mol Bi$_2$O$_3$ excess. We grew BET thin films on Pt/Ti/SiO$_2$/Si substrates by rf magnetron sputtering technique. The as-grown thin films were annealed for an hour in an oxygen atmosphere at 650 °C. For electrical measurements, the Pt electrodes of 100 nm thickness were deposited onto BET films at 400°C through a shadow mask on an area of 2×10^{-4} cm^2. The phase constitutions were characterized by a Rigaku X-ray diffractometer (D/MAX-RB) with Cu Kα radiation. The surface morphology was investigated by Scanning Electron Microscopy (SEM) (JEOL JSM-6300). The ferroelectric hysteresis loops and the leakage current were measured with the Radiant RT66A Unit. The dielectric characterizations were determined by an impendence analyzer (HP-4294A).

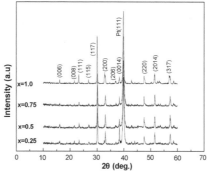

Fig.1 XRD patterns of BET films annealed at 650°C

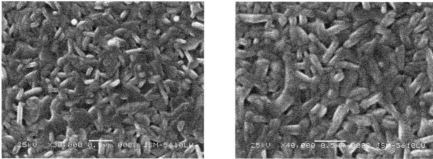

Fig. 2 SEM micrographs of the BET films with (a, left) x = 0.25 and (b, right) x = 0.75

Results and Discussions

Fig. 1 shows XRD patterns of the BET films annealed at temperature 650 °C, which indicates that all the BET films consist of single phase of a bismuth-layered structure, showing a highly (117) oriented preferential growth with a minor fraction of (00l) orientation. The results show that Er-doping did not affect the crystal orientation of the BIT film. According to the results, it is possible that the Bi-layered structure is always maintained in BET and is rather insensitive to the amount of Er. The Bi-layered perovskite structure of BIT can be maintained for such a large amount of Er doping because the radii of Bi and Er are very close, and some Bi ions are only substituted near Ti-O octahedron layers by Er ions.

The SEM micrographs of the BET thin films with x = 0.25 and 0.75 are shown in Fig. 2(a) and (b), respectively, which indicate that the BET films consist of well-developed grains with random orientation. The average grain length of the films with x = 0.25 and 0.75 are approximately 200 nm and 250 nm, respectively. The average grain diameter of the film with x = 0.75 is about 125nm, larger than that of the film with x = 0.25. It implies that the film with x = 0.75 promotes bismuth titanate grain growing greater than the film with x = 0.25.

The P-E hysteresis loops of BET thin Films at room temperature are shown in Fig.3. Hysteresis loops of BET thin Films with x=0.25 and 1.0 are distorted due to leakage current. However, well-saturated loops are measured in BET Films with x=0.5 and 0.75. The remanent polarization (P_r) and the coercive field (E_c) of the BET film with x=0.75 are 21 μC/cm^2 and 80 kV/cm, respectively. Without controlling the crystal orientation, the P_r of the polycrystalline BET film with x=0.75 can exceed 20mC/cm^2, a value larger than that of the BLT film [2] and BST films [8], and comparable to that of the BNTV film [9]. Meanwhile, the E_c is remarkably lower than that of the BNTV film (about 150 kV/cm) [9] and comparable to that of the BLT film [2]. Thus, the BET film with large P_r and low E_c seems to be more suitable for application as FRAM.

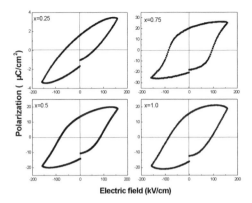

Fig. 3 P-E hysteresis loops of BET films

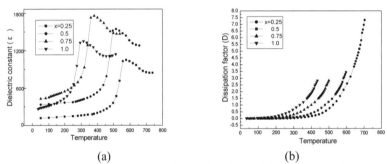

(a)   (b)

Fig.4 (a) The dielectric constant (ε) and (b) dissipation factor (D) of the BET thin films as functions of temperature measured at a frequency of 1 MHz

The dielectric constant (ε) and dissipation factor (D) of the BET thin films as functions of temperature are shown in Figure4 (a) and (b), respectively. The peaks of the dielectric constant at 640, 575, 420, and 335°C, as shown in Fig. 4(a), indicate that the Er-doping leads a shift of the Curie temperature (T_C) of paraelectric-ferroelectric phase transition of BIT from 675°C to 640, 575, 420, and 335°C for the BET films with x = 0.25, 0.5, 0.75, and 1.0, respectively. It is obviously seen that T_C decreases with Er increasing, while ε increases with Er increasing in the range of x≤ 0.75. The ε values of the BET films with x≥0.5 at room temperature are larger than 285, higher than that of the BLT films and the BTV ceramics [8, 9]. Furthermore, the broadening peaks of ε indicate that the BET films consist of a diffuse-type ferroelectric phase transition. The good ferroelectric properties of Er-doped BIT did not come from the enhancement of spontaneous polarization, because Er-doping did not give rise to a high T_C of the BIT films. Fig. 4(b) indicates that the dissipation factor of each BET film increases slowly below T_C and rises rapidly with temperature increasing beyond T_C, which confirms that paraelectric-ferroelectric phase transition of each film appears at its T_C. It should be noted that the D values of the BET films are about 3~6×10^{-3} at room temperature, one order of magnitude lower than that of the BLT films [6], comparable with that of the BTV ceramics [9]. These results indicate that Er ions doping into BIT result in a remarkable improvement in dielectric properties.

Conclusions

In summary, Er_2O_3-doped Bismuth titanate films with composition $Bi_{4-x}Er_xTi_3O_{12}$ were prepared by rf magnetron sputtering technique. XRD pattern revealed that all the films consisted of a single phase of Bi-layered perovskite structure. Er ions simultaneous substitution for Bi^{3+} ions are effective to derive enough ferroelectricity by efficient decreasing in space charge density caused by Bi^{3+}-site substitution. Er ions doping into BIT also result in a remarkable improvement in dielectric properties.

Acknowledgement

This work has supported by the Project of National Natural Science Foundation of China (grant No. 50774034).

References

[1] J. F. Scott, C. A. P. De Araujo, Ferroelectric memories. Science. 246 (1998) 1400-1404.

[2] B. H. Park, B. S. Kang, S. D. Bu, et al: Lanthanum-substituted bismuth titanate for use in non-volatile memories, Nature. 401 (1999) 682-686.

[3] H. N. Lee, D. Hesse, N. Zakharov, et al: Ferroelectric $Bi_{3.25}La_{0.75}Ti_3O_{12}$ films of uniform a-axis orientation on silicon substrates, Science. 296 (2002) 2006-2010.

[4] T. Kojima, T. Sakai, T, Watanabe, et al: large remanent polarization of $(Bi,Nd)_4Ti_3O_{12}$ epitaxial thin films grown by metalorganic chemical vapor deposition, Appl. Phys. Lett. 80 (2002) 2746-2750.

[5] Y. N0guchi, M. Miyayama, Large remanent of polarization of vanadium-doped $Bi_4Ti_3O_{12}$, Appl Phys Lett. 78(2001) 1903-1904.

[6] Y. C. Chang, D. H. Kuo: The improvement in ferroelectric performance of $(Bi_{3.15}Nd_{0.85})_4Ti_3O_{12}$ films By the addition of hydrogen peroxide in a spin-coating solution, Thin Solid Films. 515(2006) 1683-1687.

[7] T. Watanabe, H. Funakubo, M. Osada, et al: Effect of cosubstitution of La and V in $Bi_4Ti_3O_{12}$ thin films on the low-temperature deposition, Appl. Phys. Lett. 80 (2002) 2229-2233.

[8] U. Chon, K. B. Kim, H. M. Jang, et al: Fatigue-free samarium-modified bismuth titanate $(Bi_{4-x}SmxTi_3O_{12})$ film capacitors having large spontaneous polarizations, Appl. Phys. Lett. 79 (2001) 3137-3141.

[9] H. Uchida, H. Yoshikawa, I. Okada, et al: Approach for enhanced polarization of polycrystalline bismuth titanate films by Nd^{3+}/V^{5+} cosubstitution, Appl. Phys. Lett. 81 (2002) 2229-2233.

[10] M. Chen, Z. L. Liu, Y. Wang, et al: Electrical characteristics and microstructures of Sm_2O_3-doped $Bi_4Ti_3O_{12}$, Chinese Physics Letters. 21(2004) 1181-1185.

Key Engineering Materials Vol. 537 (2013) pp 122-125
© (2013) Trans Tech Publications, Switzerland
doi:10.4028/www.scientific.net/KEM.537.122

Electrical Characteristics and Microstructures of $Bi_{2.9}Pr_{0.9}Ti_3O_{12}$ and $Bi_{2.9}Pr_{0.92}Ti_{2.97}V_{0.03}O_{12}$ Thin Films

M. Chen[a], X. A. Mei, R. F. Liu, C. Q. Huang and J. Liu

School of Physics and Electronics, Hunan Institute of Science and Technology, Yueyang, 414000, China

[a]chenmin581215@yahoo.com.cn

Keywords: Ferroelectric; Dielectric; films

Abstract. $Bi_{2.9}Pr_{0.9}Ti_3O_{12}$(BPT) and $Bi_{2.9}Pr_{0.9}Ti_{2.97}V_{0.03}O_{12}$(BPTV) thin films with random orientation were fabricated on $Pt/Ti/SiO_2/Si$ substrates by rf magnetron sputtering technique. These samples had polycrystalline Bi-layered perovskite structure without preferred orientation, and consisted of well developed rod-like grains with random orientation. The experimental results indicated that Pr doping into $Bi_4Ti_3O_{12}$ (BIT) result in a remarkable improvement in ferroelectric property. The remanent polarization (P_r) and coercive field (E_c) of the BPT film were 28 $\mu C/cm^2$ and 80 kV/cm, respectively. Furthermore, V substitution improves the P_r value of the BTVT film up to 43 $\mu C/cm^2$, which is much larger than that of the BPT film.

Introduction

The most popular ferroelectric material for nonvolatile memories is lead zirconate titanate (PZT). However, the PZT thin films on platinum electrode show serious problems of degradation due to oxygen vacancies created at the interface[1]. As alternative a new class of ferroelectric based on Bi-layered perovskites has been attempted. The electrical conductivity, dielectric and optical properties of many doped (La, Nd, Sm, Pr…) bismuth titanate perovskites based on $Bi_4Ti_3O_{12}$ compound have been studied [2-5]. Recentlly, the ferroelectric property of bismuth titanate substituted by some rare-earth ions, such as La or Nd, a Bi-layered perovskite oxide with platinum electrode has received increasing attention on ferroelectric application, such as nonvolatile memory [6, 7]. These rare-earth-doped materials have many attractive properties, such as well-known fatigue-free and large values of remnant polarization. Obviously, the substitution of by rare-earth ions influences on the ferroelectric properties of this material dramatically. It is known that the role of A-site substitution is to displace the volatile Bi with rare-earth ions to suppress the A-site vacancies which are accompanied by oxygen vacancies that act as space charges. It has been reported that ferroelectric properties were improved by ion doping on A- or B-site [8, 9]. However, there has not been any report of the effects of Pr- and V-doping on the ferroelectric characteristic of $Bi_4Ti_3O_{12}$ (BIT). In the present study, $Bi_{2.9}Pr_{0.9}Ti_3O_{12}$ (BPT), $Bi_{2.9}Pr_{0.9}Ti_{2.97}V_{0.03}O_{12}$ (BPTV) films were prepared and the effects of the Bi- and Ti-site substitution by Pr and V ion on the microstructure and ferroelectric properties of BIT films were investigated.

Experimental

The BPT and BPTV targets were prepared by solid-state reaction of a mixture of Bi_2O_3 (99.9%), Pr_6O_{11} (99.95%), V_2O_5(99.9%) and TiO_2(99.9%) powder with 1% mol Bi_2O_3 excess. We grew BPT and BPTV thin films on $Pt/Ti/SiO_2/Si$ substrates by rf magnetron sputtering technique. The as-grown thin films were annealed for an hour in an oxygen atmosphere at 650 °C. For electrical measurements, the Pt electrodes of 100 nm thickness were deposited onto BPT and BPTV films at 400 °C through a shadow mask on an area of 2×10^{-4} cm^2. The phase constitutions were characterized by a Rigaku X-ray diffractometer (D/MAX-RB) with Cu Kα radiation. The surface morphology was investigated by Scanning Electron Microscopy (SEM) (JEOL JSM-6300). The ferroelectric hysteresis loops and the leakage current were measured with the Radiant RT66A Unit.

Results and Discussions

Fig. 1 shows XRD patterns of the BPT and BPTV films annealed at temperature 650 °C, which indicates that all the BPT, BPTV, and BIT films consist of single phase of a bismuth-layered structure, showing a highly (117) oriented preferential growth with a minor fraction of (00l) orientation. The results show that Pr-doping did not affect the crystal orientation of the BIT film. According to the results, it is possible that the Bi-layered structure is always maintained in the films and is rather insensitive to the amount of Pr and V. The Bi-layered perovskite structure of BIT can be maintained for such a large amount of Pr doping because the radii of Bi and Pr are very close, and some Bi ions are only substituted near Ti-O octahedron layers by Pr ions.

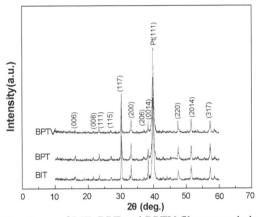

Fig.1 XRD patterns of BIT, BPT and BPTV films annealed at 650 °C

The SEM micrographs of the BPT and BPTV thin films are shown in Fig. 2(a) and (b), respectively, which indicate that the BPT and BPTV films consist of well-developed grains with random orientation. The average grain length of the BPT and BPTV films is approximately 200 nm and 250 nm, respectively. The average grain diameter of the BPTV film is about 125nm, larger than that of the BPT film. It implies that the BPTV film promotes bismuth titanate grain growing greater than the BPT film.

(a)   (b)

Fig. 2 SEM micrographs of (a) the BPT and (b) BPTV films

The P-E hysteresis loops of BPT and BPTV thin Films at room temperature are shown in Fig.3. Well-saturated loops are measured in BPT and BPTV Films. The remanent polarization (P_r) and the coercive field (E_c) of the BPT film are 28 μC/cm^2 and 80 kV/cm, respectively. Without controlling the crystal orientation, the P_r of the polycrystalline BPT film can exceed 28μC/cm^2, a value larger than that of the BLT film [6] and BST films [8], and comparable to that of the BNTV film [9]. Meanwhile, the E_c is remarkably lower than that of the BNTV film (about 150 kV/cm) [9] and comparable to that of the BLT film [6]. Furthermore, V substitution improves the P_r value of the BPTV ceramics up to 43 $\mu C/cm^2$, which is much larger than that of the BPT film. Thus, the BPTV film with large P_r and low E_c seems to be more suitable for application as FRAM.

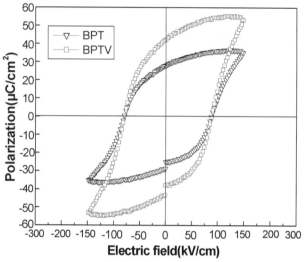

Fig.3 Fig.3 P-E hysteresis loops of BPT and BPTV films

It is known that the valence of Pr ion is 3+ and 4+, respectively, and Pr_6O_{11} can be divided into two parts by the formula:

$$Pr_6O_{11} \rightarrow Pr_2O_3 + 4PrO_2,$$

Three Pr^{4+} ions should be companied by four Bi^{3+} vacancies if Pr substituted only for Bi^{3+} ions. Therefore, it assumed that Pr^{4+} ions substitution for Bi^{3+} ions efficiently suppressed the generation of oxygen vacancies, similar to V^{5+} ions doping into BIT [9], and that Pr^{3+} ions substitution for Bi^{3+} ions, on the other hand, were similar to La^{3+} ions doping into BIT [6]. Both Pr^{3+} ions and Pr^{4+} ions simultaneous substitution for Bi^{3+} ions are effective to derive enough ferroelectricity by efficient decreasing in space charge density caused by Bi^{3+}-site substitution. This speculation is consistent with experimental results. In fact, the Pr doping into BIT led to a marked improvement in dissipation factor, dielectric permittivity, and leakage current (2.0×10^{-8} A/cm^2 at E = 90 kV/cm) properties. These experimental results imply that the Pr and V doping into BIT lead to an efficient decrease in space charge density.

Conclusions

In summary, $Bi_{2.9}Pr_{0.9}Ti_3O_{12}$ and $Bi_{2.9}Pr_{0.92}Ti_{2.97}V_{0.03}O_{12}$ Thin Films were prepared by rf magnetron sputtering technique. XRD pattern revealed that all the films consisted of a single phase of Bi-layered perovskite structure. Both Pr and V ions simultaneous substitution for Bi^{3+} and Ti^{4+} ions are effective to derive enough ferroelectricity by efficient decreasing in space charge density caused by Bi^{3+}-site and Ti^{4+}- site substitution.

Acknowledgement

This work has supported by the Project of National Natural Science Foundation of China (grant No. 50774034)

References

[1] J. F. Scott, C. A. P. De Araujo, Ferroelectric memories, Science. 246 (1998) 1400-1404.

[2] U. Chon, K. B. Kim, H. M. Jang, et al: Fatigue-free samarium-modified bismuth titanate($Bi_{4-x}SmxTi_3O_{12}$)film capacitors having large spontaneous polarizations, Appl. Phys. Lett. 79 (2001) 3137-3141.

[3] B. H. Park, B. S. Kang, S. D. Bu, et al: Lanthanum-substituted bismuth titanate for use in non-volatile memories, Nature. 401 (1999) 682-686.

[4] T. Kojima, T. Sakai, T, Watanabe, et al: large remanent polarization of $(Bi,Nd)_4Ti_3O_{12}$ epitaxial thin films grown by metalorganic chemical vapor deposition, Appl. Phys. Lett. 80 (2002) 2746-2750.

[5] H. Matssuda, S. Ito, T. Iijima, Design and ferroelectric properties of polar-axis-oriented polycrystalline $Bi_{4-x}Pr_xTi_3O_{12}$ thick films on Ir/Si substrates, Appl. Phys. Lett. 83(2003) 2003-2007.

[6] H. N. Lee, D. Hesse, N. Zakharov, et al: Ferroelectric $Bi_{3.25}La_{0.75}Ti_3O_{12}$ films of uniform a-axis orientation on silicon substrates, Science. 296 (2002) 2006-2010.

[7] Y. C. Chang, D. H. Kuo, The improvement in ferroelectric performance of $(Bi_{3.15}Nd_{0.85})_4Ti_3O_{12}$ films By the addition of hydrogen peroxide in a spin-coating solution, Thin Solid Films. 515(2006) 1683-1687.

[8] T. Watanabe, H. Funakubo, M. Osada, et al: Effect of cosubstitution of La and V in $Bi_4Ti_3O_{12}$ thin films on the low-temperature deposition, Appl. Phys. Lett. 80 (2002) 2229-2233.

[9] H. Uchida, H. Yoshikawa, I. Okada, et al: Approach for enhanced polarization of polycrystalline bismuth titanate films by Nd^{3+}/V^{5+} cosubstitution, Appl. Phys. Lett. 81 (2002) 2229-2233.

Key Engineering Materials Vol. 537 (2013) pp 126-129
© (2013) Trans Tech Publications, Switzerland
doi:10.4028/www.scientific.net/KEM.537.126

Electrical Characterization and Microstructures of Bi$_{3.3}$Tb$_{0.6}$Ti$_3$O$_{12}$and Bi$_{3.3}$Tb$_{0.6}$Ti$_{2.97}$V$_{0.03}$O$_{12}$ Thin Films

R. F. Liu[a], M. B. Sun, W. P. Ding and X. A. Mei

School of math, Hunan Institute of Science and Technology, Yueyang, 414000, China

[a]lrf_yy@yahoo.com.cn

Keywords: Ferroelectric; Dielectric; films

Abstract. Bi$_{3.3}$Tb$_{0.6}$Ti$_3$O$_{12}$(BTT), Bi$_{3.3}$Tb$_{0.6}$Ti$_{2.97}$V$_{0.03}$O$_{12}$(BTTV), and pure Bi$_4$Ti$_3$O$_{12}$ (BIT) thin films with random orientation were fabricated on Pt/Ti/SiO$_2$/Si substrates by rf magnetron sputtering technique. These samples had polycrystalline Bi-layered perovskite structure without preferred orientation, and consisted of well developed rod-like grains with random orientation. The experimental results indicated that Tb doping into BIT also result in a remarkable improvement in ferroelectric property. The remanent polarization (P$_r$) and coercive field (E$_c$) of the BTT film were 25 μC/cm^2 and 85 kV/cm, respectively. Furthermore, V substitution improves the P$_r$ value of the BTVT film up to 35 μC/cm^2, which is much larger than that of the BTT film.

Introduction

The most popular ferroelectric material for nonvolatile memories is lead zirconate titanate (PZT). However, the PZT thin films on platinum electrode show serious problems of degradation due to oxygen vacancies created at the interface[1]. As alternative a new class of ferroelectric based on Bi-layered perovskites has been attempted. The electrical conductivity, dielectric and optical properties of many doped (La, Sm, Nd...) bismuth titanate perovskites based on Bi$_4$Ti$_3$O$_{12}$ compound have been studied [2-5]. Recentlly, the ferroelectric property of bismuth titanate substituted by some rare-earth ions, such as La or Nd, a Bi-layered perovskite oxide with platinum electrode has received increasing attention on ferroelectric application, such as nonvolatile memory [6, 7]. These rare-earth-doped materials have many attractive properties, such as well-known fatigue-free and large values of remnant polarization. Obviously, the substitution of by rare-earth ions influences on the ferroelectric properties of this material dramatically. It is known that the role of A-site substitution is to displace the volatile Bi with rare-earth ions to suppress the A-site vacancies which are accompanied by oxygen vacancies that act as space charges. It has been reported that ferroelectric properties were improved by ion doping on A- or B-site [8, 9]. However, there has not been any report of the effects of Tb-doping on the ferroelectric characteristic of BIT. In the present study, Bi$_{3.3}$Tb$_{0.6}$Ti$_3$O$_{12}$ (BTT), Bi$_{3.3}$Tb$_{0.6}$Ti$_{2.97}$V$_{0.03}$O$_{12}$ (BTTV) films were prepared and the effects of the Bi-site substitution by Tb and V ion on the microstructure and ferroelectric properties of BIT-based film were investigated.

Experimental

The BTT and BTTV targets were prepared by solid-state reaction of a mixture of Bi$_2$O$_3$ (99.9%), Tb$_4$O$_7$ (99.95%), V$_2$O$_5$(99.9%) and TiO$_2$(99.9%) powder with 1% mol Bi$_2$O$_3$ excess. We grew BTT and BTTV thin films on Pt/Ti/SiO$_2$/Si substrates by rf magnetron sputtering technique. The as-grown thin films were annealed for an hour in an oxygen atmosphere at 650 °C. For electrical measurements, the Pt electrodes of 100 nm thickness were deposited onto BTT and BTTV films at 400 °C through a shadow mask on an area of 2×10^{-4} cm^2. The phase constitutions were characterized by a Rigaku X-ray diffractometer (D/MAX-RB) with Cu Kα radiation. The surface morphology was investigated by Scanning Electron Microscopy (SEM) (JEOL JSM-6300). The ferroelectric hysteresis loops and the leakage current were measured with the Radiant RT66A Unit. Raman spectra analysis was performed on a Nicolet 950 Raman Spectrometer with laser wavelength of 514nm. The Nd:YAG laser was used and a laser incidence angle optimized dielectric refractive filter was used for improving the laser power transmitting. An InGaAs detector was used with resolution of 4 cm^{-1}.

Results and Discussions

Fig. 1 shows XRD patterns of the BTT and BTTV films annealed at temperature 650 $^{\circ}$C, which indicates that all the BTT, BTTV, and BIT films consist of single phase of a bismuth-layered structure, showing a highly (117) oriented preferential growth with a minor fraction of (00l) orientation. The results show that Tb- and doping did not affect the crystal orientation of the BIT film. According to the results, it is possible that the Bi-layered structure is always maintained in the films and is rather insensitive to the amount of Tb. The Bi-layered perovskite structure of BIT can be maintained for such a large amount of Tb doping because the radii of Bi and Tb are very close, and some Bi ions are only substituted near Ti-O octahedron layers by Tb ions.

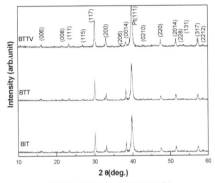

Fig.1 XRD patterns of BIT, BTT and BTTV films annealed at 650 $^{\circ}$C

The SEM micrographs of the BTT and BTTV thin films are shown in Fig. 2(a) and (b), respectively, which indicate that the BTT and BTTV films consist of well-developed grains with random orientation. The average grain length of the BTT and BTTV film are approximately 200 nm and 250 nm, respectively. The average grain diameter of the BTTV film is about 125nm, larger than that of the BTT film. It implies that the BTTV film promotes bismuth titanate grain growing greater than the BTT film.

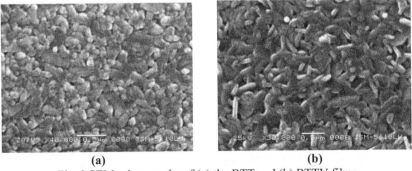

(a)   (b)

Fig. 2 SEM micrographs of (a) the BTT and (b) BTTV films

The P-E hysteresis loops of BTT and BTTV thin Films at room temperature are shown in Fig.3. Well-saturated loops are measured in BTT and BTTV Films. The remanent polarization (P$_r$) and the coercive fields (Ec) of the BTT film are 25 μC/cm^2 and 85 kV/cm, respectively. Without controlling the crystal orientation, the P_r of the polycrystalline BTT film can exceed 25mC/cm^2, a value larger than that of the BLT film [3] and BST films [8], and comparable to that of the BNTV film [9]. Meanwhile, the E_c is remarkably lower than that of the BNTV film (about 150 kV/cm) [9] and comparable to that of the BLT film [3]. Furthermore, V substitution improves the P_r value of the BTVT ceramics up to 35 µC/cm^2, which is much larger than that of the BTT film. Thus, the BTTV film with large P_r and low E_c seems to be more suitable for application as FRAM.

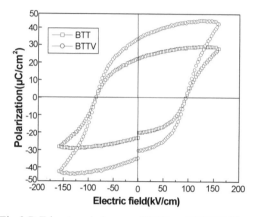

Fig.3 P-E hysteresis loops of BTT and BTTV films

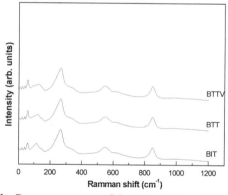

Fig. 5. The Raman spectrum of the BIT, BTT and BTTV films

The BIT, BTT and BTTV crystalline samples were examined by Raman spectroscopy to look for further short range structure information, and the results are shown in Fig. 4.The vibrational modes of BTT can be classified as internal modes of Ti-O octahedra and the lattice transitions involved in the motion of the cations. From the mass consideration of vibrational frequencies, we expect that the low-frequency modes should be dominated by the motions of heavy Bi ions. The internal modes, which stem from Ti-O octahedra, appear above 200 cm^{-1}.

The stronger Raman phonon modes appear at 62, 116, 152, 269, 551, and 851 cm^{-1}, in which the peaks of 269, 551, and 851 cm^{-1} are Ti-O octahedron modes and the range of 0-200 cm^{-1} corresponds to the vibrational modes of A-site ions. The intensity and position of the vibrational modes of BIT at 269, 551, and 851 cm^{-1} are not almost affected by Tb-doping because the radii between Bi and Tb are very close and some Bi ions are only substituted near Ti-O octahedron layers by Pr ions. Thus, Tb-doping did not apparently affect Ti-O octahedron microstructure. However, the low frequency vibrational modes are affected apparently by Tb doping into BIT. From Fig. 4 we can see clearly that

Tb-doping lead to the decrease of the intensity of 62, 116 and 152 cm^{-1} modes and the shift of the vibrational frequency of these modes from low frequency to high frequency apparently. Raman analysis shows that Tb^{3+} and Tb^{4+} ions substitution only appears in Bi-sit.

It is known that the valence of Tb ion is 3+ and 4+, respectively, and Tb_4O_7 can be divided into two parts by the formula:

$$Tb_4O_7 \rightarrow Tb_2O_3 + 2TbO_2,$$

Three Tb^{4+} ions should be companied by four Bi^{3+} vacancies if Tb substituted only for Bi^{3+} ions. Therefore, it assumed that Tb^{4+} ions substitution for Bi^{3+} ions efficiently suppressed the generation of oxygen vacancies, similar to V^{5+} ions doping into BIT [9], and that Tb^{3+} ions substitution for Bi^{3+} ions, on the other hand, were similar to La^{3+} ions doping into BIT [3]. Both Tb^{3+} ions and Tb^{4+} ions simultaneously substitution for Bi^{3+} ions are effective to derive enough ferroelectricity by efficient decreasing in space charge density caused by Bi^{3+}-site substitution. This speculation is consistent with experimental results. In fact, the Tb doping into BIT led to a marked improvement in dissipation factor, dielectric permittivity, and leakage current (10^{-8} A/cm^2 at E = 90 kV/cm) properties. These experimental results imply that the Tb and V doping into BIT lead to an efficient decrease in space charge density.

Conclusions

In summary, $Bi_{3.3}Tb_{0.6}Ti_3O_{12}$ and $Bi_{3.3}Tb_{0.6}Ti_{2.97}V_{0.03}O_{12}$ films were prepared by rf magnetron sputtering technique. XRD pattern revealed that all the films consisted of a single phase of Bi-layered perovskite structure. Both Tb and V ions simultaneous substitution for Bi^{3+} and Ti^{4+} ions are effective to derive enough ferroelectricity by efficient decreasing in space charge density caused by Bi^{3+}-site and Ti^{4+}- site substitution.Raman analysis shows that Tb ions substitution appears in A-sit.

Acknowledgement

This work has supported by the Project of National Natural Science Foundation of China (grant No. 100871061).

References

[1] J. F. Scott, C. A. P. De Araujo, Ferroelectric memories. Science. 246 (1998) 1400-1404.

[2] H. N. Lee, D. Hesse, N. Zakharov, *et al*: Ferroelectric $Bi_{3.25}La_{0.75}Ti_3O_{12}$ films of uniform *a*-axis orientation on silicon substrates, Science. 296 (2002) 2006-2010.

[3] B. H. Park, B. S. Kang, S. D. Bu, *et al*: Lanthanum-substituted bismuth titanate for use in non-volatile memories, Nature. 401 (1999) 682-686.

[4] T. Kojima, T. Sakai, T, Watanabe, *et al*: large remanent polarization of $(Bi,Nd)_4Ti_3O_{12}$ epitaxial thin films grown by metalorganic chemical vapor deposition, Appl. Phys. Lett. 80 (2002) 2746-2750.

[5] Y. N0guchi, M. Miyayama, Large remanent of polarization of vanadium-doped $Bi_4Ti_3O_{12}$, Appl Phys Lett. 78(2001) 1903-1907.

[6] Y. C. Chang, D. H. Kuo, The improvement in ferroelectric performance of $(Bi_{3.15}Nd_{0.85})_4Ti_3O_{12}$ films By the addition of hydrogen peroxide in a spin-coating solution, Thin Solid Films. 515(2006) 1683-1687.

[7] M. Chen, Z. L. Liu, Y. Wang, *et al*: Electrical characteristics and microstructures of Sm_2O_3-doped $Bi_4Ti_3O_{12}$, Chinese Physics Letters. 21(2004) 1181-1185.

[8] U. Chon, K. B. Kim, H. M. Jang, *et al*: Fatigue-free samarium-modified bismuth titanate $(Bi_{4-x}SmxTi_3O_{12})$ film capacitors having large spontaneous polarizations, Appl. Phys. Lett. 79 (2001) 3137-3141.

[9] H. Uchida, H. Yoshikawa, I. Okada, *et al*: Approach for enhanced polarization of polycrystalline bismuth titanate films by Nd^{3+}/V^{5+} cosubstitution, Appl. Phys. Lett. 81 (2002) 2229-2233.

Key Engineering Materials Vol. 537 (2013) pp 130-133
© (2013) Trans Tech Publications, Switzerland
doi:10.4028/www.scientific.net/KEM.537.130

Ferroelectric and Electrical Behavior of $Bi_{4-x}Nd_xTi_3O_{12}$ Thin Films

C. Q. Huang [a], X. B. Liu, X. A. Mei and J. Liu

School of Physics and Electronics, Hunan Institute of Science and Technology, Yueyang, 414000, China

[a]huangchongqing_yy@yahoo.com.cn

Keywords: ferroelectric; film; bismuth titanate; doping

Abstract. Nd_2O_3-doped bismuth titanate ($Bi_{4-x}Nd_xTi_3O_{12}$: BNT) thin films with random orientation were fabricated on $Pt/Ti/SiO_2/Si$ substrates by rf magnetron sputtering technique, and the structures and ferroelectric properties of the films were investigated. XRD studies indicated that all of BNT films consisted of single phase of a bismuth-layered structure with well-developed rod-like grains. The remanent polarization (P_r) and coercive field (E_c) of the BNT Film with x=0.8 were $25\mu C/cm^2$ and $55 KV/cm$, respectively. After 3×10^{10} switching cycles, 15% degradation of P_r is observed in the film.

Introduction

Bismuth titanate layered perovskite films have received much attention because of their potential for technological applications [1-3]. Bismuth titanate, $Bi_4Ti_3O_{12}$ (BIT), is a typical ferroelectric material with the spontaneous polarization lying in the a-c plane at about $4.5°$ to the a-axis and exhibiting two independently reversible components of approximate magnitude $4\mu C/cm^2$ along the c-axis and $50\mu C/cm^2$ along the a-axis [2]. In 1999, B. H. Park and colleagues first reported [3] that ferroelectric lanthanum-substituted bismuth titanate ($Bi_{3.25}La_{0.75}Ti_3O_{12}$) thin films by pulsed laser deposition exhibited a fatigue-free characteristic after 3×10^{10} read/write cycles. Therefore, this material is considered as a promising candidate for non-volatile memory devices. They suggested that the fatigue-free characteristics of $Bi_{3.25}La_{0.75}Ti_3O_{12}$ thin films result from some Bi ions being substituted near the Ti-O octahedron layers by La ions, and that there was some room for improvement by varying the amount of the La substitutes, or by substituting other rare-earth ions. Recently, these kinds of materials doped with rare earth elements attract much attention for a large remnant polarization and fatigue-free characteristic [3-6]. In order to obtain better properties, a lot of works about $Bi_4Ti_3O_{12}$ doped with the rare earth elements such as La, Pr, and Sm have been reported [7-9]. However, the effects of Bi ions substituted by rare earth ions have not been clarified. In the present work, we fabricated Nd_2O_3-doped $Bi_4Ti_3O_{12}$ thin films with rf magnetron sputtering technique，and their microstructures and electrical properties were studied in details.

Experimental

$Bi_{4-x}Nd_xTi_3O_{12}$ (BNT, x=0.0 ~ 1.0) targets were prepared by solid-state reaction of mixture of Bi_2O_3 (99.9%), Nd_2O_3 (99.95%) and TiO_2 (99.9%) powder. Raw materials of targets were mixed by ball milling with agate balls and ethanol for 24h. The powders with binder addition were pressed into discs of 50 mm in diameter and about 4mm in thickness at a pressure of 120 MPa. The discs were sintered at $1100°C$ for 2h in air, and furnace-cooled to room temperature. We grew BNT thin films (about 400 ~500nm thickness) on $Pt/Ti/SiO_2/Si$ substrates by rf magnetron sputtering technique. The deposition temperature and ambient oxygen pressure were optimized at $400 °C$ and 20 Pa, respectively. The as-grown thin films were annealed for an hour in an oxygen atmosphere at $650 °C$. For electrical measurements, the Pt electrodes of 100 nm were deposited onto BET films at $400 °C$ through a shadow mask on an area of $2 \times 10^{-4} cm^2$. The phase constitutions were characterized by Rigaku X-ray diffractometer (D/MAX-RB) with Cu Kα radiation. The surface morphology was scanned by

Scanning Electron Microscopy (SEM) (JEOL JSM-6300). The ferroelectric hysteresis loops and the leakage current were measured with the Radiant RT66A Unit. The compositions of sintered samples were determined by The JXA-8800R electron probe microanalyzer (EPMA).

Results and discussion

Fig. 1 shows XRD patterns of BNT films annealed at 650 °C. XRD studies indicate that all of the BNT films consist of single phase of a bismuth-layered structure showing a highly (117) oriented preferential growth with a minor fraction of (00*l*) orientation. The results show that Nd-doping did not affect the crystal orientation of

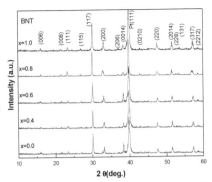

Fig.1 XRD patterns of BNT films annealed at 650 °C

the BIT film. According to the results, it is possible to think that the Bi-layered structure is always maintained in BNT and rather insensitive to the amount of Nd. The Bi-layered perovskite structure of BIT can be maintained for such high amount of Nd-doping because the radii between Bi and Nd are very close and some Bi ions are only substituted near Ti-O octahedron layers by Nd ions.

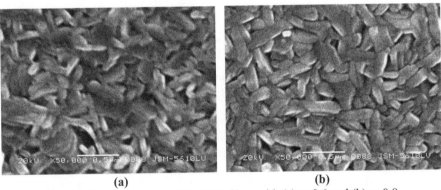

(a)   (b)

Fig.2 SEM micrographs of the BNT films with (a) x=0.6 and (b) x=0.8

The SEM images of surface morphology of BNT films with x=0.6 and 0.8 annealed at 650 °C are shown in Fig.2 which indicate that the BNT films consist of well-developed rod-like grains with random orientation. The average length and diameter of the grains of the film with x=0.8 are about 350 nm and 150 nm, respectively. It implies that the film with x=0.8 promotes bismuth titanate grain growing greater than the film with x= 0.6.

The ferroelectric hysteresis loops of the BNT (x=0.4 - 1.00) films at room temperature are shown in Fig.3. The ferroelectric hysteresis loop of the BNT film with x=0.4 is distorted due to the leakage current. However, the films with x$\geq$0.6 exhibit much improved ferroelectric hysteresis loops. It should be noted that the remanent polarization (P_r) and the coercive field (E_c) of the film with x=0.8 are 25 C/cm^2 and 55 kV/cm, respectively. The P_r value of the BNT film is about three times larger than that of the BIT, and much larger than that of $Bi_{3.25}La_{0.75}Ti_3O_{12}$ (BLT) films (16 μC/cm^2) [3]. It should be pointed out that the coercive field found for the BNT film (x=0.8) in the present work is much lower than that of $Bi_{3.5}Nd_{0.5}Ti_3O_{12}$ [10] and BLT films [3]. This lower coercive field is better for application in nonvolatile random access memory (NvRAM).

Figure 4 shows the fatigue characteristics of the BNT film with x=0.8 annealed at 650 °C. The fatigue test was done using a bipolar wave of 6 V at 50 kHz. After 3×10^{10} switching cycles, 15% degradation of P_r is observed in the film.

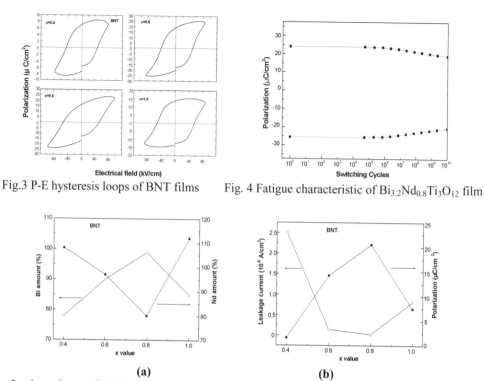

Fig.3 P-E hysteresis loops of BNT films   Fig. 4 Fatigue characteristic of $Bi_{3.2}Nd_{0.8}Ti_3O_{12}$ film

(a)   **(b)**

Fig. 5 x dependence of (a) the Bi and Nd amount, and (b) the leakage current density and the remanent polarization for BNT films

Although there are no clear explanations why such different ferroelectric properties exist in samples with different Nd amounts, the above results indicate that the electrical properties of BNT should be optimized with a proper Nd concentration. Park et al. suggested that the role of the A-site substitution is to displace the volatile Bi with La to suppress the A-site vacancies accompanied with oxygen vacancies which acted as space charge [3]. Uchida and Yoshikawa discussed the possible origin of good ferroelectric properties of V-doped BIT and pointed that the B-site substitution by high-valent cation is mainly the compensation for the defects which cause a fatigue phenomenon and strong domain pinning [7]. Watanabe et al. showed that the simultaneous substitutions for both sites are effective to derive enough ferroelectricity by accelerating the domain nucleation on passing the Curie temperature (T_C), which is lower than the deposition temperature after deposition and pinning relaxation caused by A- and B-site substitution, respectively [8]. Then, we focused on the relationship between the Nd amount and the leakage current trying to understand the reason for the difference of ferroelectric properties in BNT samples. The compositions of BNT samples were determined by EPMA. Fig. 5 (a) gives the dependence of the Bi amount and the Nd amount on the x value. The x dependence of the leakage current and the remanent polarizations at an applied field of 100 kV/cm is shown in Fig. 5 (b). Fig. 5 indicates that the leakage current of the samples decreases with the increases of Bi amount. It is noted that the amount of Bi in the BNT sample with x=0.8 is very close to the nominal composition and much higher than those of other samples, but the leakage current of the sample with x=0.8 is 1~2 order of magnitude lower than those of other samples. These results indicate that lower leakage current results from Nd preventing Bi from evaporating, and the higher leakage current is due to the evaporation of Bi during sintering. However, the Nd amount of the sample with x=0.8 is lower than those of other samples, which suggests that there might some cation vacancies at Nd sites. On the other hand, the x dependence of the Bi amount is very similar to that of P_r, demonstrating that the Bi amount is an important factor in obtaining films with large P_r. These experimental results imply that Nd ions substitution for Bi ions suppresses the generation of oxygen vacancies and reduces leakage current. In fact, for the BNT films with x=0.8, Nd doping into BIT results in a much larger P_r than that of undoped BIT and a remarkable improvement in the leakage current (2.1×10^{-7} A/cm^2 at E =100 kV/cm) properties, the loss factor (D =2.2×10^{-3}) and the dielectric permittivity (ε_s = 280) characteristics. It implies that Nd doping into BIT leads to efficient decrease in space charge density. The reason for the fatigue problem of the BNT film with x=0.8 is still uncertain, but a deficient Bi content (97.5% in Fig. 5(a)) may be a major defect to cause ferroelectric fatigue.

Conclusions

In summary, $Bi_{4-x}Nd_xTi_3O_{12}$ thin films were grown on Pt/Ti/SiO2/Si substrates by rf magnetron sputtering technique. All of the BNT films consist of single phase of a bismuth-layered structure without preferred orientation and have rod-like grains with random orientation. At an applied voltage of 100 kV/cm, the remanent polarization and the coercive field of the BET film with x=0.8 annealed at 650 °C are about 25 µC/cm^2 and 55 kV/cm, respectively. However, after 3×10^{10} switching cycles, 15% degradation of $2P_r$ is observed in the film with x=0.8. The studies on the compositions of BNT samples demonstrate that the Bi amount is an important factor in obtaining films with large P_r.

Acknowledgement

This work has supported by the Project of National Natural Science Foundation of China (grant No. 50774034),

References

[1] J. F. Scott, C. A. P. De Araujo, Ferroelectric memories. Science. 246 (1998) 1400-1404.

[2] H. N. Lee, D. Hesse, N. Zakharov, *et al*: Ferroelectric $Bi_{3.25}La_{0.75}Ti_3O_{12}$ films of uniform *a*-axis orientation on silicon substrates, Science. 296 (2002) 2006-2010.

[3] B. H. Park, B. S. Kang, S. D. Bu, *et al*: Lanthanum-substituted bismuth titanate for use in non-volatile memories, Nature. 401 (1999) 682-686.

[4] M. Chen, Z. L. Liu, Y. Wang, *et al*: Electrical characteristics and microstructures of Sm_2O_3-doped $Bi_4Ti_3O_{12}$, Chinese Physics Letters. 21(2004) 1181-1185.

[5] Y. N0guchi, M. Miyayama, Large remanent of polarization of vanadium-doped $Bi_4Ti_3O_{12}$, Appl Phys Lett. 78(2001) 1903-1907.

[6] Y. C. Chang, D. H. Kuo, The improvement in ferroelectric performance of $(Bi_{3.15}Nd_{0.85})_4Ti_3O_{12}$ films By the addition of hydrogen peroxide in a spin-coating solution, Thin Solid Films. 515(2006) 1683-1687.

[7] H. Uchida, H. Yoshikawa, I. Okada, *et al*: Approach for enhanced polarization of polycrystalline bismuth titanate films by Nd^{3+}/V^{5+} cosubstitution, Appl. Phys. Lett. 81 (2002) 2229-2233.

[8] T. Watanabe, H. Funakubo, M. Osada, *et al*: Effect of cosubstitution of La and V in $Bi_4Ti_3O_{12}$ thin films on the low-temperature deposition, Appl. Phys. Lett. 80 (2002) 2229-2233.

[9] U. Chon, K. B. Kim, H. M. Jang, *et al*: Fatigue-free samarium-modified bismuth titanate($Bi_{4-x}SmxTi_3O_{12}$)film capacitors having large spontaneous polarizations, Appl. Phys. Lett. 79 (2001) 3137-3141.

[10] T. Kojima, T. Sakai, T, Watanabe, *et al*: large remanent polarization of $(Bi,Nd)_4Ti_3O_{12}$ epitaxial thin films grown by metalorganic chemical vapor deposition, Appl. Phys. Lett. 80 (2002) 2746-2750.

[11] M. Chen, Z. L. Liu, Y. Wang, *et al*: Electrical characteristics and microstructures of Sm_2O_3-doped $Bi_4Ti_3O_{12}$, Chinese Physics Letters. 21(2004) 1181-1185.

Key Engineering Materials Vol. 537 (2013) pp 134-139
© (2013) Trans Tech Publications, Switzerland
doi:10.4028/www.scientific.net/KEM.537.134

Preparation of NASICON disk by tape casting and its CO_2 sensing properties

Jianguo Li, Xishuang Liang[a], Chengguo Yin, Fengmin Liu, Geyu Lu

State Key Laboratory on Integrated Optoelectronics Jilin University Region and College of Electronic Science and Engineering, Changchun, PR China

[a]liangxs@jlu.edu.cn

Keywords: NASICON, Tape casting, CO_2 sensor

Abstract. In this work, NASICON-type disks with the formula, $Na_3Zr_2Si_2PO_{12}$ were prepared by non-aqueous tape casting method. The effect of the dispersant on the slurry viscosity was investigated, triethanolamine was found to be an effective dispersant for NASICON slurry. The correlation between the overall conductivity and the sintering conditions (temperature and time) for the NASICON disk was also studied. Green tapes were calcined at 900°C, 1000°C, 1100°C for 6h and 12h, respectively. Results revealed that the overall conductivity increased with the increasing of the sintering temperature and decreased with the increasing of the sintering time. The segregation of resistive monoclinic ZrO_2 phase was examined to have a negative effect on the overall conductivity. The CO_2 sensor using NASICON disk and Li_2CO_3-$BaCO_3$ complex thick film was fabricated and evaluated, the sensitivity was about 82.9 mV/decade at 450°C.

Introduction

NASICON is a well known family of sodium ionic ceramic conductors with the general formula $Na_{1+x}Zr_2Si_xP_{3-x}O_{12}$ (0<x<3) [1, 2]. The highest ionic conductivity has been obtained for x=2, corresponding to a monoclinic structure [3]. NASICON was firstly suggested as solid electrolyte material for Na^+ ion-based batteries [4]. Since then the NASICON electrolyte has attracted the attention of the researchers on chemical sensors. Many NASICON-based sensors, e.g., potentiometric type CO_2 [5–9], NO_2 sensors using oxysalt auxiliary phases, amperometric type NO_2 sensor utilizing the nitrite [10–12], as well as mixed potential type sensors using oxide sensing electrode [13–17], have been investigated. In these sensors, NASICON substrates generally prepared by dry pressing method, but such preparing technique could not easily be used to prepare large and thin pellets. Contrary to the dry pressing method, tape casting process is suitable for preparing the wide and thin ceramic tapes since its low cost and precise dimensional tolerances [18, 19]. As we know tape casting has been widely used in preparing YSZ plate, researchers have established many slurry systems [20–22]. Using planar electrolyte substrate, gas sensors can be much developed in lowing power consumption, miniaturization and integrating. A few kinds of sensors can be integrated on the same platform to detect different gases. P.K. Sekhar et al. reported a combined NO_x/NH_3 sensor on the same YSZ platform [23]. D.L. West et al. reported a SO_2 sensor with one Pt and two oxide electrodes [24]. In this work, we originally used tape casting method to prepare NASICON disc, and study the correlation between the conductivity and sintering condition. In addition, the potentiometric CO_2 sensor based on NASICON prepared with tape casting process was fabricated and investigated.

Experimental

Preparation of NASICON ($Na_3Zr_2Si_2PO_{12}$) slurry. Stoichiometric amount of $ZrSiO_4$, $NaNO_3$ and $(NH_4)_2HPO_4$ were ball-milled for 12 h in ethanol, after drying at 80°C for 2 h, pre-calcined at 400°C for 1 h for removing excess ethanol and decomposing of nitrates, then sintered at 1125°C for 12 h, finally, ball-milled for another 12 h in order to get fine and homogeneous NASICON powder.

Suspensions of NASICON were prepared by mixing the NASICON powder and reagent grade solvents consisting of an azeotropic mixture of methyl ethyl ketone (MEK) and ethanol (EtOH). In order to study the effect of the dispersant (Triethanolamine, TEA) on the slurry rheology, various amounts of TEA was used in the tape casting formulations. Polyvinyl butyral (PVB) and dibutyl phthalate (DBP) were used as binder and plasticizer, respectively.

The NASICON slurry was prepared by the following two processes. At first, 40 g of NASICON powder, 50 ml of MEK, 25 ml of EtOH and different contents of TEA were mixed by ball-milling method for 12 h to break down the soft-agglomerates between particles. The viscosity of the above slurry was measured by using a viscometer (Shanghai Nirun Intelligent Technology Co.,Ltd.: SNB-1) to study the effect of the dispersant on the slurry rheology. Secondly, 8 g of PVB and 8ml DBP were added into the above slurry and ball-milled again for 24 h.

Preparation and characterization of ceramic disks. The slurry was deaired under constant stirring for 25 min for removing any entrapped air in the slurry during the ball milling. The casting was carried out by using a laboratory model tape casting unit (Beijing Orient Sun-tec CO., Ltd.: LYJ-150). The slurry was casted on tempered glass plates through double blades which were kept at a gap of 3 mm, 3.5 mm, respectively. The obtained tapes were dried at room temperature for 12 h, then at 50°C for 10 h. The green tapes were punched into small disks and sintered at 900°C, 1000°C, 1100°C for 6 h and 12 h, respectively.

In order to clarify the appropriate sintering temperature, thermogravimetric (TG)-differential scanning calorimetry (DSC) (NETZSCA: STA449 F3) measurements were carried out for the green tape of which the ratio between powder and TEA was 3g : 1 ml. Phase composition of both calcined powders and sintered disks was verified by X-ray diffraction (Rigaku wide-angle X-ray diffractometer (D/max rA, using Cu Kα radiation at wavelength λ=0.1541 nm)). Microstructure of the surface for the disk obtained at different sintering temperatures were observed by scanning electron microscopy (SEM) (FE-SEM; JEOL, JSM-7500, Japan). The density was estimated from the disks' weight and geometry. Au electrodes were formed on the polished surfaces of the disks by applying Au paste and sinering at 800°C for 20 min. Impedance measurements were performed with a Solartron SI 1287 electrochemical interface and an SI 1260 impedance/gain-phase analyzer using Zplot software from 10 to 10^6 Hz at temperatures ranging from 200°C to 350°C.

Fabrication and measurement of the electrochemical CO_2 sensor. A planar CO_2 electrochemical sensor was fabricated using a small NASICON disk and Li_2CO_3-$BaCO_3$ complex thick film. A couple of ring-shaped Au electrodes were made on the two sides of NASICON disc. Li_2CO_3-$BaCO_3$ (1:2 in molar ratio) was deposited on one side (auxiliary electrode) by melting and recrystallization, a Pt heater was attached to the other side with inorganic adhesive. This process had been described in our previous papers [25].

Gas sensing properties of the sensor were measured by a conventional static mounting method. The sample gases containing different concentrations of CO_2 were obtained by diluting pure CO_2 with air. When the CO_2 sensor was exposed to air or a sample gas at 450°C, the electromotive force (EMF) was measured with a digital multimeter (RIGOL TECHNOLOGIES, INC, DM3054, China).

Results and discussion

Effect of dispersant on slurry rheology. The role of a dispersant is to keep the primary particles separating from each other thereby preventing their agglomeration due to van der Waals attraction energy [22]. The optimum amount of dispersant is determined by searching for the lowest viscosity which gives the best degree of deagglomeration and the highest stability of particle suspension [20]. The dependence of the viscosity values on the dispersant amount is shown in Fig. 1. It can be observed that the viscosity of the slurry shows a minimum value at 5 mL / 15 g powder. It is evident that Triethanolamine is an effective dispersant for NASICON slurry.

Sintering condition of NASICON disk. In order to clarify the appropriate sintering temperature and temperature gradient, TG-DSC measurements were carried out for the green tapes. As shown in Fig. 2, mass losses of the sample were observed at the range from 100°C to 500°C, and the mass became almost constant above 500°C. The exothermal peaks in the DSC curve at 200–500°C and 630°C are ascribed to the decomposition of additive organics in the green tape and the crystallization of the NASICON phase, respectively. In order to avoid the warpage of the disk at high temperature the speed of heating up for the green tape was kept at 0.5°C/min for 100–500°C and 1°C/min for 500–1100°C.

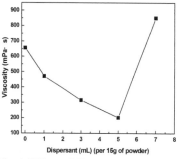

Fig. 1 Effect of dispersant on viscosity of NASICON slurry.

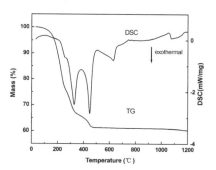

Fig. 2 The TG-DSC curves for green tape.

XRD analysis was used to measure the grain size and phase compositions of NASICON disc. As shown in Fig. 3, traces of monoclinic zirconia and monoclinic NASICON were identified in all cases, and a tetragonal $ZrSiO_4$ phase was also confirmed. Both ZrO_2 and NASICON peaks become larger with the increasing sintering temperature. The increasing of the sintering time leads to the increase of ZrO_2 peaks and the decrease of NASICON peaks. This indicated that ZrO_2 phase is being performed with consumption of NASICON. The crystallite size D of as-prepared NASICON was calculated according to the Debye-Scherrer's equation. As we can see in Table 1, the crystallite size D becomes larger with the increasing sintering temperature and time.

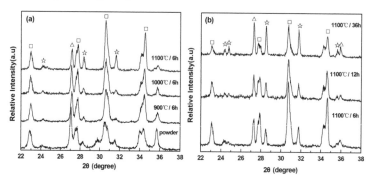

Fig. 3 XRD patterns of NASICON disks. Monoclinic zirconia (☆), zirconium silicate (Δ) and NASICON(□) are present in all cases.

Fig. 4 shows SEM images of disks sintered at different temperatures and times. From Fig. 4, it can be seen that the density of NASICON disk increases with sintering temperature and time, and measurement results were shown in Table 1. In the case of 1100°C, the density of the disk sintered for 6h, 12h and 36h is 94%, 97% and 98% of theoretical density, respectively. By comparison, it can be seen that the effect of the sintering temperature on the density is much larger than the sintering time. The presence of liquid phases could be observed when the disk was sintered at 1000°C, which became more obvious with increasing the sintering temperature. As suggested by some authors [1, 26], zirconia could be dispersed along the grain boundaries in a liquid phase. Densification and grain growth are probably associated to a liquid phase produced in sintering process [27].

Electrical properties of the disks sintered under different conditions were measured using impendence spectroscopy. Arrhenius plots of different disks are shown in Fig. 5. Generally, conductivity of NASICON materials obeys the Arrhenius law, which can be expressed as follows:

$$\sigma T = \sigma_0 \exp(-Ea / k_BT). \tag{1}$$

Here, σ is the conductivity, σ_0 is the pre-exponential factor, T is the temperature in K, Ea is the conductivity activation energy and k_B is the Boltzmann's constant. The activation energy Ea for the conduction process was extracted from the slop of the straight line plot of log (σT) against reciprocal temperature 1000/T. The magnitudes of σ and Ea for different disks are listed in Table 1. As shown in Fig. 5, the conductivity of NASICON disk increases with increasing the sintering temperature and decreases with increasing the sintering time. The maximum value of conductivity was achieved for the sample sintered at 1100°C for 12h, which was about 3.596×10^{-2} S·cm^{-1} at 250°C.

Table 1 Grain size, density, conductivity at 250°C and active energy for different disks.

Sintering Condition	Properties of NASICON Disks			
	Grain size (nm)	Density(%)	$\sigma_{250°C}$ ($10^{-3}S·cm^{-1}$)	Ea(eV)
900°C / 6h	18.9	52	7.26	0.18
900°C / 12h	21.0	56	5.97	0.18
1000°C / 6h	23.4	64	16.37	0.16
1000°C / 12h	25.6	70	11.54	0.16
1100°C / 6h	30.5	93	20.07	0.08
1100°C / 12h	37.3	97	35.96	0.08
1100°C / 36h	39.0	98	30.28	0.08

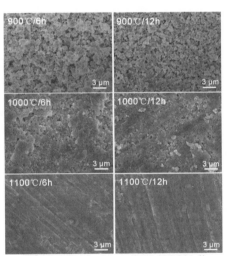

Fig. 4 SEM images of NASICON discs sintered under different conditions.

As reported, zirconia as an impure phase leads to the Zr-deficient in as-prepared NASICON, and results in the less conductivity [26]. The increase in conductivity when increasing sintering temperature could be ascribed to decrease in grain boundary density, resulting from both increasing sample density and grain size. The negative effect of zirconia on the conductivity is not so strong that can be taken into account. For the disk sintered at the same temperature, the density doesn't change much when increasing sintering time. However, increasing sintering time results in the severe segregation of a resistive phase, leading to decrease in conductivity. These results have been verified by previous XRD and density analysis.

Sensing properties of the CO_2 electrochemical sensor. The planar CO_2 electrochemical sensor combining the NASICON disks sintered at 1100°C for 12h with Li_2CO_3-$BaCO_3$ was fabricated. Fig. 6 displays the correlation between the EMF responses of the sensor and the concentration of CO_2 at 450°C. It can be observed that the EMF response is almost proportional to the logarithm of CO_2 concentration. The slope of the sensor is about 82.9 mV/decade, and the number of the reaction electronics is about 2.08 at 450°C. These results indicate that the sensing behavior of the present planar device follows the Nernst law.

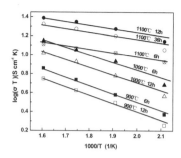

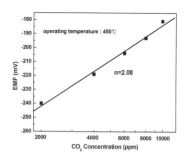

Fig. 5 Conductivity of NASICON disks as a function of inverse of temperature.

Fig. 6 CO_2 concentration vs. EMF for the sensor at 450°C.

Conclusion

Tape casting is an effective way to prepare NASICON disk with high ionic conductivity. The dispersant (TEA) plays an important role in the preparation of ceramic tape, whose content has a great effect on the viscosity of NASICON slurry. The total electrical conductivity strongly depends on density and the quantity of a resistive liquid phase. The total electrical conductivity increases with increasing the sintering temperature and decreases with increasing the sintering time. The tapes sintered at 1100°C for 12h showed highest conductivity, which was about 3.596×10^{-2} S·cm^{-1} at 250°C. The potentiometric CO_2 electrochemical sensor using NASICON disk and Li_2CO_3-$BaCO_3$ complex thick film was found to exhibit good sensing performance at 450°C.

Acknowledgement

Supported by NSFC (Nos. 61074172, 61134010, 61104203) and Program for Chang jiang Scholars and Innovative Research Team in University (No. IRT1017), and Jilin province science and technology development plan program (20106002) is gratefully acknowledged.

References

[1] H.Y.P. Hong, Crystal structures and crystal chemistry in the system $Na_{1+x}Zr_2Si_xP_{3-x}O_{12}$, Mater. Res. Bull. 11 (1976) 173-182.

[2] J.B. Goodenough, H.Y.P. Hong, J.A. Kafalas, Fast Na$^+$-ion transport in skeleton structures, Mater. Res. Bull. 11 (1976) 203-220.

[3] R.S. Gordon, G.R. Miller, E.D. Beck, J.R. Rasmunssen, Fabrication and characterization of Nasicon electrolytes, Solid State Ionics, 3/4 (1981) 243-248.

[4] P. Fabrya, J.P. Grosa, J.F. Million-Brodaza, M. Kleitza, Nasicon, Nasicon, an ionic conductor for solid-state Na$^+$-selective electrode, Sensors and Actuators 15 (1) (1988) 33-49.

[5] N. Miura, S. Yao, Y. Shimizu, N. Yamazoe, Carbon dioxide sensor using sodium ion conductor and binary carbonate auxiliary electrode, J. Electrochem. Soc. 139 (5) (1992) 1384-1388.

[6] S. Yao, Y. Shimizu, N. Miura, N. Yamazoe, Solid Electrolyte CO_2 Sensor Using Binary Carbonate Electrode, Chem. Lett. 1990 (1990) 2033-2036.

[7] S. Yao, Y. Shimizu, N. Miura, N. Yamazoe, Solid electrolyte carbon dioxide sensor using sodium ionic conductor and lithium carbonate-based auxiliary phase, Appl. Phys. A 57 (1993) 25-29.

[8] T. Kida, Y. Miyachi, K. Shimanoe, N. Yamazoe, NASICON thick film-based CO_2 sensor prepared by a sol-gel method, Sensors and Actuators B 80 (2001) 28-32.

[9] H. Dang, X. Guo, Investigation of porous counter electrode for the CO_2 sensing properties of NASICON based gas sensor, Solid State Ionics 201 (2011) 68-72.

[10] N. Miura, M. Iio, G. Lu, N. Yamazoe, Sodium ion conductor based sensor attached with $NaNO_2$ for amperometric detection of NO_2, J. Electrochem. Soc. 143 (1996) L241-L243.

[11] N. Miura, M. Ono, K. Shimanoe, N. Yamazoe, A compact solid-state amperometric sensor for detection of NO_2 in ppb range, Sensors and Actuators B 49 (1998) 101-109.

[12] N. Miura, M. Iio, G. Lu, N. Yamazoe, Solid-state amperometric NO_2 sensor using a sodium ion conductor, Sensors and Actuators B 35 (1996) 124-129.

[13] Y. Shimizu, H. Nishi, H. Suzuki, K. Maeda, Solid-state NO_x sensor combined with NASICON and Pb-Ru-based pyrochlore-type oxide electrode, Sensors and Actuators B 65 (2000) 141-143.

[14] X. Liang, Y. He et al., Solid-state potentiometric H_2S sensor combining NASICON with Pr_6O_{11}-doped SnO_2 electrode, Sensors and Actuators B 125 (2007) 544-549.

[15] X. Liang, T. Zhong, B. Quan, B. Wang, H. Guan, Solid-state potentiometric SO_2 sensor combining NASICON with V_2O_5-doped TiO_2 electrode, Sensors and Actuators B 134 (2008) 25-30.

[16] X. Liang, T. Zhong et al., Ammonia sensor based on NASICON and Cr_2O_3 electrode, Sensors and Actuators B 136 (2009) 479-483.

[17] X. Liang, G. Lu, Ti. Zhong, F. Liu, B. Quan, New type of ammonia/toluene sensor combining NASICON with a couple of oxide electrodes, Sensors and Actuators B 150 (2010) 355-359.

[18] Y. Zeng, D.L. Jiang, P. Greil, Tape casting of aqueous Al_2O_3 slurries, J. Eur. Ceram. Soc. 20 (2000) 1691-1697.

[19] K. Zhu, H. Wang, J. Qiu, J. Luo and H. Ji, Fabrication of $0.655Pb(Mg_{1/3}Nb_{2/3})O_3$-$0.345PbTiO_3$ functionally graded piezoelectric actuator by tape-casting, J. Electroceram. 27 (2011) 197-202.

[20] X.G. Capdevila, J. Folch, A. Calleja, J. Llorens, M. Segarra, F. Espiell, J.R. Morante, High-density YSZ tapes fabricated via the multi-folding lamination process, Ceramics International 35 (2009) 1219-1226.

[21] A.K. Maiti, B. Rajender, Terpineol as dispersant for tape casting yttria stabilized zirconia powder, Mater. Sci. Eng., A 333 (2002) 35-40.

[22] A. Mukherjee, B. Maiti, A. Das Sharma, R.N. Basu, H.S. Maiti, Correlation between slurry rheology, green density and sintered density of tape cast yttria stabilised zirconia, Ceramics International 27 (2001) 731-739.

[23] P.K. Sekhar et al., Application of commercial automotive sensor manufacturing methods for NO_x/NH_3 mixed potential sensors for on-board emissions control, Sensors and Actuators B, 144 (2010) 112-119.

[24] D.L. West, F.C. Montgomery, T.R. Armstrong, A technique for monitoring SO_2 in combustion exhausts: Use of a non-Nernstian sensing element in combination with an upstream catalytic filter, Sensors and Actuators B, 140 (2009) 482-489.

[25] X. Liang, S. Yang et al., Mixed-potential-type zirconia-based NO_2 sensor with high-performance three-phase boundary, Sensors and Actuators B, 158 (2011) 1-8.

[26] O. Bohnke, S. Ronchetti, D. Mazza, Conductivity measurements on nasicon and nasicon-modified materials, Solid State Ionics, 122 (1999) 127-136.

[27] R.O. Fuentes, F.M. Figueiredo, F.M.B. Marques, J.I. Franco, Influence of microstructure on the electrical properties of NASICON materials, Solid State Ionics 140 (2001) 173-179.

Key Engineering Materials Vol. 537 (2013) pp 140-143
© (2013) Trans Tech Publications, Switzerland
doi:10.4028/www.scientific.net/KEM.537.140

Structure and Optical Properties of Al$_{1-x}$Sc$_x$N Thin Films

Jing Yang, Miaomiao Cao ,Yudong Li and Yigang Chen*

Department of Electronic Information Materials, School of Materials Science and Engineering, Shanghai University, 149 Yanchang Road, Shanghai 200072, China

*yigangchen@shu.edu.cn

Keywords: Thin film; Scandium; Aluminum Nitride; Structure; Optical property.

Abstract. In this study, c-axis oriented AlN and Al$_{1-x}$Sc$_x$N films have been successfully grown on Si (100) and quartz glass by DC magnetron reactive sputtering method. The XRD patterns show that the crystal structure of the Al$_{1-x}$Sc$_x$N films is (002) orientation. The grain size and band gap energy (Eg) of the Al$_{1-x}$Sc$_x$N films decrease as the Sc concentration increases. The frequency of the E2 (high) mode observed in the Al$_{1-x}$Sc$_x$N films shows higher red shift compared to that observed in AlN film and the peak shifts to the low wave number with the increasing of Sc concentration.

Introduction

Among the semiconductor hosts, aluminum nitride (AlN), which possesses a large energy band gap of 6.2eV [1-3], has been attracting lots of interest in applications. AlN with a wurtzite crystalline structure is a superior material due to its chemical resistance, high acoustic velocity (6000m/s), thermal stability, and relatively high piezoelectric coefficient d31 (−2.65 pm/V) and d33 (5.5 pm/V) [4]. In recent years, rare-earth (RE)-doped IIIA-nitride semiconductor thin films are attracting increasing attention for use in some fields [5-9], for example, scandium doping in AlN. Scandium is a RE element no.21 in the periodic system. The first principle calculation indicated that it is possible to fabricate Sc-IIIA-N nitrides [10, 11]. Akiyama [4] has reported that Al$_{1-x}$Sc$_x$N films with a Sc concentration of 43% exhibited a four times larger piezoelectric response than pure AlN films, which is the largest piezoelectric response among the known tetrahedrally bounded semiconductors. However, Sc–Al–N is still an unexplored material system. To get a deep understanding of the nature how the scandium element affects in Al$_{1-x}$Sc$_x$N films, we studied the structure and optical properties of the films. In this work, c-axis oriented AlN thin films had been successfully grown on Si(100) substrates by DC magnetron reactive sputtering method. By Sc-doping in AlN thin films, Al$_{1-x}$Sc$_x$N alloy phase is formed. The structure and optical properties are being characterized carefully by x-ray diffraction (XRD), UV-vis spectroscopy and Raman spectroscopy, respectively.

Experiment

Al$_{1-x}$Sc$_x$N thin films were prepared on Si(100) substrates by DC magnetron reactive sputtering method. The growth conditions of the films are listed in Table 1. The aluminum sputtering targets were 60mm in diameter and 99.999% of purity. In this work, we grew Al$_{1-x}$Sc$_x$N thin films with different Sc concentrations by changing the number of Sc tips which were set on the Al target. The sputtering chamber was evacuated to a pressure below 3.5×10-4 Pa and then high-purity argon (99.999%) and nitrogen (99.999%) gases were introduced. The growth pressure was 0.7 Pa.

Results and discussion

The crystal structure and orientation of the films were determined by XRD. The film thickness of Al$_{1-x}$Sc$_x$N films is approximately about 700nm. The Sc concentration was measured by using energy dispersive X-ray spectroscopy (EDS). The XRD patterns of Al$_{1-x}$Sc$_x$N films are shown in Fig. 1.

Only a peak appears, corresponding to the (002) orientations of AlN phase. When the films are doped with Sc, the diffraction intensity of (002) peak remarkably decreases and the peak shifts to the low angle side. Generally, the AlN has the wurtzite structure while the ScN has the rock-salt

Table 1 Sputtering conditions of $Al_{1-x}Sc_xN$

Material	$Al_{1-x}Sc_xN$
Target	Al
Substrate	Si (100)
Substrate temperature (°C)	Room temperature
Power (W)	280
Gas contents	Ar:N_2=1:1
Sputtering time (H)	1
Sputtering pressure (Pa)	0.7

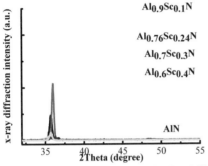

Fig. 1 X-ray diffraction spectra $Al_{1-x}Sc_xN$ films
with different Sc mole fractions

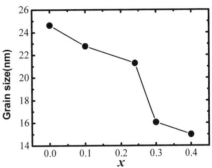

Fig. 2 The grain size of $Al_{1-x}Sc_xN$ calculated
using Scherrer's equation

(nonpolar) [12-14]. When Sc was doped into AlN films, the Sc atoms were substituted on the Al sub-lattice. It is probably that the crystal structure of $Al_{1-x}Sc_xN$ film is a hexagonal intermediate phase between wurtzite AlN and rocksalt ScN. Fig. 1 shows that with the increasing of x from 0.1 to 0.24, the diffraction intensity of (002) peak increases. While the diffraction intensity decreases with further increasing x from 0.3 to 0.4. These suggest that the crystal orientation of the $Al_{1-x}Sc_xN$ becomes lower when the Sc concentration is between 0.3 and 0.4.

The crystal size was calculated by using the Scherrer formula [15] as the following:

$$D = k\lambda / B\cos\theta \qquad (1)$$

k = 0.9; λ is the wavelength of X-ray used (0.154056 nm) and B is the FWHM of the peak.

The grain size of non-doped AlN film was calculated to be about 24.6 nm. Fig. 2 shows the relationship between x and the grain size of $Al_{1-x}Sc_xN$ films. It can be seen that the grain size of $Al_{1-x}Sc_xN$ is in the range of 14-23 nm, which is smaller than that of non-doped AlN. And with the Sc concentration increases, the grain size of $Al_{1-x}Sc_xN$ films decreases. It is reasonable to consider that Sc may cause the Sc–N phases forming and enwrapping the AlN grains, which restrict the growth of AlN grains.

The optical transmittance was obtained by UV-visible spectrometry in the range of 200-800 nm. Fig. 3 depicts the optical transmittance spectra of $Al_{1-x}Sc_xN$ thin films. The transmittance of $Al_{1-x}Sc_xN$ thin films significantly reduces compared to that of non-doped AlN films. The optical band gap, Eg, can be determined by the following formula:

$$\alpha h\upsilon = A(h\upsilon - E_g)^{1/2} \qquad (2)$$

α is a constant, Eg is the band gap energy and hυ is the photon energy.

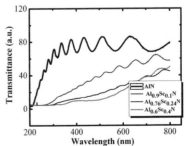

Fig. 3 Optical transmittance spectra
of $Al_{1-x}Sc_xN$ thin films

A plot of $(\alpha h\upsilon)^2$ versus $h\upsilon$ for the $Al_{1-x}Sc_xN$ films is presented in Fig.4, which shows that Eg decreases as Sc concentration increasing.

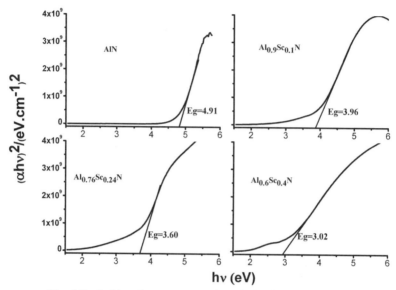

Fig. 4 Optical band gap energy values (Eg) of $Al_{1-x}Sc_xN$ films

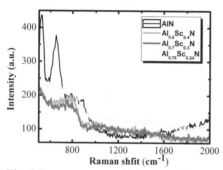

Fig. 5 Raman spectra of $Al_{1-x}Sc_xN$ film on Si(100)

The films were also investigated by using Raman spectroscopy, as shown in Fig. 5. The Raman peaks of $Al_{1-x}Sc_xN$ films at 750-800 cm-1 can be observed but they are very weak. Compared to AlN E2 (high) Raman peak, the Raman peak of $Al_{1-x}Sc_xN$ films is higher red shift. With the reference to the results of our XRD and UV-vis spectrometry, the observed peak shifting can be attributed to the change of the crystal structure.

It was reported that a red shift was observed in Raman spectra with the decrease of grain size [16]. In our cases, when AlN films are doped with Sc, the grain size and Eg decrease. These are consistent with the results of XRD and UV-vis spectrometry. The E2 (high) mode is usually being used to analyze the stress state information in the sample due to its high sensitivity to stress which affects the E2 phonon frequency [17]. An increase in the E2 phonon frequency indicates compressive stress, whereas a decrease points to tensile stress. In Fig. 5, the Raman peak of $Al_{1-x}Sc_xN$ films shift to low wave number was also found with the increasing of Sc concentration, which indicates the compressive stress in $Al_{1-x}Sc_xN$ films.

Summary

In this study, $Al_{1-x}Sc_xN$ thin films were prepared on both Si (100) and quartz glass substrates by DC magnetron reactive sputtering method. All the films show (002) preferred orientation. Through the XRD pattern, it is found that the (002) peak of $Al_{1-x}Sc_xN$ films shifts to the low angle side which is due to its crystal structure is a hexagonal intermediate phase between wurtzite AlN and rocksalt ScN. The transmittance of $Al_{1-x}Sc_xN$ thin films significantly reduces compared to that of non-doped AlN film. The grain size of $Al_{1-x}Sc_xN$ is in the range of 14-23 nm, which is remarkably smaller than that of non-doped AlN. Thus the frequency of the E2 (high) mode observed in the $Al_{1-x}Sc_xN$ film shows higher red shift compared to that observed in non-doped AlN film. And the peak shifts to low wave number with the increasing of Sc concentration which is closely related to the increasing of compressive stress in $Al_{1-x}Sc_xN$ films.

Acknowledgments

The authors thank for the financial support of Shanghai Eastern-scholar program and Shanghai Pujiang Program, China.

References

[1] J. Li, K. B. Nam, M. L. Nakarmi, J. Y. Lin and H. X. Jiang, Band-edge photoluminescence of AlN epilayers, Appl. Phys. Lett. 81 (2002) 3365.

[2] T. Onuma, S. F. Chichibu, T. Sota, et al., Exciton spectra of an AlN epitaxial film on (0001) sapphire substrate grown by low-pressure metalorganic vapor phase epitaxy, Appl. Phys. Lett. 81 (2002) 652.

[3] E. Kuokstis, J. Zhang, Q. Fareed, et al., Near-band-edge photoluminescence of wurtzite-type AlN, Appl. Phys. Lett. 81 (2002) 2755.

[4] Kamohara, Akiyama and Kuwano, Influence of polar distribution on piezoelectric response of aluminum nitride thin films, Appl. Phys. Lett. 92 (2008) 093506.

[5] A. J. Steckl and R. Birkhahn, Visible emission from Er-doped GaN grown by solid source molecular beam epitaxy, Appl. Phys. Lett. 73 (1998) 1700.

[6] A. J. Steckl, J. C. Heikenfeld, D. S. Lee, et al., Rare-Earth-Doped GaN: Growth, Properties, and Fabrication of Electroluminescent Devices, IEEE J. Sel. Top. Quantum Electron. 8 (2002) 749.

[7] R. Birkhahn, M. Garter and A. J. Steckl, Red light emission by photoluminescence and electroluminescence from Pr-doped GaN on Si substrates, Appl. Phys. Lett. 74 (1999) 2161.

[8] N. Teraguchi, A. Suzuki, Y. Nanishi, et al., Room-temperature observation of ferromagnetism in diluted magnetic semiconductor GaGdN grown by RF-molecular beam epitaxy, Solid State Commun. 122 (2002) 651-653.

[9] K. Lorenz, U. Wahl, E. Alves, et al., High-temperature annealing and optical activation of Eu-implanted GaN, Appl. Phys. Lett. 85 (2004) 2712.

[10] N. Takeuchi, First-principles calculations of the ground-state properties and stability of ScN, Phys. Rev. B 65 (2002) 045204.

[11] L. Mancera, J. A. Rodriguez and N. Takeuchi, Theoretical study of the stability of wurtzite, zinc-blende, NaCl and CsCl phases in group IIIB and IIIA nitrides, Phys. Sta. Sol. 241 (2004) 24242428.

[12] M. Akiyama, T. Kamohara, K. Kano, et al., Enhancement of piezoelectric response in scandium aluminum nitride alloy thin films prepared by dual reactive cosputtering , Adv. Mater. 21 (2009) 593-596.

[13] Y. Oussaifi, A. B. Fredj, M. Debbichi, N. Bouarissa and M. Said, Elastic properties and optical phonon frequencies of zinc-blende $Sc_xGa_{1-x}N$, Semicond. Sci. Technol. 22 (2007) 641.

[14] T. R. Paudel and W. R. L. Lambrecht, Calculated phonon band structure and density of states and interpretation of the Raman spectrum in rocksalt ScN, Phys. Rev. B 79 (2009) 085205.

[15] A. L. Patterson, The Scherrer Formula for I-Ray Particle Size Determination, Phys. Rev. 56 (1939) 978.

[16] Q. Zhao, H. Zhang, X. Xu, Z. Wang, et al., Optical properties of highly ordered AlN nanowire arrays grown on sapphire substrate, Appl. Phys. Lett. 86 (2005) 193101.

[17] M. Kuball, Raman spectroscopy of GaN, AlGaN and AlN for process and growth monitoring/control, Surf. Interface Anal. 31 (2001) 987–999.

Key Engineering Materials Vol. 537 (2013) pp 144-149
© (2013) Trans Tech Publications, Switzerland
doi:10.4028/www.scientific.net/KEM.537.144

Morphological and Optical Properties of CdS Nano-crystalline Thin Films by Chemical Bath Deposition

Liu Yi[1, a], Wei Aixiang[2,b] and Lin Xiaodong[1,c]

[1]College of Physics Science &Technology, Shenzhen University, Shenzhen, P. R. China

[2]Faculty of Materials and Energy, Guangdong University of Technology, Guangzhou, P. R. China

[a]liuy@szu.edu.cn, [b]weiax@ gdut.edu.cn, [c]linxd@szu.edu.cn

Keywords: CdS thin film, Chemical Bath Deposition, Optical constants, Spectroscopic ellipsometry

Abstract. CdS thin films were prepared by chemical bath deposition (CBD) at different ammonia concentration and different temperature using cadmium chloride hydrate, thiourea, ammonium chloride, ammonia and deionized water as precursors. The morphology structure and the optical properties of CdS thin films were characterized by scanning electron microscope (SEM), ultraviolet-visible spectra and spectroscopic ellipsometry. The results indicated that CdS thin film could be grown when ammonia concentration at the range of 0.4-1.0 mol /L and in the temperature range of 50° ~ 80 °C. Within 600-980 nm wavelength, the average value of the refractive index was found to be 1.70 and the extinction coefficient was less than 0.08.

Introduction

CdS is a technologically important material, especially for photovoltaic solar cells, sensory and optoelectronics. CdS thin films are regarded as one of the most promising materials for hetero-junction thin film solar cells. With its n-type semiconductor characteristic [1] and wide band gap (Eg = 2.44 eV), CdS has been used as the window material together with several semiconductors such as CdTe [2], InP [3] and CuInSe$_2$ [4] with 14–16% efficiency [5]. The deposition of CdS films has been explored by different techniques such as thermal evaporation [6], sputtering [7], chemical vapor deposition [8], molecular beam epitaxy [9] and chemical bath deposition [10]. Among them, chemical bath deposition (CBD) is one of best suited for thin film deposition because of simplicity, convenience, least expenses to produce uniform, adherent and reproducible large area thin films for solar related applications.

The CBD process, adapted for the preparation of CdS window layers in high efficiency solar cells, includes a cadmium salt, a complexing agent, and thiourea as sulphur source. These precursors are mixed in an alkaline aqueous solution. Deposition of CdS is based on the slow release of Cd^{2+} ions and S^{2+} ions in an aqueous bath and the subsequent condensation of these ions on substrates suitably mounted in the bath. The slow release of Cd^{2+} ions is achieved by adding a complexing agent (ligand) to the Cd salt to form some cadmium complex species, which, upon dissociation, results in the release of small concentrations of Cd^{2+} ions. The S^{2+} ions are supplied by the decomposition of thiourea. It has been shown that under a critical concentration ratio of species in solution, nano-crystalline CdS can be deposited [11]. The influence of deposition conditions, such as the solution concentration, the solution ph value and the reaction temperature, upon the films property is of great importance to improve the power conversion efficiency of solar cells. In this paper, the morphology structure and the optical properties of CdS thin films were characterized by scanning electron microscopy (SEM), ultraviolet-visible spectra and spectroscopic ellipsometry.

Experimental

In this paper, cadmium chloride hydrate, thiourea, ammonium chloride, and ammonia were used as precursors to prepare solutions of certain concentration. All the solutions were prepared in deionized water using analytical reagents. Then the solutions were mixed in different proportioning to obtain different ammonia concentration of 0.1mol/L, 0.4mol/L, 0.7mol/L and 1.0mol/L respectively. The total volume of reaction solutions was kept to 100ml by adding deionized water. CBD-CdS thin films were deposited by vertically immersing the glass substrates in an aqueous bath at certain temperature in the range of 50°C -80°C for 60 min. The concentration of the chemicals used in the bath for the preparation of the CdS thin films is shown in Table 1. The CdS films were ultrasonically cleaned in deionized water for 1 min to remove any loosely adhered particles and finally dried in air.

The film samples were evaluated in terms of morphological and optical properties. Morphology of the films was identified by a scanning electron microscope (SEM) (S-3400N). The optical transmission of the samples were obtained using a UV–vis spectrophotometer (T6, Persee) in the wavelength range of 300-800 nm with a bare glass substrate in the reference beam to eliminate the substrate contribution. And the ellipsometry spectra were obtained using a spectroscopic ellipsometer (M-2000U, J. A. Woollam) in the wavelength range of 280-980 nm with an incident angle of 70^0.

Table 1 The concentration of the chemicals used in the bath for the preparation of the CdS thin films

Ammonia concentration	$CdCl_2$ 0.1mol/L (ml)	NH_4Cl 1mol/L (ml)	Thiourea 0.5mol/L (ml)	Ammonia 25% (ml)
0.1 mol/L	5	3	5	1
0.4 mol/L	5	3	5	3
0.7 mol/L	5	3	5	5
1.0 mol/L	5	3	5	7

Note: In all cases after the mixing, deionized water was added to reach a total volume of 100ml.

Results and discussions

CBD of high quality CdS thin films is based on the control precipitation of CdS in bath. The precipitation control can be achieved by controlling the concentration of free cadmium ions. To control the free cadmium ions concentration, generally ammonia is used as a complexing agent. The cadmium salt produces free cadmium ions (Cd^{2+}) through a dissociation reaction. The cadmium ions then complex with ammonia to form the dominant complexion $Cd(NH_3)_4^{2+}$. On the surface, the $Cd(NH3)_4^{2+}$ ions react with hydroxide ions, then reacts with thiourea. Finally, CdS is formed and a new surface site is regenerated through decomposition of the adsorbed metastable complex. The overall reaction is given as follows [12]:

$$Cd(NH_3)_4^{2+} + SC(NH_2)_2 + 2OH^- \rightarrow CdS + CH_2N_2 + 2H_2O + 4NH_3 \tag{1}$$

Fig. 1 shows the SEM images for CdS thin films deposited in various ammonia concentrations on a glass substrate in an aqueous bath at 60°C for 60 min. When the ammonia concentration was 0.1mol/L, the film exhibited disordered sheet but not nano-crystalline. As the ammonia concentration increased above 0.4mol/L, nano- crystalline particle began to form on the substrate, and the grain size of the CdS films decreased. Unlike other deposition methods, the growth rate in the CBD technique cannot be kept constant during film deposition. The growth rate decays with time as the concentrations of reactants in solution decrease. The CdS film cannot grow without ammonia. The complexation of the cadmium ions was weak when the ammonia concentration was low and the concentration of the cadmium ions in the solution was relatively high, which lead to flocculating precipitate. As the ammonia concentration increased to 0.4mol/L, the CdS grew faster. Liu et. al. [13] studied the CdS growth rate dependence on ammonia concentration, and found that the growth rate was maxim when the ammonia concentration was 0.6mol/L.The redundant ammonia stabled the complexing ions, thus slowed down the growth of CdS.

Fig. 2 shows the SEM images of CdS thin films deposited in an ammonia concentration of 0.4mol/L on glass substrates in various aqueous bath temperatures for 60 min. As can be seen from the images, both films were nano-crystalline and the grain size exhibited a small difference. It can be founded that there were many pinholes on the CdS film deposited at 50°C. The increase of the bath temperature is an effective method to diminish the pinhole on the CdS films. As the bath temperature increased, the surface morphology of the CdS became more homogeneous. The grain size of CdS film increased with increasing aqueous bath temperature and then decreased when the bath temperature was larger than 60°C. During CBD deposition, the growth rate increases with increase in temperature [14]. At the beginning of deposition, the glasses are isotropic body and random nucleation occurs. With the growth of the films, CdS is inclined to orientational deposition along the former grain for the slow growth rate [15]. According to the mechanism described by Kostoglou et al. [16], CdS films may grow by the decomposition of a metastable complex, $Cd(OH)_2(NH_2)_2SC(NH_2)_2$, going with a release of NH_3. The absorbed NH_3 on the substrate would hinder the growth of CdS and cause the island growth or the hole on the films.

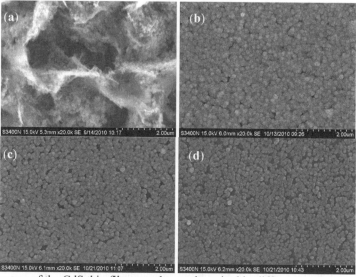

Fig. 1 SEM images of the CdS thin films on glasses deposited in different ammonia concentrations
(a) 0.1 M; (b) 0.4 M; (c) 0.7 M; (d) 1.0 M

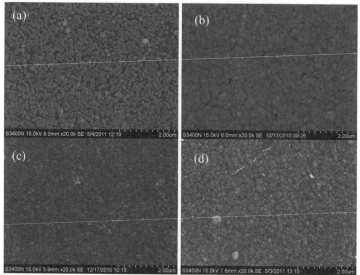

Fig. 2 SEM images of the CdS thin films on glasses deposited in different aqueous bath temperatures
(a) 50°C; (b) 60°C; (c) 70°C; (d) 80°C

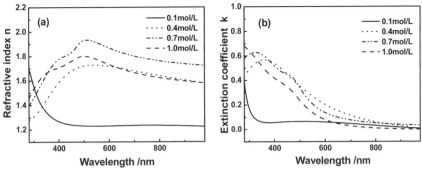

Fig. 3 Optical constants of the CdS thin films deposited on glasses at 60°C in different ammonia concentrations for 60 min. (a)Refractive indices (b) Extinction coefficients

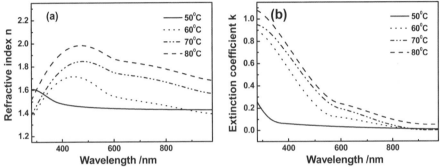

Fig. 4 Optical constants of the CdS thin films deposited on glasses in an ammonia concentration of 0.4mol/L in different aqueous bath temperatures for 60min. (a) Refractive indices. (b) Extinction coefficients.

The optical properties of the samples were investigated using spectroscopic ellipsometry and Uv-vis spectroscopy. A three-medium model of air/CdS/SLG was applied in the fitting procedure with the CdS films modeled using the empirical Cauchy model or the Tauc–Lorentz (TL) model described in Ref. [17]. The TL model is a combination of the classical Lorentz oscillator model and the Tauc model which includes parabolic bands and a constant momentum matrix element versus energy, and has been successfully applied for amorphous semiconductors [18]. Fig. 3 shows the optical constants of the CdS films deposited in different ammonia concentrations obtained by SE. The film thickness was 110nm, 101nm, 108nm and 86nm respectively.

The optical constants of the film deposited in the ammonia concentration of 0.1mol/L exhibited great difference with other samples because of the disordered sheet structure, thus the refractive indices and the extinction coefficients both were quite low. As the ammonia concentration increased above 0.4mol/L, the refractive indices increased indicating the density of the film increased. At absorption edge, refractive index n decreases sharply for Kramers-Kronig consistency. The extinction coefficient k values were low between 600nm and 980nm and increased through short wavelengths below 600 nm. The extinction coefficient of a material is directly related to its absorption characteristic. So, at long wavelengths films would have high transmittance, low absorption and low k values, or vice versa for short wavelengths.

Fig. 4 shows the optical constants of the CdS films deposited in different aqueous bath temperatures obtained by SE. The film thickness was 97nm, 101nm, 106nm and 105 nm respectively. The optical constants of the film deposited in aqueous bath temperature of 50°C exhibited different shape with other samples because of more pinholes in the film structure, thus the refractive indices and the extinction coefficients both were quite low. As the aqueous bath temperature increased, the refractive indices increased indicating the density of the film increased. The extinction coefficient k values were low between 600nm and 980nm and increased through short wavelengths below 600 nm. The k values increased with the increasing aqueous bath temperature, which may cause by the decreasing of band gap. The SE results consisted with the SEM results.

The Uv-vis spectra of the CdS thin films are shown in Fig. 5. As shown in Fig. 5(a), the average transmittance of CdS films between 500nm and 800nm deposited in different ammonia concentrations was about 80%. The absorbance of thin films increased with the increase in ammonia concentration, and the absorption edge of the films changed in the same tendency of the SE results. Using the standard expression for direct transition materials $(\alpha h v)^2 = A(h v - E_g)$, the band gap of CdS films deposited at different temperature was 2.57 eV, 2.42eV, 2.44eV and 2.52eV respectively. The CdS film deposited in an ammonia concentration of 0.1mol/L had a larger band gap than other sample because of its disordered structure. As the ammonia concentration increased above 0.4mol/L, the band gap of CdS films increased, indicating the film structure grow more compact. The results of the Uv-vis spectra were in accordance with that of SEM and SE.

The average transmittance of film deposited in different aqueous bath temperatures between 500nm and 800nm ranged from 60% to 80%, shown in Fig. 5(b). The absorbance of CdS thin films increased with the increase in aqueous bath temperature. The band gap of CdS films deposited at different temperature was found to be 2.54eV, 2.42eV, 2.40eV and 2.38eV respectively. An E_g shift toward the blue wavelength region as temperature decreases was noticed [19]. It can be proposed that at low temperature the dispersion of ion in solution due to thermal movement is reduced, and a preferential columnar growth parallel to the substrate edge appears. The decrease in band gap can be attributed to the improvement of crystallinity. Although the low bath temperature is beneficial for the crystallinity of CdS films, the rough surface contributed to big crystalline size bring the flat absorption edge, and may destroy the electric property for short current. The proper bath temperature is 60°C.

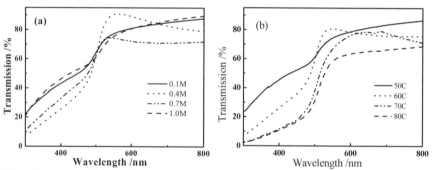

Fig. 5 Uv-vis spectra of the CdS thin films on glasses. (a) The films were deposited in different ammonia concentrations. (b) the films were deposited in different aqueous bath temperatures.

Summary

In this study, CdS thin films were prepared by chemical bath deposition (CBD) at different ammonia concentration and different temperature using cadmium chloride hydrate, thiourea, ammonium chloride, ammonia and deionized water as precursors. The morphology structure and the optical properties of CdS thin films were characterized by scanning electron microscope (SEM), ultraviolet-visible spectra and spectroscopic ellipsometry. The results indicated that CdS thin film could be grown when ammonia concentration at the range of 0.4-1.0 mol /L and in the temperature range of 50°C -80°C. The grain size of the films decreased with the increase in ammonia concentration and with the increase in aqueous bath temperature. The band gap of the films decreased with the decrease in ammonia concentration and with the increase in aqueous bath temperature. Within 600-980 nm wavelength，the average value of the refractive index was found to be 1.70 and the extinction coefficient was less than 0.07.

Acknowledgement

This work is supported by Shenzhen Key Laboratory of Sensor Technology (Grant No.SST201004) and Shenzhen Science and Technology Bureau project (Grant No.JC201005280419A).

References

[1] V.R. Shinde, S.B. Mahadik, T.P. Gujar, Supercapacitive cobalt oxide (Co_3O_4) thin films by spray pyrolysis , Appl. Surf. Sci. 252(20) (2006) 7487-7492

[2] F. Gu, C. Li, Y. Hu, L. Zhang, Synthesis and optical characterization of Co_3O_4 nanocrystals, J. Cryst. Growth. 304 (2007) 369-373.

[3] M. Verelst, T. O. Ely, C. Amiens, Synthesis and characterization of CoO, Co_3O_4, and mixed Co/CoO nanoparticules, Chem. Mater. 11(10)(1999) 2702-2708.

[4] M.M. Natile, A. Glisenti, Surface reactivity of NiO: interaction with methanol, Chem. Mater. 14 (2002) 4895-4903.

[5] J. Britt, C. Ferekides, Thin film CdS/CdTe solar cell with 15.8% efficiency, Appl. Phys. Lett. 62 (22) (1993) 2851-2852.

[6] K. Chidambaram, L. K. Malhotra, K. L. Chopra, Spray-pyrolysed cobalt black as a high temperature selective absorber, Thin Solid Films. 87 (1982) 365-371.

[7] B. S. Moon, J. H. Lee, H. Jung, Comparative studies of the properties of CdS films deposited on different substrates by R.F. sputtering , Thin Solid Films. 511-512 (2006) 299-303.

[8] E. W. Jones, V. Barrioz, S. J. C. Irvine, Towards ultra-thin CdTe solar cells using MOCVD, Thin Solid Films. 517(7) (2009)2226-2230.

[9] P. Boieriu, R.Sporken, A. Adriaens, SIMS and XPS characterization of CdS/CdTe hetero-structures grown by MBE, Nucl. Instrum. Methods Phys. Res. Sect. B. 161-163 (2000) 975-979.

[10]C. Nethravathi, S. Sen, N. Ravishankar, Ferrimagnetic nanogranular Co_3O_4 through solvo-thermal decomposition of colloidally dispersed monolayers of α-cobalt hydroxide, J. Phys. Chem. B. 109 (23) (2005) 11468-11472.

[11]A. Cortes, H.Gómez, R. E. Marotti, Grain size dependence of the bandgap in chemical bath deposited CdS thin films, Sol. Energy Mater. Sol. Cells. 82(1-2) (2004)21-34.

[12]M. Froment, M. C. Bernard, R. Corles, Study of CdS epitaxial films chemically deposited from aqueous solutions on InP single crystals, J. Electrochem. Soc. 142(8) (1995)2642-2649.

[13]Q. Q. Liu, J. H. Shi, Z. Q. Li, Morphological and stoichiometric study of chemical bath deposited CdS films by varying ammonia concentration, Physica B. 405(2010)4360–4365

[14]M. A. Martínez, C. Guillén, J. Herrero, Morphological and structural studies of CBD-CdS thin films by microscopy and diffraction techniques, Appl. Surf. Sci. 136(1998)8-16.

[15]W. Li, X. Cai, Q. Chen, Z. Zhou, Influence of growth process on the structural, optical and Electrical properties of CBD-CdS films, Mater. Lett. 59 (2005) 1-5

[16]M. Kostoglou, N. Andritsos, A.J. Karabelas, Progress towards modelling the CdS chemical bath deposition process, Thin Solid Films. 387(2001) 115-117.

[17]G. E. Jellison, F. A. Modine, Parameterization of the optical functions of amorphous materials in the interband region, Appl. Phys. Lett. 69 (1996) 371-373.

[18]G. E. Jellison, V. I. Merkulov, A. A. Puretzky, Characterization of thin-film amorphous semi-conductors using spectroscopic ellipsometry, Thin Solid Films, 377-378(2000) 68-73.

[19]O. Vogil, Y. Rodrìguez, O. Zelaya-Angel, Properties of CdS thin films chemically deposited in the presence of a magnetic field. Thin Solid Films 322 (1998) 329-333.

Key Engineering Materials Vol. 537 (2013) pp 150-154
© (2013) Trans Tech Publications, Switzerland
doi:10.4028/www.scientific.net/KEM.537.150

Preparation and Properties of ZnO/ZAO Double-Layers Thin Films on the Substrate of Glass

Guishan Liu[a], Mingjun Wang[b], Zhiqiang Hu, Yanyan Jiang, Xiaoyue Shen

Liaoning Province Key Laboratory of New Materials and Material Modification, Institute of New Energy Materials, Dalian Polytechnic University, Dalian116034, China

[a] gshanliu@dlpu.edu.cn, [b] Mingjun_Wang@163.com

Keywords: ZnO/ZAO double-layers thin film; magnetron sputtering; electrochemical deposition; CIGS solar cells

Abstract. Aluminium zinc oxide(ZAO) thin films were deposited on soda-lime-silica glass substrate by middle frequency power magnetron sputtering. Then zinc oxide (ZnO) thin films were deposited above ZAO thin films by electrochemical deposition method at different time. ZAO thin films and ZnO/ZAO double-layers thin films were characterized by X-ray diffraction (XRD) and scanning electron microscope (SEM). A four-point probe was used to determine the resistivity of the films. The optical transmittance of ZAO films and ZnO/ZAO films was measured by UV-visible spectrum. The results represent that the transmittance of ZAO/ZnO thin films decreases gradually with deposition time increasing. When the deposition time is 5 minutes, the maximum transmittance of ZnO/ZAO films reaches to 85% at wavelength from 400nm to 600nm, and the thickness and resistivity of thin film are 610nm and $2.04 \times 10^{-3} \Omega \cdot cm$, respectively. However, the thickness and resistivity are highest when the deposition time is 20 minutes, which reaches to 808nm and $1.2 \times 10^{-2} \Omega \cdot cm$. Meanwhile, the lattice constants a and c of ZAO/ZnO thin films demonstrate an expansion with deposition time increasing. In essence, good-quality double-layers thin films of ZnO/ZAO play an important role in CIGS solar cells.

Introduction

Recently, there have been growing interests in CIGS solar cells from both fundamental and applied viewpoints. In 2010, ZSW reported that CIGS solar cells with a record efficiency of 20.3% had been prepared by co-evaporation method[1]. CIGS solar cells are composed of glass and conductivity films, multilayer films which include Mo, CIGS, CdS, ZnO and ZnO:Al[2]. ZnO thin film as n-type window layer is composed of low resistance ZAO and high-resistivity i-ZnO, which is core part of built-electric field of PN junction in CIGS solar cells [3-5].

ZnO is a typical n-type metal oxide semiconductor with a wide band gap (3.37eV in room temperature) and high electron mobility ($120cm^2V^{-1}s^{-1}$)[6]. It is also a II-VI semiconductor compound, which has a much larger exciton binding energy (60meV) than GaN (~25meV) and ZnSe (22meV)[7-9]. ZnO has wurtzite structure with lattice parameters of a=0.3249nm and c= 0.5206nm[10]. It is also a good transparent conducting oxide (TCO) that can host large dopant concentrations with minimum effects on its crystal structure and transparency [11], thus using for heterojunction solar cells, transparent electrodes, surface acoustic wavedevices, optical waveguides, varistors and gas sensors, UV-LEDs (light emitting diodes) and LDs (laserdiodes), etc. [12,13].

In this paper, the ZAO/ZnO double-layers thin films was prepared by magnetron-sputtering ZAO and electrochemical-depositing ZnO, and we put more emphasis on analyzing the optical and electrical properties of ZnO/ZAO double-layers thin films to meet the demand for CIGS solar cells.

Experimental

Preparing ZAO thin films. In the same condition of time (60min) and electric current (0.12A), the ZAO thin films were deposited on soda-lime-silica glass substrate by middle frequency power supply magnetron sputtering.

Preparing ZnO thin films. The $Zn(NO_3)_2$ solution(0.005mol/L) was taken as electrolyte and the KCl solution (0.1mol/L) as auxiliary solution. The glass with as-deposited ZAO thin films was cleaned before electrochemical deposition. First of all, the glass with ZAO thin films was ultrasonically cleaned in anhydrous ethanol, acetone solution and distilled water solution for 20, 20, 10min, respectively. Secondly, the ZnO/ZAO double-layers films were prepared by electrochemical deposition with platinum sheet(purity:99.99%) as positive electrode and ZAO thin films as negative electrode. Two electrodes were parallel placed at the distance of 4.5cm. Deposition was carried out in room temperature, the electric current was 917μA and deposition time was 5, 10, 15, 20 minutes, respectively.

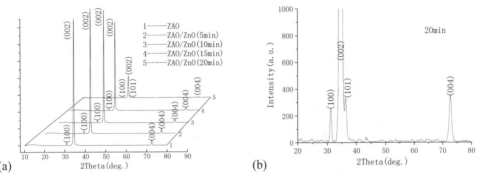

Fig.1 XRD pattern of ZAO and ZAO/ZnO thin films: (a) XRD patterns of ZAO thin films and prepared ZAO/ZnO thin films at different time; (b) Enlarge figure of 20 minutes

Results and Discussion

X-ray diffraction patterns for ZAO thin films deposited by sputtering shows a (002) preferred orientation (fig.1a). ZnO thin films are deposited above ZAO thin films by electrochemical deposition method for 5, 10, 15 and 20 minutes, respectively. A sharp peak representing a (002) orientation is observed as well for the as-deposited ZnO films. Consequently, the diffraction patterns for thin films indicated the presence of wurtzite ZnO and ZAO. However, for ZnO thin films deposited for 20 minutes, the intensity of XRD peak at (100), (002) and (004) has been found to decrease apparently (fig.1a), while additional peaks at (101) is observed that ZnO thin films have a much more relaxed structure with a greater degree of structural disorder (fig.1b).

The lattice constants a and c of ZnO and ZAO are calculated from Bragg's law:

$$n\lambda = 2d \sin\theta \qquad (1)$$

where n is the order of diffraction (usually n=1), λ is the X-ray wavelength (λ=1.5406nm) and d is the spacing between planes of given Miller indices h, k and l. In the ZnO and ZAO hexagonal structure, the plane spacing is relate to the lattice constant a, c and the Miller indices by the following relation[14]:

$$\frac{1}{d_{hkl}^2} = \frac{4}{3}\left(\frac{h^2 + hk + k^2}{a^2}\right) + \frac{l^2}{c^2} \qquad (2)$$

with the first order approximation, $n=1$

$$a = \frac{\lambda}{2\sin\theta}\sqrt{\frac{4}{3}\left(h^2 + hk + \frac{l^2}{(c/a)^2}\right)} \qquad (3)$$

$$c = \frac{\lambda}{2\sin\theta}\sqrt{\frac{4}{3(a/c)^2}(h^2 + hk + l^2)} \qquad (4)$$

Table 1 The peak position and the lattice constant of ZAO and ZAO/ZnO thin films

Thin films	Deposition time	Position of (100) peak (°)	Position of (002) peak (°)	$a(Å)$ (100)	$c(Å)$ (002)
ZAO		30.96	34.38	3.3325	5.2135
ZAO/ZnO	5min	30.96	34.38	3.3325	5.2135
	10min	30.94	34.38	3.3346	5.2135
	15min	30.90	34.32	3.3390	5.2221
	20min	30.88	34.30	3.3411	5.2241

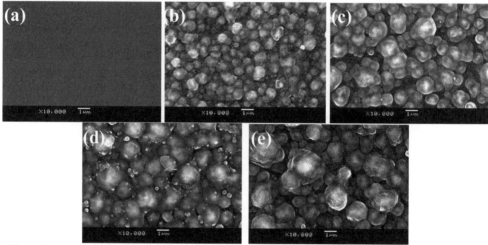

Fig.2 SEM images of ZAO thin films and prepared ZnO thin films at different deposition times
(a) ZAO thin films, (b) ZnO-5min, (c) ZnO-10min, (d) ZnO-15min, (e) ZnO-20min

for the (100) orientation, we can calculate the lattice a by

$$a = \frac{\lambda}{\sqrt{3}\sin\theta} \qquad (5)$$

for the (002) orientation, we can calculate the lattice c by

$$c = \frac{\lambda}{\sin\theta} \qquad (6)$$

The peak position and the lattice constant a and c of ZAO thin films and the as-prepared ZAO/ZnO double-layers thin films at different time are shown in Table 1. The ZnO films deposited for 5 minutes is so thin that all peaks position and the lattice constant have not been found to change by comparison with ZAO thin films. However, when ZnO thin films are deposited for 10, 15 and 20 minutes, respectively, XRD diffraction shows that the position of (100) peak and (002) peak move to the left and the lattice constant a and c demonstrates an expansion. Slightly increase for both a and c in our sample could be possible attribute to the excess Zn^{2+} or its incorporation in ZAO thin films[15].

Fig 2 demonstrates the SEM images of ZAO thin films and ZnO thin films prepared at different deposition times by electrochemical deposition method. Fig.2a shows the surface image of ZAO thin films deposited by sputtering, which reveals a structure with dense grains and smooth surface, thus we can observe the ambiguous configuration. Fig.2(b-e) shows SEM images of ZnO thin films deposited on the surface of ZAO films for 5, 10, 15 and 20 minutes, respectively. The size of grains has been found to increase as observed in XRD analysis with an increasing of deposition time. Moreover, many small grains without growing up were observed in the as-depositing film for 15 minutes. The same phenomenon was not found in additional films.

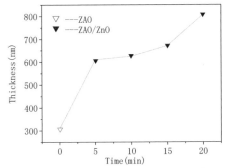

Fig.3 Thickness variation with time for
ZnO/ZAO films

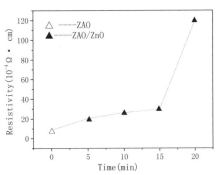

Fig.4 Resistivity variation with time for
ZnO/ZAO films

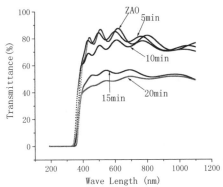

Fig.5 Transimittance as a function of
depositon time for ZnO/ZAO films

To observe the changes in electrical property and thickness of ZnO thin films, we have measured the thickness and resistivity of ZnO thin films deposited for 5, 10, 15 and 20 minutes. Fig.3 shows clearly that the thickness of ZAO substrate films is 310nm and the thickness of ZnO thin films increase nonlinearly with deposition time increasing. The measured resistivity of ZnO/ZAO thin films is shown in fig.4. The resistivity of ZAO substrate films is $8.7 \times 10^{-4} \Omega \cdot cm$ and the resistivity of ZnO films at deposition time under 15 minutes has been found to remains around $2.5 \times 10^{-3} \Omega \cdot cm$. For ZnO thin films deposited for 20 minutes, the resistivity raise sharply to $1.2 \times 10^{-2} \Omega \cdot cm$. This is due to the structural disorder as observed in XRD analysis.

The optical transmittance of ZAO thin films and ZnO / ZAO double-layers thin film is investigated by UV-vis spectra. Fig.5 shows that the sharp absorption band at ~370 nm is consistent with the ZnO band gap[16]. The optical transmittance of ZAO thin films is up to 88%, and the transmittance of ZnO/ZAO thin films decreases gradually with deposition time increasing. When the deposition time is 5minutes, the maximum transmittance of ZnO/ZAO double-layers thin films in UV light region reach to 85%. ZnO thin films for 20 minutes have a lower transmittance than those prepared at other deposition times. We deduce that the transmittance depends largely on the increases of the thin films thickness.

Conclusions

ZnO thin films have been successfully deposited on the surface of AZO films to form ZnO/ZAO double layers thin films. XRD and SEM characteristics releaved that deposition time has significant influence on the structure and morphology of thin films. The position of XRD diffraction peak of ZnO thin films move to the left and the lattice constant *a* and *c* generates an expansion with deposition time increasing. Meanwhile, the thickness of ZnO thin films increases nonlinearly with deposition time increasing, thus causing transmittance to decrease. ZnO thin films deposited for 20 minutes demonstrates a much more disorder structure resulting in the resistivity increasing sharply.

We conclude that when deposition time is 5minutes, the maximum transmittance of ZnO/ZAO films reach to 85%, and the resistivity is $2.04 \times 10^{-3} \Omega \cdot cm$. We believe that the perfect results could be obtained by optimizing the process and parameters. The high transmittance ZnO/ZAO films contribute to the further improvements in efficiency for CIGS solar cells.

Acknowledgement

The author wish to thank the Nanjing Zhaohe Photovoltaic Technology Company for our support in the funds and experimental devices.We acknowledge Mr Xu for technical support with SEM and UV-VIS measurements.

References

[1] Information on http://blog.sina.com.cn/nonferrousme

[2] M.H. Harati, J. Jia, K. Giffard, K. Pellarin, C. Hewson, A.L. David, W.M. Lau and Z.F. Ding, One-pot electrodeposition, characterization and photoactivity of stoichiometric copper indium gallium diselenide(CIGS) thin films for solar cells, Chem. Phys. 12 (2010) 15282-15290.

[3] B.N. Illy, A.C. Cruickshank, S. Schumann, *et al.*, Electrodeposition of ZnO layers for photovoltaic applications: controlling film thickness and orientation.J. Mater. Chem. 12 (2011) 12949.

[4] T. Minemoto, T. Matsui, H. Takakura, et al, Theoretical Analysis of the Effect of Conduction Band Offset of Window/CIS Layers on Performance of CIS Solar Cells Using Device Simulation. Solar Energy Materials & Solar Cells. 67 (2001) 83-88.

[5] Y.M. Xue, Y. Sun, F.Y. Li, *et al.*, Research on the Structure of the Heterojunction of the CIGS Thin Film Soler Cells. Journal of synthetic crystals. 33 (2004) 841-842.

[6] H.M. Wei, H.B. Gong, Y.Z. Wang, X.L. Hu, L. Chen, H.Y. Xu, P. Liu and B.Q. Cao, Three kinds of Cu2O/ZnO heterostructure solar cells fabricated with electrochemical deposition and their structure-related photovoltaic properties, Cryst Eng Comm. 13 (2001) 6065.

[7] C. Frederik, L. Colin, L. Neil, Y.S. Allan, N. R. Michael and H. John, Growth of ZnO thin films-experiment and theory. J. Mater. Chem. 15 (2005) 139.

[8] Z.K. Tang, G. Wang, K. Liu, et al, Room-temperature ultraviolet laser emission from self-assembled ZnO micro crystallite thin film. Appl. Phys. Letts. 72 (1998) 3270-3272.

[9] H.J. Hang, Y.F. Chen, S.K. Hong, MBE Growth of Gigh-Quality ZnO Films. Journal of Crystal Growth. 209 (200) 816-821.

[10] A.C. Rastogi, S.B. Desu, P.B. Hattacharya, R.S. Katiyar, Effect of starin Gradient on Luminescence and Electronic Properties of Pulseed Laser Deposited Zinc Oxide Thin Films.J. Electronceram. 13(2004) 345-352.

[11] D. Calestani, M. Z. Zha, L. Zanotti, M. Villani and A. Zappettini, Low temperature thermal evaporation growth of aligned ZnO nanorods on ZnO film: a growth mechanism promoted by Zn nanoclusters on polar surface, CrystEngComm. 13 (2011) 1707-1712.

[12] K.J. Omichi, N.K. Takahashi, T.K. Nakamura, *et al.*, AP-HVPE growth of ZnO with room-temperature ultraviolet emission Journal of materials, 11 (2001) 3158-3160.

[13] K.J. Omichi, K.Z. Kaiya, N.K. Takahashi, *et al.*, Growth of ZnO thin films exhibiting room-temperature ultraviolet emission by means of atmospheric pressure vapor-phase epitaxy. J. Mater. Chem. 11 (2001) 262-263.

[14] B.D. Cullity, S. Rstock, Elements of X-ray Diffraction, Prentice Hall, New Jersey,2001.

[15] O. Lupan, T. Pauporte, L. Chow, B. Viana, F. Pelle, L.K. Ono, B. Roldan Cuenya, H. Heinrich, Effects of annealing on properties of ZnO thin films prepared by electrochemical deposition in chloride medium.Applied Surface Science.256 (2010) 1895-1907.

[16] J. B. Franklin, B. Zou, P. Petrov, D. W. McComb, M. P. Ryan and M. A. McLachlan. Optimised pulsed laser deposition of ZnO thin films on transparent conducting substrate. J. Mater. Chem. 21 (2011) 8178.

Key Engineering Materials Vol. 537 (2013) pp 155-160
© (2013) Trans Tech Publications, Switzerland
doi:10.4028/www.scientific.net/KEM.537.155

Effect of Hydrolysis Modifier on the Properties of ATO Films Prepared by Spin Coating

J. A. Galaviz-Pérez[1a], Fei Chen[1b], Qiang Shen[1c], J. R. Vargas García[2d], Lianmeng Zhang[1e]

[1]State Key Laboratory of Advanced Technology for Materials Synthesis and Processing, Wuhan University of Technology, Wuhan 430070, PR China

[2]Dept of Materials and Metallurgical Eng., National Polytechnic Institute, Mexico 07300 DF Mexico

[a]jorgegalavizperez@gmail.com, [b]chenfei027@gmail.com, [c]sqqf@263.net, [d]rvargasga@ipn.mx, [e]lmzhang@whut.edu.cn

Keywords: ATO films; Acetylacetone; Optical transmittance; Spin coating

Abstract. ATO thin films have been successfully prepared by the spin coating method using an ethanol solution of $SnCl_4 \cdot 5H_2O$ and $SbCl_3$ using acetylacetone as hydrolysis modifier. ATO films from 1 to 6 layers were prepared at 300 °C as densification temperature and within an annealing temperature ranging from 450 °C to 750 °C. Films exhibited the cassiterite crystalline phase and morphology consisted in nanoparticles of about 10 nm in diameter. Acetylacetone content as well the addition of water affected the particle size and morphology, decreasing particle size and the appearance of voids. Films prepared at an acetylacetone ratio of 4 with no addition of water exhibited an optical transparency above 90% from 380 to 800 nm while resistivity was 7.99×10^{-3} $\Omega \cdot cm$. The effect of the hydrolysis modifier on the electric properties, morphology, optical transparency and microstructure has been studied

Introduction

Transparent conductive oxides (TCO´s) have been fundamental for the development of solar cells since silicon based devices have become expensive and have reached their top efficiency.[1]. Since the development of the Grätzel cell, which allows the fabrication of solar cell devices at low cost [2], the search for more efficient and cheap TCO´s has been increased. Indium tin oxide thin film is one of the most attractive TCO´s available, because it's high efficiency and elevated transparency. However, indium is an expensive material and its toxicity is an important drawback for scaled applications. The preparation of antimony doped tin oxide (ATO) have been drawn by deposition methods such as CVD [3], sputtering [4], precipitation [5] and sol-gel [6]. However the optical transparency and electric resistance of ATO films prepared by these methods have not yet reached the performance of ITO films. In chemical methods, as sol-gel, the use of additives to improve transparency and reduce electric resistivity has become common. Acetylacetone, has been successfully used to enhance the properties of TiO_2 [7] and CdS [8,9] films because it modifies the hydrolysis process and form complexes with metallic ions, leading to small particle size and selective crystalline phase formation [10]. Only two works have reported the use of acetylacetone as hydrolysis modifier for the preparation of ATO films and no systematic study on its effects on ATO films properties have been yet reported [6,11]. The present work shows the influence of acetylacetone in the properties of ATO films.

Experimental

ATO films were prepared using the spin-coating technique on quartz substrates. Solutions were synthesized using tin (IV) chloride pentahydrate ($SnCl_4 \cdot 5H_2O$, 99%) and antimony trichloride ($SbCl_3$, 99%) in proper amounts to achieve a Sb doping ratio of 10at%. Acetylacetone ($C_5H_8O_2$, 99%) was used as hydrolysis modifier in a molar ratio (acac/(Sn+Sb)) of 1 and 4. Anhydrous ethanol (C_2H_5OH, 99.7%) and deionized water were used as hydrolysis media. Procedures for the preparation of both sols and films have been reported elsewhere [6]. Films with 1, 3 and 6 layers were prepared at 450, 550, 650 and 750°C as annealing temperature. ATO solutions contents was as follows: *solution I* has an acac ratio:1 with no addition of water; *solution II* has an acac ratio:1 and water was added; *solution III* has an acac ratio:4 with no addition of water and *solution IV* has an acac ratio: 4 with addition of water. The crystal structure of ATO films was analyze by XRD (rigaku ULTIMA III, CuKα). Surface morphology was analyzed by FE-SEM (HITACHI S-4800, 10 kV). Optical and near UV was measured by UV-Vis from 200 to 800 nm in wavelenght (Shimadzu UV-2550). Electrical resistivity and Hall measurments was measured by using the van der Pawn method (Accent Optical HL5500PC).

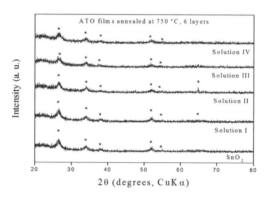

Figure 1. XRD patterns of SnO_2 and ATO films prepared using different solutions

Results and discussion

The crystallographic structure of films was determined by XRD. Fig. 1 shows the XRD patterns of 6 layered ATO films annealed at 750 °C as well as SnO_2 pattern for reference. Patterns show that films are polycrystalline SnO_2 rutile structures. No other crystalline phases are noticeable. All films exhibited the (110) and (101) SnO_2 reflections. Fig. 1 shows that onto an amorphous substrate the use of acetylacetone as hydrolysis modifier slow the crystallization process for films prepared in an acetylacetone ratio of 4, in comparison with films prepared in a ratio of 1. Due to this phenomenon films prepared by using solutions III and IV does not show the (112) reflection. Additionally, film prepared by solution II shows a more intense (112) peak due to the addition of water. Yanan et al. [7] reported this phenomenon in the preparation of TiO_2 nanoparticles by using acetylacetone, which stabilizes the titania precursor by chelating and inhibiting crystallization at 450 °C, leading to a comparatively smaller crystal size of 10 nm. Weng et al. [10] prepared PZT powders by using metal organic precursors and acetylacetone to manipulate the phase transformation process due to the formation of acetylacetone (diol) ligands with Ti and Zr atoms slowing down the crystallization process, leading to homogeneously dispersed powders formed by nanoparticles of about 20 nm in size. No phase transformation occurred in between 450 and 750 °C.

The morphology of ATO films were slightly altered by the content of acetylacetone. The films exhibited a granular morphology that consisted in agglomerates of about 20 nm in size tightly connected and leading to homogeneously dispersed pores as large as 50 nm in length. FE-SEM images (not shown) suggest the presence of nanoparticles below 10 nm in diameter (further analysis by TEM showed a particle size of 5-10 nm). Annealing temperature did not affect the ATO films morphology.

Transparency of films is highly dependent of the material properties, particle size and homogeneity of film. Fig. 2(a) to 2(c) shows the transparency in visible and near ultraviolet spectra (UV-Vis) of ATO films annealed at 650 °C. Transparency for monolayered ATO film –Fig. 2(a)- ranged between 82 to 96 % at 350 nm in wavelength. Film prepared from solution III exhibited a transparency of 97.3% at a wavelength of 369.5 nm. The lower transparency measurements were for films prepared from solutions I and III with a transparency of 90.7 and 82% respectively at 350 nm. For 3 layered ATO films –Fig. 2(b)-, the highest transparency was exhibited by film prepared from solution III in a value of 97.27% at 366.5 nm in wavelength. Films prepared from solutions I and III showed a continuous decrease in transparency, which is evident below 350 nm in wavelength, when transparency was 90.7 and 81 % respectively. The 6 layered ATO films showed a noticeable decrease in transparency, as seen in Fig. 2(c), exhibiting the highest transparency from solution III with a transparency of 92% at a wavelength of 467 nm. This film showed a different pattern in transparency due to the increase in thickness. Films prepared from solutions I, II and IV (86, 80 and 51% of transparency respectively) also changed their transparency patterns, decreasing more rapidly than 3 and monolayered films. All 6 layered ATO films also decreased in transparency for wavelengths in UV region. In comparison, the use of polymeric precursor method [13] to prepare ATO films, reported a transparency in the order of 80%, while the use of $SnCl_2·2H_2O$ and $SbCl_3$ precursors without an hydrolysis modifier [14] reported a transparency of approximately 85% within a range of 300-900 nm in wavelength. For ATO films prepared by precipitation and solution of nanoparticles by using the same precursors [15], transparency was below 90%. The difference in the transparency attained by use of acetylacetone is noticeable in Fig. 2(a) when film prepared from solution IV reached a 97.3% at 369 nm in wavelength and also sustained an excellent transparency in the UV region of 20% at 200 nm in wavelength.

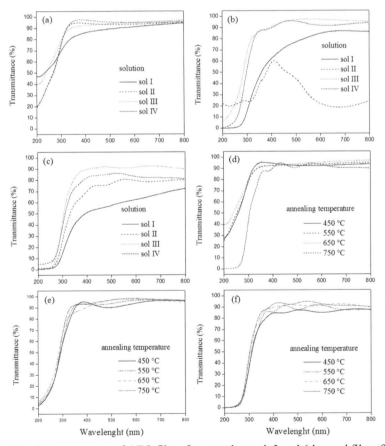

Figure 2. UV-vis transparency of ATO films for monolayered, 3 and 6 layered films for different solutions (a) to (c) and for solution IV at different annealing temperatures (d) to (f)

Fig. 2(d) to 2(f) shows the effect of annealing temperature on transparency of ATO films prepared from solution III. Monolayered ATO films exhibited noticeable changes in transparency with annealing temperature, as Fig. 2(d) shows. The transparency for these films is quite similar but the maximum transparency attained for film annealed at 450 °C is 95.45% at 360 nm. Then transparency decrease to 91.75% at the same wavelength for 550 and 650 °C. When the annealing temperature is 750 °C, transparency decreased to 87.2% at the same wavelength. Figure 2(e) shows that optical transparency results for 3 layered ATO films are similar for films at different annealing temperature with a small shift in the wavelength and significant change for the maximum transparency. For the film annealed at 450 °C, transparency is 95.3% at 386 in wavelength. At 550 °C as annealing temperature, the maximum transparency is 93.72% at 380 in wavelength. At 650 °C is 81.2% (325nm) and at 750 °C is 93% at 360 nm. For 6 layered ATO films transparency exhibited undulatory response, showed by Fig. 2(f) possibly related with films thickness and morphology. Film transparency at annealing temperature of 750 °C exhibited the highest transparency of 93.74% at 422 nm. In this study, there is no reason to relate the presence of acetylacetone in the post-annealed films because the complete decomposition of acetylacetone has been observed by TGA between 380 and 460 °C [10]. Thus, the change in transparency of films prepared by using acetylacetone is related to the effect of acetylacetone on the grain growth. As discussed above, the addition of acetylacetone limits the grain and crystal growth, leading to homogeneously dispersed crystalline nanoparticles no strongly influenced by the annealing temperature. The enhance in transparency of ATO films by using acetylacetone is explained by its ability to form stable ligands by tautomerization with metallic ions [9] and also by its hydrolysis stabilization properties that lead to well dispersed complexes into the sol and consequently leading to nanostructured homogeneous films [16].The formation of ATO nanoparticles as small as 5 nm by controlled hydrolysis and the low crystallization rate due the acetylacetone content allowed to keep a small particle size with no significative change with temperature.

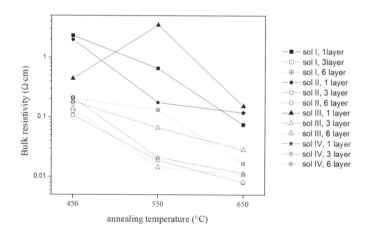

Figure 3. Bulk resistivity for monolayered, 3 and 6 layered ATO films prepared using different solutions at 450, 550 and 650 °C as annealing temperature

Hall measurements were performed in order to analyze the effect of the addition of acetylacetone on the electrical properties of films. Fig. 3 shows the resistivity of ATO films annealed at 450, 550 and 650 °C. The resistivity of ATO films exhibits a noticeable decrease with annealing temperature in almost one order of magnitude. The number of layers also diminish the resistivity of films due to the increase in thickness. The increase in acetilacetone ratio from 1 to 4 and the addition of water affects notoriously the electrical resistivity of films. The 6 layered ATO films annealed at 650 °C exhibited the lowest electric resistivity (7.99×10^{-3} $\Omega \cdot$cm) for film prepared from solution I with 3 layers, followed by film prepared from solution II with 6 layers (8.23×10^{-3} Ω·cm). These results are similar to those reported by Elangovan et al. [17] and Liu et al. [5], however, optical transmittance in this study is elevated in comparison to these reports and comparatively high for films prepared by other techniques [18–20].

Conclusions

Transparent ATO films were prepared by addition of acetylacetone as hydrolysis modifier by the sol gel process. The optical transmittance of films was high as to 97% in the visible region and up to 45% in the ultraviolet (200-40 nm) region. Acetylacetone is an excellent hydrolysis modifier in alcoholic solution either for chlorinated or metalorganic precursors. The complexing mechanism of acetylacetone on metallic ions limits the particle and crystalline size leading to homogenous and smooth films, no noticeable affected by annealing temperature. The homogeneity of film and small grain size allows an elevated optical transmittance.

Acknowledgements

This research was funded by the International Science & Technology Cooperation Program of China (No. 2011DFA52650). Author J. A. Galaviz-Pérez wish to thank to National Council of Science and Technology of Mexico (CONACYT) for the support for the postdoctoral stage.

References

[1]   K.L. Chopra, P.D. Paulson, V. Dutta, Thin films solar cells: an overview, Prog. Photovolt: Res. Appl. 92 (2004) 69-92.

[2]   N. Cherepy, G. Smestad, M. Grätzel, J. Zhang, Ultrafast electron injection: implications for a photoelectrochemical cell utilizing an anthocyanin dye-sensitized TiO_2 nanocrystalline electrode, J. Phys. Chem. B 101 (1997) 9342-9351.

[3]   A. de Graaf, J. van Deelen, P. Poodt, T. van Mol, K. Spee, F. Grob, A. Kuypers, Development of atmospheric pressure CVD processes for high quality transparent conductive oxides, Energy Procedia 2 (2010) 41-48.

[4]   Y. Muto, N. Oka, N. Tsukamoto, Y. Iwabuchi, H. Kotsubo, Y. Shigesato, High-rate deposition of Sb-doped SnO_2 films by reactive sputtering using the impedance control method, Thin Solid Films 520 (2011) 1178-1181.

[5]   S-m. Liu, W-y. Ding, W-p. Chai, Influence of Sb doping on crystal structure and electrical property of SnO_2 nanoparticles prepared by chemical coprecipitation, Physica B 406 (2011) 2303-2307.

[6]   H. Deng, J. Kong, P. Yang, Optical and structural characteristics of Sb-doped SnO_2 thin films grown on Si (111) substrates by Sol-Gel technique, J. Mater. Sci.: Mater. Electron. 20 (2009) 1078-1082.

[7]   Y. Fu, Z. Jin, Y. Ni, H. Du, T. Wang, Microstructure, optical and optoelectrical properties of mesoporous nc-TiO_2 films by hydrolysis-limited sol – gel process with different inhibitors, Thin Solid Films 517 (2009) 5634-5640.

[8]   A. Apolinar-Uribe, M. C. Acosta-Enriquez, M. A. Quevedo-Lopez, R. Ramirez-Bon, A. De Leon, S. J. Castillo, Acetylacetone as complexing agent for CdS thin films grow chemical bath deposition, Chalcogenide Letters 8 (2011) 77-82.

[9]   A. de Leon, M. C. Acosta-Enríquez, A. F. Jalbout, A. Apolinar-Iribe, S. J. Castillo, The role of tautomerization in acetylacetone as a complexing agent: Theoretical perspectives, Journal of Molecular Structure: THEOCHEM 957 (2010) 90-93.

[10] L. Weng, X. Bao, K. Sagoe-crentsil, Effect of acetylacetone on the preparation of PZT materials in sol-gel processing, Mat. Sci. Eng. B 96 (2002) 307-312.

[11] M. Guglielmi, E. Menegazzo, S. Materiali, M. Paolizzi, I. G. Spa, G. Gasparro, D. Ganz, J. P. Utz, M. A. Aegerter, I. Stadtwald, C. Pascual, A. Duran, H. X. Willems, M. VAN Bommel, L. B. Uttgenbach, L. Costa, Sol-Gel Deposited Sb-Doped Tin Oxide Films, J. Sol-gel Sci. Technol. 683 (1998) 679-683.

[12] Y. Li, R. M. Almeida, Elimination of porosity in heavily rare-earth doped sol–gel derived silicate glass films, J. Sol-gel Sci. Technol. 61 (2011) 332-339.

[13] T. R. Giraldi, M. T. Escote, M. I. B. Bernardi, V. Bouquet, E. R. Leite, E. Longo, J. A. Varela, Effect of Thickness on the Electrical and Optical Properties of Sb Doped SnO_2 (ATO) Thin Films, J. Electroceram. 13 (2004) 159-165.

[14] D. Zhang, L. Tao, Z. Deng, J. Zhang, L. Chen, Microstructure and electrical properties of antimony-doped tin oxide thin film deposited by sol-gel process, Mat. Chem. Phys. 100 (2006) 275-280.

[15] Y-J. Lin, C-J. Wu, The properties of antimony-doped tin oxide thin films from the sol-gel process, Surf. Coat. Technol. 88 (1996) 239-247.

[16] W. Hamd, A. Boulle, E. Thune, R. Guinebretiere, A new way to prepare tin oxide precursor polymeric gels, J. Sol-gel Sci. Technol. 55 (2010) 15-18.

[17] E. Elangovan, K. Ramamurthi, Studies on optical properties of polycrystalline SnO_2: Sb thin films prepared using $SnCl_2$ precursor, Crys. Res. Technol. 784 (2003) 779-784.

[18] M. Vishwas, K. Narasimha Rao, K. V. Arjuna Gowda, R. P. S. Chakradhar, Effect of sintering on optical, structural and photoluminescence properties of ZnO thin films prepared by sol-gel process, Spectrochimica acta A, Mol. and Biomol. Spectroscopy 77 (2010) 330-333.

[19] K. Ravichandran, P. Philominathan, Fabrication of antimony doped tin oxide (ATO) films by an inexpensive, simplified spray technique using perfume atomizer, Mat. Lett. 62 (2008) 2980-2983.

[20] H. Bisht, H-T. Eun, A. Mehrtens, M. Aegerter, Comparison of spray pyrolyzed FTO, ATO and ITO coatings for flat and bent glass substrates, Thin solid films 351 (1999) 109-114.

Key Engineering Materials Vol. 537 (2013) pp 161-164
© (2013) Trans Tech Publications, Switzerland
doi:10.4028/www.scientific.net/KEM.537.161

Determination of Thickness of PZO Films Deposited on ITO Glass Substrates

Xuejiao Li, Cheng Zhang[a] and Na Zhang

School of Materials Science and Engineering, Shanghai Institute of Technology, Shanghai 200235, China

[a]czhang@sit.edu.cn

Keywords: PbZrO₃; film thickness; UV-vis; microhardness.

Abstract. PZO (PbZrO$_3$) coatings with different thicknesses were deposited onto Indium Tin Oxide ITO glass substrates at room temperature by magnetron sputtering technique. UV-Vis absorption spectra method and microhardness testing method were used to measure the thickness of coating. It was proved that the measuring results of film thickness by two kinds of methods were equivalent, and either one method can be alternatively used to determine the thickness of deposited films.

Introduction

Film thickness is an important parameter in determining the efficiency of optical devices and it is closely related to the structure of the films. Thus, the testing of the film thickness is particularly important in film preparation. Recently, lots of measurement techniques have been developed to measure film thickness, fast and efficiently. Among which, non-destructive measurement technique is used widely because it gives no damage on the film[1]. There are several non-destructive thickness measurement techniques including X-ray radiation [2], ultrasonic [3] and optical methods such as reflection, transmission and interference measurement of light [4]. These techniques can be used to measure thickness of polymer films, optical coatings and photo resist layers [5].

The choice of measurement technique depends on the material and the desired measurement resolution. However, film thickness is usually numbers of micron or lower, conventional X-ray diffraction method for thin films usually provides weak diffraction peak that might lead to a larger measurement error. Ultrasonic method is not suitable to measure thickness of cross films.

UV-Vis spectrophotometer is a simple and familiar experimental facility. Film thickness can be calculated from optical constant as a by-product of UV-Vis spectrophotometer testing. This method provides a simple and NDT way for thin films. In addition, microhardness test method is also used to measure the thickness, for its characteristics of fast, simple, low cost, wide measurement range, high precision, and macro damage-free, which is suitable for many kinds of materials such as ceramics, metals and so on.

In this paper, PZO coatings with different thicknesses were deposited onto ITO glass substrates at room temperature by magnetron sputtering technique. UV-Vis spectrophotometer (UV1102) and Vickers microhardness meter (HXS-1000A) were then used to measure the thicknesses of such films, and measuring results were compared.

Experimental

PZO films were deposited on the ITO glass substrate from a PZO target (ϕ76×5 mm) of 3N5 purity by a PVD magnetron sputtering system in a 99.99% pure argon atmosphere. To make the glass surfaces clean, the ITO glass substrates (10×10×1.1 mm) were ultrasonically washed in acetone, ethanol, and deionic water for 20 minutes in sequence. Then the substrates were dried by N$_2$ flow before being placed into sputtering chamber. Before deposition, chamber was exhausted to a pressure below 5×10^{-5} Torr by vacuum pump and molecular pump. The distance between the substrate and the target was 100 mm. The sputtering power was 100W and the working gas pressure was 2.4×10^{-2} Torr. First, a procedure of pre-sputtering 5 minutes should be done to remove impurities of target, then the samples with sputtering time of 1 h, 1.5 h, 2 h, 2.5 h and 3 h were prepared with the substrates keeping at room temperature.

The as-deposited film specimens were scanned by using a UV-Vis spectrophotometer over the wavelength range 300–800 nm with 1nm increment. For microhardness testing, a rhombic pole was pressed into the surface of substrate and film by microhardness meter, respectively, whose applied loading is 200g. The indentation impression was the average of the horizontal and vertical dimensions.

In order to determine the influence of different sputtering time on the film, we choose PZO coatings with sputtering power of 100W as reference object. The crystal structure of the deposited films was analyzed by theX-ray diffraction (XRD). XRD patterns of the films were determined with D/Max2500X, Rigaku Corporation, Cu Kα radiation.

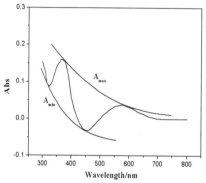
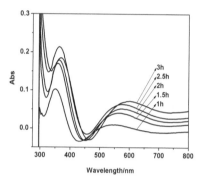

Fig. 1 The absorbance spectrum of PZO thin film   Fig. 2 UV-Vis spectrum of PZO films with different sputtering time

Results and discussion

Fig. 1 shows the absorbance spectrum scanned via UV-Vis spectrophotometer, and envelope line with the highest and lowest points is fitted. The complex index of refraction (N) and index of refraction (n) were calculated by equation(1) and (2), respectively. Ultimately, the thicknesses of the films were gained by equation (3).

$$N = 2n_s (10^{A_{max}} - 10^{A_{min}}) + \frac{n_s^2 + 1}{2} \tag{1}$$

$$n = [N + (N^2 - n_s^2)^{0.5}]^{0.5} \tag{2}$$

$$d = \frac{\lambda_1 \lambda_2}{2[n_1(\lambda_1)\lambda_2 - n_2(\lambda_2)\lambda_1]} \tag{3}$$

where N is complex index of refraction, n is index of refraction, n_s is the refractive index of the glass substrate. A_{max} and A_{min} are fitted by envelope line drawing with the highest and lowest points. λ_1, λ_2 are corresponding to wavelength of the adjacent maximum (or the minimum) absorbance.

Fig. 2 shows UV-Vis spectrum of PZO films with different sputtering time. It is obviously shown that a nearly linear enhancement of absorbance was probed with the sputtering time increasing, which could provide a good evidence for the stepwise and regular growth of the PZO films. The facts mentioned above could demonstrate that the UV-Vis spectra could indirectly predict the thicknesses of PZO films.

Fig. 3 shows the indenter of Vickers microhardness meter, 2α=136°, the pressed depth h has the relationship as follows equation (4) with impression dimension d.

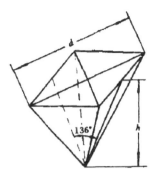

Fig. 3 Image of indenter of Vickers
microhardness meter

Fig. 4 Impression image

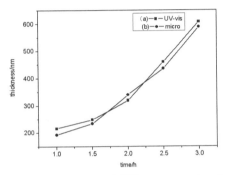

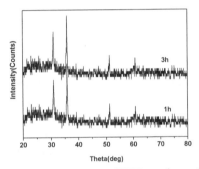

Fig. 6 Film thickness of different sputtering time

Fig. 7 XRD parrerns of PZO coatings of
1h and 3h

$$h = \frac{d}{2(2)^{1/2} \tan \dfrac{136°}{2}} = \frac{d}{7.000603} \tag{4}$$

For hard substrate, the hardness is far harder than the film. The indenter is pressed onto the film surface, if the thin film is removed, the impression dimension of the substrate is almost the same with film substrate. Therefore, the film thickness film can be calculated by equation (5).

$$h = \frac{D - d}{7.000603} \tag{5}$$

where D stands for impression dimension of thin film surface, and d stands for impression dimension of substrate surface.

Fig. 4 shows the impression image revealed by optical microscope, the image is rhombic as described above, thus, impression dimension d can be read by horizontal and vertical dimensions.

The comparison of experiment results on film thickness measured by two different methods for different PZO thin films or sputtering time is shown in Fig. 6. It is clear that the film thickness increases linearly with the increasing in sputtering time while deposition. While sputtered longer than two hours, the curve can be considered as the straight line and the result can help us predict the coating thickness of other time. Besides, the two growth curve are almost adjacent which means that the measurement of film by the two methods has no big deviation.

In order to detect the microstructure, XRD has been employed. Fig. 7 shows that the XRD patterns of PZO coatings sputtered by 1h and 3h have the same XRD profile. It sees that sputtering time have no effect to the structure of the deposited films.

Conclusions

In conclusion, the UV-Vis absorption spectra method and the microhardness testing can be used to measure film thickness. By comparison the testing results on film thickness of two different methods, it is confirmed that both are equivalent, and either one method can be alternatively used to determine the thickness of deposited films. And the XRD measurement showed that sputtering time have no effect to the structure of the deposited films.

Ackonwledgements

This work is supported by Foundation of Science and Technology Commission of Shanghai Municipality (No.11dz0501600), and Shanghai Leading Academic Discipline Project (Project No: J51504).

References

[1] F. Mitsugi, A. Matsuoka, Y. Umeda, Development of thickness measurement program for transparent conducting oxide, Thin Solid Films. 518 (2010) 6330–6333.

[2] J. Pei, F.L. Degertekin, B.V. Honein, B.T. Khuri-Yakub, K.C. Saraswa, In-situ thin film thickness measurement using ultrasonics waves, Proc. Int. Ultrasonic Symp. (1994) 1237–1240.

[3] J. Lhotka, R. Kuzel, G. Cappuccio, V. Valvod, Thickness determination of thin polycrystalline film by grazing incidence X-ray diffraction, Surf. Coat. Technol. 148 (2001) 96–101.

[4] K.H. Chen, C.C. Hsu, D.C. Su, Appl. Phys. B 77 (2003) 839.

[5] G. Cox, C.J.R. Sheppard, et al. Measurement of thin coatings in the confocal microscope, Micron 32 (2001) 701–705.

Key Engineering Materials Vol. 537 (2013) pp 165-168
© (2013) Trans Tech Publications, Switzerland
doi:10.4028/www.scientific.net/KEM.537.165

Electrochemical Performance of V_2O_5 Nano-Porous Aerogel Film

Lifeng Zhang, Guangming Wu[a], Guohua Gao and Huiyu Yang

Pohl Institute of Solid State Physics, Shanghai Key Laboratory of Special Artificial Microstructure and Technology, Tongji University, NO.1239 Siping Road, Shanghai 200092.China

[a] wugm@tongji.edu.cn

Keywords: aerogel; film; li-ion battery; electrochemical

Abstract. Nano-porous V_2O_5 aerogel films were prepared by dip-coating vanadium oxide sol onto ITO substrate, using V_2O_5 powder, Benz alcohol, Isopropanol as precursor materials. The nano pores were characterized by scaning electron microscope (SEM). The electrochemical properties were investigated by chronopotenyiometry (CP) and cyclic voltammograms (CV). Results showed that this porous V_2O_5 aerogel film exhibited good cycling stability with initial discharge capacity of 143mAh/g and 128 mAh/g after 18 cycles, staying 89.5 % of the initial discharge capacity, at a charge/discharge current density of 200 mA/g.

Introduction

Lithium ion battery was one of the most promising technologies in modern society due to their extensive applications in a wide range of areas including portable electronic devices, electric vehicles, and implantable medical devices[1,2]. As the importance of cathode for lithium ion battery, the development of high electrochemical performance cathode materials is key to the production of a new generation of battery[3]. Among cathode candidates, V_2O_5 materials have aroused special interest[3,4] for lithium ion insertion due to their layered structure, high energy density, high capacity, controllable fabrication method, multiple valence states[5]. But industrial application of traditional V_2O_5 is limited, because of the capacity degradation. So various nanostructured vanadium pentoxide are prepared to obtain high electrochemical performance: sub-microbelt[6], nanofibers[7], nanowires[8]. Comparing with the other microstructure, nano-porous structure may be the most controllable, and its porous structure could significantly improve the battery life and increase Li^+ diffusion rate.

To surpass the properties of crystalline V_2O_5, V_2O_5 aerogels are attracting increasing interests for their high surface, three-dimensional network, nano-porous structure[9,10]. The nano-porous structure could be permeated by electrolyte and its slim solid–phase structure can shorten the length of the ion diffusion, which is much better than crystalline V_2O_5 materials[11]. Short distances of lithium-ion transportation may significantly increase the rate of lithium insertion/removal and high surface area may permit a high contact area with the electrolyte and a high lithium-ion flux across the interface as sequence[12].

In this paper, V_2O_5 nano-porous aerogel films were prepared by dip-coaing method using the sol derived from V_2O_5 powder. The surface morphology of pore size was characterized by SEM. Chronopotentiometric (CP) tests and cyclic voltammograms (CV) were further employed to investigate the electrochemical performance of the aerogel film.

Experimental

V_2O_5 sol was prepared via sol-gel method as follows. V_2O_5 powder (99.6%, Kefeng Ltd), benzyl alcohol (99.0%, Sinopharm) and isopropyl alcohol (99.7%, Sinopharm) were firstly stirred for 20 minutes at room temperature with a certain mol ratio. Then the mixture was refluxed at 110°C for 4 hours. A limpid yellow sol was obtained after filtration. The resultant sol was subjected to a aged process at 40°C in a thermostat. Before depositing the film, the sol was centrifuged at 1500 r/min for half an hour. V_2O_5 films were then prepared by dip-coating the sol onto Indium-Tin Oxide (ITO) conducting glass substrates in the saturated acetone atmosphere. Then the films were quickly dipped into acetone to exchange the residual water in the film, at least were drawn out of acetone and dipped into cyclohexane to replace acetone in the film with cyclohexane for several times. After drying under ambient conditions for 12 h, the films were then annealed at 300°C in air atmosphere for 3 hours.

The surface morphology of these films were characterized by SEM(Philips-XL-30FEG). Electrochemical measurements were carried out with a three-electrode system, in an argon-filled glove box where H_2O and O_2 consentration were less than 1 ppm, using 1M $LiPF_6$ in the mixed solution of ethylene carbonate (EC), methyl carbonate (EMC) and dethyl ethylene carbonate (DEC) (1:1:1 by vol) as the electrolyte. Lithium plates were employed to work as both the counter electrode and the reference electrodes. Electrochemical tests were cycled on a battery test system Land CT2001A (Wuhan Jinnuo Ltd.) between 1.5 and 4 V versus Li/Li^+ at a current density of 200 mA g^{-1}. Cyclic voltammentry (CV) was performed with CHI660C electrochemical workstation system (Shanghai Chenhua Ltd.) between 1.5 and 4 V at a scanning rate of 5mV s^{-1}. All tests were performed at room temperature.

Fig. 1 SEM image of V_2O_5 aerogel film

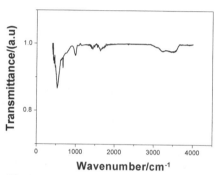

Fig.2 FTIR spectra of V_2O_5 aerogel film

Results and discussion

Fig.1 showed the SEM observation of the surface morphology of the V_2O_5 aerogel film annealed at 300°C for 3 h. The film showed a typical smooth surface morphology, with pores in range of 40-200 nm, which resulted from the solvent exchange procedure by cyclohexane. The solvent exchanging process may prevent pore collapse and retain the nano-porous structure of aerogel film.

Fig.2 showed the FTIR spectra of the synthesized V_2O_5 aerogel film. The bands at 667 and 989 cm^{-1} were ascribed to the stretching vibration of V-O-V (683-658 cm^{-1})[6] and V=O(1044-972cm^{-1}), respectively[13], indicating that the main component of the film was V_2O_5.

The cyclic performance of the aerogel film was investigated in the voltage range of 1.5-4.0 V vs Li/Li^+ at a current density of 200 mA g^{-1} and the results were shown in Fig.3. The initial discharge capacity was 143 mAh g^{-1}, and after 18 cycles, the discharge capacity was 128 mAh g^{-1}, indicating a good capacity stability of 89.5% for the V_2O_5 material.

Fig.4 showed the CV curves of the films before and after 18 cyclic tests. The oxidation peak in the profile of the aerogel film was around 2.7 V and the reduction peak was around 2.4 V. That was typical cyclic valtammetric property of sol-gel derived amorphous V_2O_5 material[14]. The aerogel film before cyclic test has a couple of redox peaks at 2.9 and 2.3 V, as shown in Fig.4, which corresponds to the lithium deintercalation/intercalation process. And after 18 cycles, there was almost no change in peak positions and the shape of the CV curve. It should be noted that the area of the CV curve did not change, which agreed well with the cycling stability of the above charge/discharge curves. It concluds that the V_2O_5 aerogel film exhibited excellent cycling stability. This may own to the porous aerogel structure, which made the electrolyte entering the cavities in the nanoporous electrodes and contacting the "inner surface" of the electrode[15].

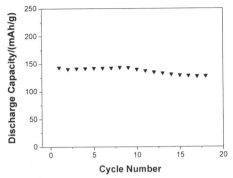

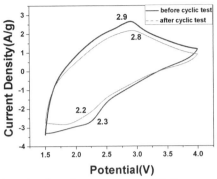

Fig.3 Discharge capacity of V$_2$O$_5$ aerogel film. The measurements were carried out in a potential window between 1.5 and 4 V vs Li/Li$^+$ at a current density of 200 mA g^{-1}

Fig.4 Cyclic voltammograms of V$_2$O$_5$ aerogel film before cyclic test (solid line) and after 18 cycles (dash line), with a scan rate of 5 mV s^{-1} in the potential range of 1.5-4 V vs Li/Li$^+$

Conclusion

V$_2$O$_5$ aerogel films were successfully synthesized through a sol-gel method. The surface nanostructure of the aerogel films were shown by SEM to be composed of nano pores in range of 40-200 nm, resulted from the solven exchange. The FTIR spectra showed the component of the film was V$_2$O$_5$. Electrochemical tests demonstrated that this V$_2$O$_5$ aerogel film had good reversibility with the initial charge/discharge capacity of 143 mAh g^{-1} and maintaining 128 mAh g^{-1} after 18 cycles at the current density of 200 mA g^{-1}. The high capacity, good cyclic stability can be ascribed to the nano-porous structure [11] having a higher surface area, shorter ion diffusion way, better contacting with electrolyte.

Acknowledgement

This work was supported financially by National Natural Science Foundation of China (Granted No. 51072137, No. 50802064, NO.51102183), Shanghai Committee of Science and Technology (10JC1414800) and Key Projects in the National Science & Technology Pillar Program (2009BAC62B02), the Doctor Subject Fund of Education Ministry of China under Grant No. 20100072110054, the Shanghai Postdoctoral Sustentation Fund, China(Grant No. 11R21416000), and the National Science Foundation for Post-doctoral Scientists of China (Grant No. 20100480619)

References

[1] G.A.Nazri, G.Pistoia, Lithium Batteries Science and Technology, Kluwer Academic, Boston, 2004

[2] M.Armond, J.M.Torosoon, Building better batteries, Nature. 451 (2008) 652-657.

[3] Y.Liu, M.Clark, Q.F.Zhang, D.Yu.D, L.J.Liu, G.Z.Cao, V$_2$O$_5$ ano-electrodes with high power and energy densities for thin film li-ion batteries, Advanced Energy Materials, 1 (2011) 194-202.

[4] Z.Chen, V.Augustyn, J.Wen, Y.W.Zhang, M.Q.Shen, B.Dunn, Y.F.Lu, High-performance supercapacitors based on intertwined CNT/V$_2$O$_5$ nanowire nanocomposites, Adv.Mater. 23 (2011) 791-795.

[5] W.Dong, J.S.Sakamoto, D.bruce, Electrochemical properties of vanadium oxide aerogels and aerogel nanocomposites, Sci.Technol.Adv.Mater. 26 (2003) 641-644.

[6] X.Ren, Y.Jiang, P.Zhang, J.Liu, Q.Zhang, Preparation and electrochemical properties of V$_2$O$_5$ submicron-belts synthesized by a sol-gel H$_2$O$_2$ route, J.Sol-Gel Sci.Technol. 51 (2009) 133-138.

[7]   C.M.Ban, N.A.Chernova, M.S.Whittingham, Electrospun nano-vanadum pentoxide cathode, Electrochem.Commun. 11 (2009) 522-525.

[8]   C.K.Chan, H.L.Peng, R.D.Twesten, K.Jarausch, X.F.Zhang, Y.Cui, Fast completely reversible Li insertion in vanadium pentoxide nanoribbons, Nano Lett. 7 (2007) 490-495.

[9]   G.M.Pajonk, Catalytic aerogels, Catal.Taday, 35 (1997) 319-337.

[10] E.Nappi, Aerogel and its application to rich detectors, Nucl.Phys.B. 61 (1998) 270-276.

[11] K.Xiao, G.Wu, J.Shen, D.Xie, B.Zhou ,Preparation and electrochemical properties of vanadium pentoxide aerogel film derived at the ambient pressure, Materials Chemistry and Physics, 100 (2006) 26-30.

[12] P.Bruce, B.Scrosati, J.Tarascon, Nanomaterials for Rechargeable Lithium Batteries, Angew.Chem.Int.Ed. 47 (2008) 2930-2946.

[13] M.A.Carreon, V.V.Guliants, Macroporous Vanadium Phosphorus Oxide Phases Displaying Three-Dimensional Arrays of Spherical Voids, Chem.Mater. 14 (2002) 2670-2675.

[14] N.Ozer, Electrochemical properties of sol-gel deposited vanadium pentoxide films. Thin Solid Films. 305 (1997) 80-87.

[15] L.Kavan, K.Kratochilova, M.Gratzel, Study of nanocrystalline TiO_2 (anatase) electrode in the accumulation regime, J.Electroanal.Chem.Soc. 394 (1995) 93-102

Key Engineering Materials Vol. 537 (2013) pp 169-173
© (2013) Trans Tech Publications, Switzerland
doi:10.4028/www.scientific.net/KEM.537.169

A Composite Film of V_2O_5 and Multiwall Carbon Nanotubes as Cathode Materials for Lithium-ion Batteries

Xiaowei Zhou[a], Guangming Wu[b], Guohua Gao, Huiyu Yang, Jichao Wang

Pohl Institute of Solid State Physics, Shanghai Key Laboratory of Special Artificial Microstructure Materials and Technology, Tongji University, Shanghai 200092, PR China

[a] magnificent1@163.com; [b] wugm@tongji.edu.cn (Corresponding author)

Key words: Sol-gel technique; V_2O_5 thin-film cathode; multiwall carbon nanotubes; electrochemical performance.

Abstract: In this paper, V_2O_5-Multiwall Carbon Nanotubes (MWCNTs) composite film is prepared through sol-gel technique and dip-coating method. MWCNTs pretreated by mixed acid, H_2O_2 and cheap V_2O_5 powder are used as raw materials. The morphology, structure and components of the composite film are characterized by corresponding testing instruments. The final electrochemical measurements indicate that: the V_2O_5-MWCNTs composite film as cathode materials for lithium batteries exhibits better electrochemical performance than ordinary V_2O_5 film.

Introduction

In recent years, much attention has been focused on the research and development of lithium-ion batteries owing to the growing demand for electronic devices. The low power and current requirements of the electronic devices have resulted in the appearance of thin-film batteries. The energy storage capacity of the thin-film battery is mainly dependent on the cathode materials used. The most widely used commercial cathode material is $LiCoO_2$ [1] which exhibits a specific capacity of about 140 mAh g^{-1}. Among various kinds of transitional metal oxides, V_2O_5 displays the most attractive theoretical capacity of about 450 mAh g^{-1}. Although V_2O_5 offers high theoretical capacity, the poor electronic conductivity [2], low diffusion rate of lithium-ion [3] and obvious structural transformation with lithium-ion insertion prevent us from realizing the full capacity and commercial utilization of V_2O_5 thin-film cathode. In the present work, we synthesize a composite film of V_2O_5 and multiwall carbon nanotubes through the dip-coating method. As we know, Multiwall Carbon Nanotubes (MWCNTs) discovered in 1991 [4] possess very good electrical conductivity and high thermochemical stability. Our aim is to improve the electronic conductivity of V_2O_5 thin-film cathode and enhance the diffusion rate of lithium-ion by compounding with MWCNTs. Through further electrochemical measurements, we notice that compounding with MWCNTs helps to enhance the electronic conductivity and lithium-ion diffusion rate of V_2O_5 film, resulting in the significant improvement in electrochemical properties of electrodes.

Experimental section

The rare materials are commercial V_2O_5 powder, hydrogen peroxide (H_2O_2, 30%) and Multiwall Carbon Nanotubes (MWCNTs). The synthetic procedure for V_2O_5-MWCNTs composite film is described as follows: Firstly, 0.5 g of raw MWCNTs is pretreated by 40 ml mix-acid (volume ratio, H_2SO_4:HNO_3 = 3:1) at 40 °C for 2 h under ultrasonic vibration condition. Through repeated filtration and rinsing, the well-dispersed MWCNTs are obtained. After that, a mixture of 1 g V_2O_5 powder and 0.2 g pretreated MWCNTs are slowly added in 40 ml of hydrogen peroxide solution (volume concentration, 30%). The exothermic reaction takes place during the reactive process, leading to the release of oxygen gas and the formation of V vanadium peroxide [5]. When a black suspension forms, 40 ml alcohol is poured into it. And then, the whole V_2O_5 sol-MWCNTs mixed system is continuously stirred for 6 h and aged for 3 days at 25 °C. Finally, V_2O_5-MWCNTs composite film is prepared by dip-coating method, using ITO conductive glass as the substrate and the suspension mentioned above as precursor. The resulting V_2O_5-MWCNTs film is sintered in air at 420 °C. For the purpose of comparison, common V_2O_5 film is also prepared by the same synthetic process except for the adding of MWCNTs.

Scanning Electron Microscopy (SEM, Philips-XL-30FEG) is performed to demonstrate the MWCNTs. Atomic Force Microscopy (AFM) is employed to observe the surface morphology of V_2O_5 film and V_2O_5-MWCNTs composite film. X-ray powder diffraction (XRD) pattern was obtained by using a RigataD/max-C diffractometer with Cu Kα radiation source (λ=1.5406 Å). The diffraction data is recorded in the angle range from 10° to 60°. The infrared spectroscopy measurements are carried out in the range of 400–4000 cm^{-1} using a Bruker-TENSOR27 FTIR spectrometer. The Raman spectra are obtained on the Horiba Jobin Yvon LABRAM-HR800 laser micro-Raman spectrometer.

The V_2O_5/V_2O_5-MWCNTs films are used as the working electrodes, a microporous film (Celgard 2500) as the separator, and lithium metal as the counter and reference electrode. The electrolyte is 1 M LiPF$_6$ dissolved in ethylene carbonate (EC)/dimethyl carbonate (DMC) (1:1 volume ratio). The three-electrode testing system is assembled in an argon-filled glove box. Electrochemical impedance spectroscopy (EIS) measurements are carried out in the frequency range between 100 kHz and 0.01 Hz at a state of charge of 3 V, and the amplitude of the AC signal is 5 mV. Cyclic voltammetry (CV) is conducted over a potential range from 1.5 to 4.0 V with a scanning rate of 5 mV s^{-1} using CHI660C (Chenghua, Shanghai) electrochemical workstation.

Results and discussion

The SEM image of MWCNTs pretreated by mixed-acid is presented in Fig. 1. Based on some relevant reports, Mixed-acid treatment is conducive to the dispersion of raw MWCNTs and can promote the uniform compounding of MWCNTs with V_2O_5 sol. Fig. 2 gives the AFM images of V_2O_5 film and V_2O_5-MWCNTs composite film in Fig. 2 (a) and Fig. 2 (b), respectively. It can be seen that V_2O_5-MWCNTs film displays more rough surface (about 80 nm surface relief) than V_2O_5 film (about 20 nm surface relief). The substantial surface relief for V_2O_5-MWCNTs film is due to the adding of MWCNTs and helps to provide more active sites of lithium-ion insertion [6].

Fig. 1. SEM image of MWCNTs pretreated by mixed-acid.

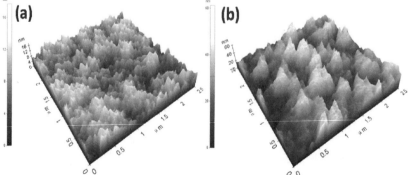

Fig. 2. AFM images of V_2O_5 film (a) and V_2O_5-MWCNTs composite film (b).

The XRD patterns of V_2O_5, V_2O_5-MWCNTs composite and MWCNTs are illustrated in Fig. 3. MWCNTs exhibit two characteristic peaks (002) and (004) centered at 25.8° and 42.8° respectively [7]. Both V_2O_5 and V_2O_5-MWCNTs composite show a series of characteristic diffraction peaks (2 0 0), (0 0 1), (1 0 1), (1 1 0), (4 0 0), etc. However, the (110) diffraction peak of V_2O_5-MWCNTs composite at about 26° has stronger relative intensity compared to that of V_2O_5. This is due to the impact of (002) peak of MWCNTs.

FTIR spectra of V_2O_5 film and V_2O_5-MWCNTs film are given in Fig. 4. As we can see, both of the two samples display the absorption peaks at 819, 1008, 1620 and 3421 cm^{-1} corresponding to the vibration of doubly coordinated oxygen (bridge oxygen) bonds, the stretching vibration of terminal oxygen bonds (V=O), the H-O-H bending and O-H stretching vibration modes of H_2O molecules within the samples respectively. The absorption peaks at 468 and 521 cm^{-1} for V_2O_5 film can be assigned to the asymmetric and symmetric stretching vibrations of triply coordinated oxygen (chain oxygen) bonds, but these two peaks merge into a peak at 501 cm^{-1} for V_2O_5-MWCNTs film. Besides, the newly appeared peak at 1569 cm^{-1} is associated with C=C vibrations in MWCNTs [8].

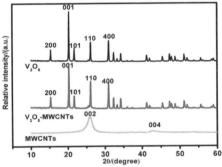

Fig. 3. XRD diffraction patterns of V_2O_5, V_2O_5-MWCNTs composite and MWCNTs.

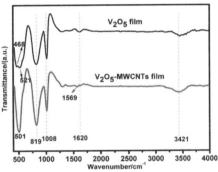

Fig. 4. FTIR spectra of V_2O_5 film and V_2O_5-MWCNTs composite film.

As shown in Fig. 5, for both of the samples: the two low-frequency Raman peaks at 142 and 192 cm^{-1} are attributed to the chain translation which is strongly associated with the layered structure; the three peaks centered at 283, 407 and 483 cm^{-1} are related to the bending and stretching vibrations of the V=O bonds; the Raman peaks appeared at 521, 691 and 995 cm^{-1} can be ascribed to the stretching modes of chain oxygen (V_3-O), bridge oxygen (V_2-O) and terminal oxygen (V=O), respectively [9].

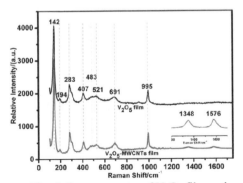

Fig. 5. Raman spectra of V_2O_5 film and V_2O_5-MWCNTs film (the inset is the longitudinally magnified Raman spectrum for V_2O_5-MWCNTs film at 1200-1700 cm^{-1}).

Especially, the weak D-mode of MWCNTs can be found at 1348 cm^{-1} in Raman spectrum of V_2O_5-MWCNTs film, followed by the G-mode at 1576 cm^{-1} [10]. For clarity, the Raman spectrum for V_2O_5-MWCNTs film at 1200-1700 cm^{-1} is longitudinally magnified (inset in Fig. 5).

EIS and CV are conducted to investigate the effect of MWCNTs on the electrochemical properties of V_2O_5 film. The Nyquist plots of V_2O_5 film and V_2O_5-MWCNTs film are depicted in Fig. 6 (a). All the plots display a upward curve in the high-frequency region and a sloped line in the low-frequency region, which are associated with the charge transfer reaction at the electrode/electrolyte interface and lithium-ion diffusion in the cathode materials, respectively. For V_2O_5 film, lithium-ion diffusion corresponds to the frequency ranged from 10 Hz to 0.01 Hz, and for V_2O_5-MWCNTs film, from 1 Hz to 0.01 Hz. In order to obtain the specific situation about lithium-ion diffusion, the plots of the imaginary Warburg impedance against $\omega^{-1/2}$ for both samples are illustrated in Fig. 6 (b) (ω represents the frequency). We can approximately get a straight line by connecting the relevant dots for each sample. The slope of the straight line is known as Warburg coefficient (W_A). According to some reports [11], the smaller the Warburg coefficient is, and the better the lithium-ion diffusion is. By calculations, W_A of V_2O_5-MWCNTs film is 57.73 which is much smaller than that (86.67) of V_2O_5 film. It indicates that V_2O_5-MWCNTs film exhibits better lithium-ion diffusion property.

The initial three CVs of both the samples are given in Fig. 7. The downward reduction and upward oxidation peaks stand for the lithium-ion insertion and extraction process in cathode films. It is obvious that V_2O_5-MWCNTs film (Fig. 7 b) exhibits larger peak area and better cycling performance than V_2O_5 film (Fig. 7 a). This concludes that the compounding of MWCNTs improves the specific capacity of V_2O_5 film and enhances its cyclic reversibility due to the good electrical conductivity and thermochemical stability of MWCNTs.

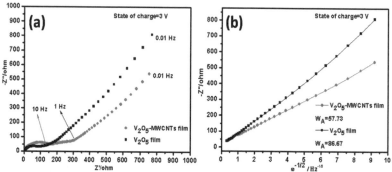

Fig. 6. The Nyquist plots (a) of V_2O_5 film and V_2O_5-MWCNTs film. The plots of the imaginary Warburg impedance against $\omega^{-1/2}$ for the both samples.

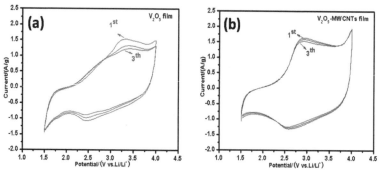

Fig. 7. The initial CVs of V_2O_5 film (a) and V_2O_5-MWCNTs film (b) at the potential range of 1.5-4 V with a scanning rate of 5 mV s^{-1}.

Conclusions

In this paper, V_2O_5-MWCNTs composite film is successfully synthesized through a simple sol-gel technique and the subsequent dip-coating method. The relevant film samples are characterized by AFM, XRD, FTIR and Raman spectra. The final electrochemical measurements indicate that this V_2O_5-MWCNTs film exhibits better electrochemical performance compared to common V_2O_5 film when used as cathode materials for lithium-ion batteries.

Acknowledgements

This research is financially supported by National Natural Science Foundation of China (Granted No. 51072137, No. 50802064, NO.51102183), the Doctor Subject Fund of Education Ministry of China under Grant No. 20100072110054, the National Science Foundation for Post-doctoral Scientists of China (Grant No. 20100480619), Shanghai Committee of Science and Technology (10JC1414800) and Key Projects in the National Science & Technology Pillar Program (2009BAC62B02), the Shanghai Postdoctoral Sustentation Fund, China (Grant No. 11R21416000), and Specialized Research Fund for the Doctoral Program of Higher Education.

References

[1] J. Cho, C.S. Kim, S.I. Yoo, Improvement of structural stability of $LiCoO_2$ cathode during electrochemical cycling by sol-gel coating of SnO2, Electrochem. Solid State Lett. 3 (2000) 362–365.

[2] J. Muster, G.T. Kim, V. Krstic, J.G. Park, Y.W. Park, S. Roth, M. Burghard, Electrical transport through individual vanadium pentoxide nanowires, Adv. Mater. 12 (2000) 420–424.

[3] T. Watanabe, Y. Ikeda, T. Ono, M. Hibino, M. Hosoda, K. Sakai, T. Kudo, Characterization of vanadium oxide sol as a starting material for high rate intercalation cathodes, Solid State Ionics 151 (2002) 313–320.

[4] S. Iijima, Helical microtubules of graphitic carbon, Nature 354 (1991) 56-58.

[5] B. Alonso, J. Livage, Synthesis of vanadium oxide gels from peroxovanadic acid solutions: a ^{51}V NMR study, J. Solid State Chem. 148 (1999) 16-19.

[6] X.W. Zhou. G.M. Wu, G.H. Gao, The synthesis, characterization and electrochemical properties of multi-wall carbon nanotube-induced vanadium oxide nanosheet composite as a novel cathode material for lithium ion batteries, Electrochim. Acta 74 (2012) 32-38.

[7] Q.L. Li, Y. Ye, D.X. Zhao, W. Zhang, Y. Zhang, Preparation and characterization of $CNTs–SrFe_{12}O_{19}$ composites, J. Alloy. Compd. 509 (2011) 1777–1780.

[8] F. Zhao, H.Y. Duan, W.G. Wang, Synthesis and characterization of magnetic Fe/CNTs composites with controllable Fe nanoparticle concentration, Physica B 407 (2012) 2495–2499.

[9] X.Q. Liu, C.M. Huang, J.W. Qiu, The effect of thermal annealing and laser irradiation on the microstructure of vanadium oxide nanotubes, Appl. Surf. Sci. 253 (2006) 2747-2751.

[10] S.H. Lee, P.S. Alegaonkar, U.H. Lee, A.S. Berdinsky, J.B. Yoo, Formation of buried-layer CNTs in porous SiO_2 templates, Diam. Relat. Mater. 16 (2007) 326–333.

[11] C.J. Cui, G.M. Wu, H.Y Yang, A new high-performance cathode material for rechargeable lithium- ion batteries: Polypyrrole/vanadium oxide nanotubes, Electrochim. Acta 55 (2010) 8870–8875.

Key Engineering Materials Vol. 537 (2013) pp 174-178
© (2013) Trans Tech Publications, Switzerland
doi:10.4028/www.scientific.net/KEM.537.174

Characterization and Electrochemical Performances of Vanadium Oxide Films Doped with Metal Ions

Wang Jichao, Wu Guangming[a], Gao Guohua, and Zhou Xiaowei

Pohl Institute of Solid State Physics, Tongji University, Shanghai, China

[a] wugm@tongji.edu.cn

Keywords: vanadium oxide film; sol-gel process; doping; diffusion rate;

Abstract. Vanadium oxide films were prepared via the sol–gel process and dip coating method, using V_2O_5 as raw materials and H_2O_2 (volume fraction 30) as the solvent. Mn and Ni ions were added to vanadium oxide sol to prepare doping vanadium oxide films. The films were characterized by atomic force microscopy, FT-IR, X-ray diffraction and electrochemical techniques. The add-on of Metal ions will not affect the morphology of the vanadium oxide films, but change the valence of vanadium ion and vanadium oxide crystal phase. Furthermore, cyclic voltammetry curves show that metal ions doping vanadium oxide films exhibit reversible electrochemical reaction. But electrochemical impedance spectroscopy indicates pure vanadium oxide film has a better diffusion rate.

Introduction

Vanadium oxide thin films with excellent physical and chemical properties have drawed significant attention over the past decades. Its layered structure, good chemical and thermal stability and excellent electrochromic properties have made it a promising material for applications such as electrochromic devices and reversible cathode materials for Li batteries. [1] Vanadium oxide films have been studied extensively. Reactive-sputtering is a common process for deposition of vanadium oxides.[2-4] In the present work, vanadium films prepared by sputtering have been characterized by several methods and the electrochemical lithium intercalation in vanadium oxide film has been examined.[5] But these methods are too complex and difficult to control.

The aim of this paper was to prepare thin films of vanadium oxide by sol-gel technologies, suitable for intercalation applications. Also, vanadium oxide films doped with metal ions, such as Mn and Ni, will be studied.

Experimental

Analytically pure V_2O_5 as raw materials, H_2O_2 (volume fraction 30) as the solvent, the sol was prepared by adding 1g V_2O_5 into 50 mL H_2O_2. H_2O_2 was stirred by magnetic stirrer in the beaker which is in a condensing unit, because the reaction will produce a lot of heat. V_2O_5 powder was added slowly into the beaker of H_2O_2 solvent, while stirring constantly. Solution color gradually becomes deeper, and air bubbles, until the reddish brown. No precipitation, transparent solution was gotten after continue stirring for 30 minutes. This is the derived precursor solution after filtering, placed in the 0 °C under the conditions for use. A certain amount of precursor solution was taken for pure vanadium oxide films coating, by adding 5 times the ethanol in accordance with the volume ratio of 1 to 5. A certain amount of manganese chloride and nickel chloride were added to the same percentages of water alcohol vanadium oxide solution for metal ions doping films coating.

ITO conductive glass was used as substrate, firstly soaked in acetone for 10 to 15 minutes, using distilled water washing and ultrasonic cleaning, and then evaporated to dryness after ethanol ultrasonic cleaning spare. Conductive glass was dipped at a slow speed (8 cm per minute) into the vanadium oxide sol or vanadium oxide sol mixed with metal ions, stand for 1 minute, pull out at the same speed. Then the vanadium oxide films or doping films were gotten; wet films at room temperature were dried for 24 h. Then the thin films on ITO conductive glass substrate were treated in muffle furnace under atmospheric pressure at 300 °C, the time was 120 minutes, at a heating rate of 3 degrees Celsius per minute, as positive electrodes.

Characterization

The oxide thin films produced were analyzed by AFM, FT-IR, and XRD. AFM imaging was performed with a Molecular Imaging instrument operating in intermittent contact mode (tapping) in air. AFM images were acquired in topographic mode. A silicon cantilever with a force constant of 1.2–3.5 N·m^{-1} at a resonance frequency of about 68 kHz was employed. The fourier transform infrared (FT-IR) absorption spectroscopy measurements were carried out using a Bruker-TENSOR27 FTIR spectrometer in the range of 400–4000 cm^{-1} to determine the bonding nature. X-ray diffraction (XRD) patterns were obtained by using a RigataD/max-C diffractometer with CuK$_\alpha$ radiation source (λ= 1.5406 Å) in the range of 10-60° to identify the phase and the structure of the synthesized materials.

Electrochemical cells with the oxide films as the working electrode and two Li metal foils as the reference and counter electrode were used for electrochemical measurements. The electrolyte was 1M LiPF6 dissolved in ethylenecarbonate(EC) / dimethylcarbonate(DMC) / diethylcarbonate (DEC) (1:1:1 in volume). During the experiments, the cell was maintained inside an argon-filled glove box with moisture content and oxygen levels less than 1ppm. The electrochemical tests were performed using CHI660C electrochemical workstation in the potential range of 1.5–4.0V (vs. Li/Li+). Electrochemical impedance spectroscopy (EIS) measurements were carried out in the frequency range between 100 kHz and 0.01 Hz at a charge potential of 3.0V, and the amplitude of the AC signal was 5mV. All tests were performed at room temperature.

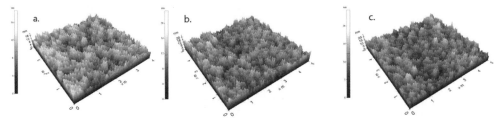

Fig. 1. AFM images of pure vanadium oxide films (a), vanadium oxide films doped with Mn ions (b) and vanadium oxide films doped with Ni ions (c)

Results and discussions

The AFM images of samples are shown in Fig. 1. It can be seen that the use of the AFM allows us to observe the surface morphology of the films and to give a schematic representation of their formation. The images clearly show the well synthesized layers of vanadium oxide films. Fig. 1 shows a pseudo-three- dimensional AFM images. The scanned area was 4 × 4 or 5 × 5 μm. The z-axis of Fig. 1 is expanded to evidence the surface structures. Surface roughness is below 20 nm. The AFM pictures appear completely flat. This indicates that adding of the metal ions, such as Mn ions and Ni ions, will not affect the film formation.

The FTIR spectra of the films are presented in Fig. 2. IR spectra for the pure vanadium oxide film showed absorption bands at 742 and 489 cm^{-1}, which are assigned to V-O-V deformation mode of VO$_2$[6]. These bands of vanadium oxide films after doping shifted toward the higher energy side. The shifted bands were observed from 536 and 771 to 505 and 781 cm^{-1}, respectively, which are assigned to V-O-V stretching

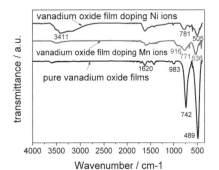

Fig. 2. FTIR spectra of vanadium oxide films, vanadium oxide films doped with Mn and Ni ions

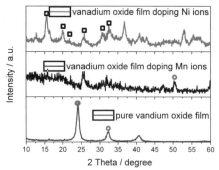

Fig. 3. XRD patterns of vanadium oxide films, vanadium oxide films doped with Mn and Ni ions. Black square and red circle denote for V_2O_5 [1] and V_3O_7 diffraction peaks, respectively.

mode of V_2O_5. All absorption bands were broad for vanadium oxide films doped with Mn and Ni ions. These bands were sharpened for pure vanadium oxide films. The broadening of the absorption bands for doping vanadium oxide films indicates that the distribution of the bond length between vanadium ion and oxygen is broad.[7] From these results, we concluded that valence state of vanadium ion in the pure vanadium oxide films is lower than V(V), and it is V(IV). And the valence state is V(V) for doping vanadium oxide films. Maybe the metal ions, such as Mn, Ni ions, get the electrons easily, and vanadium ion will be oxidated quickly when heat treatment.

Absorption bands at 3400 and 1620 cm^{-1}, which are assigned to an O-H stretching mode and an H-O-H deformation mode, were also observed for the films. This is due to the deliquescence of the films. These bands are more significant for vanadium oxide films doped with Mn ions and Ni ions than pure vanadium oxide film, because additives of manganese chloride and nickel chloride are easy to absorb moisture to hydrate.

The XRD patterns of vanadium oxide films, vanadium oxide films doped with Mn and Ni ions are shown in Fig. 3, respectively. The pure vanadium oxide film was crystallized. Diffraction peaks assigned to V_3O_7 were observed. V_3O_7 is a mixed oxide and one of the intermediates of oxidizing from V(IV) to V(V) oxide. It consists of VO_2[8] and V_2O_5[7]. The vanadium oxide film doped with Mn ions was with not significant diffraction peak, less diffraction peaks of V_2O_5 were observed. However, diffraction peaks assigned to a mixture of V_3O_7 and V_2O_5 were observed for the films doped with Ni ions. From the X-ray diffraction data, we conclude that manganese ions and nickel ions will affect the crystal structure of the films. This corresponds to the FTIR results.

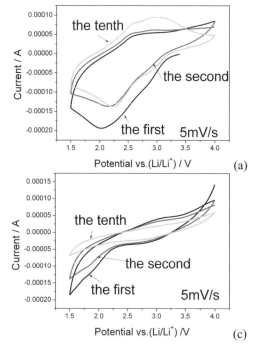

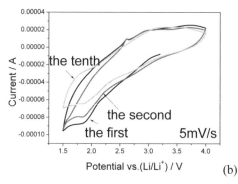

Fig. 4. The CVs of pure vanadium oxide films (a), Mn ions doping vanadium oxide films (b) and Ni ions doping vanadium oxide films (c)

The first, second, and tenth cyclic voltammetry curves of vanadium oxide films (pure, doping Mn and Ni ions) are presented in Fig. 4 (a, b, c). Comparing the cycle for the three samples, it is clear that the pure vanadium oxide film has good oxide crystallinity, as indicated by the better defined peaks in the tenth cyclic voltammetry curve which exhibit a pair of redox peaks that correspond to lithium-ions insertion/extraction process in cathode materials. The first cyclic voltammetry curve for pure vanadium oxide film shows a big broad peak around 2.0V and only very weak peak at 2.5 V. The cathodic peak at 2.0 V shifted towards to 2.25 V after first cycling. The cyclic voltammetry curve for vanadium oxide film doped with Mn ions shows weak peaks at 2.6, 1.9 V and diffuse peaks at 2.6,2.55, 1.6 V. There are only two broad peaks at 3.4, 2.8 V for vanadium oxide film doped with Ni ions. This is because the peaks were interrelated with crystal phase transition. The pure vanadium oxide film has single crystal structure, so the peaks are identified.

It can be observed that the redox current of first cycle for pure vanadium oxide film is much higher than that of the tenth cycle, and the potential positions of the cathodic and anodic peaks changed. These results indicate that the sample has an irreversible change. The samples doped with metal ions didn't change, which maybe attributed to Mn and Ni ions which stabilized the structures.

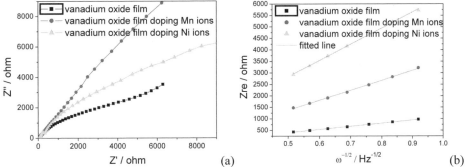

Fig. 5. The Nyquist plots of vanadium oxide films (a) and the plot of the Warburg impedance Zre against $\omega^{-1/2}$ (b)

The typical Nyquist plots of the samples are given in Fig. 5a, respectively. The kinetic characteristics of lithium insertion into VO_x were deduced from impedance measurements.[9] A typical impedance spectrum obtained at room temperature should display a depressed semicircle in high-frequency region and a sloped line in low-frequency region, which correspond to the charge transfer reaction at the electrode/electrolyte interface and Warburg impedance associated with lithium-ions diffusion in the cathode materials, respectively[9]. It is well known that the smaller the resistance of the electrode materials, the higher the electronic conductivity.

From Fig. 5, it can be seen that there are no significant semicircles in high-frequency region, the slope of vanadium oxide film doped with Mn ions is larger than the other two slopes. Pure vanadium oxide film has the smallest slope. Therefore, the small impedance of vanadium oxide film doped with Mn ions can be attributed to the higher electronic conductivity due to the adding of Mn ions.

The kinetics study of lithium-ions diffusion is another important aspect for electrochemical impedance spectroscopy (EIS). The plots of the Warburg impedance Zre as a function of $\omega^{-1/2}$ for the films are given in Fig. 5b, respectively. According to the Warburg equation[9], the values of the apparent chemical diffusion coefficient D_{Li} are related to the slope of the straight line Zre vs.$\omega^{-1/2}$. The smaller the slope of the straight line, the larger the apparent chemical diffusion coefficient of the electrode. It can be seen from Fig.5b that the slope of the straight line for pure vanadium oxide film is lower than that of vanadium oxide films doped with metal ions, indicating pure vanadium oxide film has a bigger diffusion rate.

Summary

Vanadium oxide films were prepared by sol–gel process. Mn and Ni ions were added to vanadium oxide sol to prepare metal ions doping vanadium oxide films. Metal ions will not affect the morphology of the vanadium oxide films, the surface roughness is below 20nm. But the valence of vanadium ion is IV for vanadium oxide film and V for vanadium oxide films doped with metal ions. Vanadium oxide crystal phase will be changed to a mixed phase for V_3O_7 and V_2O_5. Cyclic voltammetry curves show that doping vanadium oxide films own reversible electrochemical reaction. But electrochemical impedance spectroscopy indicates that pure vanadium oxide film has a larger diffusion rate.

References

[1] S. Beke, A review of the growth of V2O5 films from 1885 to 2010, Thin Solid Films. 519 (2011) 1761-1771.

[2] T. Wang, Y.D. Jiang, H. Yu, Z.M. Wu, H.N. Zhao, Target voltage behaviour of a vanadium-oxide thin film during reactive magnetron sputtering, Chinese Phys B. 20 (2011)

[3] H. Yu, Y.D. Jiang, T. Wang, Z.M. Wu, J.S. Yu, X.B. Wei, Modeling for calculation of vanadium oxide film composition in reactive-sputtering process, J Vac Sci Technol A. 28 (2010) 466-471.

[4] T. Wang, Y.D. Jiang, H. Yu, Z.M. Wu, Analysis of titanium and vanadium oxide thin film by method of reactive co-sputtering, P Soc Photo-Opt Ins. 7658 (2010)

[5] N. Kumagai, H. Kitamoto, M. Baba, S. DurandVidal, D. Devilliers, H. Groult, Intercalation of lithium in rf-sputtered vanadium oxide film as an electrode material for lithium-ion batteries, J Appl Electrochem. 28 (1998) 41-48.

[6] H.-N. Cui, V. Teixeira, L.-J. Meng, R. Wang, J.-Y. Gao, E. Fortunato, Thermochromic properties of vanadium oxide films prepared by dc reactive magnetron sputtering, Thin Solid Films. 516 (2008) 1484-1488.

[7] S. Deki, Y. Aoi, Y. Miyake, A. Gotoh, A. Kajinami, Novel wet process for preparation of vanadium oxide thin film, Mater Res Bull. 31 (1996) 1399-1406.

[8] T.H. Yang, C.M. Jin, R. Aggarwal, R.J. Narayan, J. Narayan, On growth of epitaxial vanadium oxide thin film on sapphire (0001), J Mater Res. 25 (2010) 422-426.

[9] R. Lindstrom, V. Maurice, H. Groult, L. Perrigaud, S. Zanna, C. Cohen, P. Marcus, Li-intercalation behaviour of vanadium oxide thin film prepared by thermal oxidation of vanadium metal, Electrochim Acta. 51 (2006) 5001-5011.

Key Engineering Materials Vol. 537 (2013) pp 179-183
© (2013) Trans Tech Publications, Switzerland
doi:10.4028/www.scientific.net/KEM.537.179

Mechanism of Annealing Treatment Effect on the Gasochromic Properties of WO$_3$ Bulks and Films

JIN Xiaobo, WU Guangming[*], GAO Guohua, FENG Wei, ZHANG zenghai, SHEN Jun

Pohl Institute of Solid State Physics, Tongji University, Shanghai, 200092, China

[*]wugm@tongji.edu.cn

Keywords: sol-gel; WO$_3$; gasochromism; annealing treatment

Abstract. Nanostructural WO$_3$ materials exhibit excellent gasochormism performance under the action of the hydrogen gas. In this paper, we study the mechanism of annealing treatment effect on the gasochromic properties both of the WO$_3$ bulks and WO$_3$ films. WO$_3$ bulks were prepared from WO$_3$ sol and dried under ambient pressure, which was previously synthesized via sol-gel method. The WO$_3$ films were prepared using dip-coating method. UV-visible spectroscopy, Raman spectroscopy, Fourier transforms infrared spectroscopy (FT-IR) and X-ray diffraction spectra (XRD) were utilized to investigate components, structures and gasochromic properties. And we find that the mechanism of the thermal treatment on the gasochromic performance depends on H$^+$ diffusion velocity which largely relies on the structural water content.

Introduction

Nanostructural tungsten trioxide (WO$_3$) shows excellent properties in controlling the transmittance of optical spectrum from visible to near-infrared spectrum [1]. Thus it has broad application prospect in the field of display devices, gas sensors, and smart windows [2~4].

The electrochromic property of WO$_3$ films was found firstly in the 60s of 20th century [5]. Afterwards, the photochromism, thermalchromism and gasochromism properties were studied extensively. The gasochromism smart window, duo to the facile and simple construction, has attracted much attention in the recent years [6~8]. It consists of a set double layer glasses with gasochromic material attached on the inner faces.

The nanostructual WO$_3$ can be economically synthesized through sol-gel method [9], which is also simpler than magnetron sputtering, vacuum vapor deposition and chemical vapor deposition.

The gasochromic properties strongly depend on the microstructure, which may easily be converted by post-treatment. In this paper, we study the effect of heat treatment on the structure of PdCl$_2$-doped WO$_3$ films, and its visible-infrared transmittance switching performance. The selected annealing temperatures are 50°C, 150°C, 250°C, 350°C and 450°C. As the nano-sized WO$_3$ bulks have asimilar gasochromic effect, we make a targeted comparison between the films and bulks for further study of the gasochromic mechanism.

Experimental

The WO$_3$ sol was prepared via sol-gel method. 20ml H$_2$O$_2$ (30%) was added to 20ml ethanol (98%), stirring with 5g metal tungsten powder (99.8%) dissolved gradually. After the drastic reaction, continued to stir for 30 min. Then the mixture was centrifuged in a speed of 1500 rpm for 30 min. The buff sol was obtained after eliminating big particles. A certain amount of ethanol was added to the solution and 2 hours reflux proceeded afterward at the temperature of 80 °C. Finally, the 0.3mol/L WO$_3$-ethanol sol was prepared. To dope with catalyst, PdCl$_2$ was added to the sol to produce the W: Pd molar ratio of 50:1. The yellow WO$_3$ bulks were obtained from drying WO$_3$ sol for 3-day, after which PdCl$_2$ was dropped on the surface. The bulks were ground into powders when analyzing the FT-IR spectra and XRD patterns.

The thin films were prepared on the glasses slides and silicon chips by dip-coating method, with the pulling speed of 10cm/min.

For the H_2 exposure, we used 10% H_2 in N_2 atmosphere for 10mins, while the recovery in pure O_2 for 30mins.

UV-visible spectroscopy (V-570, Jasco Inc.) measurement was carried out to investigate the gasochromic depth of WO_3 films in the wavelength range from 200nm to 2000nm. Fourier transform infrared spectroscopy (FT-IR, TENSOR27, Bruker Optik Gmbh) was taken to analyze IR absorption of WO_3 films and powders, with the range from 400cm^{-1} to 4000cm^{-1}. Raman spectra (HORIBA800) were used for investigating the presence of groups and components. X-ray diffraction spectra (XRD, D/MAX2550, Cu Kα) were taken to analyze the crystalline structure of films and powders.

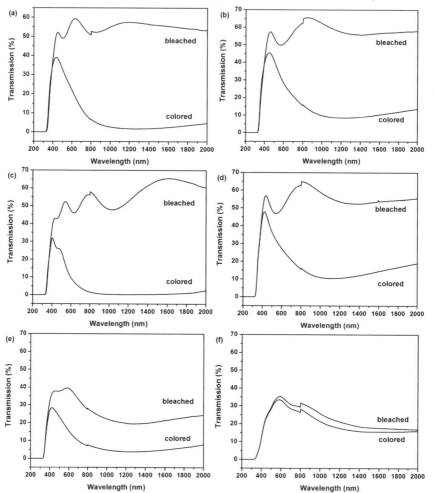

Fig. 1 Transmission spectra for bleached and colored Pd doped WO_3 as deposited film (a) and films annealed at the temperature of 50 °C (b), 150 °C (c), 250 °C (d), 350 °C (e) and 450 °C (f)

Results and Discussion

For the as-deposited film (Fig. 1a), the average transmittance of the bleached state in the visible and near-infrared range is nearly up to 60%, while the colored state lowers to 5%. For the 50 °C (Fig. 1b) and 150°C (Fig. 1c) treatment films, the change of transmittances from bleached state to colored state approximately reach 50%. The transmittances hardly change when the temperature is up to 450°C (Fig. 1f).

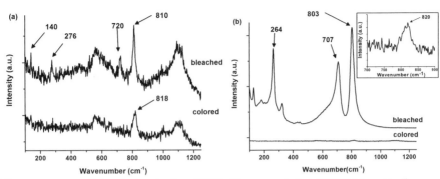

Fig. 2 Raman spectra of bleached and colored WO$_3$ films annealed at 450 °C (a) and bulks annealed at 650°C (b).

However, for the WO$_3$ bulks, we find that they distinctly color after all these annealing treatments, event at the temperature of 650 °C.

To identify the gasochromic phenomena of WO$_3$ bulks, Raman spectra are used to examine the structure changes. The film was selected the one which are annealed at 450 °C, which exhibit the worst gasochromic property. And the bulk is chosen the one annealed at 650°C, which is the highest temperature we can reach. Fig. 2 shows the Raman spectra of bleached and colored WO$_3$ films and bulks annealed at 450 °C (a) and 650 °C (b), respectively. The coloration of bulks WO$_3$ is observed from the drastically changes of Raman peaks, while the films with poor gasochromic performance show unconspicuous structural changes. The characteristic peaks at 720 cm^{-1} and 810 cm^{-1} of films are assigned to O-W-O stretching vibration, while 276 cm^{-1} is assigned to O-W-O bending vibration [10]. The shapes of the bleached and colored films annealed at 450 °C appear similar, which corresponds with the color change. On the contrary, the bulks behave much different in the bleached and colored state. The O-W-O vibration (264cm^{-1}, 707 cm^{-1}, and 803 cm^{-1}) sharply decrease in the colored state, which suggests that accompanying with the dissociated H^{+} interact with the lattice, oxygen vacancy is caused. The inset image shows the wave number shift of the main O-W-O absorption peak from 803 cm^{-1} to 820 cm^{-1}, which does a similar red shift as 450 °C treated films (from 810 cm^{-1} to 818 cm^{-1}).

Fig. 3 shows the FT-IR spectra of WO$_3$ films annealed at gradient temperatures from 50°C to 450°C and as-deposited film. The 3420 cm^{-1} and 1620 cm^{-1} absorptions are assigned to O-H-O stretching vibration and bending vibration, respectively. The 1428 cm^{-1} peak is assigned to W-OH bond vibration, which is produced by the decomposition of the surface absorbed water. The 800 cm^{-1} absorption peak is assigned to stretching vibration of peroxide groups W-O-O-W. The 640 cm^{-1} peak is assigned to the asymmetrical stretching vibration of corner-shared W-O-W bond. These appearances correspond with literatures [11~13]. At temperature under 50 °C sees the 980 cm^{-1} W=O absorption peak which is characteristic at room temperature condition. High temperature annealing cause the condensation of the films and the break of W=O, switching the vibration to W-O-W bond.

The peaks caused by water molecular obviously reduce along with the annealing temperature increasing. The 3420 and 1620 cm^{-1} peak nearly disappears when the temperature is over 350 °C. The 800 cm^{-1} peak also decreases when the temperature rise, which may account for the transformation to the WO$_6$ cyclic groups. While the 640 cm^{-1} peak growths with the rising of temperature, but the position of the peak shifts to the higher wave number. These imply that the water content and crystallization of WO$_3$ film largely influence the gasochromic performance.

Fig. 4 gives the comparison of the components and groups of films and bulks after 450 °C treatment. The main differences are obviously noticed. The absorption peaks assigned to O-H vibration (3420cm^{-1} and 1620cm^{-1}) do not disappear in the bulks. The other difference appears in the 980cm^{-1}, which corresponds to the W=O absorption. This bond does not totally break in the bulks and no obvious W-O-W absorption peak appears which corresponds to orthorhombic phase of WO$_3$.

Different form other crystal structures, an orthorhombic WO_3 unit consists a W=O bond and a structural H_2O molecule which shows a large binding energy and high thermal stability. As we know the physically absorbed water plays an important role in the transmission of H^+ [14]. While in the films, the water evaporates thoroughly at high temperature, intensively decelerating the H^+ diffusion in the films.

Fig. 5 shows the XRD patterns of WO_3 thin films. The temperature under 350 °C show the amorphous structure as no obvious diffraction peak appears. When it rises to 450 °C, diffraction peaks sharply emerge, which indicates the crystallization of the films [15]. Furthermore, the spectrum demonstrates the two main crystal phases, orthorhombic phase and triclinic phase.

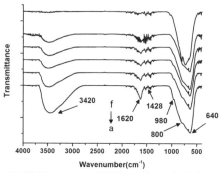

Fig. 3 FT-IR spectra of WO_3 as-deposited film (a) and films annealed at 50 °C (b), 150 °C (c), 250 °C (d), 350 °C (e) and 450 °C (f)

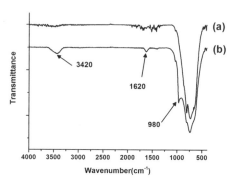

Fig. 4 FT-IR spectra of WO_3 film (a) and bulks (b) annealed at 450 °C

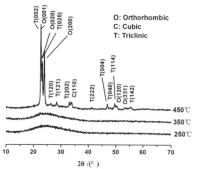

Fig. 5 XRD patterns of WO_3 thin films annealed at different temperatures

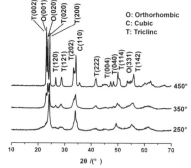

Fig. 6 XRD patterns of WO_3 powder after annealing treatment

Fig. 6 shows the XRD patterns of WO_3 powders. Although the 250 °C and 350 °C treated bulks appear lower diffraction intensity to the 450 °C treated ones, the diffraction peaks are still obvious, which differ from the films of 250 °C and 350 °C condition. The bulks treated at high temperature show multicrystalline structure, where more orthorhombic structures can be detected than that in films. As the bulks preserve excellent gasochromic performance after high temperature annealing, this may suggest that the crystal structure is not a curial factor for gasochromic properties. Considering the IR discussion, the mechanism of the thermal treatment on the gasochromic performance depends on H^+ diffusion velocity. Increasing the content of structural water molecules will improve the H^+ diffusion ability. Though most crystalline WO_3 exhibit a condensed structure without any structural water molecules, the gasochromic behavior can be enhanced if enough orthorhombic phases remain in films or bulks.

Conclusion

The components and crystalline structures of Pd doped WO_3 films and bulks treated at different annealing temperatures were systematically characterized. The hydroxyl groups drastically reduce in the films with the temperature rising. While WO_3 bulks have good heat resistance, as they contain relatively more water content than films under the same annealing treatment. The XRD patterns also indicate that the color properties have little to do with the crystalline structure. In consideration of the different drying and solvent removal process, we suppose the mechanism of the thermal treatment on the gasochromic performance depends on H^+ diffusion movability which relies on structural water content. The gasochromic behavior can be enhanced by increasing the content of structural water molecules such as remaining enough orthorhombic phases in films or bulks.

Acknowledgement

This work was supported financially by National Natural Science Foundation of China (Granted No. 51072137, No. 50802064, NO.51102183), Shanghai Committee of Science and Technology (10JC1414800) and Key Projects in the National Science & Technology Pillar Program (2009BAC62B02), the Doctor Subject Fund of Education Ministry of China under Grant No. 20100072110054, the Shanghai Postdoctoral Sustentation Fund, China (Grant No. 11R21416000), the National Science Foundation for Post-doctoral Scientists of China (Grant No. 20100480619), and Specialized Research Fund for the Doctoral Program of Higher Education (20100072110054).

References

[1] L. Zhuang, X. Q. Xu, H. Shen, A study on the gasochromic properties of WO_3 thin films, Surface and Coatings Technology. 167(2000) 217-220

[2] C. G. Granqvist, Electrochromic tungsten oxide films: Review of progress 1993–1998, Solar Energy Materials and Solar Cells. 60 (2000) 201-262

[3] J. L. Solis, S. Saukko, L. Kish, Semiconductor gas sensors based on nanostructured tungsten oxide, Thin Solid Films. 391 (2) (2001) 255-260

[4] A. Geog, W. Graf, R. Neumann, Stability of gasochromic WO_3 films, Solar Energy Materials and Solar Cells. 63 (2000) 165-176

[5] S. K. Deb, Electrochromism in tungsten oxide film, Appl. Opt. 58(1969) 190-196

[6] V. Wittwer, M. Datz, I. Ell, Gasochromic windows, Solar Energy Mater Solar Cells. 84 (2004) 305

[7] C. G. Granqvist, Transparent conductors as solar energy materials: A panoramic review, Solar Energy Mater Solar Cells. 91(17) (2007) 1529

[8] R. B. Goldner, R. D. Rauh, Electrochromic materials for controlled radiant energy transfer in buildings, Solar Energy Mater. 11(3) (1984) 177

[9] A. Georg, W. Graf, R. Neumann, Mechanism of the gasochromic coloration of porous WO_3 films, Solid State Ionics. 127 (2000) 319-328

[10] J. Z. Ou, M. H. Yaacob, In situ Raman spectroscopy of H_2 interaction with WO_3 films, Phys. Chem. Chem. Phys. 13(2011) 7330-7339

[11] N. Sharma, M. Deepa, FTIR and absorption edge studies on tungsten oxide based precursor materials synthesized by sol–gel technique, Journal of Non-Crystalline Solids. 306 (2002) 129-137

[12] U. O. Krasovec, B. Orel, A. Georg, The gasochromic properties of sol–gel WO_3 films with sputtered Pt catalyst, Solar Energy. 68 (6) (2000) 541-551

[13] N. Ozer, C. M. Lampert, Electrochromic performance of sol-gel deposited WO_3–V_2O_5 films, Thin Solid Films. 349 (1/2) (1999) 205-211

[14] B.H. Loo, J.N. Yao, H.D. Coble, A Raman microprobe study of the electrochromic and photochromic thin films of molybdenum trioxide and tungsten trioxide, Appl. Surf. Sci. 81(1994) 175-181

[15] K. Nishio, T. Sei, T. Tsuchiya, Preparation of electrochromic tungsten oxide thin film by sol-gel process, Journal of the Ceramic Society of Japan. 107 (3) (1999) 199-205

Key Engineering Materials Vol. 537 (2013) pp 184-188
© (2013) Trans Tech Publications, Switzerland
doi:10.4028/www.scientific.net/KEM.537.184

Structural Study of WO_3 and MoO_3 Compound Films in H_2 Gasochromism

Zenghai ZHANG, Guangming WU [*], Guohua GAO, Wei FENG, Xiaobo JIN, Jun SHEN

Pohl Institute of Solid State Physics, Tongji University, Shanghai, 200092, China

[*]wugm@tongji.edu.cn

Keywords: WO_3; MoO_3; compound films; FTIR; Raman; gasochromism.

Abstract: Sol-gel technique was used to prepare disordered tungsten oxide and molybdenum oxide sols. A series ratio of W:Mo compound sols were obtained via mettalic powder co-peroxided with H_2O_2 as precursors in ethonal. Compound films were made by dip-coating method. Fourier Transform Infrared Spectroscopy, Raman Spectroscopy were taken to characterize the structure of these compound films. Uv-visible Spectroscopy was used to test the gasochromic property. The results showed the gasochromics property was much different from that of pure tungsten oxide and molybdenum oxide sol-gel thin films. The effect was origined from the structrue alteration, which was not due to the spectrum superposition but the co-reaction of W and Mo.

Introduction

As a special transitional metal oxide, tungsten oxide is widely studied in electrochromism, photochromism, gasochromism, gas sensing, thermalchromism, photon catalyst, et al. Researches on gasochromic property of tungsten oxide thin films used for smart windows have attained some available achievements [1, 2]. The molybdenum trioxide is similar to tungsten oxide in properties. Molybdenum trioxide has a lower coloration efficiency compared with tungsten oxide in gasochromism, and the closer position of its optical absorption peak to the human eye sensitivity peak makes this material very attractive for many applications. Synthesis of such nanostructure films via sol-gel method is more economical than other techniques, such as vacuum vapor deposition, magnetron sputtering or chemical vaporous deposition [3].

There is not a common concept to describe the mechanism of gasochromic coloration, including the double injection model and the oxygen deficiency model. In this paper, it is reported the synthesis methods, microstructure characterizations and gasochromics property of W-Mo compound thin films. The study focuses on the structure changes of different compound ratios of W: Mo.

Experimental

Two synthesizing ways of W-Mo compound nano-films were taken to produce various ratios solutions by sol-gel method. 35.1g W metallic powder and 17.55g Mo metallic powder were used to prepare pure tungsten oxide and molybdenum oxide solution, which had consisted of series ratio of W:Mo solution mixture system that contained 1:0, 8:1, 6:1, 3:1, 2:1, 1:1 and 0:1. Metal powders (tungsten and molybdenum), H_2O_2 and ethanol as solvent reacted in the molar proportion according to [4]. After centrifugal and reflux processes, a series 0.3 mol/L solutions in molar concentration were prepared. $PdCl_2$ was added into sols as gasochromic catalyst in a certain proportion [2]. Then, nanostructure thin films were made on silicon slices and glass slides by dip-coating method, as shown in Table 1.

Table 1 W-Mo compound thin films prepared with varying molar proportions in S systems

1:0	8:1	6:1	3:1	2:1	1:1	0:1
W	S_{81}	S_{61}	S_{31}	S_{21}	S_{11}	Mo

The samples obtained were investigated by Fourier Transform Infrared Spectroscopy (FTIR, Tensor27, Bruker Optik Gmbh) in 400-4000 cm^{-1} and Raman Spectra (HORIBA JOBIN YVON HR800) in 200-1200 cm^{-1} for characters on microstructure.

Results & Discussion

Fig.1 shows the gasochromic (coloring and bleaching, with the wavelength of 700 nm) curves of compound films, including S_{11}, S_{21}, S_{31}, S_{61} and S_{81}. It can be seen that gasochromics responses of all the samples are within 10s and coloring states can be achieved in 120s to 280s except that of S_{11}. There is at least one coloring- bleaching recycle within 1000s. The gasochromic property of S81 shows an advantage in recycling compared with other samples, but is absolutly an disadvantage compared with WO$_3$ in recycling stability [4]. MoO$_3$ addition to WO$_3$ has not improve the gasochromic response or recycling stability. It was analysized that there are a small quantitive co-edge structure and a large number of co-corner structure in the compound materials. Co-corner structure dose not act benefit for gasochromic property optimization based on WO$_3$ thin films. However, during the coloring-bleaching recycle process, M_3O_{13} (M=W,Mo) in co-edge style turns to MO_6 (M=W,Mo) in co-corner style. In the process of transformation, the micro-structure of WO$_3$ has changed from loose and disordered structure to dense, partially ordered and crystallined structure, which hinder gasous moleculars from contacting with compound cluster of WO$_3$-MoO$_3$ and atomic hydrogen from adsorpting and desorpting. This limits the surfacial reaction and causes the declination of coloring-bleaching speed based on WO$_3$ thin films [5].

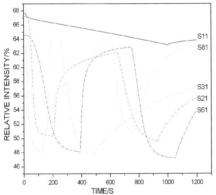

Fig.1 Samples of compound films in gasochromic property test

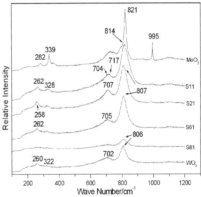

Fig.2 Raman spectra of WO$_3$, MoO$_3$ and compound films

It is observed that MoO$_3$ films, shown in Fig.1 demonstrates a slow gasochromic responsing speed, which needs about 5-10 minites and 3-5 hours to reach the saturated state in coloring. The coloring process is as slow as bleaching process. However, the electrochromic property of MoO$_3$ is better than gasochromics [6], which is assumed that drive force of reaction is different. H_xMoO_3 is molybdenum bronze structure, which is similar to tungsten bronze structure. But it is difficult to produce H_xMoO_3 and that results in a slow gasochromic speed. Obviously, coloring speed *of* W/Mo compound films has been promoted compared with pure MoO$_3$, different from the bleaching speed. This phenomenon is not agree with oxygen vacancy theory, because the diffusion of oxygen vacancies in bleaching process is similar with, or even faster than that in coloring process, for the production of vacancies in bleaching is not needed. Actually this phenomenon can be explained by double-injection model, which demonstrates that the absorption and injection of hydrogen atom is depended on its absorption energy.

We can see the Raman spectra in Fig.2 and find absorbed peaks at 806 cm^{-1} and 702 cm^{-1}. They are considered as stretching viberation of bridge oxygen in WO_3, peak at 322 cm^{-1} is due to viberation of $\gamma(W-OH_2)$, and peak at 260 cm^{-1} is the characteristic absorbed peak of $\gamma(W-O-W)$ [7]. For that there isn't apparently absorbed peaks of $\gamma(W=O)$, this structure is defined as monocline or hexagonal phase of WO_3. After the adulteration of MoO_3, all the compound films except S_{81} can well maintain absorbed peaks at or near 702 cm^{-1} and 806 cm^{-1} of WO_3 spectrum: 705 cm^{-1}, 707 cm^{-1}, 704 cm^{-1} and 717 cm^{-1} resulted from the new composited characteristic absorbed peak of the co-action with MoO_3 adultered into WO_3.

Fig.3 shows FT-IR spectrum of pure WO_3, S_{81} and MoO_3 in the range of 400-4000 cm^{-1}. Viberation peaks of Water molecular can be seen: the absorption peaks at 3490 cm^{-1} and 1626 cm^{-1} are seperately attributed to the stretching and bending viberation of O-H-O, respectively. For WO_3, peak at 981 cm^{-1} is corresponding to stretching viberation bond of $\gamma(W=O)$; peaks at 802 cm^{-1} and 652 cm^{-1} are corresponding to bending viberation of W-O-W and W_3-O in W_3O_{13} co-edge structure, peak at 633 cm^{-1} (not marked) is due to antisymmetric stretching viberation of W-O-W in WO_6 co-corner structure. Combined with Raman spectra, it can be concluded that the compound film is a state of mixture phases with the main part of co-corner hexatomic ring and small part of $WO_3 \cdot xH_2O$ and monocline phase [7].

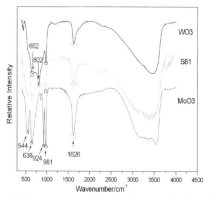
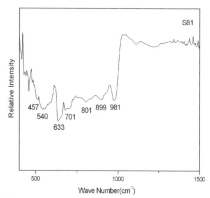

Fig.3 FT-IR spectra of WO_3, S_{81} and MoO_3 films   Fig.4 Short range of S_{81} in FT-IR spectrum

It is analyzed that, after adulteration of MoO_3, in spectrum of S_{81} (Fig.4), peak at 981 cm^{-1} maintains. Absorbed peak at 802 cm^{-1} also maintains indicating that the co-edge structure has not been destroyed. Except characteristic peaks of WO_3, MoO_3 and water in the compound films, peaks at 899 cm^{-1} is analysized as peaks compromised at 864 cm^{-1} and 924 cm^{-1} of MoO_3 under the action of WO_3. It is considered that peaks at 701, 697, 680 cm^{-1} are new peaks of WO_3-MoO_3 compound structure[8].

There are reactions between groups because of composition of W and Mo. The reactions makes absorbed peaks redshift, with middle peaks arising between the peaks at 864 cm^{-1} and 802 cm^{-1} (respectly representing Mo-O-Mo and W-O-W). The result is consistent with T. Ivanova et al.[6], who applied CVD method to prepare WO_3-MoO_3 compound films and assumed that structure of $Mo_xW_{1-x}O_3$ had been produced. Preliminary work[8] showed that peak at 958 cm^{-1} was caused by stretching viberation of $\gamma(Mo=O)$ end-bond in $MoO_3 \cdot xH_2O$. So M=O (M=W, Mo) mainly dominates the structure after MoO_3 adulterated,including spectrum of S_{81}.

It is found that the compound films acts an absolutely better gasochromic process than pure MoO_3, when molar ratio of W:Mo equals 8:1. After the compound style formed with adulteration of MoO_3, the mixture phases with hexagonal and monocline states arised. Action of WO_3 improved coloring speed of MoO_3, and the compound films performed well in cycle stability. Based on the double-injection theory, premilinary view is that [9,10] hydrogen gas formed H atoms and electrons after contaction with catalyst and diffuses into the film structure and reacts with WO_3 first. Then they begin to react with MoO_3 with the proceeding of gas quantity increasing. That is different from the pure MoO_3 gasochromic process. The possible reason is that reaction of H and WO_3 reduced the required potential energy of MoO_3 [11,12]. Therefore, the responsing speed is increased.

Conclusion

This paper introduced a preparation process of WO_3-MoO_3 compound films, tested the gasochromic property, characterized the structure, and analysized the theorical explanation. From the FT-IR and Raman spectra, the WO_3-MoO_3 compound film contains mainly hexagonal phase and small part of monocline phase. Some new peaks (701, 697, 680 cm^{-1}) arised and were caused by interaction of W and Mo. The gasochromics property was much different from that of pure tungsten oxide and molybdenum oxide sol-gel thin films. The reason was structrue alteration, which was not due to the spectrum superposition but the co-reaction of W and Mo. The results showed that WO_3 could enhance the gasochromic process of MoO_3 with the potential energy reduced. However, the compound structure verified the double-injection theory, and that prepared well for WO_3 to expand the research in improve smart windows application.

Acknowledgement

This work was supported financially by National Natural Science Foundation of China (Granted No. 51072137, No. 50802064，NO.51102183), Shanghai Committee of Science and Technology (10JC1414800) and Key Projects in the National Science & Technology Pillar Program (2009BAC62B02), the Doctor Subject Fund of Education Ministry of China under Grant No. 20100072110054, the Shanghai Postdoctoral Sustentation Fund, China(Grant No. 11R21416000), the National Science Foundation for Post-doctoral Scientists of China (Grant No. 20100480619)，and Specialized Research Fund for the Doctoral Program of Higher Education （20100072110054）.

References

[1] A. Georg, W. Graf, R. Neumann, V. Wittwer, Mechanism of the gasochromic coloration of porous WO films, Solid State Ionics. 127 (2000) 319–328

[2] S.H. Lee, H.M. Cheong, P Liu, Raman spectroscopic studies of gasochromic a-WO3 thin films, Electrochimica Acta. 46 (2001) 1995-1999

[3] M. Zayat, R. Reisfeld, H. Miniti, Gasochromic Effect in Platinum-Doped Tungsten Trioxide Films Prepared by the Sol-Gel Method, J. Sol-Gel Sci. Tech., 11 (1998) 161-168

[4] J. Shi, G. Wu, J. Shen, et al., Gasochromic Properties of WO3-MoO3 Composite Films Prepared by Sol-Gel Process, Rare metal materials and enginnering. 39, Suppl.2 (2010) 40-43

[5] M.F. Daniel, B.Desbat, Infrared and Raman study of WO3 tungsten trioxides and WO3·xH2O tungsten trioxide hydrates, J.Solid State Chem. 67 (1987) 235-247

[6] T. Ivanova, K.A. Gesheva, G. Popkirov, Electrochromic properties of atmospheric CVD MoO3 and MoO3-WO3 films and their application in electrochromic devices, Mater. Sci. Eng. B, 119 (2005) 232-239

[7] S. Morandi, G. Ghiotti, A. Chiorino, E. Comini. FT-IR and UV-Vis-NIR characterisation of pureand mixed MoO3 and WO3 thin films. Thin Solid Films. 2005,490:74 – 80

[8] Z. Zhang, G. Wu, G. Gao, Influence of MoO3 addition on the gasochromism of WO3 thin films. 7th Int. Conf. Thin Film Physics and Applications, Proc. of SPIE. 7995 (2010), 79951O

[9] N. Miyata, S. Akiyshi, Preparation and electrochromic properties of RF-sputtered molybdenum oxide films, J. Appl. Phys., 58 (1985) 1651-1655

[10] L. Zhou, J. Zhu, $Mo_xW_{1-x}O_3 \cdot 0.33H_2O$ Solid Solutions with Tunable Band Gaps, J. Phys. Chem. C., 114 (2010) 20947-20954

[11] L. Seguin, M. Figlarz I, R. Cavagnat, *et al.*, Infrared and Raman spectra of MoO_3 molybdenum trioxides and $MoO_3 \cdot xH_2O$ molybdenum trioxide hydrates, Spectrochimica Acta Part A. 51 (1995) 1323-1344

[12] L. F. Zhu, J. C. She, J. Y. Luo, Study of Physical and Chemical Processes of H_2 Sensing of Pt-Coated WO_3 Nanowire Films, J. Phys. Chem. C., 114 (2010) 15504-15509

Key Engineering Materials Vol. 537 (2013) pp 189-192
© (2013) Trans Tech Publications, Switzerland
doi:10.4028/www.scientific.net/KEM.537.189

The Process Optimization and Structural Analysis of Gasochromic Thin Films Derived by Peroxopolytungstic Acid

Wei Feng, Guangming Wu*, Guohua Gao, Zenghai Zhang

Pohl Institute of Solid State Physics, Shanghai Key Laboratory of Special Artificial, Microstructure Materials and Technology, Tongji University, Shanghai 200092, China

*wugm@tongji.edu.cn

Keywords: Sol-Gel; Gasochromic effect; Durability

Abstract. The sol-gel method with a combination of dip-coating process was employed to prepare peroxopolytungstic acid gasochromic thin films. The influence of preparation process on the structural and gasochromic durability was then discussed. We found that hydrogen peroxide content shown a significant impact on the structure of tungsten oxide thin films, which directly determined the gasochromic durability.

Introduction

As a typical intercalation compound, tungsten trioxide has been extensively studied for its interesting optical, electrical, structural and defect properties, such as metal-insulator transitions [1] and superconductivity at very low temperature [2], especially the applications of gas sensors [3] and smart glazing [4, 5].

Although, electrochromic properties[6-11] of WO_3 is very famous, we focus on its gasochromic properties in this paper. Gasochromic effect is based on spill-over effect[12], which can by obtained by exposure films to atomic hydrogen provided by a thin layer of catalyst (Pt or Pd) which is contact with WO_3 thin films[13] or incorporate the catalyst inside the WO_3 thin films[14]. WO_3 thin films can be prepared by several methods, including sputtering[13], electron beam[15], sol-gel[14], etc. Within those methods, sol-gel processing offers many advantages over traditional vacuum deposition techniques, such as the high performance and low cost. Despite that a lot of work has been done over several decades on peroxopolytungstic acid (W-PTA) derived gasochromic thin films, several contradictions still exist in the interpretation of experimental results, especially in the degradation of durability, which is a fatal problem of this material.

In this paper, through the optimum preparation process, we investigate the IR spectroscopic properties of WO_3 thin films to elucidate the relationship between structure and durability. Then we make a preliminary conclusions on the mechanism leading to degradation of WO_3 gasochromic durability.

Experimental

Preparation of W-PTA Sols and gasochromic films. W-PTA sols were prepared according to kudo[16] et al, then improved by many other adherents. Firstly, 21 g of W powders (99.8%) was reacted with a certain amount of H_2O_2 (30%) and 80 ml EtOH. The mixture was stirred until all the tungsten had dissolved before centrifuged. After the addition of appropriate EtOH, the sols was refluxed at 80°C until the color turned to transparent orange.

All glasses were cleaned with detergent soap, alcohol and de-ionized water before coating. All films were deposited by dip-coating from either freshly prepared sols, or from sols that $PdCl_2$ was added into (named $PdCl_2$-doped WO_3). For the first case, two membranes were needed, one is WO_3 membrane, and then is $PdCl_2$ (named WO_3 / $PdCl_2$).

Instrumental and Measuring Techniques. For IR analysis the films were deposited on Si wafers. Infrared spectra were using a FT-IR (TENSOR2, Bruker OpTik Gmbh, spectra rang $4000 \sim 400$ cm^{-1}). Films were colored and bleached by switched flushing with 4% H_2/Ar gas mixture or O_2 gas at 2

L/min speed and 0.15 Mpa. The transmittance measurement of the films at colored and bleached state was carried out in-suit with UV/Vis/NIR spectrophotometer (V-570, Jasco Inc). Especially, the coloring / bleaching kinetic was determined by measuring the monochromatic transmittance at $\lambda=1000$ nm of the films.

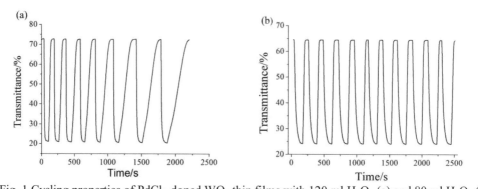

Fig. 1 Cycling properties of PdCl$_2$-doped WO$_3$ thin films with 120 ml H$_2$O$_2$ (a) and 80 ml H$_2$O$_2$ (b)

Table 1: IR vibrational band frequencies (in cm^{-1}) and band assignment of films

Wave number (cm^{-1})	Band assignment
3450	
3240	OH-O
3070	
1630	δ H$_2$O
971	ν W=O terminal
803	ν W-O-W edge-shared
640	ν W-O-W corner-shared

Results and Discussion

Fig.1 show the in-suit transmittance response of the PdCl$_2$-doped WO$_3$ thin films with different content of hydrogen peroxide plotted against the time during alternate exposures to H$_2$ / Ar mixture or O$_2$ gas. For each cycling, we maintained the same transmission states (colored and bleached state) and then recorded the response and recovery time. A degradation progress is obvious that in 120 ml H$_2$O$_2$ case, where the switched time increased with the increase of the cycle number, from 140 s for the first time to 420 s for the 7th cycle. However, in case of 80 ml H$_2$O$_2$, the gasochromic switching durability is stable, and about 150 s for each cycle.

As known, IR spectra can be efficiently to detect structure. The structure of W-PTA is supposed to be made of peroxo polytungstate anions linked together by hydrogen bonding [12,14,17]. Those anions belong to a kind of deformed kegging-liked structure, was assessed by Nanba et al[18], in which a six-membered ring of corner-shared (WO$_6$ or WO$_5$(O$_2$)) polyhedra is sandwiched by two edge-shared 3-membered (W$_3$O$_{13}$) rings. The characteristic peaks of the samples are listed in Table 1. Those peaks are basically the same with Krasovec and Orel's sample[12, 19]. As shown in Fig.2, the main difference between the FT-IR spectra of the films prepared drom different H$_2$O$_2$

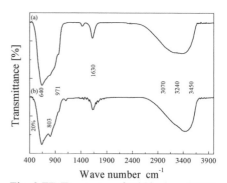

Fig. 2 FT-IR spectra of PdCl$_2$-doped WO$_3$ thin films with 120 ml H$_2$O$_2$ (a) and 80 ml H$_2$O$_2$ (b)

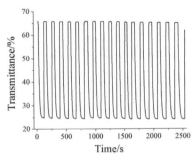

Fig. 3 Cycling properties of WO₃ / PdCl₂ thin films with 80 ml H₂O₂

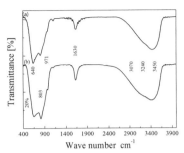

Fig. 4 FT-IR transmittance spectra of PdCl₂-doped WO₃ (a) and WO₃ / PdCl₂ (b) thin films with 80 ml H₂O₂

content is the sharpness and the multiplicity of bands in the range 971 ~ 560 cm^{-1}, especially in the band intensity associated with 640 and 803 cm^{-1}, which belongs to the corner-shared and edge-shared W-O-W stretching modes, respectively. Obviously, the less H₂O₂ was used, the stronger the peak intensity of 803 cm^{-1}, and the weaker of 640 cm^{-1} can be detected. In addition, a shoulder band at 971 cm^{-1} indicates the presence of terminal W=O groups[20]. The water vibrations in FT-IR spectra of all samples appearing as a triplet with frequencies at 3450 cm^{-1}, 3240 cm^{-1}, and 3070 cm^{-1}.

In previous work [21], we have proven that the edge-shared W-O-W will translate into corner-shared structure during coloring / bleaching switched, which is the main reason for the performance degradation of pour WO₃ gasochromic thin films. This transition will cause the microstructure of films changing from disordered porous to partial ordered and dense, as corner-shared W-O-W is more dense than edge-shared structure, which hinders the contact between H₂ gas or O₂ gas with WO₃ clusters. Moreover, it will restrict the surface reaction to slow down the switched speed. Therefore, we can draw a simple conclusion: the structure of edge-sharing W-O-W is benefit for gasochromic durability.

To further prove that point, we measured the cycling properties of WO₃/PdCl₂ films with 80 ml H₂O₂, as shown in Fig.3. Compared with Fig.2 (b), the gasochromic preference is excellent, without any attenuation. Additionally, a faster response and recovery switching are attained. Switch time is less than 100s, approximately 67% of the PdCl₂-doped WO₃ films at the same lever of H₂O₂ content.

As shown in Fig.4, there is only a tiny difference between FT-IR spectra of the PdCl₂-doped WO₃ thin films and WO₃ / PdCl₂ thin films, except the band intensity at 640 and 803 cm^{-1}. The influence of PdCl₂ doping on the structure of WO₃ thin films cause the same transmition of W-O-W from edge-shared to corner-shared structure as overmuch H₂O₂ do. Comparing with cycling performance (Fig.1(a) and Fig.1(b), Fig.1(b) and Fig.3) and FT-IR spectra (Fig.2, Fig.4), it is evident that the structure of edge-sharing W-O-W is benefit for gasochromic durability.

Conclusion

WO₃ thin films deposited using a dip-coating method from peroxopolytungstic acid sols exhibit good gasochromic properties reflected by fast coloring / bleaching switch. Through the optimization of preparation process, we found that hydrogen peroxide content has a significant impact on the structure of tungsten oxide films. Additionally, repetitive gasochromic cycling and FT-IR spectra revealed that the edge-shared W-O-W structure directly determined the gasochromic durability. Therefore, more work needs to be done to enhance the edge-shared W-O-W structure, as so to meet the durability requirement.

Acknowledgements

This work was supported financially by National Natural Science Foundation of China (Granted No. 51072137, No. 50802064, NO.51102183), Shanghai Committee of Science and Technology (10JC1414800) and Key Projects in the National Science & Technology Pillar Program

(2009BAC62B02), the Doctor Subject Fund of Education Ministry of China under Grant No. 20100072110054, the Shanghai Postdoctoral Sustentation Fund, China(Grant No. 11R21416000), and the National Science Foundation for Post-doctoral Scientists of China (Grant No. 20100480619) and Specialized Research Fund for the Doctoral Program of Higher Education (20100072110054).

References

[1] L. Whittaker, T.L. Wu, C.J. Patridge, *et al.*, Distinctive finite size effects on the phase diagram and metal-insulator transitions of tungsten-doped vanadium (IV) oxide, J Mater Chem. 1 (2011) 5580-5592.

[2] S. Reich, G. Leitus, R. Popovitz-Biro, A. Goldbourt, S. Vega, A Possible 2D H_xWO_3 Superconductor with a T (c) of 120 K, J Supercond Nov Magn. 2 (2009) 343-346.

[3] L.F. Zhu, J.C. She, J.Y. Luo, S.Z. Deng, J. Chen, X.W. Ji, and N.S. Xu, Self-heated hydrogen gas sensors based on Pt-coated $W_{18}O_{49}$ nanowire networks with high sensitivity, good selectivity and low power consumption, Sensor Actuat B-Chem. 53 (2011) 354-360.

[4] A. Chen, N.W. Gao, Z.Y. Zhu, Y.Q. Han, and M.Q. Wu, Some aspects affecting transmittance spectra of composite smart film WO_3, Ieee T Compon Pack T. 22 (1999) 17-20.

[5] C.G. Granqvist, A. Azens, A. Hjelm, L. Kullman, G.A. Niklasson, G. Vaivars, Recent advances in electrochromics for smart windows applications, Solar Energy. 63 (1998) 199-216.

[6] A. Karuppasamy, and A. Subrahmanyam, Studies on electrochromic smart windows based on titanium doped WO_3 thin films, Thin Solid Films. 516 (2007) 175-178.

[7] H.N. Cui, M.F. Costa, V. Teixeira, I. Porqueras, and E. Bertran, Electrochromic tungsten oxide multilayer thin films for use in smart windows, Optics for the Quality of Life. 4829 (2003) 817-818

[8] N.I. Jaksic, and C. Salahifar, A feasibility study of electrochromic windows in vehicles, Sol Energ Mat Sol C. 79 (2003) 409-423.

[9] S.H. Lee, C.E. Tracy, G. Jorgensen, J.R. Pitts, and S.K. Deb, Cyclic environmental testing of electrochromic window devices, Electrochim Acta. 46 (2001) 2237-2242.

[10] R. R. David, Electrochromic windows: an overview, Electrochimica Acta 44 (1999) 3165-3176.

[11] C.M. Lampert, Electrochromic materials and devices for energy efficient windows, Sol Energ Mat Sol C. 11 (1984) 1-27.

[12] B. Orel, N. Groselj, U.O. Krasovec, R. Jese, and A. Georg, IR spectroscopic investigations of gasochromic and electrochromic sol-gel - Derived peroxotungstic acid/ormosil composite and crystalline WO_3 films, J Sol-Gel Sci Techn. 24 (2002) 5-22.

[13] D. Schweiger, A. Georg, W. Graf, and V. Wittwer, Examination of the kinetics and performance of a catalytically switching (gasochromic) device, Sol Energ Mat Sol C. 54 (1998) 99-108.

[14] B. Orel, N. Groselj, U.O. Krasovec, *et al.*, Gasochromic effect of palladium doped peroxo-polytungstic acid films prepared by the sol-gel route, Sensor Actuat B-Chem. 50 (1998) 234-245.

[15] .Y. Luo, S.Z. Deng, Y.T. Tao, *et al.*, Evidence of Localized Water Molecules and Their Role in the Gasochromic Effect of WO_3 Nanowire Films, J Phys Chem C .113 (2009) 15877-15881.

[16] T. Kudo, H. Okamoto, K. Matsumoto, *et al.*, Peroxopolytungstic acids synthesised by direct reaction of tungsten or tungsten carbide with hydrogen peroxide, Inorg Chim Acta. 111 (1986) 27-28.

[17] M. Zayat, R. Reisfeld, H. Minti, B. Orel, and F. Svegl, Gasochromic effect in platinum-doped tungsten trioxide films prepared by the sol-gel method, J Sol-Gel Sci Techn. 11 (1998) 161-168.

[18] T. Nanba, S. Takano, I. Yasui, and T. Kudo, Structural study of peroxopolytungstic acid prepared from metallic tungsten and hydrogen peroxide, J Solid State Chem. 90 (1991) 47-53.

[19] U.O. Krasovec, B. Orel, A. Georg, and V. Wittwer, The gasochromic properties of sol-gel WO_3 films with sputtered Pt catalyst, Sol Energy. 68 (2000) 541-551.

[20] B. Orel, U.O. Krasovec, N. Groselj, *et al.*, Gasochromic behavior of sol-gel derived Pd doped peroxopolytungstic acid (W-PTA) nano-composite films, J Sol-Gel Sci Techn. 14 (1999) 291-308.

[21] T. Liang, G.M. Wu, G.H. Gao, J.D. Wu, Z.H. Zhang, J. Shen, B. Zhou, Gasochromic Properties of Stable WO_3-SiO_2 Composite Films, Mater Rev 24 (2010) 50-54.

Key Engineering Materials Vol. 537 (2013) pp 193-196
© (2013) Trans Tech Publications, Switzerland
doi:10.4028/www.scientific.net/KEM.537.193

Improvement of n/i Interface Layer Properties in Microcrystalline Silicon Solar Cell

Xiangbo Zeng* [a], Jinyan Li [b], Xiaobing Xie [c], Ping Yang [d], Hao Li [e], Haibo Xiao [f], Xiaodong Zhang [g] and Qiming Wang [h]

State Key Laboratory on Integrated Optoelectronics, Institute of Semiconductors, Chinese Academy of Sciences, Beijing 100083, China

[a] xbzeng@semi.ac.cn (corresponding author), [b] lijy@semi.ac.cn, [c] xbxie@semi.ac.cn, [d] yangping10@semi.ac.cn, [e] lihao2010@semi.ac.cn, [f] hbxiao03@sina.com, [g] xdzhang@126.com, [h] qmwang@semi.ac.cn

Keywords: microcrystalline silicon, hydrogen plasma treatment, solar cell.

Abstract. Properties of n-i interface are critical for hydrogenated microcrystalline silicon (µc-Si:H) substrate-type (n–i–p) solar cell as it affects carrier collection, which is visible in the red response. Here, we report a remarkable improvement in visible-infrared responses upon hydrogen plasma treatment (HPT) of n/i interface. We demonstrate that hydrogen plasma treatment in the initial stage of a µc-Si:H i layer growth affects the red response of µc-Si:H solar cell. At the optimal deposition condition, 18% higher short-circuit current density was obtained than its count part without using HPT.

Introduction

Hydrogenated microcrystalline silicon (µc-Si:H) is a good absorber material for thin-film solar cells[1,2].To achieve optical enhancement, µc-Si:H layers often have to be deposited on substrates with intentional roughness and the current density Jsc can be increased[3,4], However, it has been reported that the Jsc increase was accompanied by a reduction of open-circuit voltage Voc and fill factor FF and that substrate might result in a detrimental effect[5,6] on the performance of solar cells. [7,8,9] Therefore the balance between optical gain and reduction of Voc and FF losses is needed for the µc-Si:H solar cell [10,11].

On the other hand, the *n* layer is the first deposited layer in nip configuration [6,12] and is affected by the substrate roughness. The *n* layer influences the growth of the subsequent layers such as intrinsic layer in µc-Si:H solar cell. Due to the incubation effect [13] and homogeneous growth process, depositing a high-quality intrinsic µc-Si:H thin layer at initial growth[14,15] onto an n-type silicon layer is tricky. It was reported that the optoelectronic properties of the intrinsic bulk layer in µc-Si:H solar cell could be controlled by adjusting the starting procedure of film growth [14].

In this paper we report a method of alternately deposition a thin µc-Si:H layer and hydrogen plasma treatment (ADHT) [16,17,18] to fabricate n/i interface of µc-Si:H nip solar cell. Our results exhibit that the performance of µc-Si:H substrate solar cel is greatly improved by ADHT in visible-infrared responses.

Experimental

First we investigated how the hydrogen plasma treatment n/i interface affects the external quantum efficiency (QE) of the solar cell .Then we try to figure out the origin of difference in QE. Lastly we applied our new method to a µc-Si:H solar cell at optimized deposition condition.

The deposition system in this paper was a plasma system with a capacitively coupled electrode and the glow discharge was produced by a 13.56 MHz rf power source through a matching network for doped layer (p or n type). For the deposition of intrinsic µc-Si:H (i) layers we used a capacitively coupled very high frequency glow discharge (VHF-GD) plasma reactor at an excitation frequency of

60 MHz. The typical cell structure used in this study was Indium Tin Oxide (ITO)/p/i/n/stainless steel (SS) whose area was 0.07 cm^2. The p, i, and n layers were fabricated in a separated chamber deposition respectively. The detailed fabrication conditions of the n a-Si:H layer ,i µc-Si:H layer and hydrogen plasma treatment are shown in Table1. The i-layer was deposited from pure SiH_4 diluted by hydrogen, whose thickness was about 550 nm. The n-layer was a amorphous silicon that was deposited from a $SiH_4/PH_3/H_2$ mixed gas (thickness about 40 nm). An anti-reflective coating layer ITO and back reflector Ag or Al, Zinc Oxide (ZnO) were prepared by magnetron sputtering method. The solar cell performance was investigated by current–voltage measurements under AM 1.5 illumination through an orifice which corresponds to the active area 0.07 cm^2 using a Xeon lamp solar simulator and external quantum efficiency measurements.

Table 1 Typical deposition condition of n/i µc-Si interface layer

	n- layer	i-layer (initial a-Si:H deposition)	hydrogen plasma treatment
Gas flow rate[SCCM]	SiH_4--3 H_2--100 PH_3--5	SiH_4--3 H_2--60	H_2--60
Pressure [pa]	200	127	121
power density [mW/cm^2]	10 (13.56 MHZ)	50 (60 MHZ)	50 (60 MHZ)
Deposition time [s]	240	300	120
T_s [°C]	250	220	220

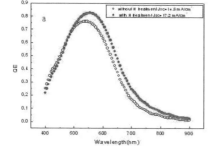 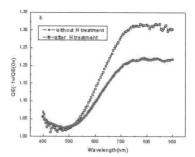

Figure1 QE for µc-Si nip solar cell: (a) At 0V without hydrogen treatment (open circle) and with hydrogen treatment (solid circle); (b) Reverse-bias QE ratio QE(-1V)/QE(0v) without hydrogen treatment (open square) and with hydrogen treatment (solid circle)

Table 2 Defect density at intrinsic silicon layer for nip solar cell from capacitance-voltage measurement

	Without hydrogen treatment (cm^{-3})	With hydrogen treatment(cm^{-3})
Defect density in intrinsic layer in µc-Si nip solar cell N_{Def}	2.37×10^{16}	1.25×10^{16}

Results and discussion

Fig.1 (a) shows the QE of µc-Si:H solar cell. Open circles refer to the µc-Si:H solar cell deposited by conventional method, while the solid circle refers to its counterpart deposited by ADHT method. It can be seen that the long wavelength spectral response is improved by ADHT (solid circles). Fig.1(b) shows the normalized spectra QE(-1V)/QE(0V). QE(-1V) denotes the QE with an applied reverse bias voltage of -1V, QE(0V) represents the QE in the short circuit condition. Fig1(b) exhibits that for

ADHT sample the ratio QE(-1V)/QE(0v) decreases to 1.2 (solid circles) at long wavelength (>700 nm) while for conventional sample the ratio QE(-1V)/QE(0V) is 1.3 (open square). The reduction in QE(-1V)/QE(0 V) indicates lower recombination possibility in the μc-Si:H layer[19] because the long-wavelength QE is mainly dominated by collection of the photogenerated carriers in μc-Si:H layer ,. Correspondingly the Jsc was increase to 17.2 mA/cm^2 with ADHT compared to 14.5 mA/cm^2 of its counterpart. Similar reports show that the p-i-n amorphous silicon solar cell performance improved by the applied ADHT to p/i buffer layer [20]. For both cases the film is improved at it initial stage by ADHT.

To explore this improvement reason we measured the capacitance-voltage to get the defect density N_{def} at intrinsic Si layer as shown in Table 2. It can be seen that N_{def} reduced from 2.37×10^{16} cm^{-3} to 1.25×10^{16} cm^{-3} and decreased 47%. We suggest that hydrogen plasma treatment decrease the thickness of incubation layer at n/i interface and N_{def} effectively, which reduce the carriers recombination possibility and thus improve the QE at longer wavelength. It is the structural relaxation of the Si matrix that results in N_{def} reduction. The structural relaxation of the a might be mediated by penetration of hydrogen atoms into the initial μc-Si:H region[21] and the insertion of H atom into strained Si matrix [22] ,which causes a reduction of the strain energy of the Si lattice [23]and thus decrease the defect density N_{def} of the μc-Si:H layer.

Finally we applied the ADHT technique to a μc-Si nip single junction solar cell on SS with Ag/ZnO as back reflector. QE for such a μc-Si substrate solar cells is shown in Fig. 2 and short current density as 24.7mA/cm2 is obtained under xenon lamp illumination (calibrated 100 mW/cm^2.). QE (λ = 900nm) reaches 25%. Fig. 3 exhibits the current-voltage characterization under Xeon lamp illumination.

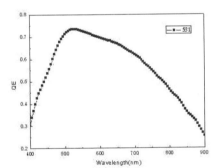

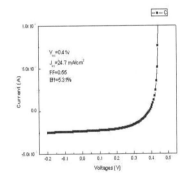

Figure 2 QE at 0 v for μc-Si nip solar cell on SS with Ag/ZnO as back reflector

Figure 3 Current-voltage (J-V) measurements under Xeon lamp illumination

Summary

To improve the conversion efficiencies of μc-Si single junction cells, new type regime to fabricate n/i interface layer were developed. Significantly high long wavelength spectral response (> 550 nm) was attained for μc-Si solar cells. The fabrication of new n/i interface layer by ADHT method resulted in improving the μc-Si film properties and the solar cell performance. As a result, 18% higher short-circuit current density was obtained than its count part without using ADHT at the n/i interface.

Acknowledgments

This work was financially supported by National High Technology Research and Development Program (863 Program) of China (No.2011AA050504) and the Knowledge Innovation Program of the Chinese Academy of Sciences (No. 1KGCX2-YW-383-1).

References

[1] J. Meier, R. Flückiger, H. Keppner, and A. Shah, Complete microcrystalline pin solar cell - Crystalline or amorphous cell behavior, Appl. Phys. Lett. 65 (1994) 860-862.

[2] A.V. Shah, J. Meier, E. Vallat-Sauvain, N. Wyrsch, U. Kroll, C. Droz, U. Graf,Material and solar cell research in microcrystalline silicon, Sol. Energy Mater. Sol. Cells 78 (2003) 469–491

[3] R. H. Franken, R. L. Stolk, H. Li, et al., Understanding light trapping by light scattering textured back electrodes in thin film n-i-p-type silicon solar cells, J. Appl. Phys. 102(2007)014503.

[4] J. Meier, S. Dubail, S. Golay, et al., Microcrystalline silicon and the impact on micromorph tandem solar cells Sol. Energy Mater. Sol. Cells 74 (2002) 457-467

[5] J. Bailat, D. Dominé, R. Schlüchter, et al., High-efficiency P-I-N microcrystalline and micromorph thin film silicon solar cells deposited on LPCVD ZnO coated glass substrates, IEEE 4 WCPEC (2006)1533.

[6] Y. Nasuno, M. Kondo, A. Matsuda, Microcrystalline silicon thin-film solar cells prepared at low temperature using PECVD, Sol. Energy Mater. Sol. Cells 74 (2002) 497-503.

[7] J. Bailat, E. Vallat-Sauvain, L. Feitknecht, et al., Influence of substrate on the microstructure of microcrystalline silicon layers and cells, J. Non-Crystalline Solids 299–302 (2002) 1219-1223

[8] E. Vallat-Sauvain, J. Bailat, J. Meier, et al., Influence of the substrate's surface morphology and chemical nature on the nucleation and growth of microcrystalline silicon, Thin Solid Films 485 (2005) 77-81

[9] M. Python, E. V.Sauvain , J. Bailat, D. Domine´, L. Fesquet, A Shah, C. Ballif,Relation between substrate surface morphology, *J.Non-Cryst. Solids* **354** (2008) 2258–2262

[10] B.T. Li, R.H. Franken, J.K. Rath, etal., Structural defects caused by a rough substrate and their influence on the performance of hydrogenated nano-crystalline silicon n–i–p solar cells, *Sol. Energy Mater. Sol. Cells*, 93 (2009) 338-349

[11] T. Söderström, F.J. Haug, V. Terrazzoni-Daudrix, et al., N/I buffer layer for substrate microcrystalline thin film silicon solar cell, J. Appl. Phys.104 (2008)104505

[12] H. Ihara and H. Nozaki, Improvement of Hydrogenated amorphous silicon n-i-p diode performance by H_2 plasma treatment for i/p interface, Jpn. J. Appl. Phys 29 (1990) L2159

[13] J.H. Zhou, K. Ikuta, T. Yasuda, et al., Growth of amorphous-layer-free microcrystalline silicon on insulating glass substrates by plasma-enhanced chemical vapor deposition, Appl. Phys. Lett. 71(1997) 1534

[14] J. Chantana, Y. Tsutsui, Y. Sobajima, et al., Importance of Starting Procedure for Film Growth in Substrate-Type Microcrystalline-Silicon Solar Cells, Jpn. J. Appl. Phys.50 (2011) 045806

[15] F. Köhler, S. Schicho, B. Wolfrum, et al., Gradient etching of silicon-based thin films for depth-resolved measurements: The example of Raman crystallinity, Thin Solid Films 520 (2012) 2605–2608

[16] Y. Ashida, M. Koyama, K. Miyachi, et al., Properties and stability of a-Si:H films by alternately repeating deposition and hydrogen plasma treatment, IEEE 22(1991)1352

[17] S. Ishihara, D. He, and I. Shimizu, Structure of Polycrystalline silicon thin film fabricated from Fluorinated precursors by layer –by-Layer technique, Jpn. J. Appl. Phys., Part 1, 33 (1994)513.

[18] N. Layadi, P. Roca i Cabarrocas, and B. Drevillon, Reai-time spectroscopic ellipsometry study of the growth of amorphous and microcrystalline silicon thin films prepared by alternating silicon deposition and hydrogen plasma treatment, Phys. Rev. B 52 (1995) 5136

[19] T. Toyama , H. Okamoto,Structural and electrical studies of plasma-deposited polycrystalline silicon thin-films for photovoltaic application, Solar Energy 80(2006).658–666

[20] H. Tanaka, N. Ishiguro, T. Miyashita, et al., Improvement of p-I buffer layer properties by hydrongen plasma treatment and its applications to pin a-Si:H solar cells", Proc. 23th IEEE PVSC (1993) 811

[21] K. Saitoh, M. Kondo, M. Fukawa, et al., Role of the hydrogen plasma treatment in layer-by-layer deposition of microcrystalline silicon, Appl. Phys. Lett. 71(1997) 3403-3405

[22] S. Sriraman, S.Agarwal, E. S. Aydil,D Maroudas, Mechanism of hydrogen-induced crystallization of amorphous silicon,Nature 418(2002) .62-65

[23] I. Kaiser, N. H. Nickel, and W. Fuhs,Hydrogen-mediated structural changes of amorphous and microcrystalline silicon, Phys. Rev. B 58(1998)R1718-1721

Key Engineering Materials Vol. 537 (2013) pp 197-200
© (2013) Trans Tech Publications, Switzerland
doi:10.4028/www.scientific.net/KEM.537.197

The Study of Microcrystalline Silicon Thin Films Prepared by PECVD

Chunya Li[a], Hao Zhang[b], Jun Li[c], Xifeng Li[d], Jianhua Zhang[e]

Key Laboratory of Advanced Display and System Applications, Ministry of Education, Shanghai University, Shanghai 200072, China

[a]lichunya2006@163.com, [b]zhkeylab@shu.edu.cn, [c]lijun_yt@shu.edu.cn, [d]xifeng.li@hotmail.com, [e]jhzhang@staff.shu.edu.cn

Keywords: Microcrystalline silicon thin film, PECVD

Abstract. Under different growth conditions, microcrystalline silicon thin films are deposited successfully on glass substrates by the double-frequency plasma enhanced chemical vapor deposition (PECVD). We report the systematic investigation of the effect of process parameters (hydrogen dilution, substrate temperature, forward power, reaction pressure, et al.) on the growth characteristics of microcrystalline silicon thin films. Raman scattering spectra are used to analyze the crystalline condition of the films and the experimental results. Optimizing the process parameters, the highest crystalline volume fraction of microcrystalline silicon films was achieved. It is found that the crystalline volume fraction of microcrystalline silicon films reaches 72.2% at the reaction pressure of 450 Pa, H_2/SiH_4 flow ratio of 800sccm/10sccm, power of 400 W and substrate temperature of 350 °C.

Introduction

Microcrystalline silicon (μc-Si) thin-film transistors (TFTs) have been studied for applications in electronic devices, including active matrix organic light-emitting devices (OLEDs) and sensor arrays [1-2]. Several deposition techniques have been used to deposit microcrystalline silicon thin films, including hot wire chemical vapor deposition (HWCVD) [3-4], High-Density Microwave Plasma [5-6], and conventional plasma enhanced chemical vapor deposition (PECVD) [7-10]. PECVD is widely used due to its high potential to prepare high quality materials uniformly on a large area substrate.

In this paper, in order to achieve the high-deposition-rate and high quality μc-Si thin film, double-frequency technique (27.12 MHz) was introduced in PECVD equipment. Microcrystalline silicon thin films were deposited successfully on glass substrates by PECVD. Raman scattering spectra were used to analyze the crystalline condition of the films and the experimental results.

Experimental

Microcrystalline silicon thin films were deposited on the corning glass by plasma enhanced chemical vapor deposition (PECVD) method. The radio frequency is 27.12 MHz. In order to analyze the effect of process parameters (hydrogen dilution, substrate temperature, forward power, reaction pressure, et al.) on the crystalline fraction volume, different microcrystalline silicon thin films were deposited and characterized. Microcrystalline silicon thin film was produced form the decomposition of a gas mixture consisting of SiH_4 and H_2, the SiH_4 gas flow is kept at 10 sccm, and H_2 gas flow ranged from 400 sccm to 1000 sccm. The applied power varied from 100 W to 600 W.

The Raman spectra of the films were obtained by a LabRam HR-800 micro-Raman spectrometer. All of the tests are performed at room temperature.

Result and discussion

Effect of hydrogen dilution. In order to study hydrogen dilution ratio of microcrystalline silicon thin films, a series of different concentration of hydrogen dilution of microcrystalline silicon thin films were prepared by of SiH_4 flow is maintained at 10 sccm, hydrogen flow 400 sccm changes to 1000 sccm. The substrate temperature was 350 °C, the reaction pressure was 450 Pa, and the forward power was 400 W.

Figure 1 shows Raman spectra of microcrystalline silicon films under different hydrogen dilution.

Raman spectroscopy was employed to estimate the crystalline volume fraction (Xc) of the μc-Si active layer according to the formula [11-12]

$$X_C = \frac{I_{520} + I_{510}}{I_{520} + I_{510} + I_{480}}$$

(1)

where I_{480}, I_{510}, and I_{520} are the areas of the corresponding Gaussian peaks used to fit the measured Raman spectra near 480 cm^{-1} (associated with the amorphous phase), 510 cm^{-1} and 520 cm^{-1} (associated with the crystalline phase), respectively.

The crystallization rate of microcrystalline silicon thin film changes when hydrogen flow changes from 400sccm to 1000sccm. The detailed data is shown in Figure 2. It is seen that the hydrogen dilution have an important impact on the crystalline volume fraction of microcrystalline silicon films.

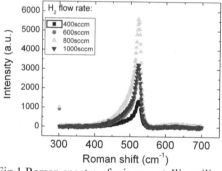

Fig.1 Raman spectra of microcrystalline silicon films under different hydrogen dilution.

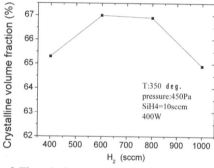

Fig.2 The relationship of hydrogen dilution and crystalline fraction volume

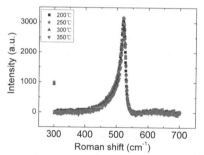

Fig. 3 Raman spectra of microcrystalline silicon films at different substrate temperatures

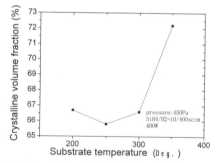

Fig. 4 The relationship of substrate temperature and crystalline volume fraction

Effect of substrate temperature. The reaction pressure is 450 Pa, H_2/SiH_4 flow ratio is 800sccm/10sccm, and power is 400 W. We prepared a series of different substrate temperature ceramics silicon thin film samples. Figure 3 shows the Raman spectra of different substrate temperature growth of microcrystalline silicon thin film. The calculation shows that, when the substrate temperature from 200 °C to 350 °C, the crystalline volume fraction with the rise of the substrate temperature increases, the crystalline volume fraction change from 66.7% to 72.2%, shown in Figure 4.

Effect of forward power. In order to study the impact of the RF power density on the nature of the microcrystalline silicon thin film, the reaction pressure is 450 Pa, H_2/SiH_4 flow ratio is 800sccm/10sccm, substrate temperature is 350°C, we prepared a series of different RF power microcrystalline silicon thin films, the RF power density changes from 100 W to 600 W. Figure 5 Raman spectra of microcrystalline silicon films prepared under different RF power. After Gaussian fitting of the Raman spectra, we have come to the RF power density of 100W, 200 W, 300 W and 600 W, the crystalline volume fraction of microcrystalline silicon films were 62.5%, 67.4%, 67.3% and 60.6%, more detailed results shown in Figure 6. Be seen as the power density increases, the crystallization of microcrystalline silicon thin film is the first to increase and then decrease.

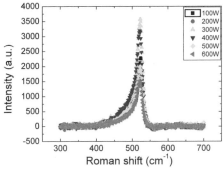

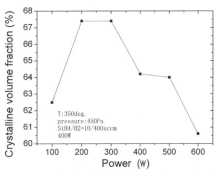

Fig. 5 Raman spectra of microcrystalline silicon films under different forward powers

Fig. 6 The relationship of forward power and crystalline volume fraction

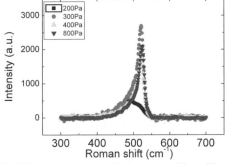

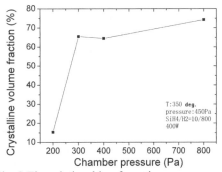

Fig. 7 Raman spectra of microcrystalline silicon films under different reaction pressure

Fig. 8 The relationship of reaction pressure and crystalline volume fraction

Effect of reaction pressure. In order to study the reaction gas pressure of microcrystalline silicon thin films, H_2/SiH_4 traffic than the conditions under for 800sccm /10sccm power density of 400 W, substrate temperature of 350°C, we were prepared under different reaction gas pressure the a series of microcrystalline silicon thin film samples, the reaction gas pressure by 200 Pa changes to 800 Pa. Figure 7 shows the Raman spectra of microcrystalline silicon thin film growth under different reaction pressure. When the gas pressure varies from 200 Pa to 800 Pa, the crystalline volume fraction of μc-Si film first increases quickly then leads to steady. The relationship of the crystalline volume fraction and the gas pressure is shown in Figure 8.

Summary

Microcrystalline silicon thin films are deposited successfully on glass substrates by the double-frequency plasma enhanced chemical vapor deposition (PECVD). We report the systematic investigation of the effect of process parameters (hydrogen dilution, substrate temperature, forward power, reaction pressure, et al.) on the growth characteristics of microcrystalline silicon thin films. Optimizing the process parameters, the highest crystalline volume fraction of microcrystalline silicon films was achieved. It is found that the crystalline volume fraction of microcrystalline silicon films reaches 72.2% at the reaction pressure of 450 Pa, H_2/SiH_4 flow ratio of 800sccm/10sccm, power of 400 W and substrate temperature of 350 °C.

Acknowledgements

This work was financially supported by the Science and Technology Commission of Shanghai Municipality (Grant No. 10DZ1100102), by national natural science foundation of china (Grant No. 61006005) and by excellent youth project funding of Shanghai Municipal Education Commission.

References

[1] R.A. Street, Thin-Film Transistors, Adv. Mater. 21 (2009) 2007–2022

[2] D. Knipp, R.A. Street, H. Stiebig, et al., Vertically Integrated Amorphous Silicon Color Sensor Arrays, IEEE Trans.Electron. 53 (2006) 1551-1558.

[3] Y. Mai, S. Klein, R. Carius, et al., Open circuit voltage improvement of high-deposition-rate microcrystalline silicon solar cells by hot wire interface layers, Appl. Phys. Lett. 87 (2005) 1-3.

[4] S. Klein, F. Finger, R. Carius, *et al.,* Intrinsic microcrystalline silicon prepared by hot-wire chemical vapour deposition for thin film solar cells, Thin Solid Films. 430 (2003) 202-207.

[5] H. Jia, H. Shirai, In Situ Study On the Growth of Microcrystalline Silicon Film Using the High-Density Microwave Plasma for Si Thin Film Solar Cells, Thin Solid Films. 506 (2006) 27-32.

[6] Y. Sakuma, L. Haiping, H. Ueyama, et al., High-Density Microwave Plasma for High-Rate and Low-Temperature Deposition of Silicon Thin Film, Vacuum 59 (2000) 266-276.

[7] L.H. Guo, R.M. Lin, Studies on the formation of microcrystalline silicon with PECVD under low and high working pressure, Thin Solid Films. 376 (2000) 249-254.

[8] P.Roca i Cabarrocas, R.Brenot, P.Bulkin, et al., Stable microcrystalline silicon thin-film transistors produced by the layer-by-layer technique, J. Appl. Phys. 86 (1999) 7079-7082.

[9] E.Amanatides, D. Mataras and D.E.Rapakoulias, Deposition rate optimization in SiH_4/H_2 PECVD of hydrogenated microcrystalline silicon, Thin Solid Films. 383 (2001) 15-18.

[10] R.Platz and S.Wagner, Intrinsic microcrystalline silicon by plasma-enhanced chemical vapor deposition from dichlorosilane, Appl. Phys. Lett. 73 (1998) 1236-1238.

[11] J. K. Rath, R. H. Franken, A. Gordijn, and R. E. Schropp, et al., Growth mechanism of microcrystalline silicon at high pressure conditions, J. Non Cryst. Solids. 338 (2004) 56-60.

[12] V. G. Golubev, V. Y. Davydov, A. V. Medvedev, and A. B. Pevtsov, et al., Raman scattering spectra and electrical conductivity of thin silicon films with a mixed amorphous-nanocrystalline phase composition: determination of the nanocrystalline volume fraction, Phys. Solid State. 39 (1997) 1197-1201.

Key Engineering Materials Vol. 537 (2013) pp 201-204
© (2013) Trans Tech Publications, Switzerland
doi:10.4028/www.scientific.net/KEM.537.201

Annealing Effect on Reversible Photochromic Properties of Ag@TiO$_2$ Nanocomposite Film

ZUO Juan[1,2]

[1] Department of Materials Science and Engineering, Xiamen University of Technology, Xiamen, China

[2] Department of Interface Chemistry and Surface Engineering, Max-Planck-Institut für Eisenforschung GmbH, Düsseldorf, Germany

zuojuan@xmut.edu.cn

Keywords: Ag@TiO$_2$, Nanocomposite film, RF magnetron sputtering, Annealing, Photochromic.

Abstract. Large-scale uniform Ag@TiO$_2$ films was prepared by RF magnetron sputtering in pure Ar plasma using polycrystalline TiO$_2$ semiconductor sintered target. The effect of annealing on the photochromic properties was studied to obtain a better understanding of the interaction of the structure. Ultraviolet-visible absorption and scanning electron microscopy were performed to investigate the possibility of tailoring the structure with consequent modification of the optical properties. Ag nanoparticles were formed between TiO$_2$ films after annealing the samples with Ag film structure. The annealed Ag@TiO$_2$ films present a photochromic property in comparison with the as-prepared samples. Such nanocomposite films can be used as smart windows, high density multiwavelength optical memory and rewritable electronic paper.

Introduction

Ag@TiO$_2$ nanocomposite films have recently received increasing attention due to their potential applications, such as solar energy conversion [1], photocatalysis, heat mirror [2], antibacterial coatings [3] and optical components[4-9]. These properties result from quantum size effects of the embedded nanoparticles in the TiO$_2$ matrix, and interface and/or surface effects between nano-particles and the TiO$_2$ matrix. TiO$_2$ as the matrix can create a new type of functional nanocomposite where the interaction between the nanoparticles and matrix can be utilized, like the reversible photochromism properties of Ag embedded in TiO$_2$ matrix [4, 6-8]. Photochromic materials can change their colors reversibly in response to UV and visible light. They can be applied to smart windows, high density multiwavelength optical memory and rewritable electronic paper[4, 6-8].

Ag@TiO$_2$ nanocomposite film are traditionally prepared via sol-gel method[3], photo-reduction of Ag under UV irradiation on nanoporous TiO$_2$ film by spin-coating[5, 6], layer-by-layer RF-magnetron deposition [8] and hybrid RF-sputtering/sol-gel routes[10]. RF magnetron sputtering is one of the most feasible methods due to its inherent versatility and the capability of obtaining a homogeneous surface coverage at low temperatures. The main advantage of RF magnetron sputtering is the capability to produce metal nanoparticles with desired size, shape and distribution by a proper choice of process parameters such as RF-power, pressure, substrate temperature, deposition time and annealing process[11-13]. In previous works by other groups, a series of Ag embedded TiO$_2$ film with different metal atomic contents were prepared by co-sputtering method[14] or layer-by-layer deposition using metallic Ti and Ag as targets in 100% O$_2$ or Ar & O$_2$ mixture gas [8]. The as-prepared nanocomposite films generally consist of amorphous or crystalline metal nanoparticles in amorphous TiO$_2$ matrix. The subsequently annealing process induced the agglomeration of nanoparticles and crystallization of TiO$_2$ matrix. The main inconvenience of these methods is the poor stability of the thin Ag layer in the reactive O$_2$ plasma (specially the Ag nanopartices) when TiO$_2$ films are deposited on an Ag layer, e.g. the oxidation of Ag[2]. Thus, the optical properties of the Ag layer can be easily damaged during the reactive plasma process, e.g. degradation, oxidation and tarnishing[15, 16]. A simple method to deposit TiO$_2$ film by RF-sputtering in pure Ar plasma using

TiO_2 semiconductor target, instead of the conventional Ti metal target were developed in order to prevent that. Large-scale uniform $Ag@TiO_2$ multilayer films were prepared layer-by-layer on substrates. These nanocomposite coatings exhibit reversible photochromic behavior and are expected to have a wide range of application.

Experimental

Preparation. Pure polycrystalline TiO_2 sintered target (99.9%) and Ag metal targets (99.99%) were used to prepare the films under high vacuum condition. The targets were powered by a RF generator at the frequency of 13.56 MHz. Arand O_2 were used as plasma gas. Indium tin oxide glass (≤ 20Ohm/sq.) and quartz glass were used as substrates.

Characterization. The film thickness was controlled by QCM and calibrated afterwards by spectroscopic ellipsometry. Scanning electron microscopy (SEM) characterization was performed with a LEO 1550VP Field-Emission Scanning Electron Microscope. UV-Vis spectra measurements were measured with a Lambda 800 spectrophotometer. Films for photochromic measurement with spot areas of approximate 2×2 cm^2 were irradiated with a red helium-neon laser (632.8nm, 15mW) in ambient air in a dark room. The UV light source is a 48W lamp (365nm).

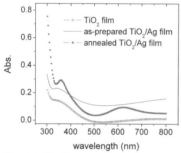

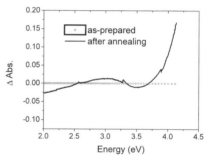

Fig. 1 UV-Vis spectra of TiO_2 films, as-prepared and annealed TiO_2/Ag films.

Fig. 2 Differential absorption spectra (Δ absorbance) after red light irradiation for 10 minutes of the as-prepared and annealed TiO_2/Ag/TiO_2 films.

Results and Discussions

The optical properties of TiO_2 films and TiO_2/Ag films were investigated by UV-Vis absorption spectrum as shown in Fig. 1. TiO_2 films deposited by RF magnetron sputtering at room temperature are generally amorphous[11, 13].The bandgap of TiO_2 films are around 3.2 eV and hence they have almost zero absorption in the range of visible light. The curve in Fig. 1 from TiO_2 film shows almost zero absorption in the range of 500-800 nm and has a rapid increase from 450 nm till 300 nm which can be assigned to the bandgap absorption. The second layer of TiO_2 film results in the increase of absorption in the whole UV-vis range, but no other new peaks appear. After annealing at 500°C for one hour, the absorption displays a characteristic surface plasmon resonance (SPR) of Ag nanoparticles centered at around 618 nm for the island-like film. The electrons are trapped in individual clusters or islands and cannot move freely like in a closed film. The electromagnetic field of an incoming light wave can excite surface plasmons which lead to absorption.

The TiO_2/Ag/TiO_2 multilayer film was irradiated with a laser at wavelength of 632.8 nm for 10 min. The absorbance of irradiated films measured by UV-Vis spectroscopy was subtracted by absorbance of the non-irradiated films. The as-prepared film shows no change before and after irradiation of UV light as shown in Fig. 2. However, an absorption decrease at the laser wavelength by the annealed samples is observed in the differential absorption spectrum. This indicates the nanocomposite films with SPR resulting from Ag nanoparticles after annealing have photochromic properties. This results from the effect of spectral hole burning by laser irradiation and its reversal by UV irradiation. The color of the nanocomposite film initially depends on the structure, size, shape and arrangement of the embedded Ag nanoparticles in the TiO_2 matrix, but changes under monochromic visible light approximately to the same color. By irradiation with UV light, the colored nanocomposite film can turn back to its initial color. Tatsuma et. al. [4, 6, 7] proposed a possible mechanism for the photochromic process by irradiation with monochromic and UV light. In this model, the Ag nanoparticles with a plasmon resonance matching the incident light frequency adsorb the energy which excites the electrons in the Ag and emitted them into TiO_2 conduction band. Subsequently, the electrons are captured by adsorbed oxygen on the TiO_2 surface, resulting in oxidation of the Ag nanoparticles to Ag^+ and in consequence an absorption decrease at the corresponding light wavelength.

The effect of the annealing on the reversible photochromic properties of $Ag@TiO_2$ nanocomposite film was investigated. Coinage metals (Au, Ag, and Cu) deposited on the metal oxide by RF-sputtering has three major morphological groups. (see e.g. [10, 13, 17]): (1) cluster-like system where metal crystallites have a narrow size distribution and spherical shape. (2) island-like system where metal crystallites are partially interconnected between each other and their geometry changes progressively from the typical spherical-like shape of isolated clusters to oblate or prolate particles resulting from cluster agglomeration. (3) Continuous films where all particles are interconnected to form a coating of variable thickness on the oxide surface. Generally, this nanosystem with well tailored structural and optical properties can be obtained as a function of the applied RF-power, total pressure, deposition time and heat treatment [13]. The film structure of Ag was obtained on TiO2 film by adjusting the at DC potential 164V, $w = 100$ W, $p = 0.01$ mbar, $t = 30s$ as shown in Fig. 3 (a). The films were annealed at $500^{\circ}C$ in N_2 atmosphere for one hour. The top view is shown in Fig. 3 (b). The Ag films start to aggregate along the surface during annealing and thus large nanoparticles 40-100 nm in size on TiO_2 film were formed. These annealed Ag has large distance between nanoparticles. The interface energy could play a crucial role in determining the final shape and size of Ag clusters after annealing. Ag aggregation can minimizes the interface energy by decreasing the contact area between the two materials Ag and TiO_2 and thus formed nanoparticles with lower surface-to-volume ratio.

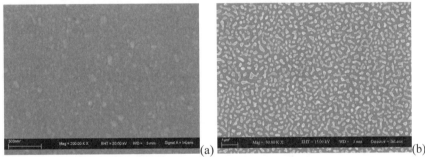

(a)   (b)

Fig. 3 SEM images of (a) as prepared and (b) annealed TiO_2/Ag film.

Conclusion

$Ag@TiO_2$ nanocomposite film were deposited by RF magnetron sputtering under soft synthesis conditions, with particular attention to annealing effect on the film morphology and optical properties. Controlled variations of these deposition parameters were performed with the aim of tailoring the system structure and optical properties as a function of processing conditions. After annealing the films the Ag film between two TiO_2 films tended to aggregated along the interface and thus decrease thermal expansion mismatch stresses between the TiO_2 film and the Ag film. The formation of island structure can minimize the interface energy between the two materials. The comparison of as-prepared and annealed films after UV irradiation indicates the nanocomposite films after annealing have photochromic properties, which means the SPR resulting from Ag nanoparticles is necessary for such properties.

Compared to the conventional method of photocatalytically deposited Ag on nanoporous anatase TiO_2 film to prepare photochromic materials, the Ag nanoparticles have higher density and dense packed structure with controllable size, shape and distribution prepared by RF-magnetron sputtering. Moreover, the pure applied Ar plasma in our work are believed to avoid the oxidation of Ag in the interface between TiO_2 and Ag films during TiO_2 deposition and thus the films are expected to have better optical properties and long-term stability.

Acknowledgement

The author would like to thank the financial support from the International Max-Planck-Research School (SurMat), the Nature Science Foundation of Fujian (No. 2011J05143) and the Education Department of Fujian Province (No. JK2010053).

References

[1] S. Ito, T. Takeuchi, T. Katayama, *et al.* Conductive and transparent multilayer films for low-temperature-sintered mesoporous TiO_2 electrodes of dye-sensitized solar cells. Chem. Mater.,15(2003):2824-2828.

[2] A. Romanyuk and P. Oelhafen. Formation and electronic structure of TiO2-Ag interface. Sol. Energy Mater. Sol. Cells,91(2007):1051-1054.

[3] L. Z. Zhang, J. C. Yu, H. Y. Yip, *et al.* Ambient light reduction strategy to synthesize silver nanoparticles and silver-coated TiO2 with enhanced photocatalytic and bactericidal activities. Langmuir,19(2003):10372-10380.

[4] K. Naoi, Y. Ohko and T. Tatsuma. TiO2 films loaded with silver nanoparticles: Control of multicolor photochromic behavior. J. Am. Chem. Soc.,126(2004):3664-3668.

[5] L. L. Bao, S. M. Mahurin and S. Dai. Controlled layer-by-layer formation of ultrathin TiO2 on silver island films via a surface sol-gel method for surface-enhanced Raman scattering measurement. Anal. Chem.,76(2004):4531-4536.

[6] K. Kawahara, K. Suzuki, Y. Ohka, *et al.* Electron transport in silver-semiconductor nano-composite films exhibiting multicolor photochromism. Phys. Chem. Chem. Phys.,7 (2005): 3851-3855.

[7] K. L. Kelly and K. Yamashita. Nanostructure of silver metal produced photocatalytically in TiO_2 films and the mechanism of the resulting photochromic behavior. J. Phys. Chem. B,110 (2006): 7743-7749.

[8] J. Okumu, C. Dahmen, A. N. Sprafke, M. Luysberg, G. von Plessen and M. Wuttig. Photochromic silver nanoparticles fabricated by sputter deposition. J. Appl. Phys.,97(2005):094305.

[9] C. Y. Wang, C. Y. Liu, X. B. Yan, *et al.* Investigation on the behavior of porphyrins at the surface of the colloidal silver particles. J. Photochem. Photobiol., A,104(1997):159-163.

[10] L. Armelao, D. Barreca, G. Bottaro, *et al.* Rational Design of Ag/TiO2 Nanosystems by a Combined RF-Sputtering/Sol-Gel Approach. ChemPhysChem,10(2009):3249-3259.

[11] J. Zuo, P. Keil, G. Grundmeier. Synthesis and Characterization of photochromic Ag-embedded TiO_2 nanocomposite thin films by non-reactive RF-magnetron sputter deposition. Appl. Surf. Sci.,258 (2012): 7231-7237.

[12] H. B. Liao, R. F. Xiao, J. S. Fu, *et al.* Large third-order optical nonlinearity in Au:SiO2 composite films near the percolation threshold. Appl. Phys. Lett.,70(1997):1-3.

[13] J. Zuo. Deposition of Ag nanostructures on TiO2 thin films by RF magnetron sputtering. Appl. Surf. Sci.,256(2010):7096-7101.

[14] Q. Q. Wang, S. F. Wang, W. T. Hang and Q. H. Gong. Optical resonant absorption and third-order nonlinearity of (Au,Ag)-TiO2 granular composite films. J. Phys. D: Appl. Phys.,38(2005):389-391.

[15] R. J. Martin-Palma and J. M. Martinez-Duart. Ni-Cr passivation of very thin Ag films for low-emissivity multilayer coatings. J. Vac. Sci. Technol., A,17(1999):3449-3451.

[16] R. C. Ross, R. Sherman, R. A. Bunger and S. J. Nadel. Plasma Oxidation of Silver and Zinc in Low- Emissivity Stacks. Solar Energy Materials,19(1989):55-65.

[17] D. Barreca, A. Gasparotto, E. Tondello, *et al.*. Influence of process parameters on the morphology of Au/SiO₂ nanocomposites synthesized by radio-frequency sputtering. J. Appl. Phys.,96 (2004): 1655-1665.

Key Engineering Materials Vol. 537 (2013) pp 205-208
© (2013) Trans Tech Publications, Switzerland
doi:10.4028/www.scientific.net/KEM.537.205

Preparation at Low-Temperature and Characterization of TiO$_2$ Film Used for Solar Cells

Wang Xiaoqiang[1, a], Li Yueliu[1], Wang Jianmin[1], Guo Jing[1], Li Mingya[1]

[1]School of Resources and Materials, Northeastern University at Qinhuangdao Branch, Qinhuangdao 066004, P. R. China

[a]wangxq18@gmail.com(corresponding author)

Keywords: TiO$_2$ nanotube arrays, Anodic oxidation, Ti mesh, Photoelectric properties.

Abstract. In this work, nanocrystalline TiO$_2$ powder was prepared by the sol-gel method via tetrabutyl titanate as raw material, non-ion surfactant TO8 as a template. Then the nanocrystalline TiO$_2$ thin film electrodes which were coated on FTO glass substrates via the slurry consisting of TiO$_2$ powder prepared by us and the trabutyl titanate precursor were successfully prepared by using a simple and convenient hydrothermal method at low temperature. The structure and morphology of powders and films were characterized by X-ray diffraction, scanning electric microscopy. The influence factors on the samples were discussed. The photoelectric properties of cells assembled by the films were measured. The results show that, at 25°C and under 1000W/m^2 light intensity, open voltage is 708mV, Jsc is 14.648mA/cm^2, fill factor is 53.788, the conversion efficiency is 5.5988%.

Introduction

The substrate of the traditional dye-sensitized solar cell (DSSC) is the conductive glass substrate. Instead of conductive glass substrate, the flexible DSSC assemble with flexible conductive plastic film has many advantages, such as light weight, good flexibility, impact resistance, low cost various shapes or surface design and so on, which can be continuously produced in large area via roll-in-roll or fast coating technology. This would reduce the cost of production and make the DSSC has stronger competitiveness. In recent years, it's become a new hot spot in the field of DSSC research [1-3]. At present, the photoelectric conversion efficiency of flexible DSSC is about 1%-7% [4-7]. In order to ensure the best physical contact and electrical contact between particle and particle of semiconductor film or between particle and the substrate, in general, the TiO$_2$ films need high heat sintering. But the flexible substrate can't withstand high temperature which limits its applied range.

In this paper, TiO$_2$ nano powders with good dispersancy were prepared by sol-gel method with non-ion surface active agent TO8 as the template. The TiO$_2$ film was coated on the FTO substrate via the the the slurry consiting of TiO$_2$ powder prepared by us and the retrabutyl titanate precursor. Then the mesoporous TiO$_2$ film was obtained by hydrothermal method at low temperature. At last, the cell was assembled and the photoelectric properties were characterized.

Experimental

Put 100mL absolute alcohol into beaker, 1.7 ml tetrabutyl titanate was slowly added at room temperature. After stirring, the surfactants aqueous solution TO8 was added, after stirring 10min, the beaker was moved into water bath and reacted in 2h at 60°C constant temperature. After centrifugal separation and washing in several times via ethanol and deionized water, the product was dried under natural conditions. The annealing of dried product was carried out at a rate of 5°C/min, 300°C for 1h, 550°C for 1h. Then the nanometer TiO$_2$ powders were obtained.

Blend 40.2ml solution of tetrabutyl titanate in n-butanol (concentration is 1mol/L) and 0.53g self-made TiO$_2$ powders, then paste with good viscosity was obtained through thorough stirring for 2h. After coating the paste on the FTO glass substrate, the film was put into the airtight container full of deionized water. Then the mesoporous TiO$_2$ electrode was obtained by oil bath for 4h at 100°C.

After heat treatment of electrode in drying oven for 1h at 100 °C, the film was immediately immersed into the solution of N719 in ethanol (5×10^{-4} mol/L), soaking time is 12h. At last, the dye-sentisized cell was assembled.

The crystal structures of powders and film were characterized by X-ray diffraction (DX-2500 diffractomter, Cu Kα λ=0.15405nm). The microstructures of the TiO$_2$ powders and film were analyzed by using scanning electron microscopy (Hitachi S-4800). The photocurrent (I) and photovoltage (V) of the cell were measured with an active area of 0.36cm^2 using a simulated sunlight at AM-1.5 produed by 150W San-Ei Solar Simulator.

Results and discussion

Crystal structure. Fig. 1 are the XRD patterns of the annealed TiO$_2$ powders synthesized by sol-gel method for different dosage of surfactant TO8 which is 0.7ml, 0.8ml, 0.9ml in respective. It can be seen, after annealing at 500 °C, the diffraction peaks appeared at 25.281°, 37.800°, 48.049°, 55.060° and 62.688° which is corresponding to anatase (101), (004), (200), (211) and (204) crystal face. In addition, there is no obvious difference between the samples which TO8 amount is 0.8ml and 0.9ml in respectively, but the only small difference appears when the TO8 amount is 0.7ml. It indicates that the impurity phase of the annealed samples decreased with the addition of surfactant.

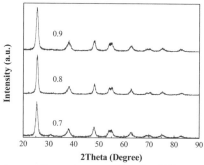

Fig. 1 The XRD patterns of samples-annealed for different dosage of surfactant

Micro-morphology. Fig.2 is the SEM images of TiO$_2$ powders prepared by sol-gel method for different dosages of surfactant. When the amount of the surfactant TO8 is 0.7ml, the particle size is relatively coarse. But when the amounts of TO8 are 0.8 and 0.9ml, the distribution of the particle size is uniform and the size is about 800nm. Contrast Fig.2 (a) and (b), it is found that the level of cluster phenomenon is way down when the amount of TO8 is 0.8ml.

Fig. 2 The SEM images of TiO$_2$ powders prepared by sol-gel method for different dosages of TO8
(a) 0.7ml TO8, (b) 0.8ml TO8, (c) 0.9ml TO8

Microstructure and morphology of TiO$_2$ film. Fig. 3 and Fig.4 is the XRD pattern and SEM image of the TiO$_2$ film prepared by hydrothermal method. According to XRD pattern, it can be seen that there is the obvious diffraction peak that exhibit anatase structure. It indicates that the anatase structure of TiO$_2$ has been obtained by the hydrothermal method at low-temperature. The SEM image shows the morphology of the film. The surface of the film is smooth and the structure is dense. It indicates that the TiO$_2$ film with good morphology and structure was prepared by hydrothermal method.

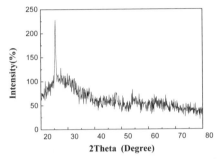

Fig. 3 The XRD pattern of TiO$_2$ film prepared By   Fig. 4 The SEM image of TiO$_2$ film prepared by
the hydrothermal method   the hydrothermal method

I-V characterization. In this work, the solar cell with TiO$_2$ film prepared by hydrothermal method was assembled. The negative electrode is platinum film and sensitization dye is N719. The I-V characteristics of the cell are shown in Fig. 5. The operating voltage of the cell is about 708mV, the short-circuit current (J$_{sc}$) is about 14.685mA/cm^2, the fill factor is about 53.788%, the conversion efficiency is about 5.5988%.

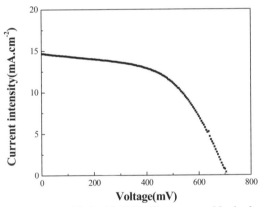

Fig. 8 I-V curves of cell assembled with TiO$_2$ film prepared by hydrothermal method

Summary

Nanocrystalline TiO$_2$ powder was synthesized by the sol-gel method based on the non-ion surfactant TO8 as a template. TiO$_2$ film electrodes which were coated on FTO glass substrates were prepared in success by adopting the slurry consisting of TiO$_2$ powder and the trabutyl titanate precursor. The influencing rule of experimental operating parameters on the structure and morphology of powders and films were analyzed. The dosage of TO8 is the key factor. The test result about the photoelectric properties of cell assembled by the film shows that the conversion efficiency is 5.5988%.

Acknowledgments

This work was supported by the Fundamental Research Funds for the Central Universities (NO. N100423006, NO. N110823001).

References

[1] T. Kado, M. Yamaguchi, Y. Yamada, et al. Low Temperature Preparation of Nano-porous TiO$_2$ Layers for Plastic Dye-sensitized Solar Cells, Chem. Lett., 15 (2003) 1056-1062.

[2] C. Longo, J. Freitas, M. A. DePaoli, Performance and Stability of TiO$_2$/Dye Solar Cells Assembled with Flexible Electrodes and a Polymer Electrolyte, J. Photochem. Photobiol. A: Chem., 23 (2003) 33-40.

[3] T. Yamaguchi, N. Tobe, D. Matsumoto, H. Arakawa, Highly Efficient Plastic Substrate Dye-sensitized Solar Cells using a Compression Method for Preparation of TiO$_2$ Photoelectrodes, J. Chem. Commun., 18 (2007) 4767-4768.

[4] H. Lindstrom, A. Holmberg, E. Magnusson, et al., Optimization of Dye-sensitized Solar Cells Prepared by Compression Method, J. Nano. Lett., 125 (2002) 11-15.

[5] D. S. Zhang, J. A. Downing, F. J. Knorr, et al., Room-temperature Preparation of Nanocrystalline TiO$_2$ Films and the Influence of Surface Properties on Dye-sensitized Solar Energy Conversion, J. Phys Chem B., 110 (2006) 21890-21898.

[6] J. Nemoto, M. Sakata, T. Hoshi, et al., All-plastic Dye-sensitized Solar Cell using a Polysaccharide Film Containing Excess Redox Electrolyte Solution, J. Electroanal Chem., 62 (2007) 23-40.

[7] Y. Takeshi, T. Nobuyuki, et al. Highly Efficient Plastic Substrate Dye-sensitized Solar Cells using a Compression Method for Preparation of TiO$_2$ Photoelectrodes, Chem Commun., 32 (2007) 4767-4783.

Key Engineering Materials Vol. 537 (2013) pp 209-213
© (2013) Trans Tech Publications, Switzerland
doi:10.4028/www.scientific.net/KEM.537.209

Study on Properties of Thick-Film Front Silver Electrodes for Silicon Solar Cells

Ming Fu[1, a], Gonglei Jin[1], Xiao Ding[1], Lin Fan[2] and Dong Chen[2]

[1]Department of Electronic Science and Technology, Huazhong University of Science and Technology, Wuhan 430074, P.R. China

[2]Wuhan Supernano Optoelec Technology Co. Ltd., Wuhan 430206, P.R. China

[a]fuming@mail.hust.edu.cn

Keywords: Silicon Solar Cells, Thick-Film Silver Electrodes, Ohmic Contact, Silver Pastes

Abstract. The front electrode is usually made by the screen printing thick-film silver pastes and the high-temperature firing process in industrial production of silicon solar cells. This paper analyzed the ohmic contact mechanism of thick-film front silver electrodes and studied the microstructure of Ag-Si interface by SEM. The paste samples, used to form front silver electrodes of silicon solar cells, were prepared. Thick-film silver electrodes were printed on silicon wafers with different sheet resistances, and the relationships between the sheet resistances and the contact properties were investigated by changing the firing temperature. By adding right amount of phosphorus compounds to the silver paste, the effects of the donor-doping (N-doping) concentrations on the series resistance of cells were studied. The experimental results show that firing temperature is critical to the Ag-Si ohmic contact, particularly when the silver pastes are designed for the wafers with high sheet resistance and the right amount of N-doping addition in the paste may decrease the series resistances of solar cells.

Introduction

In industrial production of monocrystal silicon solar cells and polysilicon solar cells, the front electrodes are usually made by thick-film Ag pastes by using screen-printing and high-temperature firing process. Research on high-performance front Ag pastes is one of the high lights in recent silicon solar cells techniques. Currently, the properties of front silver pastes have a wide promotion space in industrial production. At the same time, there are many reports on the mechanism of the front Ag electrodes, which put forward some structure models and theoretical assumptions [1-4]. These models can explain some phenomena and experimental results but still need further supplement to meet the practical applications. This paper studied the mechanism and characteristics of the front electrodes based upon some experiments and the micro structures by SEM. Some meaningful results to improve the conversion efficiency of solar cells were proposed.

Mechanism Analysis

Structure analysis of Si solar cells. The structure of silicon solar cells is a parallel multi-layer structure, which stems from its special manufacturing process [3]. The production process of the solar cells is shown as below: the p-type silicon substrate (about 200μm thick) is used, after cleaning and texture etching of the wafer, a undulating n-type silicon layer (less than 0.6μm thick) is made on the surface of p-type silicon substrate, by diffusion of liquid phosphorus source. Then the surface of n-type silicon layer is covered with a silicon nitride antireflection coating (about 70 ~ 80 nm thick) by PECVD. Finally, the back silver electrodes, back surface field aluminium film and the front silver electrodes are fabricated by screen printing and firing process. Fig. 1 (left) is the SEM photograph of cell cross section. Fig. 1 (right) is the schematic structure of SiN_x antireflection coating. There are some differences between monocrystal silicon cells and polysilicon cells in the surface microstructure due to the different texture etchings [5], but the electrical parameters are basically consistent, and the test results of electrode properties also show the same results. To ensure the credibility of the experimental conclusions, the experiments are all based on monocrystal silicon wafers in this paper.

The N-doping concentration decreased rapidly from the surface to the interior on the aforementioned n-type silicon layer (about 0.6μm thick), and the more shallow the junction depth is, the bigger the sheet resistances become. In this experiment, the sheet resistances of silicon wafers were chosed as 45Ω/□, 60Ω/□, 75Ω/□ and 90Ω/□, respectively.

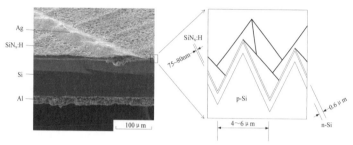

Fig. 1 SEM image of cell cross section (left) and schematic of antireflection coating (right)

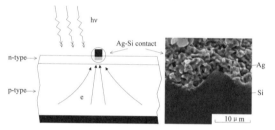

Fig. 2 Front Ag-Si contact of Silicon Solar Cell

Ag-Si ohmic contact of front silver electrodes. For the metal-semiconductor interface, the atomic energy level changed gradually. There is a transition area from semiconductors to metals because of the continuous transformation from covalent bond of the semiconductor to metal bond of the mental. That is to say, metal-semiconductor interface is the inter diffusion zone of the two kinds of atoms, also known as the metal-semiconductor alloyed region. There are two types of metal-semiconductor alloyed zone: one type is the atoms of metal and semiconductor mutual diffuse and dissolve with each other, but this process does not form any new compounds; the other type is that metal and semiconductor will form new compounds after a high-temperature chemical reaction. Ag-Si interface belongs to the first type.

Silver and silicon will inter infiltration after contacting with each other at high temperature (about 700°C ~ 800°C), a small amount of silicon atoms spread to the metal silver, and a small amount of silver atoms diffuse into silicon [6,7]. So a continuous transition and differential interface environment formed in the vertical direction with the silicon surface. Fig. 2 is the schematic and SEM image of front Ag-Si contact of silicon solar cells. Due to the differences of the contact and diffusion between the silver film and the silicon surface in different parts, Ag-Si contact forms and conductive mechanism of the front-side silver electrodes can not be discribed by only using a single model. Normally, several current conduction mechanism may be valid in the mean time for the front silver electrodes. In the n-type shallow junction region, the silver contacts with the high doping level silicon. As the firing deepens, the N-doping concentration of n-type silicon contacted with the silver will decrease.

According to the metal-semiconductor contact theory, ohmic contact closely relates to two factors, one is the barrier height $(q\phi_{Bn})$ between metal silver and semiconductor silicon, the other is the doping concentration (E_{00}) of semiconductor silicon. When the doping concentration is low, the changes of contact resistance (R_c) subject to Eq. 1, the lower the barrier height is, the less the contact resistance becomes. When the doping concentration is moderate, the changes of contact resistance (R_c) subject to Eq. 2, the contact resistance is determined by both barrier height and doping concentration. When the doping concentration is at a high level, the changes of contact resistance (R_c) subject to Eq. 3, in this case, the contact resistance is determined by doping concentration.

$$R_c \propto \exp(q\phi_{Bn} / kT). \tag{1}$$

$$R_c \propto \exp[q\phi_{Bn} / E_{00}\coth(E_{00}/kT)]. \tag{2}$$

$$R_c \propto \exp(q\phi_{Bn} / E_{00}). \tag{3}$$

Obviously, the high doping concentration and the low barrier height are the necessary condition for excellent ohmic contact.

Functional analysis of front silver electrodes. Normally, the front silver pastes for silicon solar cells consist of four parts: conductive silver power, organic carrier, glass frits [4,7] and some additives for improvement of paste properties. Front silver paste formed conductive grids on lighting-receiving side of solar cells to collect photo-generated current after screen-printed and fired as shown in Fig. 3. As Fig. 3 shows, the series resistance (R_s) of silicon solar cells consists of five parts: metal silver bar line resistance (R_{silver}); front Ag-Si contact resistance (R_c); n-type silicon sheet resistance (R_{sheet}); p-type silicon bulk resistance (R_{bulk}) and back Al-Si contact resistance (R_{BSF}). $R_s = R_{silver} + R_c + R_{sheet} + R_{bulk} + R_{BSF}$. Front silver line critically affects R_{silver} and R_c, R_c is the most important part of R_s, so Ag-Si contact directly affect R_s.

During the firing period, the silver paste will corrode the silicon nitride antireflection coating and seep into n-type silicon surface, about $0.1 \sim 0.3\mu m$ below. The permeation degree can be controlled by adjusting the silver paste formula and sintering process parameters. Too shallow and too deep Ag-Si contact junction will cause adverse impact on the properties of cells characterized by a lower efficiency and a higher R_s.

In summary, the excellent front silver conductors must meet three properties: a higher ratio of line height to line width, low line resistivity and the good ohmic contact. The paste formula and firing process create the prerequisites for ohmic contact undoubtedly.

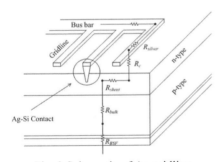

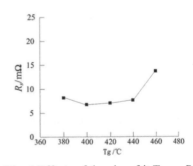

Fig. 3 Schematic of Ag gridline   Fig. 4 Effects of the glass frit Tg on R_s

Experimental

Preparation of thick-film electrodes. The silver powers, glass frits, organic carrier, and some additives were blended at appropriate proportion and mixed well, then the mixture was grinded by a three-roll miller. The viscosity of the paste is 200-250Pa·S. In the experiments, monocrystal silicon wafers with size of 125 mm×125mm were used. By using the 280 mesh screen, the silver paste was printed on the front side of silicon wafers with an antireflection coating. After dried in an oven at 200°C for 10min, wafers were fired in a infrared furnace. The silver electrodes were made with the line width of 100μm. The microstructures of thick-film silver electrodes were analyzed by SEM and the electric properties of cells were tested by the solar simulator.

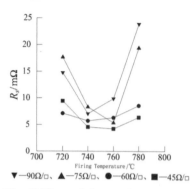

Fig. 5 Effects of sheet resistances on R_s

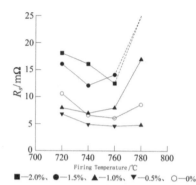

Fig. 6 Effects of the N-type doping content on R_s

Effects of glass transition temperature (Tg) on R_s. In the test, Pb-B-Si glass frit was adopted for silver paste, and the Tg of glass frit was chosen as 380°C, 400°C, 420°C, 440°C and 460°C, respectively. The sheet resistances of silicon wafers were chosen as 45Ω/□. The cells were fired at the same temperature (750°C). Effects of the glass frit Tg on R_s is shown in Fig.4.

Effects of sheet resistances on R_s. The sheet resistances of silicon wafers were chosen as 45Ω/□, 60Ω/□, 75Ω/□ and 90Ω/□, respectively. The front silver grids were printed by using the same silver paste. Wafers were sintered at 720°C, 740°C, 760°C and 780°C, respectively. R_s of samples were measured and the average R_s were calculated (5 pieces for each sample). The effects of sheet resistance on R_s is shown as in Fig. 5.

Effects of phosphorus doping concentrations on R_s. To realize the heavy doping effects of grid lines, à small amount of phosphorus compounds were added to the pastes. The proportions of phosphorus were 0%, 0.5%, 1.0%, 1.5%, 2.0%, respectively. In the experiment, the sheet resistance of the silicon wafers was 45Ω/□, after screen-printing, wafers were fired at 720°C, 740°C, 760°C and 780°C, respectively. R_s of samples were measured and the average R_s were calculated (5 pieces for each sample). The effects of N-type doping content on R_s is shown as in Fig. 6. The dotted line represents the results of the resistance is more than 100mΩ.

Results and Discussions

Fig. 4 shows that the series resistance (R_s) is greatly affected by the transition temperature Tg of glass frit added in the paste. When Tg is between 400°C and 440°C, R_s is smaller. When Tg is higher than 460°C, R_s increased significantly. The glass frit plays a major role of firing aids and film-forming aids in thick-film conductors, and has a great influence on firing process, line conductivity and adhesion of films. The glass frit also acts the main role to corrosive the SiN_x antireflection coating during firing process. When Tg is too low, the glass frit will melt the SiN_x film at the lower temperature, and the sliver will permeating deeper at emitter region while firing, this will make R_s become higher. When Tg is too high, the glass frit will melt the SiN_x film inadequately, the Ag-Si mutual diffusion cannot achieve the higher degree, which will also make R_s increase.

From Fig. 5 we know that the best firing temperature region is between 740°C and 760°C for different wafers with different sheet resistances. The smaller the sheet resistance is, the smaller the series resistances become. When the firing temperature is too high or too low, R_s will increase. R_s will increase rapidly when the temperature is over 760°C, this is because the diffusion depth of the silver atoms in silicon wafer increases as the firing temperature rise. In the emitter region of the wafer, the further from the surface, the lower of the phosphorus concentrations, making Ag-Si contact resistance increases. When firing peak temperature is too high, the emitter region can be burn through by the front silver conductors, which makes the contact resistance increases greatly. The silicon wafers with the sheet resistance of 45Ω/□, was less influenced by over firing, when firing temperature is over

760°C, R_s only slightly increased. This is because the emitter region of wafers with 45Ω/□ sheet resistance is deeper. When the firing peak temperature is below 740°C, R_s also obviously increase, especially for the shallow junction wafers with the sheet resistance of 75Ω/□ and 90Ω/□.Too low firing temperature causes R_s increase, the main reason is the corrosion of glass frit in the pastes to the antireflective coating is inadequate, and Ag-Si cannot form fine ohmic contact. According to the theoretical analysis above, the phosphorus atoms concentration is lower in shallow junction (high sheet resistances) cells, and it is more difficult to form Ag-Si ohmic contact. So when the sheet resistance of silicon wafer is higher, the firing process of front thick-film silver electrodes requires more strict control.

From Fig. 6 we know that the characteristics of R_s will be improved by adding 0.5% phosphorus compound. When the addition is excessive (more than 1.0%), the properties of R_s become worse obviously. A moderate amount of phosphorus doping concentration will form heavy doping in emitter region while firing, which can improve the concentration of carriers, and make Ag-Si ohimc contact easily. When phosphorus doping is excessive, interstitial impurity will form in emitter region, which will reduce the concentration of carriers and make Ag-Si ohmic contact become worse.

Conclusions

(1) Based on analyzing the structure and mechanism of the front silver electrodes, approaches of making solar cells with low R_s is proposed. Reasonable silver paste formula and appropriate firing process are necessary for making cells with small R_s.

(2) The Tg of glass frit has a direct influence upon the ohmic contact properties of thick-film Ag electrodes, the glass frit with suitable transition temperature for silver pastes shoud be chosen.

(3) The surface sheet resistance of silicon wafers has great influence on firing process. The cells with higher sheet resistances have a more narrow firing window.

(4) The suitable amount of phosphorus addition can form heavy donor doping when firing, which can reduce the R_s of the cells.

References

[1] G. Schubert, F. Huster, P. Fath, Physical understanding of printed thick-film front contacts of crystalline Si solar cells—Review of existing models and recent developments, Sol. Energy Mater. Sol. Cells 90 (2006) 3399-3406.

[2] L.K. Cheng, L. Liang, Z.G. Li, Nano-Ag colloids assisted tunneling mechanism for current conduction in front contact of crystalline Si solar cells, Photovoltaic Specialists Conference 34 (2009) 002344-002348.

[3] Z.G. Li, L. Liang, A.S. Lonkin, B.M. Fish, M.E. Lewittes, L.K. Cheng, K.R. Mikeska, Microstructural comparison of silicon solar cells' front-side Ag contact and the evolution of current conduction mechanisms, J. Applied Phys. 110 (2011) 074304.

[4] K.K. Hong, S.B. Cho, J.S. You, J.W. Jeong, S.M. Bea, J.Y. Huh, Mechanism for the formation of Ag crystallites in the Ag thick-film contacts of crystalline Si solar cells, Sol. Energy Mater. Sol. Cells 93 (2009) 898-904.

[5] D. Pysch, A. Mette, A. Filipovic, S.W. Glunz, Comprehensive analysis of advanced solar cell contacts consisting of printed fine-line seed layers thickened by silver plating, Progress in Photovoltaics: Research and Applications 17 (2009) 101–114.

[6] S.B. Cho, K.K. Hong, J.Y. Huh, H.J. Park, J.W. Jeong, Role of the ambient oxygen on the silver thick-film contact formation for crystalline silicon solar cells, Current Applied Phys. 10 (2010) S222–S225.

[7] M.M. Hilali, S. Sridharan, C. Khadilkar, A. Shaikh, A. Rohatgi, S. Kim, Effect of glass frit chemistry on the physical and electrical properties of thick-film Ag contacts for silicon solar cells, J. Elec. Mater. 35 (2006) 2041-2047

Key Engineering Materials Vol. 537 (2013) pp 214-219
© (2013) Trans Tech Publications, Switzerland
doi:10.4028/www.scientific.net/KEM.537.214

Preparation and Photocatalytic Activity of Shape- and Size-Controlled TiO₂ Films on Silica Glass Slides

Rufen Chen[1,a], Xinxin Jia[2,b], Xiangmin Meng[3,c]

[1]College of Chemistry and Material Science, Hebei Normal University, Shijiazhuang, 050024, P.R.China

[2]School of Biomedicine, Bejing city university, Bejing, 100094, P.R.China

[3]Key Laboratory of Photochemical Conversion and Optoelectronic Materials, TIPC, Chinese Academy of Sciences, Bejing, 100190, P.R.China

[a] rufenchen7@gmail.com (corresponding author); [b] xinxinjia66@163.com, [c] xmmeng@mail.ipc.ac.cn

Keywords: TiO₂ film; Adjustable Size; Crystal morphology; Photocatalytic activity

Abstract. TiO₂ films consisting of rod-like to sphere-like TiO₂ particles on glass slides were synthesized by assembly technique. The results showed that the shape and size of TiO₂ particles could be manipulated using different concentrations of polyethylene glycol 20000(PEG). By increasing of PEG, the shapes of the TiO₂ particles transformed from rod-like to sphere-like, the size of TiO₂ particles became gradually smaller. The size became bigger when an excess amount of PEG was added. With the adding of PEG, the amount of the TiO₂, the hydroxyl content, and the rutile phase content on the surface of TiO₂ films increased, respectively. The photocatalytic activity of TiO₂ films added PEG was higher than that of unadded samples.

Introduction

Titanium dioxide (TiO₂) is one of the most attractive materials in the experimental investigation due to its scientific and technological importance [1]. Especially, TiO₂ photocatalysts in thin film form have potential application for the decomposition of organic pollutants [2]. A number of methods have been employed to fabricate TiO₂ films, including vacuum evaporation, magnetron sputtering, ion beam technique, and sol-gel process [3-5]. Among these methods, sol-gel technique presents many advantages [5]. The photocatalytic activity of TiO₂ films depends on not only the phase and the porosity of the coatings but also their shapes and sizes [6]. Although shape- and size-controlled TiO₂ particles on substrates have been prepared through doping ions or using certain templates [7], it is still a great challenge to further develop simple and low-cost approaches for the synthesis of homogeneous TiO₂ films with variable shapes and sizes.

In this paper, we described a simple route to synthesize TiO₂ films on silica glass substrates adopting the assembly technique. The sodium dodecyl sulfate (SDS) was used as self-assembled medium (SAM). The shape and size of TiO₂ particles were manipulated using different concentrations of PEG during preparation. The effects of PEG on the shape, size, and photocatalytic activity of the resultant samples were investigated.

Experimental

Preparation of TiO₂ films. As the carriers, glass slides with a suitable size (45 × 20 mm) were cleaned in the solution (mixture of 70% H_2SO_4: 30% H_2O_2, 80ml) at 80°C. The SAM with SDS anionic micelle was then deposited on the glass slides surface by immersing the glass slides into the solution of SDS (7 mmol/L, 100 mL) for 2h. After filtration, the carriers with SAM were collected, which were then washed with distilled water and dried in an oven at 100 °C for 2h. The TiO₂ colloids were prepared as follow: the titanium tetrochloride (5 mol/L, 10 mL) was dropped with stirring into aqueous hydrochloric acid (4mol/L, 100 mL) in a 250mL beaker, and then various amount of PEG (0,

2, 5, 10, 15 and 20 at %) were added with stirring to the above TiO_2 colloids. The mixtures were stirred for 1h at the room temperature. The carriers with SAMs were immersed in the above TiO_2 colloids containing different concentrations of PEG. Then, the mixtures were heated by water bath at 80 ° C for 2 h. TiO_2 colloidal carry a position surface at pH < 6, electrostatic interactions between the TiO_2 colloid and the SDS micelle. The hot colloids were cooled and aged at the room temperature. After being washed with distilled water and dried in the oven at $100^{\circ}C$, the substrates coated with TiO_2 gel films were obtained. The TiO_2 coating films on glass slides could be obtained after calcining at 550 °C for 2 h.

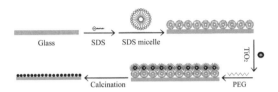

Fig.1 Schematic illustration of preparing TiO_2 films by the assembly technique

Photocatalytic activity. The photoatalytic activity of samples was evaluated by the degradation rate of methyl orange. The concentration of methyl orange was maintained at 10^{-6} mol/L. The photocatalytic measurements were carried out in a 100 mL pyrex photoreactor at room temperature. The reactor was placed at a fixed distance of 10 cm from the lamp. The solution of methyl orange (100 mL) and the photocatalyst were sonicated in a photoreactor before irradiation. The catalysts were agitated for 1h in methyl orange solution in the absence of light to attain the equilibrium adsorption on the catalyst surface. UV irradiation was carried out using a 150 W high-pressure Hg-lamp. At regular intervals, samples were collected and filtrated, and the filtrates were analyzed employing UV-visible spectrophotometer at its characteristic absorption band (462 nm).

Characterization. The structures of the samples were characterized by X-ray diffractiometer (Bruker D8 advance) with Cu Kα radiation. The morphologies and element content of samples were measured by a scanning electron microscopy (SEM Japan S-570) equipped with EDS (Kevex Sigma TM Quasar, USA). The compositions of coated samples were analyzed by X-ray photoelectron spectroscopy (XPS ESCALAB MK II). An MgKα X-ray source (E=1253.6 ev) was used. The analyzer was operated at 50ev for survey spectra. Spectroscopic analyses of samples were performed using an UV-visible spectrophotometer (type UV160A).

Results and discussion

Surface characterization. The surface morphologies of the TiO_2 coating films prepared from precursor solutions containing different amount of PEG (A: 0 %; B: 2 %; C: 5 %; D: 10 %; E: 15 %; F: 20 %) are shown at Fig.2. When PEG has not been added to the precursor solution, the shape of TiO_2 particles is rod-like (Fig.2A). With the increasing of PEG, the ratio of axial lines and radial lines of the particles decreases from about 4:1 (Fig.2A) to 1:1(Fig. 2D), the sphere-like particles appearance gradually. When the molar ratio reaches 20%, the uniform TiO_2 sphere microparticles are obtained. This can be because the molecules of PEG can be absorbed on different faces of TiO_2 through intermolecular forces. The surface energy of the crystalline faces that have absorbed PEG decreases so their growth is restrained, leading to the growth of sphere-like particles. Thus, the shape of TiO_2 particles transforms from rod-like to sphere-like.

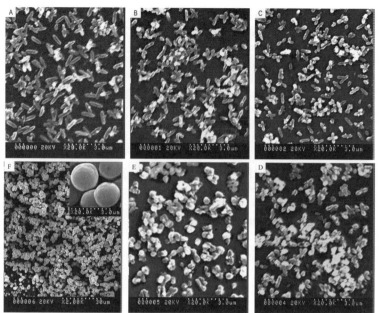

Fig.2 Surface morphologies of the TiO_2 films on glass slides prepared from precursor solutions containing PEG: (A) 0 %; (B) 2 %; (C) 5 %; (D) 10 %; (E) 15 %; (F) 20 %

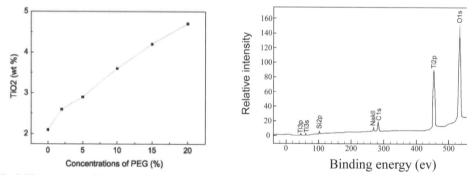

Fig.3 The amount of TiO_2 (wt %) with various Fig.4 XPS survey spectra for surface of TiO_2 films prepared concentrations of PEG from precursor solution containing 15 % PEG

    In addition, the size of the TiO_2 particles varies gradually with the increasing of PEG. The size is big when PEG is not added. After the small amount of PEG is added, the size becomes smaller. Then the size becomes larger when the large amounts of PEG are added. The diameter of the TiO_2 particles is about 2um on average (Fig.2F) when the concentration of PEG reaches 20%. When no PEG is added, TiO_2 particles have higher surface energy so that it is easier to reunite each other and the consequence is that the TiO_2 particles on glass slides become bigger. PEG is a kind of nonionic surfactant, which can reduce the surface tension of the solution and make the TiO_2 particles have a better decentralization. So, while a small amount of PEG is added to the precursor solution, TiO_2 particles will growth homogeneously on the glass slides and the size of the TiO_2 particles becomes smaller gradually with the role of PEG increasing. On the other hand, the viscosity of colloids becomes greater with the increasing of PEG. As a result, the colloidal particles that have formed nucleus are easier to reunite, and then bigger particles when an excess amount of PEG is used.

Fig.3 shows the amount of TiO_2 (wt %) on the surface of glass slides with various concentrations of PEG. It is obviously that the amount of TiO_2 (wt %) increases with the increasing concentrations of PEG. It is because that the viscosity of colloids becomes stronger in this period which causes the TiO_2 colloidal particles moves slowly little by little. So the amount of TiO_2 increases with the adding of PEG.

Fig.4 shows the XPS survey spectrum for the surface of TiO_2 films prepared from the precursor solution containing 15 % PEG. XPS spectrum reveals characteristic peaks from Ti, O, C, Na and Si. The photoelectron peak for Ti2p appears clearly at a binding energy, $E_b = 458.5$ eV, O1s at $E_b = 531.1$ eV, C1s at $E_b = 284.5$ eV. The Nakll and Si photoelectron peaks are at binding energies 269.2 and 103.3eV, respectively. The XPS peak for Na and Si are observed in the spectrum, implying that some chemical reaction occur at the interface between the film and the glass substrate [8].

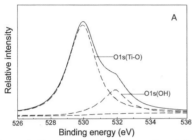
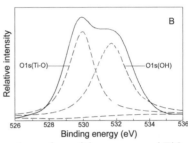

Fig.5 High resolution XPS of the O1s region, taken on the surface of the as-prepared TiO_2 films from precursor solutions containing (A) 0% and (B) 15% PEG

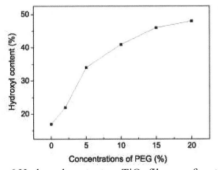
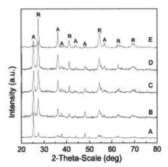

Fig.6 Hydroxyl content on TiO_2 films as function of the amount of added PEG in precursor solution

Fig.7 XRD diagrams of TiO_2 films with different concentrations of PEG: (A) 0 %; (B) 5%; (C) 10 %; (D) 15%; (E) 20%; (A: anatase; R: rutile)

Fig.5 shows the high resolution XPS spectra of the O1s region, taken on the surface of the as-prepared TiO_2 films from precursor solutions containing PEG(A: 0.0 %; B: 15 %). The O1s region is decomposed into two peaks, as shown in Fig.5. The main peak is attributed to the contribution of Ti-O in TiO_2. The other is related to the hydroxyl groups from the adsorbed H_2O. Although some H_2O is physically adsorbed on the surface of TiO_2 films, the physically adsorbed H_2O on TiO_2 is easily desorbed under theultra- high vacuum condition of the XPS system. So, Hydroxyl group existed in the films can be attributed to the chemically adsorbed H_2O. The results reveal that the hydroxyl content of TiO_2 films added with PEG is higher than that without PEG.

Fig.6 shows the hydroxyl content on TiO_2 films as a function of the amount of the added PEG in precursor solution, where the hydroxyl content represents the ratio of the area of O1s (OH) peak to the total area of all the 2 kinds oxygen peaks. The hydroxyl content increases with the increasing of PEG concentration.

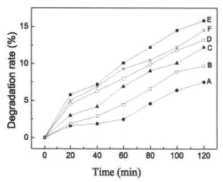

Fig.8 Degradation rate of methyl orange with different concentrations of PEG (A) 0 %; (B) 2 %; (C) 5 %; (D) 10 %; (E)15 %; (F) 20 %

XRD study. Fig.7 shows the XRD patterns of the TiO_2 coating films prepared from precursor solutions containing different amounts of PEG. The results of XRD analysis show that mixed crystalline phases of anatase and rutile are formed for all samples. The relative content of rutile phase increases with the increasing of PEG. This can be due to the fact that the PEG promotes the phase transformation of anatase to rutile in the films. A possible explanation is that the decomposition of PEG has a positive effect on the phase transformation of anatase to rutile. With increasing of PEG, the heat of decomposition of PEG is sufficient for the anatase to rutile phase transformation [9]. So the relative content of rutile phase increases with increasing of the PEG. The data is shown in table 1.

Photocatalytic activity. Fig.8 shows the degradation rate of methyl orange caused by TiO_2 films prepared from precursor solutions with different amounts of PEG. It can be seen that the catalytic activity of TiO_2 films enhances with the increasing of PEG up to 15%. The main results from the following factors: Firstly, the small amount of PEG is added, the size becomes smaller and the shapes of the TiO_2 particles transform from rod-like to sphere-like, leading to the increasing of the surface area of TiO_2 particles. The large surface area is an important contribution to its excellent activity because it can absorb more oxygen and molecular water. Secondly, the hydroxyl content increases with increasing of the PEG(as shown in Fig.6), the increase of the hydroxyl on the surface of TiO_2 is beneficial to the enhancement of photocatalytic activity [10]. And the third factor is that the amount of the TiO_2 increases on the surface of glass slides with the increasing concentrations of PEG (as shown in Fig.3), leading to the increasing of the photocatalytic activity. However, the catalytic activity of samples decreases when the amount of PEG added becomes too larger (> 15 %). The reason could be that when PEG is added too much, the size of TiO_2 particles becomes bigger. The increasing of the size may lead to the reducing of the surface area of TiO_2 particles. Meanwhile, the result of XRD shows that the relative content of the rutile phase increases with increasing of PEG. Some researchers reported that the catalytic activity of anatase TiO_2 was better than that of rutile TiO_2 [11]. In our experiment, the optimal amount of PEG is about 15%. And further work will be proceeding in our laboratory.

Table 1 The relative content of rutile in the mixed phases with increasing of the PEG.

The concentrations of PEG (%)	0 5 10 15 20
The relative content of rutile in the mixed phases (%)	12.5 47.6 51.9 59.6 88.7

Conclusion

The shape- and size-controlled TiO_2 films on glass slides were prepared by the assembly technique. The shapes (rod-like and sphere-like) and sizes of TiO_2 particles could be controlled by adjusting the concentrations of PEG. With the increasing of PEG, the rutile phase content of TiO_2 films increased, and the amount of TiO_2 and the hydroxyl on the surface of films increased,too. The photocatalytic activity of TiO_2 films added PEG was higher than that of unadded samples. The result of our experiment will be very important for the understanding and further development of environmental purification materials.

Acknowledgment

This work was supported by the National Natural Science Foundation of China (21077031), the Key Laboratory of Photochemical Conversion and Optoelectronic Materials, TIPC, CAS (PCOM201110) for financial support.

References

[1] H. Song , M.Dai, Y.-T. Guo, Y.-J.Zhang, Preparation of composite TiO_2–Al_2O_3 supported nickel phosphide hydrotreating catalysts and catalytic activity for hydrodesulfurization of dibenzothiophene, Fuel Process. Technol. 96 (2012) 228–236.

[2] P. Pansila, N. Witit-anun, S. Chaiyakun. Influence of sputtering power on structure and photocatalyst properties of DC magnetron sputtered TiO_2 thin film, Procedia Engineering 32(2012) 862 – 867.

[3] F.Zhao, X.Cui, B.Wang, and J.G.Hou, Preparation and characterization of C54 $TiSi_2$ nanoislands on Si(111) by laser deposition of TiO_2, Appl. Surf. Sci. 253 (2006)2785-2791.

[4] R.Suryanarayanan, V.M.Naik, P.Kharel, P.Talgala, and R.Naik, Ferromagnetism at 300K in spin- coated films of Co doped anatase and rutile-TiO_2, Solid State Commun. 133(2005) 439-443.

[5] J. Zhou, Y. Cheng, J.Yu. Preparation and characterization of visible-light-driven plasmonic photocatalyst Ag/AgCl/TiO_2 nanocomposite thin films. J. Photoch. Photobio. A 223 (2011)82–87.

[6] M.J.Uddin, F.Cesano, F.Bonino, S.Bordiga, G.Spoto, D.Scarano, and A. Zecchina, Photoactive TiO_2 films on cellulose fibres: synthesis and characterization, J. Photoch. Photobio. A 189（2007) 286-294.

[7] S.Liang, M.Chen, Q.J. Xue, Y.L.Qi, and Chen J.M., Site selective micro-patterned rutile TiO_2 film through a seed layer deposition, J.Colloid Interf. Sci. 311 (2007) 194-202.

[8] J.Yu, X.Zhao, Q.Zhao, Effect of surface structure on photocatalytic activity of TiO_2 thin films prepared by sol-gel method, Thin Solid Films 379 (2000) 7-14.

[9] J. C.Yu, J. F.Xiong, B.Cheng, S.W Liu., Fabrication and characterization of Ag–TiO_2 multiphase nanocomposite thin films with enhanced photocatalytic activity, Appl. Catal. B: Environ. 60 (2005) 211-221.

[10] Q.Fang, M.Meier, J. J.Yu, Z. M.Wang, J. Y. Zhang, J. X.Wu, FTIR and XPS investigation of Er-doped SiO_2-TiO_2 films, Mater. Sci. Eng. B 105 (2003) 209-213.

[11] D. S. Seo, H. Kim, Synthesis and characterization of TiO_2 nanocrystalline powder prepared by homogeneous precipitation using urea, J. Mater. Res. 18(2003)571-577.

Key Engineering Materials Vol. 537 (2013) pp 220-223
© (2013) Trans Tech Publications, Switzerland
doi:10.4028/www.scientific.net/KEM.537.220

Preparation and Characterization of TiO$_2$ Film on Glass Flake

Yi Guo[a], Shiguo Du, Jun Yan, Bin Wang

Shijiazhuang Mechanical Engineering College, Shijiazhuang, Hebei province, P.R.China

[a] yangli666555@126.com

Keywords: Glass flake, TiO$_2$, Film.

Abstract. A TiO$_2$ film was prepared on glass flake with Ti(OBu)$_4$ as a main material by peptization and reflux method at 80°C in acidic abundant aqueous solution. The surface topography and surface phase structure were characterized by scanning electron microscope, energy spectrum, X-ray diffractometer, X-ray photoelectron spectroscope, and the bonding strength between film and glass flake was researched by ultrasonic vibration. The results indicated that the TiO$_2$ film, the main phase was anatase and a little rutile, coated on the glass flake uniformly. Along with growth of TiO$_2$ on surface of glass flake, the formation maybe base on two different mechanics, at the earlier stage, TiO$_2$ coupled surface of glass flake with Si-O-Ti chemical bond, but with physical deposition at final.

Introduction

Glass flake is a packing of great predominance, large length thick ratio, when filled in to the resin, it will array like the scale by the effect of gravity to form a laminar structure. The thickness of glass flake is 2-5μm, grain size usually is 0.4-3mm, so in theory, the 1mm thick coating can contain 20-80 layers of glass flake, so the corrosion medium can not penetrate to the surface of glass flake directly, unless along the long and flexuose line.

Recently, for improving the resistance of glass flake coat, Yuan Jianjun[1]retreated glass flake with KH560 coupling agent, the results showed that a layer of organic molecule was formed on the glass flake surface after modified, and the interface was optimized, the impermeability and the mechanical properties of composites were also improved.

Ni Nannan[2] coated the surface of glass flake with polyaniline, prepared a conductive epoxy anti-corrosion coating with this composite material. Through test, the anti-corrosion of the coating with this composite material was better than the coating with glass flake, and it had good electrical conductivity. Japanese scholars coated the micron-sized glass flake with Zn and Al, then mixed glass flake into epoxy resin. Glass flake could prevent the intrusion of rainwater and salt, Zn could prevent corrosion, and Al could effectively stop the ageing of coating due to ultraviolet rays.

A TiO$_2$ film was prepared on surface of glass flake by peptization and reflux method at 80°C in this paper, and discussed the growth mechanization of TiO$_2$ on the surface of glass flake by ultra-sonic oscillation.

Experimental

Preparation of sample. The main reagents in experimental were Ti(OBu)$_4$(chemically pure), C$_2$H$_5$OH, C$_5$H$_8$O$_2$(ACAC, analytically pure), SnCl$_4$(analytically pure), BTA(analytically pure), glass flake; KH-550(chemically pure).

The KH-550 weighed 1% of the powder was added into the absolute alcohol with enough glass flake on electromagnetic stirring, then glass flake was separated from the solution after stirring 30 minutes, and cleaned 2-3 times by absolute alcohol, at last the glass flake was dried at 60°C was spare sample.

A white sediment was got after distilled water of 10mL was dropped into the absolute alcohol with ACAC of 0.4mL and Ti(OBu)$_4$ of 3mL, then on the acutely stirring a certain amount of concentrated nitric acid was dropped into the solution, the white sediment disappeared or dissolved, so the sol sample was prepared[3].

The sol maxed with glass flake was shifted to three necked flask, and stirrer was turned on. After refluxing 50 minutes on 80°C, turned off the stirrer, then the powder was separated, and cleaned in turn with absolute alcohol and distilled water, the TiO_2 film was prepared on surface of glass flake after being dried on 65°C. The TiO_2 powder got in the previous preparation process with no glass flake was called idle sample.

Characterization of sample. The morphology of TiO_2 film on surface of glass flake was observed by S-4800 scanning electron microscope. D8ADVANCE X-ray diffractometer and EDX attached to SEM were used to analyze the elemental composition of the surface of powder. The adhesion of TiO_2 film and glass flake was detected by ultrasonic cleaner.

Fig.1 SEM photos of (a) glass flake and (b) glass flake with TiO_2 film

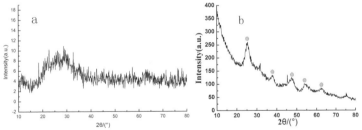

Fig.2 XRD patterns of (a) glass flake with TiO_2 film and (b) TiO_2 powder

Results and discussion

Surface Topography of Compound Particle. Fig.1 was photos of TiO_2 film on surface of glass flake and blank sample.

Fig.1 was the SEM photographs of the glass flake and the TiO_2 film on the surface of glass flake. On the surface of original glass flake, there were lots of small chippings produced remaining in the production and transportation process of the glass flake, these small chippings maybe effect TiO_2 film formation on the surface of glass flake. Preparation test proved that if glass flake was not retreated by KH-550, it was difficult to form a TiO_2 film on the surface of glass flake, because KH-550 was a silane coupling with amino group that could attract Ti^{4+} in the solution, so the TiO_2 film was easy to form on the surface. Fig.1b showed the surface of glass flake was covered by a homogeneous films, and the film was smooth, non-cracks, fewer chippings.

XRD analysis. Fig.2a was the X-ray diffractometer of glass flake coated TiO_2 film, because glass flake was non crystalline, and the peak of TiO_2 did not show in the XRD of glass flake coated TiO_2 film, the reason was that the content of TiO_2 was so few that it was covered by the non crystalline glass flake.

For confirming the crystal of TiO_2 on the surface of glass flake, a idle experiment was designed. XRD of TiO_2 sample was Fig.2b, the weak diffraction peaks appeared on the position of 25.5°, 37.3°, 47.9°, 54.1° and 61.1°, so the XRD showed the crystalline of TiO_2 film on the surface of glass flake was mainly anatase, but the high background showed there was amorphous TiO_2 existing.

Elements composition and phase analysis. Fig.3 was the EDS of TiO_2 film on the surface of glass flake.

Comparing Fig.3a and Fig.3b, the different was the Ti element appearing, showed that the film on surface of glass flake contained Ti. For further explanation of element composition of film, XPS was given as Fig.4.

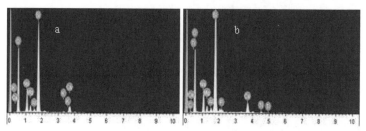

Fig.3 EDS of (a) glass flake and (b) glass flake with TiO_2 film

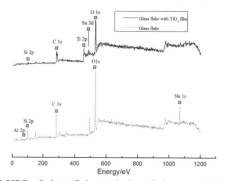

Fig.4 XPS of glass flake and glass flake with TiO_2 film

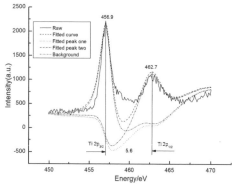

Fig.5 Fitting XPS spectra of Ti element on glass flake with TiO_2 film

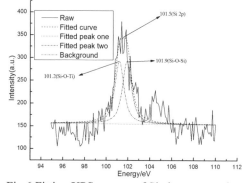

Fig.6 Fitting XPS spectra of Si element on glass flake with TiO_2 film

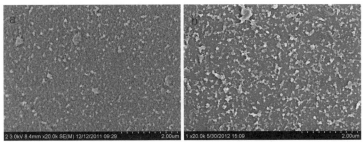

Fig.7 SEM photos of glass flake with TiO_2 film (a) before and (b) after ultrasonic vibration

The two curves were got in a wide scanning of glass flake and glass flake with TiO_2 film. The content of Ti of glass flake with film increased relative to the glass flake. The elements on the surface glass flake with film were Si, C, Ti, O, Sn. Through overlapping peak resolving to narrow spectrum of Ti (Fig.5) and analyzing with standard card, the valent state of Ti was +4, valent state of Sn was +4.0°

The Si(101.4) peak was divided to two peaks on the position of 101.2 and 101.9(Fig.6), according to reference[4], these two peaks were Si-O-Si bond(101.9) and Si-O-Ti bond(101.2). It showed that TiO_2 films and glass flake connected with chemical bond.

For evaluating the bonding intensity of TiO_2 film and glass flake, the glass flake coated TiO_2 with absolute ethyl alcohol was put into ultrasonic cleaner, and vibrated for 1minute on electric powder of 100W, then observed by stereoscan photograph. The photos are given in Fig. 7.

Comparing with two photos, through ultrasonic vibration for 1minute, the TiO_2 film dropped partly from the surface of glass flake, it showed the bond intensity of TiO_2 particles that did not drop and glass flake was higher than the dropped, and TiO_2 non-dropped and glass flake bonded with chemical bond, Si-O-Ti, the dropped TiO_2 just deposited on the surface of glass flake. So the process of TiO_2 film on surface of glass flake was surmised as follows, firstly, the nano-TiO_2 particle bonded glass flake with Si-O-Ti to form the islands, with the growth of islands, the islands was bigger and bigger, finally, TiO_2 deposition filled in interspace between islands.

Conclusion

TiO_2 film, the main phase was anatase and a little rutile, was prepared on the surface of glass flake by peptization and reflux method at 80°C. The SEM photos showed that a TiO_2 film covered the surface of glass flake, and XPS indicated that the TiO_2 bonded glass flake with Si-O-Ti. Through ultrasonic vibration for 1minute, TiO_2 film dropped partly from the surface of glass flake, so process of growth of TiO_2 film maybe base on two different mechanics, firstly, the TiO_2 might form islands on the surface of glass flake on the function of Si-O-Ti bond, then just physical deposition.

References

[1] J.J. Yuan, Z.E. Liu, M.J. Xue, A study of the interface optimization of glass -flake/epoxy-resin composites, Journal of Materials Engineering. 5(1997)23-25, in Chinese.

[2] N.N. Ni, M.Q.Jia, Preparation and characterization of conductive polyaniline coated glass flake, PAINT & COATINGS INDUSTRY. 40(2010)19-23, in Chinese.

[3] J. Yan, H.P. Cui, B. Wang and et al, Preparation of nano-TiO_2 coated micro copper powder by heterogeneous nucleation method, Powder Metallurgy Technology. 25(2007)430-434, 439, in Chinese.

[4] G.J. Ji, Z.M. Shi, AFM and XPS study of glass surface coated with Titantia nano films by sol-gel method, Chinese Physics Letters. 27(2010)179-183.

Key Engineering Materials Vol. 537 (2013) pp 224-228
© (2013) Trans Tech Publications, Switzerland
doi:10.4028/www.scientific.net/KEM.537.224

Structural and Optical Properties of Pulse Laser Deposited TiO₂ Thin Films

Liu Yi[1, a], Huang Hongmo[2,b] and Lin Xiaodong[1,c]

¹College of Physics Science and Technology Shenzhen University, Shenzhen. P. R. China

²Tongcheng Teachers College, Tongcheng, P. R. China

ᵃliuy@szu.edu.cn, ᵇhhmd@163.com, ᶜlinxd@szu.edu.cn

Keywords: Titanium dioxide, Pulse laser deposition, Optical constants, Spectroscopic ellipsometry

Abstract. TiO_2 thin films were prepared on quartz glasses by pulsed laser deposition (PLD) using a KrF laser excimer. The crystalline structure was characterized by X-ray diffraction, and the optical properties of the films were investigated using spectroscopic ellipsometry and UV-vis spectra respectively. The effects of the PLD conditions, including substrate temperature and O_2 pressure on the crystalline structure and the optical properties of the films were investigated. The results indicated that there are a suitable substrate temperature and an O_2 pressure which is favorable for the synthesis of anatase-type TiO_2.

Introduction

TiO_2 is an important n-type wide band gap II–VI semiconductor with high refractive index and high dielectric constant. TiO_2 and doped TiO_2 thin films have drawn a great deal of attention in recent years due to their wide application in dye-sensitized solar cells, photo catalyst and potential transparent conducting oxides [1-3]. TiO_2 thin films can have three structural forms anatase (tetragonal), rutile (tetragonal) and brookite (orthorhombic). Many studies reveals that film structure affect the photo-catalytic activity as well as the power conversion efficiency of solar cells [4,5]. Therefore, the control of the film structure is very important for applications of TiO_2 films.

TiO_2 thin films have been prepared by many growth techniques, such as RF-magnetron sputtering [6], metal–organic chemical vapor deposition (MOCVD) [7] and sol–gel process [8]. Among all the reported techniques of TiO_2 film deposition, pulsed laser deposition (PLD) provides the precise manipulation of the properties of thin films of TiO_2 by controlling the deposition parameters [9,10]. Spectroscopic ellipsometry (SE) is a fast and non-invasive method to characterize macro- and microscopic structural properties such as layer thickness, surface roughness, interface property, and homogeneity, as well as optical and electronic properties such as refractive index, band-gap, and oscillator parameters, which provide indirect information on the chemical structure [11]. In this study, TiO_2 thin films were prepared on quartz glass substrates by PLD, under different O_2 pressure and temperature. The dependence of the structural and optical properties of the film samples on the deposition conditions was investigated.

Experimental

In this paper, TiO_2 films were deposited on quartz glass substrates by PLD. A KrF excimer laser (λ=248nm) beam was focused on a ceramic target by a quartz lens. The laser was operated with a repetition frequency of 20 Hz and pulse duration of 20 ns. The single pulse energy was set to 180mJ. A sintered pellet of TiO_2 (99.99%) was used as the target for the film deposition. This pellet had a diameter of 49 mm and a thickness of 5 mm. The distance from the target to the substrate was approximately 50 mm. A 20mm×20mm square quartz glass substrate placed parallel to the target was fixed on a substrate holder and rotated during the deposition to ensure the uniformity of the films. The deposition chamber was initially evacuated by using a turbo molecular pump to achieve a base pressure of approximately 5×10^{-4} Pa. It was then backfilled with O_2 to obtain a suitable pressure.

The film samples were evaluated in terms of structural and optical properties. Crystalline phases in the films were identified by a Cu Kα X-ray diffractometer (XRD) (D8 ADVANCE, BRUKER AXS GMBH) at scanning step 0.02° and scanning rate 0.5°/min. The optical transmission and ellipsometry spectra of the samples were obtained by using a UV–vis spectrophotometer (Lamda 950, PerkinElmer) in the wavelength range of 200-800 nm with a bare quartz glass substrate in the reference beam to eliminate the substrate contribution, and a spectroscopic ellipsometer (M-2000U, J. A. Woollam) in the wavelength range of 280-980 nm.

The system operations, including data acquisition and scanning were fully and automatically controlled by a computer. The incident angle was set to be 70°. SE measures the complex reflectance ratio ρ, between the reflection coefficients of σ_p and σ_s for the light polarized parallel and perpendicular, respectively, to the plane of incidence. By analyzing the state of the polarization of reflected light, one determines two parameters: $\tan\Psi$ and $\cos\Delta$, which are related to by: [12]

$$\rho = \sigma_p/\sigma_s = \tan\Psi \cdot \exp(i\Delta) \tag{1}$$

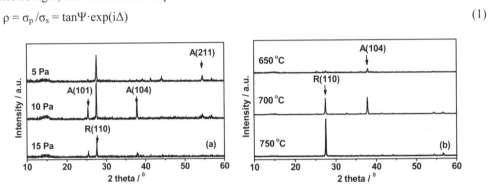

Fig. 1 XRD patterns of the TiO_2 thin films on glasses. (a) The films were deposited at 680°C in different O_2 pressures for 30 min. (b) the films were deposited at 10 Pa with different substrate temperature for 30min.

Results and discussions

The X-ray diffraction patterns for TiO_2 thin films deposited at various O_2 pressures on quartz glass substrates at 680°C is shown in Fig. 1(a). The film deposited at ambient vacuum exhibited no obvious diffraction peaks, which is not shown in this paper, indicating the formation of an amorphous structure or very small crystals. The film deposited at 5 Pa of O_2 exhibited a diffraction peak attributed to the anatase (211) in addition to the rutile (110) as the main fraction. The film deposited at 10 Pa exhibited diffraction peaks due to anatase (101), (004) and (204) and the rutile (110) peak. Thus, with the increase in the O_2 pressure to 10 Pa during deposition, the anatase phase was considered to form along with the rutile phase. The films deposited at 15 Pa of O_2 pressures exhibited the anatase (101) peak and the rutile (110) peak. The anatase fraction increased with the O_2 pressures and reached the maximum value at 10 Pa, then decreased. The full width at half maximum was 0.18° for the anatase (104) peak of the film deposited at 10 Pa.

Fig. 1(b) shows the X-ray diffraction patterns for TiO_2 thin films deposited at various substrate temperatures at 10 Pa of O_2 pressure. The film deposited at substrate temperature below 600°C exhibited no obvious diffraction peaks, which is not shown in this paper. The film deposited at 650°C exhibited a weak diffraction peak attributed to the anatase (104), indicating the formation of small crystals. The film deposited at 700°C exhibited both the rutile (110) and the anatase (104) at almost same peak intensity, which represents the sample was in a mixed phase of rutile and anatase. As the substrate temperature increased to 750°C, the anatase fraction transformed into the rutile phase completely since rutile is thermodynamically stable at high temperatures and is consequently the most common form of TiO_2. The grain size of the deposited TiO_2 thin films ranged from 40nm to 65nm calculated by Scherrer equation.

Anatase is a metastable phase that can be formed at relatively low temperatures and relatively high O_2 pressures. Too high O_2 pressure may increase the collision opportunities of plasma ions, therefore decreases the kinetic energy of plasma ions as well as the crystallinity of deposited films. The temperature and O_2 pressure during deposition were suggested to be the key parameters for growing TiO_2 thin films with an anatase structure [13].

The optical properties of the samples were investigated using spectroscopic ellipsometry and Uv-vis spectroscopy. A four-medium model of air/surface roughness layer(SR)/TiO_2/SiO_2 was applied in the fitting procedure with the TiO_2 films modeled using the empirical Cauchy model or the Tauc–Lorentz (TL) model described in Ref. [14]. The TL model is a combination of the classical Lorentz oscillator model and the Tauc model which includes parabolic bands and a constant momentum matrix element versus energy, and has been successfully applied for amorphous semiconductors [15].

Typical SE spectra of TiO_2 films on quartz glass substrates in the wavelength range of 280–980nm is shown in Fig. 2. The experimental data were analyzed by using the TL model, and a good fit was found between the model and experimental data. However, there still were some deviations on fitted Ψ values. Films produced by PLD do not have uniform and homogeneous surfaces. So, the deviations of Ψ values may result from the PLD technique, roughness, grain boundaries and morphology of the samples. Also, the rough surface and backsides of the samples may affect the reflection of light. Besides, the grain boundaries and morphologies would depolarize the incident polarized light, resulting deviated experimental ellipsometric parameters and deteriorating the SE fitting [16].

Fig. 3 shows the optical constants of the film deposited in different O_2 pressure obtained by SE. The film thickness was 879nm, 722nm and 710nm respectively, and the surface roughness was 21nm, 14nm and 20nm. The refractive indices of the films increased with the O_2 pressure between 400nm and 900nm, indicating the film density increases with the O_2 pressure. On the other hand, the absorption edge obtained from the extinction coefficients exhibited a small blue-shift as the O_2

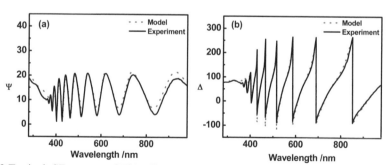

Fig. 2 Typical SE spectra of TiO_2 film deposited on glass at 680°C in 10 Pa O_2 pressure

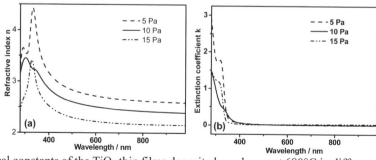

Fig. 3 Optical constants of the TiO_2 thin films deposited on glasses at 680°C in different O_2 pressures for 30 min. (a)Refractive indices (b) Extinction coefficients

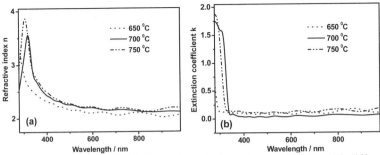

Fig. 4 Optical constants of the TiO$_2$ thin films deposited on glasses at 10 Pa with different substrate temperature for 30min. (a) Refractive indices. (b) Extinction coefficients.

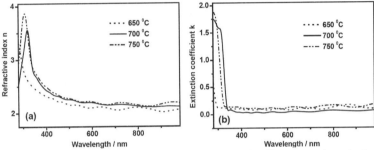

Fig. 5 Uv-vis spectra of the TiO$_2$ thin films on glasses. (a) The films were deposited at 700°C in different O$_2$ pressures for 30 min. (b) the films were deposited at 10 Pa with different substrate temperature for 30min.

pressure increasing. Anatase is recognized as a wide-band gap semiconductor with an energy gap of 3.2 eV [17], which is a little larger than 3.02 eV of rutile. The blue shift was caused by increasing anatase fraction.

Fig. 4 shows the optical constants of the film deposited at different substrate temperature obtained by SE. The film thickness was 843nm, 769nm and 743nm respectively, and the surface roughness was 21nm, 30nm and 16nm. The refractive indices of the films between 400nm and 900nm were not zero and slightly increased with the substrate temperature. On the contrary, the absorption edge obtained from the extinction coefficients slightly red-shifted as the substrate temperature increasing. The red shift accorded with the transformation from anatase to rutile of the films, which was in agreement with the XRD results.

The Uv-vis spectra of the TiO$_2$ thin films are shown in Fig. 5. As shown in Fig. 5(a), the average transmittance of TiO$_2$ film deposited at different oxygen pressures was about 80%, which was similar to Kao and Chen [17]. Using the standard expression for direct transition materials $(\alpha h\nu)^2 = A(h\nu-E_g)$, the optical band gaps of films deposited in different oxygen pressures were 3.06eV, 3.18eV and 3.12eV respectively. The absorbance of TiO$_2$ thin films decreased with the increase in oxygen pressure, and the absorption edge of the films changed in the same tendency of the SE results. The average transmittance of TiO$_2$ film deposited at different substrate temperatures ranged from 40% to 20%, shown in Fig. 5(b), which was much less than that of the oxygen pressure series. This was partly due to the quite large film thickness and the absorbent film. The optical band gaps of films deposited at different substrate temperatures were 3.06eV, 2.94eV and 2.90eV respectively. The absorbance of TiO$_2$ thin films increased with the increase in substrate temperature. The degeneration of optical transmission was accompanied by the rutile phase-transformation. This further confirms formation of rutile phase at higher deposited temperatures in accordance with the XRD results (Fig. 1(b)).

Summary

In this study, we deposited TiO_2 thin films on quartz glass substrates by pulsed laser deposition and examined the dependence of their crystalline and optical properties on the deposition conditions. The anatase fraction increased with the O_2 pressures and reached the maximum value at 10 Pa, then decreased. As the substrate temperature increased, the anatase grains grew up and accompanied by phase transformation from anatase to rutile. There are a suitable substrate temperature and an O_2 pressure which is in favor of the synthesis of anatase-type TiO_2.

Acknowledgement

This work is supported by Shenzhen Key Laboratory of Sensor Technology (Grant No.SST201004) and Shenzhen Science and Technology Bureau project (Grant No.JC201005280419A).

References

[1] A. Maldotti, A. Molinari, R. Amadelli, Photocatalysis with organized systems for the oxo-functionalization of hydrocarbons by O_2, Chem.Rev.102(2002)3811-3836.

[2] M. Grätzel, Photoelectrochemical cells, Nature 414(2001)338-344.

[3] Y. Furubayashi, T. Hitosugi, T. Hasegawa, A transparent metal: Nb-doped anatase TiO2, Appl. Phys. Lett. 86 (2005) 252101-252103.

[4] K. Hara, K. Miyamoto, Y. Abe, Electron transport in coumarin-dye-sensitized nanocrystalline TiO_2 electrodes, J. Phys. Chem. B. 109 (2005) 23776-23778.

[5] D. Yoo, I. Kim, S. Kim, Effects of annealing temperature and method on structural and optical properties of TiO_2 films prepared by RF magnetron sputtering at room temperature, Appl. Surf. Sci. 253 (2007) 3888-3892.

[6] H. Kikuchi, M. Kitano, M. Takeuchi, Extending the photoresponse of TiO_2 to the visible light region: photoelectrochemical behavior of TiO_2 thin films prepared by the radio frequency magnetron sputtering deposition method, J. Phys. Chem. B. 110 (11) (2006) 5537-5541

[7] A.L. Linsebigler, G.Q. Lu, J. T. Yates, CO photooxidation on TiO2(110), J. Phys. Chem. 100(16) (1996) 6631-6636.

[8] C. Anderson, A.J. Bard, Improved photocatalytic activity and characterization of mixed TiO_2/SiO_2 and TiO_2/Al_2O_3 materials, J. Phys. Chem. B. 101(14) (1997) 2611-2616.

[9] J. H. Kim, S. Lee, H.S. Im, The effect of target density and its morphology on TiO_2 thin films grown on Si(100) by PLD, Appl. Surf. Sci. 151 (1999) 6-16.

[10] D. Luca, D. Macovei, C.M. Teodorescu, Characterization of titania thin films prepared by reactive pulsed-laser ablation, Surf. Sci. 600 (2006) 4342-4346.

[11] M. Gioti, D. Papadimitriou, S. Logothetidis, Diamond Relat. Mater. 9(2000) 741-744.

[12] M. Schubert, Generalized ellipsometry and complex optical systems, Thin Solid Films. 313-314 (1998) 323-332

[13] D. Wicaksana, A. Kobayashi, A. Kinbara, Process effects on structural properties of TiO_2 thin films by reactive sputtering J. Vac. Sci. Technol. A. 10 (1992) 1479-1482.

[14] G.E. Jellison, F.A. Modine, Parameterization of the optical functions of amorphous materials in the interband region, Appl. Phys. Lett. 69 (1996) 371-373.

[15] G. E. Jellison, V. I. Merkulov, A. A. Puretzky, Characterization of thin-film amorphous semi-conductors using spectroscopic ellipsometry, Thin Solid Films. 377-378(2000) 68-73.

[16] J. A. Wollam Company, The user manual of variable angle spectroscopic ellipsometry, J. A. Wollam Company, Lincoln, Neb., 2000, pp2-50

[17] Y. R. Park, K. J. Kim, Structural and optical properties of rutile and anatase TiO_2 thin films: Effects of Co doping, Thin Solid Films. 484 (2005) 34-38.

[18] M.C. Kao, H.Z. Chen, S.L. Young, The effects of the thickness of TiO_2 films on the performance of dye-sensitized solar cells, Thin Solid Films. 517 (2009)5096-5099

Key Engineering Materials Vol. 537 (2013) pp 229-233
© (2013) Trans Tech Publications, Switzerland
doi:10.4028/www.scientific.net/KEM.537.229

Low Temperature Preparation and Microstructure of TiO$_2$ Nanostructured Thin films Coated Short Glass Fibers by Biomimetic Synthesis Technology

Ming Qiu Wang[a], Jun Yan[b], Shi Guo Du, Yi Guo, Hong Guang Li, Hai Ping Cui and Hao Qin

The 3rd Department, Mechanical Engineering College, Shijiazhuang 050003, PR China

[a]mqwang1514@163.com, [b]yan-junjun@263.net

Keywords: Biomimetic Synthesis Technology, Composite Particles, Nanostructured Thin Films, Short Glass Fiber, TiO$_2$ Nanoparticles

Abstract. By using organic amine as the soft template, TiO$_2$ nanostructured thin films were deposited on the surface of short glass fibers (SGFs) by biomimetic synthesis technology at low temperature (85°C). The properties of composite particles, including surface morphology, the phase composition of the coating layer and microstructure were characterized by SEM, XRD, XPS and Raman. The results show that the functional layers containing NH$_2$ and OH groups can be formed through a kind of organic amine, which can induce the bio-mineralization of nano-TiO$_2$ on the SGFs surfaces. XRD shows that TiO$_2$ thin films uniformly coated on SGFs surfaces were mainly anatase. The surface roughness of SGFs was remarkably increased after coating. Compared with uncoated SGFs as filler, the abrasive loss of composite coatings decreased to 47 percent of that coatings filled with uncoated SGFs, when the composite fiber powders were filled in wear-resistant coatings based on the silicone modified epoxy resins. Therefore, anti-wear property of coatings was notably enhanced.

Introduction

Fibers reinforced polymer composites have been widely used as structural materials[1]. Among the different fillers for resins, short glass fibers (SGFs) have been extensively used owing to their qualities of higher thermal conductivity, higher modulus, lower price and well-distributed internal stress in the SGFs-filled matrix. However, SGFs are usually of smooth surface which have disadvantageous effects on the mechanical properties of composites as its weak interfacial action with the polymer matrix. In order to improve the interfacial action between the SGFs and the polymer matrix as well as to optimize the mechanical properties, kinds of approaches, such as including coupling agent treatment[2], plasma[3], and graft polymerization[4] have been developed in most of the previous studies. However, these methods did not radically eliminate the disadvantageous effects of smooth surface.

In the present work, the approach of biomimetic synthesis [5] was firstly used in the SGFs surfaces. We prepared TiO$_2$/SGFs by using organic amine as the soft template in an acidic aqueous solution at low temperature to develop the advantages by combining SGFs and TiO$_2$ nanoparticles.

Experimental

Materials. Tetrabutyl titanate (Ti(OBu)$_4$, AR), tin(II) chloride (SnCl$_2$·2H$_2$O, AR), acetylacetonat (acac, CR), Short glass fibers (SGFs, the glass composed of Si, Na, Ca, etc, are broken and calcined by high temperature after which the said fibers are obtained). Silicone modified epoxy resin (SEP, m(silicone):m(epoxy=0.2:1)), polyamide. The structure of organic amine as follows.

Preparation of SGFs/TiO$_2$ composite powders and the polymer matrix composite. TiO$_2$/SGFs composite powders were prepared by the following procedure. Firstly, SGFs were pretreated with organic amine 1% concentration in a simple ethanol system for 30 minutes. Subsequently, the pretreated samples were cleaned fully with anhydrous ethanol and dried at 65 °C in desiccators. 5.0 g pretreated SGFs, 0.1 g SnCl$_2$·2H$_2$O and 80.0 mL C$_2$H$_5$OH are put into the flask. 3.6 mL Ti(OBu)$_4$ were dissolved in 20.0 mL C$_2$H$_5$OH, and then 6.0 mL H$_2$O was added with vigorous stirring. A limpid yellow sol was formed by dropwise addition of hydrochloric acid into the precipitate. Then, pour the sol into the flask and the mixture was refluxed at 85°C for 50 minutes. Subsequently, the as-prepared samples were cleaned carefully with anhydrous ethanol and deionized water. The idling samples were fabricated by refluxing directly the sol without adding SGFs, with all other conditions unchanged.

The cure process of all polymer composites as follows. The as-prepared TiO$_2$/SGFs after pre-treatment with 0.5 wt.%-2.0 wt.% KH-560 ethanol solution were dried at 85°C for 1 hour in an air circulated oven. A measured amount of the formulated epoxy resin dissolved with component solvent, then 20% as-prepared composite powders, 2% other additives were mixed with the formulated epoxy resin in weight ratios, which remarked Group A. Polyamide, which remarked Group B, is used as a curing agent. Mixing B with A on the basis of stoichiometric ratio of m(A):m(B)=1:0.5, all samples were left for a cure at 45°C for 4 hours and 65°C for 8 hours in desiccators.

Characterization. The measurements were carried out at 25±1°C. The phases presented in the as-prepared powders, its chemical composition, and microstructure were determined by X-ray diffraction (XRD; Rigaku RAD-C; Cu-Kα, 40 kV, 150 mA), X-ray photoelectron spectroscopy (XPS; ESCA System; PHI1600X), respectively. The crystal structure of TiO$_2$ was determined by Raman spectroscopy (JOBINYV-1000). Scanning electron micrograph (SEM) of the samples were taken on HITACHI (S-4800) scanning electron microscope under a gold film. Wear-resistance testing was carried out on worn of apparatus (QMH, Tianjin, China), load (5N) and 500 turn under the Chinese national standards of GB 1768-79.

Results and discussion

SEM analysis. Fig. 1 shows the SEM images of the original SGFs (Fig. 1a and 1b), TiO$_2$/SGFs powders (Fig. 1c and 1d), respectively. It can be seen from Fig. 1a and 1b that the original SGFs have a smooth surface with about 10 μm width. Fig. 1c and 1d illustrate the SEM images of TiO$_2$/SGFs powders prepared by biomimetic synthesis technology. From Fig. 1d, we can see that the surface of the TiO$_2$/SGFs (Fig. 1c and 1d) became more rough than that of the original SGFs. TiO$_2$ nanoparticles of size about 50 nm are formed compactly and homogeneously on the SGFs surface. And agglomeration of TiO$_2$ nanoparticles can also be seen. It is clear that the utilization of organic amine has visibly promoted the formation and growth of TiO$_2$ nanoparticles.

Fig.1 SEM of the original SGFs (a, b) and TiO$_2$/SGFs (c, d)

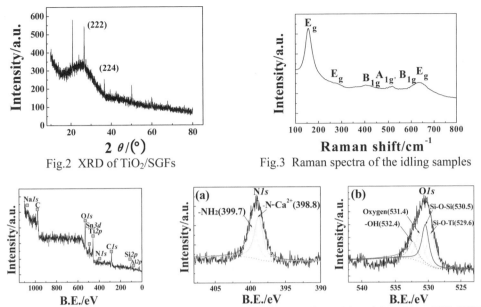

Fig.2 XRD of TiO$_2$/SGFs   Fig.3 Raman spectra of the idling samples

Fig.4 XPS spectra of TiO$_2$/SGFs   Fig.5 Fitting XPS spectra of N$1s$ (a) and O$1s$ (b) on TiO$_2$/SGFs

X-ray diffraction and Raman analysis. XRD was performed to confirm the crystal structure of the coating layers on the composite powders surface. The XRD patterns of the as-prepared samples are shown in Fig. 2. It can be found that peaks emerge at 2θ=23.3, 30.2, which indicates that the composition of SGFs was mainly SiO$_2$. The analysis of the diffraction peaks for composite powders reveals the presence of SGFs, while no obvious peaks indicated the existence of TiO$_2$. The reason is that the concentration of the loaded TiO$_2$ nanoparticles is relatively low[6].

In order to confirm further the crystal of TiO$_2$, the idling samples were made by directly refluxing the sol without SGFs. The Raman spectroscopy was shown in Fig. 3. The typical peaks appeared at 155, 403, 515 and 639cm^{-1} stand for Raman oscillating mode E$_g$ (155, 633 cm^{-1}), B$_{1g}$ (403cm^{-1}) and A$_{1g}$ (515 cm^{-1}) separately, which can further confirm that the structure of TiO$_2$ loaded on the surface of SGFs is mainly of anatase type[7]. This result was in accordance with that of XRD analysis.

XPS analysis. XPS technique was employed to determine the surface chemical composition of the representative as-obtained samples. Fig.4 showed the XPS spectra of typical TiO$_2$/SGFs spectrum. As shown in Fig.4, the peaks of Na$1s$, C$1s$, O$1s$, Ti$2p$, N$1s$, Sn$3d$ and Si$2p$ appeared on the surface of the TiO$_2$/SGFs particles, showing that the surface mainly contained elements of Na, C, O, Ti, N, Sn and Si. The existence of N$1s$ is further confirmed that the functional layers are formed on the SGFs surface. In the preparation process, the SGFs were cleaned several times with anhydrous ethanol and deionized water, therefore, the organic amine through physical absorption is prone to be washed by solvent. It is indicated that the organic base have been reacted with the hydroxyl on the SGFs surface, thus, the functional layers are gradually formed.

In order to confirm the composition of functional layers, the spectrum of N$1s$ and O$1s$ are further analyzed by XPS. Fig.5a showed the spectra of N$1s$. As shown in Fig.5a, the peaks of the N$1s$ spectrum of TiO$_2$/SGFs, can be fitted with two peaks, in which the higher binding energy peaks (399.7 eV) are assigned to -NH$_2$, and the lower binding energy peaks (398.8 eV) to coordination bonds between nitrogen atom and calcium ions with unoccupied orbit, and this implies that the interaction between SGFs and organic amine is chemical absorption.

The peak of O$1s$ at BE of about 531eV in Fig. 5b demonstrates the presence of oxygen atom on the surface of SGFs. The O$1s$ spectrum shown in Fig. 5b can be fitted with four peaks at BEs of 532.4 eV, 531.4 eV, 530.5 eV and 529.6 eV for -OH, absorption oxygen, Si-O-Si and Ti-O-Si, respectively.

Wear-resistance properties of compound materials based on silicone modified epoxy resins. Table 1 compares the variations of the weight loss against different fillers. Totally, the weight loss of composites filled with TiO$_2$/SGFs decreased to 47% of SGFs, that is to say, the surface nano-structured modification of SGFs is helpful for the enhancement of wear-resistance in the coatings, which were mainly due to the composite particles have a better interfacial action with the polymer matrix when used as filler materials[8]. Therefore, fewer debris was peeled off from the composites. On the other hand, TiO$_2$ nanoparticles formed on the surface of SGFs may work effectively like "ball effect"[9], thus, synergistic effect of nano-TiO$_2$ together with SGFs may be expected. Although wear data of SGFs/polymer system are not available in the literature according to the authors' knowledge, the result of short glass fibers filled PA6-PU composites can be cited as Ref. [10]. It is indicated that the wear-resistance were relatively lower at filler content of 5 wt.% than that of composites. Therefore, it is preliminarily concluded that the combination between TiO$_2$ nanoparticles and SGFs can enhance the wear-resistance of the coatings when filled in the matrix. Though the surface roughness of SGFs can increase by the alkaline pre-washing, the synergistic effect of TiO$_2$ nanoparticles together with SGFs couldn't develop if only through enhancing its roughness.

Table 1 Abrasive loss of fillers filled in the matrix (silicone modified epoxy resins, SEP)

Samples	Coatings filled with SGFs	Coatings filled with TiO$_2$/SGFs
Abrasive loss/mg	18.4	9.7

Conclusions

(1) TiO$_2$ nanostructured thin films have been prepared on the surface of short glass fibers (SGFs) by biomimetic synthesis technology at low temperature (85 °C). The SEM images show that the TiO$_2$ nanoparticles of size about 50nm have been formed compact and homogeneous layers on the SGFs surface, the main phase of which was anatase type. And the chemical bond of Ti-O-Si, which has been confirmed by XPS, is existed at the interface. The functional layers containing -NH$_2$ and -OH can induce the heterogeneous nucleation on the SGFs surface.

(2) Compared to SGFs/SEP composites, the use of the TiO$_2$/SGFs in SEP improves largely the wear-resistance of composite materials at the same filling content. It is provide a new method to design the wear-resistance coatings by filling in the as-prepared composite powders.

Acknowledgements

The financial support provided by the Project (50842054) of National Natural Science Foundation of China and Foundation of Mechanical Engineering College(YJJXM11016) for this work were gratefully acknowledged.

References

[1] N. Mohan, S. Natarajan, S.P. KumareshBabu, Abrasive wear behaviour of hard powders filled glass fabric-epoxy hybrid composites, Mater. Des. 32 (2011) 1704-1709.

[2] Y.J. Song, L.J. Huang, P. Zhu, et al. Friction and wear of coupling agent treated glass fiber modified polyimide composites, J. Mater. Eng. 2 (2009) 58-62.

[3] M. Norkhairunnisa, A.B. Azhar, C.W. Shyang, Effects of organo-montmorillonite on the mechanical and morphological properties of epoxy/glass fiber composites, Polym. Int. 56 (2007) 512-517.

[4] N.M.L.R. Mondadori, C.R. Nunes, A.J. Zattera, Relationship between processing method and microstructural and mechanical properties of poly(ethylene terephthalate)/short glass fiber composites, J. App. Poly. Sci. 109 (2008) 3266-3 274.

[5] J.N. Cha, G.D. Stucky, D.E. Morse, Biomimetic synthesis of ordered silica structures mediated by block copolypeptides, Nature, 403 (2000) 289-292.

[6] Z.Y. Ma, S.X. She, S.X. Min, Sythesis and characterization of Cr^{3+} doped TiO_2 composite nanoparticles, J. Lanzhou Univ. 41 (2005) 64-67.

[7] S.P.S. Porto, P.A. Fleury, T.C. Damen. Raman Spectra of TiO_2, MgF_2, ZnF_2, FeF_2, and MnF_2, Phys. Rev. 154 (1967) 522-526.

[8] S.R. Chauhan, A. Kumar, I. Singh, Study on friction and sliding wear behavior of woven S-glass fiber reinforced vinylester composites manufactured with different comonomers, J. Mater. Sci. 44 (2009) 6338-6347.

[9] S Bahadur, C Sunkara, Effect of transfer film structure, composition and bonding on the tribological behavior of polyphenylene sulfide filled with nanoparticles of TiO_2, ZnO, CuO and SiC, Wear, 258 (2005) 1411-1421.

[10] D.X. Li, X. Deng, J. Wang, Mechanical and tribological properties of polyamide 6-polyurethane block copolymer reinforced with short glass fibers, Wear, 269 (2010) 262-268.

Key Engineering Materials Vol. 537 (2013) pp 234-237
© (2013) Trans Tech Publications, Switzerland
doi:10.4028/www.scientific.net/KEM.537.234

Structure Characterization and Hydrophilic Property of S+Ce Co-doped TiO₂ Nanocomposite Film

Difa Xu[1,a], Haibo Wang[2, b], Xiangchao Zhang[1,c], Shiying Zhang[1,d]

[1] Institute of New Materials, Department of Biological and Environmental Science,Changsha University, Changsha,410022, China

[2] Institute of Technology East China Jiao Tong University, Nanchang Economic and Technical Development Zone,Nanchang, 330100 China

[a] xudifa@ccsu.cn, [b]whbcsu@126.com, [c]csuzxc@126.com, [d] cdzhangshiying@163.com

Keywords: Titanium dioxide (TiO₂), thin film, photocatalysis, dope, hydrophilic

Abstract. The S+Ce co-doped TiO₂ nanocomposite films deposited on glass substrate had been synthesized by the sol–gel dip-coating method. The as-synthesized samples were characterized using X-ray photoelectron spectroscopy (XPS), atomic force microscopy (AFM) and ultraviolet-visible (UV-vis) absorption spectra analysis technologies. The surface morphology and surface chemical composition of the S+Ce co-doped TiO₂ nanocomposite film had been primarily investigated. The results shows that the properties of doped TiO₂ thin films with different ions have close relations with the intrinsical properties of S and Ce doped ions, the absorption edge shifted towards visible light region and the water contact angle of the surface of the nanocomposite films with the water droplet was only 6°, indicating that the S+Ce co-doped TiO₂ nanocomposite film showed promising applications in the self-cleaning and other potential fields.

Introduction

Titanium dioxide (TiO₂) is one of important functional inorganic materials. Since Fujishima[1] discovered the photocatalytic splitting of water on a TiO₂ electrode under ultraviolet light in 1972, enormous efforts have been devoted to the research of TiO₂ materials due to its excellent optical, electrical, photocatalytic and thermal properties, which led to promising applications in the fields of photovoltaics, photocatalysis, sensors, medicine et al[2]. However, it is well known that there are two typical drawbacks in practical application. Firstly, only the ultraviolet part of the solar irradiation (amounting to ~4% of the incoming solar energy on the earth's surface) could be absorbed by TiO₂, Secondly, a low rate of electron transfer to oxygen and a high rate of recombination between excited electron–hole pairs. In order to resolve the above listed problems, continuous efforts have been made to improve the properties of TiO₂ thin film, for example, noble metal loading, metal ion doping, anion doping, dye sensitization, composite semiconductors, metal ion-implantation, etc[3,4].

The development of photocatalysts under visible light irradiation is one of the major goals for enhancing the efficient utilization of solar energy and realizing the practical industrialization, which is also one of the challenging tasks in the field of photocatalysis. Choi [5] performed a systematic study of TiO₂ nanoparticles doped with 21 metal ions by the sol-gel method and found the presence of metal ion dopants significantly influenced the photoreactivity, charge carrier recombination rates, and interfacial electrontransfer rates. Metal ion doping on TiO₂ can expand its photo-response to visible region through formation of impurity energy levels. However, the effect of red shift is negligible and doped ions tend to become recombination centers. Therefore, the benefit of metal ion doping is limited. Qualitatively, anion doping, such as nitrogen doping and sulfur doping, is more effective than metal ion doping for red shift.

In this paper, the S+Ce co-doped TiO₂ nanocomposite films deposited on glass substrate had been synthesized by the sol–gel dip-coating method. The crystalline structure, surface morphology and surface chemical composition of the S+Ce co-doped TiO₂ nanocomposite film had been primarily investigated by XPS, AFM and UV-vis analysis technologies. Hydrophilic properties were evaluated from the measurements of the water contact angles on the surface of the samples.

Experimental

The precursor solution of S+Ce co-doped TiO_2 sol was synthesized by a sol–gel method, using titanium tetrabutyloxide(TTBO), H_2NCSNH_2, $CeCl_3 \cdot 7H_2O$, diethanolamine(DEA) and anhydrous alcohol as starting materials. In typical synthesis process, 17.02 ml TTBO and 4.8 ml DEA were mixed with 48.28 ml EtOH, after the sol being magneticly stirred for 30 min at room temperature, 0.0025 mol H_2NCSNH_2 and $CeCl_3 \cdot 7H_2O$ dissolved in 20 ml ethanol were added to the sol drop by drop keeping vigorous stirring for 30min. Afterwards, the mixed solution was added dropwise into another mixture consisting of 0.9 ml deionized water and 10 ml EtOH under roughly stirring to hydrolyze. After continuously stirring for 2 h, a transparent sol of 5% S+Ce co-doped TiO_2 in molar concentration was obtained, which was then aged for 24 h and served for film preparation. The S+Ce co-doped TiO_2 thin film was prepared by a dip-coating method. Prior to the coating process, microscope slides glass substrates were ultrasonically cleaned. The gel film was obtained by dipping the substrate in the precursor solution bath and pulled upwards with a constant speed of 4 cm/min to keep uniform thickness of the film. After a layer was deposited, the glass was treated at 100°C for 5min, and then coated another layer. The thickness of the film was adjusted by dip-coating cycles. The films were coated with six layers and then were annealed at 500°C for 2h.

The X-ray photoelectron spectroscopy (XPS) analysis was performed on a VG ESCALAB MK-II spectrometer equipped with an Mg Kα (1253.6 eV) monochromator X-ray source with a power of 240 W. The test chamber pressure was maintained below 5×10^{-7} Torr during spectral acquisition. The XPS binding energy (BE) was internally referenced to the C 1s peak (BE=284.6 eV). The surface morphology was observed using a NT-MDT atomic force microscopy (AFM) with silicon probe tapping contact mode. The UV–visible absorption spectra of the sample was performed using a UV–visible spectrophotometer (UV-2450). Hydrophilic properties of glasses were evaluated from the measurements of the water contact angles on the surface using a JJC-I contact angle instrument.

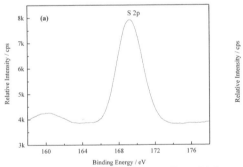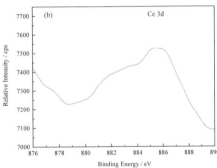

Fig. 1 S+Ce co-doped TiO_2 thin films high-resolution XPS spectra(a) S 2p and (b) Ce 3d

Results and discussion

To investigate the states of the doping ions, we applied XPS to record the S 2p and Ce 3d XPS spectra of the S+Ce co-doped TiO_2 nanocomposite film as shown in Figure 1.

According to the previous studies[6], the strong peak at 169.5 eV could be assigned to S^{6+} (SO_4^{2-}) states. These sulfate ions can form S,O and O-S-O bonds on the TiO_2 surface, creating unbalanced charge on Ti and vacancies/defects in the titania network. As a result, the introduction of sulfate ions will bring strong acidity on the TiO_2 surface. It Accordingly, it can be concluded that the contribution of the doping sulfur atoms to the photocatalytic activity originates mainly from the introduction of acidic sites on the surface of the co-doped TiO_2. The XPS of Ce 3d is quite complicated due to the hybridization of Ce 4f and O 2p electrons. It was reported that the Ce 3d spectra could be assigned to two sets of spin–orbital multiplets, i.e., $3d_{5/2}$ and $3d_{3/2}$. [7] The Ce 3d spectrum of sample basically denotes a mixture of Ce^{3+}/Ce^{4+} oxidation states giving rise to a myriad of peaks, indicating that the surface of the sample is not fully oxidized.

Two- and three-dimensional AFM images of S+Ce co-doped TiO_2 nanocomposite film deposited on glass substrate are shown in Fig. 2. As shown in Fig. 2, the nanoparticles of S+Ce co-doped TiO_2 nanocomposite film exhibit spherical shape. The surface of S+Ce co-doped TiO_2 nanocomposite film was smooth, with the maximum protuberance about 111.2 nm and the size of TiO_2 nanoparticles was 63.6 nm in average. The image based root mean square roughness for the TiO_2 films is 11.3 nm.

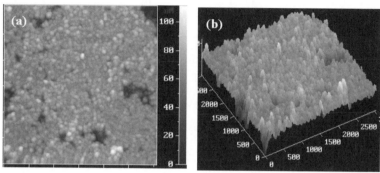

Fig. 2 AFM images of S+Ce co-doped TiO_2 thin films: (a) Two and (b) three dimensional

Fig. 3 gives the UV-vis absorption spectra of several doped TiO_2 samples. Compared to the spectrum of pure anatase TiO_2, the absorption edge was shifted to the lower energy region in the spectra of all the doped TiO_2 materials. Pure TiO_2 has no ability to respond to visible light, whereas the absorption edges of S+Ce co-doped TiO_2 nanocomposite film are shifted into the visible region. The absorption data were fitted to following equation for indirect band-gap transitions[8]: $ahv = A (hv-E_g)^2$. A quantitative evaluation of the band gap energy (E_g) can be performed by plotting $(ahv)^{1/2}$ versus hv and extrapolated from linear part of the curve as shown in the Fig. 3b.The calculated band gap (E_g) value of the TiO_2,S doped, Ce doped and S+Ce co-doped TiO_2 thin films are determined to be 3.20 eV, 2.63 eV, 2.51 eV and 2.36 eV, respectively. We prefer to believe that the red-shifted absorption dominantly originates from the oxygen vacancies created during the codoping process.[9] The generation of these oxygen vacancies is due to the bulk and surface defects induced by the codoping effect. The codoping with two elements can increase the number of oxygen vacancies in TiO_2 and thus make the absorption more red-shifted.

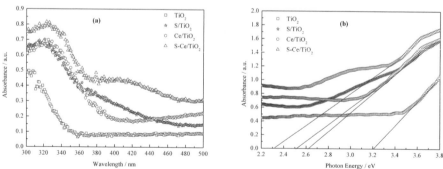

Fig. 3 TiO_2 films (a)UV-vis absorption spectra of and (b) $(ahv)^{1/2}$ versus hv for indirect transition

To investigate the hydrophilic properties of samples, the contact angles of water droplets with the surface of thin films were measured under ambient conditions, the results are shown in Fig. 4. It can be obviously seen that the water contact angle of the synthesized thin film for Ce doped, S doped and S+Ce co-doped thin films are 43°, 25° and 6°, respectively. It is interesting that the S+Ce co-doped TiO_2 nanocomposite film exhibits super-hydrophilicity, which indicates the films would be widely application in the self-cleaning and other potential fields.

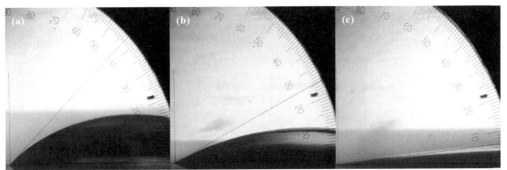

Fig.4 The contact angles (a)Ce doped;(b)S doped and (c) S+Ce co-doped TiO_2 thin films

Summary

The S+Ce co-doped TiO_2 nanocomposite films deposited on glass substrate had been synthesized by the sol–gel dip-coating method. The surface morphology and surface chemical composition of the samples had been primarily investigated. The S+Ce co-doped TiO_2 nanocomposite film exhibit spherical shape with the maximum protuberance about 111.2 nm and the size of TiO_2 nanoparticles was 63.6 nm in average. The calculated band gap (E_g) value of the TiO_2 thin film is about 2.36 eV, the water contact angles for the thin film was only about 6°.

Acknowledgements

This work was supported by the Program for the National Natural Science Foundation of China (51102026) and Aid program for Science and Technology Innovative Research Team in Higher Educational Instituions of Hunan Province.

References

[1] A. Fujishima, A. Honda. Nature, Electrochemical photolysis of water at a semiconductor electrode, Nature, 238 (1972) 37-38.

[2] A. Kubacka, M. F. Garcia, G. Colon, Advanced nanoarchitectures for solar photocatalytic applications, Chem. Rev., 112 (2012) 1555-1585.

[3] O.Carp, C. L.Huisman, A.Reller, Photoinduced reactivity of titanium dioxide, Progress in Solid State Chemistry, 32 (2004) 33-177

[4] H. Tong, S. X. Ouyang, Y. P. Bi, N. Umezawa, M. Oshikiri, J. H. Ye, Nano-photocatalytic materials: possibilities and challenges, Adv. Mater., 24 (2012) 229-251.

[5] W.Chio, A.Termin, M. R.Hoffman, The role of metal ion dopants in quantum-sized TiO_2, J. Phys. Chem., 98(1994) 13669-13679.

[6] Y. Wang, Y. Wang, Y. Meng, H. Ding,Y. Shan, A highly efficient visible-light-activated photocatalyst based on bismuth- and sulfur-codoped TiO_2, J. Phys. Chem. C, 112 (2008) 6620-6626.

[7] C. Liu, X. Tang, C. Mo, Z. Qiang, Characterization and activity of visible-light-driven TiO_2 photocatalyst codoped with nitrogen and cerium, J.Solid State Chem. 181 (2008) 913–919

[8] T. Zhang, H. Peter, H. Huang, Sintering study on commercial CeO_2 powder with small amount of MnO_2 doping, Materials Letters, 57 (2002) 507–512.

[9] M. S. Andrew, S. M. Nie, Semiconductor nanocrystals: structure, properties, and band gap engineering, Accounts of Chemical Research,43 (2010) 190-200.

Key Engineering Materials Vol. 537 (2013) pp 238-242
© (2013) Trans Tech Publications, Switzerland
doi:10.4028/www.scientific.net/KEM.537.238

Synthesis and Characterization of Graphene Sheets as an Anode Material for Lithium-Ion Batteries

L. Dong, Wei Ren, L. Dong, D.J. Li[a]

College of Physics and Electronic Information Science, Tianjin Normal University, Tianjin, 300387, P.R. China

[a]dejunli@mail.tjnu.edu.cn (corresponding author)

Keywords: Nano electrode material, Lithium ion battery, anode, graphene sheets.

Abstract. The graphene sheets were obtained using rapid thermal expansion in nitrogen atmosphere from graphene oxide sheets prepared from graphite powder through improved Hummers' method. The structure and morphology of the graphene sheets as anode materials with different dimensions were characterized through X-ray diffraction (XRD), Fourier transform infrared spectroscopy (FTIR), Scanning electron microscope (SEM). The electrochemical performances were evaluated in coin-type cells versus metallic lithium. The results showed that graphene as the two dimensional nanomaterials possessed more advantages in microstructure and better Li-ions intercalation performances than the traditional materials.

Introduction

Graphene has been paid significant attention several years [1]. Graphene is the name given to a flat monolayer of carbon atoms tightly packed into a two-dimensional (2D) honeycomb lattice. It can be stacked to form 3D graphite, rolled to form 1D nanotubes, and wrapped to form 0D fullerenes [1]. Recently it attracted great interests for both fundamental and applied research since its discovery by mechanical exfoliation. Graphene has unique chemical and physical properties, such as high chemical stability, excellent conductivity which is better than silver, as the highest conductive substance known at room temperature, and the easiness of functionalization and production. These unique physical and chemical properties lead graphene as a perfect base material for applications in electric device, biosensors, supercapacitors, and even application in biotechnology. In particular, the graphene have superior electrical conductivities than graphitic carbon, high surface areas of over 2600 m^2/g, and a broad electrochemical window that would be very advantageous for application in energy technologies [2]. Single graphene sheets derived from splitting graphite oxide have been successfully prepared by Schniepp through thermal exfoliation in 2006 [3]. In recent years, graphene has more and more applications in the battery materials field not only its improved performance by nanodispersion and nanostructuring, but also better electrochemical reactions as well as emerged new materials from graphene combined with materials.

In this work, we prepared the graphene by the heat-treatment of graphene oxide under Ar, and studied the structure of graphene systematically by various characterization techniques. The study influence of dimensions of graphene sheets as LIBs' anode on their electrochemical properties was also investigated.

Experimental

Graphene was first prepared by the oxidation of graphite powder using the improved synthesis of graphite oxide [4]. First, a 9:1 mixture of concentrated H_2SO_4/H_3PO_4 (180:20 mL) was added to a mixture of graphite flakes (3.0g, 1 wt equiv) and $KMnO_4$ (12.0g, 4 wt equiv) in the room temperature. After a slight exotherm to nearly 40°C the reaction was heated to 50°C and stirred for 24 h. Then the reaction was cooled to 20°C and poured into ice (300 mL) with 30% H_2O_2 (20 mL). After that the mixture' color became golden yellow from turquoise, the mixture was centrifuged (12000 rpm for 20 min), and the supernatant was decanted away. The remaining solid material was then washed in succession with 5% HCL until the SO_4^{2+} cannot be checked. After washing in succession with deionized water until the pH value of the filtrate was nearly neutral, suspension material was dried overnight at room temperature to get graphite oxide. Finally, graphene was obtained after the dried graphite oxide was heated at 1050°C for 30 s under Ar environment.

The structure and morphologies of graphene were analyzed by XRD (Rigaku D/MAX 2500 X), FTIR (IRAffinity-1), and SEM (Hitachi S-5200). The electrochemical performances of the graphene as anodes were measured as follow. Composite anode made up of active substance of graphene (90 wt. %) and polyvinylidene fluoride binder (PVDF, 10 wt. %) was obtained in N-methylpyrrolidinone (NMP) solvent. The electrolyte was composed of 1 M $LiPF_6$ dissolved in ethylene carbonate (EC) and diethyl carbonate (DEC) with a volume ratio of 1:1. After drying 24 h under vacuum at 100°C, the anode was assembled into CR-2032-type coin half-cells using lithium foil as cathode in glove box (M/Braun, moisture and oxygen concentration <1 ppm). Electrochemical performance was tested galvanostatically with a cell test instrument (Arbin, BT-2000). The electrochemical window was 0.01 to 2.0V. Cyclic voltammetry tests were carried out on a versatile electrochemical workstation (CHI, 660D) at a scan rate of 0.1 mVs^{-1} within a potential range of 0.01–3.0 V (vs. Li^+/Li). Electrochemical impedance spectroscopy (EIS) measurements of the electrodes were also performed on the electrochemical workstation (CHI, 660D) using the frequency response analysis applying a sine wave with amplitude of 5.0 mV over the frequency range from 100 kHz to 0.001 Hz.

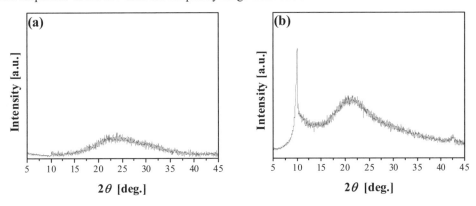

Fig 1. XRD patterns of graphene (a) and graphene oxide (b).

Result and discussion

Fig.1 shows the XRD patterns of graphite oxide and graphene sheets. XRD pattern of graphite oxide exhibits a characteristic peak at 11^0, corresponding to the (0 0 1) diffraction peak of graphite oxide. The d-spacing of graphite oxide increased to 0.779nm from 0.335nm of graphite powder, which is ascribed to the oxide-induced O-containing functional groups and inserted H_2O molecules [5] that can be confirmed by FTIR in Fig. 2. These results suggest that the graphite powder has been completely oxidized. After rapid heat treatment, the diffraction peak of graphite oxide disappears and the graphene displays a weak, broad diffraction peak of (002) diffraction at about 23° (Fig.1(b)), indicating a typical amorphous structure with a layer-layer distance (d-spacing) of 0.379 nm calculated according to Bragg's equation. This compares with a d-spacing of 0.335 nm for graphite. Annealing the graphite oxide at 1050℃ with Ar for 30 s shifted the diffraction peak to 23°(Fig. 1a), nearly identical to that of graphite but less crystallized, indicating that the graphite oxide has successfully exfoliated to graphene sheets.

FT-IR spectrum of graphene oxide exhibits representative peaks at 3435, 1730, 1630, 1235 and 1061 cm^{-1} corresponding to O-H stretch, C=O stretch, aromatic C=C and O–H bending, epoxy C–O stretch and alkoxy C–O stretch, as shown in Fig.2. These evidences indicate that the original extended conjugated π-orbital systems of the graphite were opened and oxygen-containing functional groups were inserted into carbon during the oxidation process of the graphite powder with $KMnO_4$ in the concentrated H_2SO_4/H_3PO_4. After heated at 1050℃ for 30 s with Ar, oxygen-containing functional groups derived from the intensive oxidation were negative, which can be proved by the XRD in Fig.1. While a new peak at $1565cm^{-1}$ reflects the skeletal vibration of graphene sheets [6]. These results indicate that the graphite oxide have been successfully exfoliated to graphene by thermal exfoliation. The FT-IR spectrum of graphene and pristine graphite are very similar, which means the oxygen groups have been largely stripped.

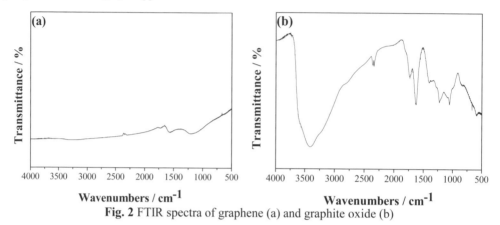

Fig. 2 FTIR spectra of graphene (a) and graphite oxide (b)

Fig. 3 SEM images of graphene (a). TEM of images of graphene (b)

Fig. 3 shows the SEM and TEM pictures of graphene. The image revealed that most of the graphite oxide has been effectively exfoliated to ultrathin sheets with wavy structures. This structure suggests that the intrinsic ripples of graphene might develop into wavy structures in the macroscopic scale. In respect of the graphene, one may consider that it could be easy to form a porous structure in the macroscopic scale due to the curled and wrinkled graphene sheets stacked structure [7]. Fig 3(b) demonstrated the two-dimensional ultrathin flexible structure of graphene. Corrugations and scrolling also can be detected.

Fig. 4 shows the first three discharge/charge profiles of the prepared graphene sheets at the current density of 0.1C. It is obviously that the voltage curve of the first discharge process was different from the second one. The graphene sheets with fewer layers shows an irreversible capacity at about 0.5-0.75V in the first cycle, which could be attributed to the formation of a solid electrolyte interface (SEI). The first discharge and charge capacities of graphene sheets are as high as 1471mAh g^{-1} and 350mAh g^{-1}, respectively. The possible reason is that the larger surface area and curled morphology of graphene sheets with fewer layers can provide more lithium insertion active sites [8]. Higher reversible capacities of graphene are benefit from their microstructures. It's well known that the disordered carbons can yield higher capacity values than graphite [9-11], and the graphene can be considered as a very disordered carbon. Otherwise, we find that the graphene sheets electrode exhibits a broad electrochemical window (0.01–2.8 V) as a function of lithium capacity and large voltage hysteresis between discharge and charge voltage curves, which is different from graphite and similar to the disordered carbons [8-11]. Although the exact insertion/extraction mechanism of lithium ions in graphene sheets is not clear yet, the insertion/extraction mechanisms of lithium ions in graphene sheets and graphite should be different [12]. The reversible specific capacity of the prepared graphene sheets reduced to 384mAh g^{-1} in the second cycle, but it was still maintained at 315mAh g^{-1} in the third cycle. It should be noted the graphene sheets we obtained are not single layer and all surfaces are exposed but also agglomerated and overlapped (Fig. 3a).

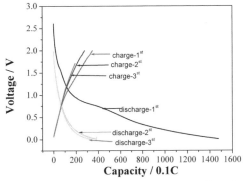

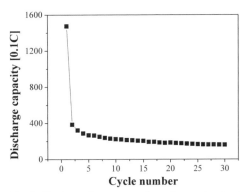

Fig.4. First three discharge/charge profiles of graphene sheets at the current density of 0.1c

Fig.5. Cycle performance of graphene sheets at the current densities of 0.1C

Cycle performance of graphene sheets at 0.1C are shown in Fig. 5. After 30 cycles, the reversible capacity is still maintained at 200mAhg^{-1} for graphene sheets with fewer layers, which was much higher than that of previously reported graphene nanosheets (290mAhg^{-1} after 20 cycles at a low current density of 50mAg^{-1}) [13]. These results demonstrated that the prepared graphene sheets with fewer layers have an intensive potential as a candidate of anode materials with high reversible capacity, good cycle performance and high rate discharge/charge capability.

Conclusions

The advantages are shown using the improved method for producing GO. The protocol for running the reaction does not involve a large exotherm and produces no toxic gas. The high quality graphene sheets with a curled morphology consisting of a thin wrinkled paper-like structure, fewer layers and a large specific surface area were prepared using a thermal exfoliation method. The electrochemical performance testing showed that the first reversible specific capacity of the prepared graphene sheets was as high as 350mAh g^{-1} at a current density of 100mAg^{-1}. Due to its two dimensional layer structure, graphene could bind more lithium ions and possesses a better cycle performance for the lithium storage.

Acknowledgements

This work was supported by National Natural Science Foundation of China (51272176, 11075116), National Basic Research Program of China (973 Program, 012CB933604), and the Key Laboratory of Beam Technology and Material Modification of the Ministry of Education, Beijing Normal University, China.

References

[1] A. K Geim, K. S. Novoselov, The rise of graphene Nature Materials 6 (2007) 183-191.

[2] Allen MJ, Tung VC, Kaner RB, Honeycomb Carbon: A Review of Graphene, Chem. Rev, 110 (2010) 132–145.

[3] H.C. Schniepp, J.L. Li, M.J. McAllister, Functionalized Single Graphene Sheets Derived from Splitting Graphite Oxide, J. Phys. Chem. B 110 (2006) 8535-8539.

[4] D.C. Marcano, D.V. Kosynkin, Improved Synthesis of Graphene Oxide, ACS Nano, 4.8.(2010) 4806–4814.

[5] D. Pan, S. Wang, B. Zhao, Li Storage Properties of Disordered Graphene Nanosheets, Chem. Mater. 21 (2009) 3136-3142.

[6] P. Guo, H. Song, X. Chen, Electrochemical performance of graphene nanosheets as anode material for lithium-ion batteries, Electrochem. Commun. 11 (2009) 1320-1324.

[7] A. Fasolino, J.H. Los, M.I. Katsnelson, Intrinsic ripples in graphene. Nat. Mater. 6 (2007) 858-861.

[8] Y. Wu, C. Jiang, C. Wan, E. Tsuchida, Effects of catalytic oxidation on the electrochemical performance of common natural graphite as an anode material for lithium ion batteries, Electrochem. Commun. 2 (2000) 272-275.

[9] T. Zheng, J. S. Xue, and J. R. Dahn, Lithium Insertion in Hydrogen-Containing Carbonaceous Materials, Chemistry of Material. 8 (1996) 389-393

[10] L. Li, E. Liu, Y.Yang, Nitrogen-containing carbons prepared from polyaniline as anode materials for lithium secondary batteries, Materials Letters. 64 (2010) 2115-2117.

[11] A. Concheso, R. Santamaría, R. Menéndez, Electrochemical improvement of low-temperature petroleum cokes by chemical oxidation with H_2O_2 for their use as anodes in lithium ion batteries, Electrochim. Acta 52 (2006) 1281-1289.

[12] P. Lian, X. Zhu, S. Liang, Large reversible capacity of high quality graphene sheets as an anode material for lithium-ion batteries Electrochimica Acta 55 (2010) 3909-3914

[13] J.L. Tirado, G.F. Ortiz, I. Honma, Tin–carbon composites as anodic material in Li-ion batteries obtained by copyrolysis of petroleum vacuum residue and SnO_2 Carbon. 7 (2007) 1396-1409.

Key Engineering Materials Vol. 537 (2013) pp 243-246
© (2013) Trans Tech Publications, Switzerland
doi:10.4028/www.scientific.net/KEM.537.243

High Performance of YBCO Films Prepared by Fluorine-free MOD Method and a Direct Annealing Process

Xiangxue Xu[a], Wanchen Xu[b], Yueling Bai[*, c], Chuanbin Cai[d] and Jianhui Fang[*,e]

Department of Chemistry, College of Sciences, Shanghai University, Shanghai 200444, P. R. China

[a]nannan@shu.edu.cn, [b]wanchengb@163.com, [c]yuelingbai@shu.edu.cn, [d]cbcai@staff.shu.edu.cn, [e]jhfang@shu.edu.cn

Keywords: YBCO films, Fluorine-free MOD, One-step heat treatment, Xylene.

Abstract. Chemical solution-deposited (CSD) is a cost-effective non-vacuum method for YBCO coated conductor fabrication. We developed a new fluorine-free metal–organic deposition (MOD) method with metal acetates, propionic acid and xylene as the starting materials. Using this non-fluorine MOD method, we were able to get high performance YBCO superconducting films within a shortened heat treatment time, which was reduced by at least 5 h in comparison with that for the pyrolysis-annealing non-fluorine MOD process. Superconducting property, with a critical current density (J_c) over 0.55 MA/cm^2 at 77K , self-field has been obtained for 380 nm epitaxial YBCO thin films on (00l) LaAlO$_3$ single crystal substrate. Owing to the low price of starting materials and the shorter heat treatment time, fluorine-free MOD method is a very effective and cost-cutting process.

Introduction

YBCO high-temperature superconductor films with high J_c have been given more and more attention in recent years due to great promise applications on various superconducting devices such as cable, transformer, generators, and motors [1]. As an alternative to the relatively high cost and complexity of vacuum deposition techniques, Chemical solution deposition (CSD), especially metal–organic deposition (MOD) method is considered to be a cost-effective non-vacuum technique for producing coated conductors [2–8].

Metal-organic deposition using trifluoroacetates (TFA-MOD) method is one of the most promising ways for producing coated conductors due to high quality of YBCO films at a low cost. However, the conventional TFA-MOD method has many drawbacks, such as a long calcining time and harmful HF gas formed during the process [9,10], which makes fluorine-free MOD method interested for the researchers because of the good environmental compatibility and commercial-scale application of coated conductors [11,12].

In this work, we report the preparation of high performance YBCO films on (00l) LAO single crystal substrate by fluorine-free MOD method with metal acetates, propionic acid and xylene as the starting materials. The microstructure and superconducting properties of the YBCO thin films were investigated.

Experimental

YBCO films were prepared on (00l) LaAlO$_3$ (LAO) single crystal substrate by fluorine free MOD method using metal acetates as starting materials. The fluorine free precursor A was prepared by dissolving yttrium, barium, and copper acetates in propionic acid with a molar ratio of Y: Ba: Cu=1:2:3.15, 0.3 mL xylene was added into the solution subsequently and then ultrasounded for 80 min. A transparent dark green solution diluted by methanol was obtained with a total cation concentration of 1.5 M. Precursor B was prepared by the same procedure with A except no xylene. The precursors were then coated on LAO substrate by dip coating with a speed of 50 mm/min followed by the thermal decomposition, firing process, and oxygenation annealing.

Heat treatment schedule for YBCO films was shown in Fig. 1, the coated precursor films on LAO substrate were firstly dried at 120 °C for 3 min under O_2 atmosphere. Then the film was heated up to 300 °C in humid O_2 atmosphere at the rate of 5 °C /min, followed by heat treatment up to 840 °C at the rate of 20 °C /min under humid Ar/O_2 mixture gas. YBCO film experienced the partial melting process at 840 °C for 5 min under Ar/O_2 mixture gas before YBCO phase transform to form tetragonal phase at relatively lower temperature (780°C) for 90 min and finally go through the oxygenation annealing at 450 °C for 90 min to form YBCO orthogonal phase which shows superconducting properties. The humid ambient was obtained by passing the dry gas through a water reservoir in which the humidity can be controlled by the water temperature.

The phases of samples were determined by X-ray diffraction (XRD) using CuKα radiation by a D2700 XRD diffractormeter. The surface morphology and film thickness of samples were observed by a scanning electron microscope (SEM: JEOL JSM-6700F). Laser Micro-Raman spectrometer (INVIA) was used to detect the orientation of YBCO phase. The J_c value was derived from M (B) curves at different magnetic field using the Bean critical state model.

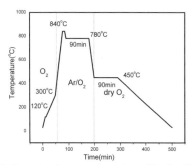 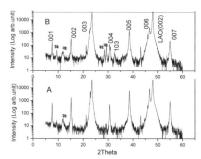

Fig. 1 One-step heat treatment profile for YBCO film A and B.   Fig. 2 The XRD patterns of YBCO films prepared by precursors A and B, respectively.

Results and Discussions

Fig. 1 shows the one-step heat treatment technique used in this work. The coated films were partial melted at 840°C for 5 min followed by down to 780°C for 90 min for the films crystallization, and then annealed at 450°C in pure O_2 for 90 min. In the conventional pyrolysis-annealing process, one or more pyrolysis processes at 400-500°C are easy to form the stable $BaCO_3$ phase which will deteriorate the film surface. So one-step heat treatment here was introduced to prepare high performance YBCO superconducting film [13]. Furthermore, the precursors were prepared without the evaporation process which greatly shorten the preparation time, so many different solvents such as acid, methanol remain in the solution were removed by drying at 120°C. Partial melting process was introduced to reduce the random and a-axis oriented phases, leading to a purer c-axis oriented YBCO phase produced [14]. The $BaCO_3$ phase can convert into the $Ba(OH)_2$ phase by controlling the water vapor pressure, which can be used as the barium source to react with CuO and Y_2O_3 to generate the YBCO phase [15].

The XRD patterns of YBCO film A and B prepared by the partial melting processes are shown in Fig. 2. The main peaks of film A and B are almost consistent, but in the case of B, a polycrystalline peak (103) is observed, and there are still several unidentified peaks which indicates the phase of film B is not pure, while in the case of A, the polycrystalline peak (103) almost disappears, and there are almost no peak of impurity phases. Furthermore, film A and B does not obviously show peak YBCO(200), indicating there is no or a small amount of a-axis oriented YBCO grains in both film A and B. According to the results, we can conclude that the purer c-axis oriented YBCO grains and better texture of film A was obtained, which illustrates the addition of xylene is very important because it greatly improved the performance of YBCO films. The current density results show the value of J_c for film A is over 0.55 MA/cm^2 at 77 K, self-field while film B shows no J_c.

The surface morphology of YBCO films A and B are shown in Fig. 3. Both film A and B are a microstructure almost without pores and obviously stripe-like a-axis YBCO grains observed on the film surface, but film A looks much better and denser than that of B. For film A, high boiling point xylene is added which maybe preserve the solid skeleton formed along the c-axis orientation to some extent. Apart from the high boiling point role, xylene may also be the stress-relaxing reagent just like PVP did [16]. The thickness of YBCO films examined here was 380 nm as determined by SEM cross-sectional imaging of the films.

YBCO grains grow epitaxially by heterogeneous nucleation, in which the substrate can be a catalyst for nucleation reported by Chen et al [17]. The Gibbs free energy of forming the spherical nuclei with small intersection angle θ between the substrate and the coated film is smaller than that with large intersection angle θ. The addition of xylene may enhance the wettability between the precursor film and substrate which further promote the epitaxial growth of the YBCO film.

To further investigate the crystal orientation, impurities, and oxygen content of the films, the non-destructive micro-Raman spectroscopy was performed on the films A and B. The r value of Raman intensity ratio of O(2,3)-B_{1g} and O(4)-A_g represents the ratio of c-axis and a-axis oriented YBCO grains of the films. For film A, the r value is 1.72, while it is only 1.50 for B which indicates xylene plays an important role on promoting the c-axis YBCO grains growth. As shown in Fig. 4, the peak for O(4) appears at 501 cm^{-1} in both films which represent both films are optimally doped to form $YBa_2Cu_3O_{6.95}$ which may have the superconductivity. A broad Raman peak at 560-580 cm^{-1} was found in both films indicating a cation disorder of the barium and yttrium atom in the YBCO structure. In addition, no evident Raman peaks at 220-250 and 590-600 cm^{-1} for Cu(1) and O(1) respectively imply the infinite Cu(1)-O(1) metal chains keeping very well and no presence of broken defects [18].

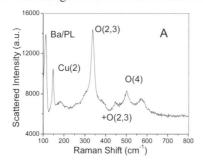 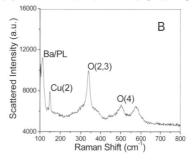

Fig. 3 SEM images of YBCO films fabricated by precursor A (left) and B (right), respectively.

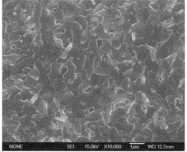 

Fig. 4 Micro-Raman spectroscopy for YBCO films prepared by precursor A and B, respectively.

Conclusions

High performance YBCO film by fluorine-free MOD approach was successfully fabricated on (00l) LAO single crystal substrate. By employing the one-step heat treatment, the procedure to fabricate the YBCO superconducting films was shortened by at least 5 h. The results of XRD, SEM and micro-Raman spectroscopy indicate high quality c-axis oriented YBCO film could be obtained by using xylene as one precursor material, and the critical current density of the film is over 0.55 MA/cm^2 at 77K, self-field.

Acknowledgement

This work was supported by the Key Project of Shanghai Science and Technology (11D1100300), Shanghai University, the Excellent Young Teacher Foundation of Shanghai Municipal Education Commission, and the Innovation Foundation of Shanghai University. We also thanked the help from the Analytical and Testing Center of Shanghai University.

References

[1] K. Tsukada, M. Sohma, I. Yamaguchi, H. Matsui, T. Kumagai, T. Manabe, Enhanced flux-pinning in fluorine-free MOD YBCO films by chemical doping, J. Physica C 470 (2010) 1449-1451.

[2] B.J. Kim, S.W. Lim, H.J. Kim, G.W. Hong, H.G. Lee, New MOD solution for the preparation of high Jc REBCO superconducting films, J. Physica C 445-448 (2006)582-586.

[3] T. Manabe, W. Kondo, S. Mizuta, T. Kumagai, Preparation of high-Jc Ba$_2$YCu$_3$O$_{7-y}$ films on SrTiO$_3$ (100) substrates by the dipping-pyrolysis process at 750°C, J. Appl. Phys. 30 (1991) 1641-1643.

[4] Y. Xu, A. Goyal, K. Leonard, P. Martin, High performance YBCO films by the hybrid of non-fluorine yttrium and copper salts with Ba-TFA, Physica C. 421 (2005) 67-72.

[5] Y. Tokunaga, T. Honjo, T. Izumi, *et al.*, Advanced TFA-MOD process of high critical current YBCO films for coated conductors, Cryogenics. 44 (2004) 817-822.

[6] M.W. Rupich, YBCO coated conductors by an MOD/RABiTS process, IEEE. Trans. Appl. Supercond. 13 (2003) 2458-2461.

[7] T. Araki, I. Hirabayashi, T. Niwa, A large volume reduction and calcining profile for large-area YBa$_2$Cu$_3$O$_{7-x}$ film by metalorganic deposition using trifluoroacetates, Supercond. Sci. Technol. 17 (2004) 135-139.

[8] C. Apetrii, H. Schlorb, M. Falter, *et al.*, YBCO thin films prepared by fluorine-free polymer-based chemical solution deposition, IEEE. Trans. Appl. Supercond. 15 (2005) 2642-2644.

[9] T. Araki, I. Hirabayashi, Review of a chemical approach to YBa$_2$Cu$_3$O$_{7-x}$ -coated superconductors-metalorganic deposition using trifluoroacetates, Supercond. Sci. Technol. 16 (2003) R71–R94.

[10] T. Puig, J. C. Gonzalez, *et al.*, The influence of growth conditions on the microstructure and critical currents of TFA-MOD YBa$_2$Cu$_3$O$_7$ films, Supercond. Sci. Technol. 18 (2005) 1141–1150.

[11] D. E. Wesolowski, Y. R. Patta, M. J. Cima, Conversion behavior comparison of TFA-MOD and non-fluorine solution-deposited YBCO films, J. Physica C. 409 (2009) 766–773.

[12] Y. Xu, A. Goyal, J. Lian, *et al.*, Preparation of YBCO films on CeO2-Buffered (001) YSZ substrates by a non-fluorine MOD method, J. Am. Ceram. Soc. 87 (2004) 1669–1676.

[13] Y. Chen, F. Yan, G. Zhao, G. Qu, L. Lei, Fluorine-free sol-gel preparation of YBa$_2$Cu$_3$O$_{7-x}$ superconducting films by a direct annealing process, J. Alloys and Compounds 505 (2010) 640-644.

[14] J.W. Lee, Y.S. Joo, S.H. Moon, S.I. Yoo, Fabrication of high-quality epitaxial YBCO films prepared by fluorine-free MOD, J. IEEE Trans. Appl. Supercond, vol. 21, pp. 2767-2770, 2011.

[15] L. Lei, G.Y. Zhao, J.J. Zhao, H. Xu, Water-vapor-controlled reaction for fabrication of YBCO films by fluorine-free sol-gel process, J. IEEE Trans. Appl. Supercond, vol. 20, pp. 2286-2293, 2010.

[16] H. Kozuka, M. Kajimura, *et al.*, Crack-free, thick ceramic coating films via non-repetitive dip-coating using polyvinylpyrrolidone as stress-relaxing agent, J. Sol-Gel. Sci. Tech, 19 (2000) 205-209

[17] J.Z. Chen, Modern Crystal Chemistry, second ed., Higher Education Press, Beijing, 2004.

[18] J. C. G. Gonzalez, Coated Conductors and Chemical Solution Growth of YBCO Films: A Micro-Raman Spectroscopy Study, Barcelona, 2005.

Key Engineering Materials Vol. 537 (2013) pp 247-251
© (2013) Trans Tech Publications, Switzerland
doi:10.4028/www.scientific.net/KEM.537.247

Synthesis and Characterization of Nanostructured Co(OH)$_2$ and Co$_3$O$_4$ Thin Films by A Simple Solution Growth Process

Yu-Xin Liu[a], Qun-Yan Li[b], Zuo-Ren Nie, Qi Wei

College of Materials Science and Engineering, Beijing University of Technology, 100022 China

[a] chengzi@emails.bjut.edu.cn; [b] qyli@bjut.edu.cn

Keywords: Co(OH)$_2$, Co$_3$O$_4$ thin film, Magnetic measurement, Nanostructure

Abstract. Nanostructured Co(OH)$_2$ thin films were synthesized by a simple solution growth process. In experiment, F$^-$ and NH$_3$ were used as Co^{2+} ordination agents, and hydroxyl ions were supplied by ammonia hydroxide solution for the hydrolysis reaction. The results showed that the solution pH and F/Co ratio had great influences on the morphologies of the thin films. The Co(OH)$_2$ thin films were constructed with lots of Co(OH)$_2$ nanoflakes. Nanostructured Co$_3$O$_4$ thin films were prepared by annealing the Co(OH)$_2$ thin films at 400°C for 2 h. The magnetic properties of Co$_3$O$_4$ thin films were also investigated.

Introduction

In recent years, nano-sized Co$_3$O$_4$ has attracted much interest due to its much intriguing magnetic, optical, electronic and electrochemical properties [1-9] and potential applications in various fields such as electrochemical capacitors[4], solar energy absorbers[5,6], solid-state sensors[7], and magnetic materials[8-11]. Especially, nanostructured Co$_3$O$_4$, as an important magnetic material, has been researched in the past few years.

It is well known that the properties of magnetic material depend on their morphologies and crystallinity, such as crystallite sizes, shapes and orientations. Co$_3$O$_4$ structures with different morphologies have been prepared using various methods such as the thermal decomposition [12], chemical spray pyrolysis [13], microwave irradiation[14], and the traditional sol-gel method [15]. Some of these synthetic methods solved many drawbacks of bulk phase due to the quantum size effect of the nanometer-scale Co$_3$O$_4$. However, the microstructures of Co$_3$O$_4$ nano materials synthesized by these methods may not be controlled well during the preparation. In this work, a simple solution growth process [16] is reported to prepare nanostructured Co(OH)$_2$ thin films, using F$^-$ and NH$_3$ as Co^{2+} coordination agents and ammonia hydroxides solution as OH$^-$ supplier. The relationships between the experimental factors and microstructure of the Co(OH)$_2$ thin films are discussed in detail. Moreover, Co$_3$O$_4$ thin films are prepared by thermal decomposition of Co(OH)$_2$ thin films, and the magnetic properties of the prepared Co$_3$O$_4$ thin films were investigated.

Experimental

In a typical experiment, 0.03mol/L cobalt sulfate solutions were prepared by dissolving cobalt sulfate (CoSO$_4$·7H$_2$O) into water. Next 40 wt% hydrofluoric acid was added into CoSO$_4$ aqueous solution and its amount depended on the expected atomic ratio of F/Co, 4:1,5:1,6:1,8:1. The mixed solution pH was adjusted to 8.80~9.00 by adding 25 wt% ammonia hydroxide solution. Then the glass substrate, activated and washed by nitric acid, ethanol and distilled water successively, was immersed into the reactive solution and the reaction system was kept in the water bath at 60°C for 4 hours. At the last, the substrate with Co(OH)$_2$ thin film was washed with distilled water, and further dried at room temperature. The Co$_3$O$_4$ thin films were obtained by annealing the Co(OH)$_2$ thin films at 400°C for 2 h. In addition, all reagents used in the experiment were analytical grade without further purification.

The crystal structure of the prepared thin film was determined by X-ray diffraction (XRD, X' Pert Pro MPD diffractometer) using Cu kα radiation. The morphology of the films was observed by scanning electron microscope (SEM, FEI Quanta 200 ESEM) and transmission electron microscope

(TEM, JEOL 2010). Differential scanning calorimetric analysis (DSC) and thermogravimetric analysis (TG) were conducted on simultaneous TG-DTA/DSC apparatus (STA449C/1/G). N_2 adsorption was determined by BET (Brunauer-Emmett-Teller) measurements using a surface area analyzer (Micromeritics ASAP2020). Magnetic measurements were carried out with the Quantum Design Model 6000 vibrating sample magnetometer (VSM) option for the physical property measurement system (PPMS).

Results and Discussion

The deposition mechanism for the $Co(OH)_2$ thin film formation can be described with the following formulas.

$$Co^{2+} + x\,F^- \leftrightarrow [CoF_x]^{(x-2)-}. \tag{1}$$

$$Co^{2+} + y\,NH_3H_2O \leftrightarrow [Co(NH_3)_y]^{2+} + y\,H_2O. \tag{2}$$

$$NH_3H_2O \leftrightarrow NH_4^+ + OH^-. \tag{3}$$

$$[CoF_x]^{(x-2)-} + x\,OH^- \leftrightarrow [Co(OH)_x]^{(x-2)-} + x\,F^-. \tag{4}$$

$$[Co(NH_3)_y]^{2+} + y\,OH^- \leftrightarrow [Co(OH)_y]^{(y-2)-} + y\,NH_3. \tag{5}$$

First, Co^{2+} ions are coordinated with F^- ions to form $[CoF_x]^{(x-2)-}$ shown in formula (1), and the other free Co^{2+} ions in the solution are coordinated with NH_3H_2O to form $[Co(NH_3)y]^{2+}$ according to the formula (2). At the same time, the produced $[CoF_x]^{(x-2)-}$ and $[Co(NH_3)y]^{2+}$ ions occur the hydrolysis reaction in the solution, shown by the formulas (4),(5). The OH^- concentration increases with the increasing ammonia hydroxide. As a result, the $[CoF_x]^{(x-2)-}$ and $[Co(NH_3)_y]^{2+}$ ions are accelerated to hydrolyze into $[Co(OH)_x]^{(x-2)-}$ and $[Co(OH)_y]^{(y-2)-}$. Then, the $Co(OH)_2$ films are deposited on the substrate from some condensation reactions between $[Co(OH)_x]^{(x-2)-}$ and $[Co(OH)_y]^{(y-2)-}$ ions. In follow-up research, it was found that no cobalt hydroxide films would be deposited on the glass substrate, if no F^- ions were added to the experimental system, or the ammonia hydroxide solution was replaced with other metallic alkali solutions like the sodium hydroxide solution. It suggests that F^- ions and NH_3, as Co^{2+} ordination agents, play an important role in the solution growth process. The mechanism of coordination reaction makes the hydrolysis more slower and controllable. It can be explained that the two nickel complex species of $[CoF_x]^{(x-2)-}$ and $[Co(NH_3)y]^{2+}$ have different hydrolysis sequences. The slower and more controllable hydrolysis is favorable to the $Co(OH)_2$ thin film growth.

The SEM images of $Co(OH)_2$ thin films, deposited with different F/Co ratios when the solution pH is 8.90, are shown in the Figure 1. (a),(b),(c). It is seen that the $Co(OH)_2$ thin films are constructed with $Co(OH)_2$ nanoflakes, which are interconnected with each other in the films. When the F/Co ratio is 4:1, the $Co(OH)_2$ nanoflakes are long and complete, but there are lots of open pores in the thin films and the number of the nanoflakes is few. When the F/Co ratio is 6:1, the $Co(OH)_2$ nanoflakes grow up less completely, but their quantity increases sharply. The $Co(OH)_2$ thin films also become denser with the increase of the F/Co ratio. When the F/Co ratio increases to 8:1, the quantity of $Co(OH)_2$ nanoflakes is so small that a continuous film cannot be gained on a substrate Apparently, the F/Co ratio have a great influence on the morphologies of the $Co(OH)_2$ thin films. It can be explained that the F/Co ratio controls the ratio of the two cobalt complex species of $[CoF_x]^{(x-2)-}$ and $[Co(NH_3)y]^{2+}$, which provide the $Co(OH)_2$ nuclei by the hydrolysis reactions in the solution. However, the relationship between the special microstructures of the thin films and the F/Co ratio needs a further explanation.

The morphologies of $Co(OH)_2$ thin films, obtained with different solution pH when the F/Co ratio is 6:1, are also shown in the Figure 2. (a),(b),(c). The solution pH denoting OH^- concentration is the other important factors to influence the morphology of the $Co(OH)_2$ thin films. When the pH is 8.80, the hydrolysis in the solution is not complete and only few $Co(OH)_2$ nuclei have flaky structures. The

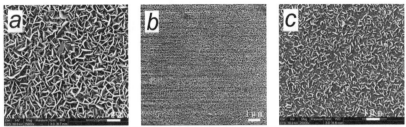

Fig. 1 SEM images of the Co(OH)$_2$ thin films deposited with different F/Co ratio at the same solution pH 8.90 : (a) 4:1; (b) 6:1; (c) 8:1

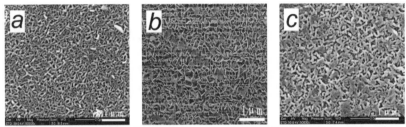

Fig.2 SEM images of the Co(OH)$_2$ thin films deposited with different solution pH at the same F/Co ratio 6:1: (a) 8.80; (b) 8.90; (c) 9.00

Co(OH)$_2$ nanoflakes have small lengths of about 200nm.The hydrolysis is found to be more complete and the flaky structures of Co(OH)$_2$ become more intact when the solution pH increases to 8.90. The Co(OH)$_2$ nanoflakes size becomes large with the largest lengths of about 450nm. With the increase of solution pH, the Co(OH)$_2$ nanoflakes don't continue to grow up, but the Co(OH)$_2$ cores decrease and the flaky structures become less complete. The lengths of Co(OH)$_2$ nanoflakes decrease to 270nm. It is because that the supersaturation degree is so high that much free Co(OH)$_2$ precipitation forms in the solution and a continuous film cannot be gained on a substrate. The thickness of the film obtained at different pH when the F/Co ratio is 6:1 is also researched (Fig. 2 d). When the solution pH increases from 8.80 to 8.90, the film growth rate increases gradually because of the more complete hydrolysis and more cobalt hydroxide nuclei, and consequently the film thickness increases rapidly from 97nm to 530nm. When the solution pH is 8.90, the nucleation is inhibited and a larger number of nuclei have the chance to grow up completely. The influence of the solution pH to the morphology of Co(OH)$_2$ thin films can be explained by the competition between two processes: nucleation and growth. When the solution pH decreases, the nucleation becomes dominant and leads to the increase of small cobalt hydroxide nuclei.

Fig. 3 is the TEM image of the sample with solution pH 8.90 and the F/Co ratio 6:1. It is apparent that the Co(OH)$_2$ thin film is constructed with Co(OH)$_2$ nanoflakes. The BET surface area of the film sample is 26.37 m^2/g, a large surface area value suggesting the porous microstructure of the Co(OH)$_2$ thin film.

Figure 4. a shows the thermal behavior of the Co(OH)$_2$ thin film when the solution is 8.90 and the F/Co ratio is 6:1. The temperature ranges of the endothermic peaks in the DSC cure fit very well with those of weight loss in the TG cure. The slight weight loss observed initially between 30°C and 80°C was attributed to the evaporation of the free water in the Co(OH)$_2$ thin films. The obvious endothermic peak located at 120°C, which

Fig.3 TEM image of the Co(OH)$_2$ film deposited with solution pH 8.90 and the F/Co ratio 6:1

indicated a transformation of β- $Co(OH)_2$ to Co_3O_4. The biggest weight loss of 11.5% was observed over the temperature range 100-160°C in the TG curve, which is in good agreement with the theoretical weight loss value (13.6%) due to the decomposition of $Co(OH)_2$. A slight endothermic peak is also observed at 620°C, suggesting that Co_3O_4 is decomposed to CoO.

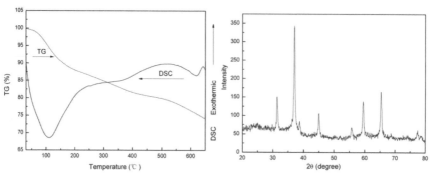

Fig. 4 TG-DSC curves of $Co(OH)_2$ thin film (a, left); XRD pattern of Co_3O_4 thin films (b, right)

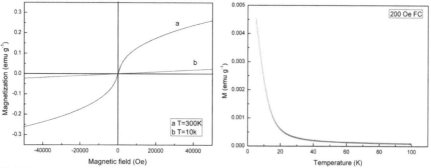

Fig. 5 M-H curve of synthesized Co_3O_4 thin films at (a) 300K and (b)10K (a, left); M-T curves of reflux synthesized Co_3O_4 thin films with 200Oe applied field (b, right)

According to the TG-DSC curves of cobalt hydroxide, the thin films were annealed at 220°C for 2 h to obtain Co_3O_4 thin films. Figure 4. b is the XRD pattern of the Co_3O_4 thin films, which is almost the same as t the standard PDF data (JCPDS file no. 42-1467). Moreover, there are no peaks of the $Co(OH)_2$ in the pattern, suggesting that the $Co(OH)_2$ thin films were completely transformed into Co_3O_4 thin films.

To investigate the magnetic properties of the obtained Co_3O_4 thin films with the solution pH 8.90 and the F/Co ratio 6:1, magnetic hysteresis measurement of Co_3O_4 was carried out in an applied magnetic field at 300K and 10K. Fig. 5 a shows the isothermal hysteresis loops of the Co_3O_4 thin films. It can be seen that saturation is not reached up to the maximum even in the presence of 50kOe magnetic field. The hysteresis loops at above the critical temperature indicate that the sample exhibits a paramagnetic behavior. However, the hysteresis loops at 10K below the critical temperature indicate that the sample exhibits a ferromagnetic behavior with a significant canting still present at 50kOe.

Temperature dependence of magnetization of Co_3O_4 thin films in 200Oe field-cooled (FC) case is shown in Fig. 5 b. The magnetization slightly increases when the temperature decreases from 100K to 50K and strongly increase between 50K and 20K. Below 30K, the magnetization sharply increases by decreasing temperature. This sharp increase in magnetization indicates that the transition occurs at around 20K. The Neel temperature of Co_3O_4 bulk crystal is reported as $T_N = 40K$ [17]. This change of transition temperature may be ascribed to smaller Co_3O_4 nanoflakes size of the thin films.

Conclusions

Nanostructured Co(OH)$_2$ thin films have been successfully prepared by a simple solution growth process with different solution pH and different F/Co ratio. The SEM images showed that the solution pH and F/Co ratio had a significant influence on the morphology and the thickness of the thin films by controlling the quantity and growth rate of the Co(OH)$_2$ nuclei. The Co$_3$O$_4$ thin films were also obtained by annealing the Co(OH)$_2$ thin films at 220°C for 2 h. The Co$_3$O$_4$ thin films exhibit a paramagnetic behavior and a ferromagnetic behavior at 300K and 5K respectively. The Neel temperature of Co$_3$O$_4$ thin films, different from the bulk value, is observed to be 20K because of the nano-sized nanoflakes.

Acknowledgements

This project was financially supported by National 863 Plan project (No. 2009AA03Z213) and National Science Foundation of China (No. 21171014).

References

[1] C.Y. Ma, Z. Mu, J. J. Li, *et al.*, Mesoporous Co$_3$O$_4$ and Au/Co$_3$O$_4$ catalysts for low-temperature oxidation of trace ethylene, J. Am. Chem. Soc. 132 (2010) 2608–2613.

[2] B. Varghese, Y.S. Zhang, L. Dai, V.B.C. Tan, C.T. Lim, C.H. Sow, Structure-mechanical property of individual cobalt oxide nanowires, Nano Lett. 8 (2008) 3226–3232.

[3] L. H. Hu, Q. Peng, Y. D. Li, Selective synthesis of Co$_3$O$_4$ nanocrystal with different shape and crystal plane effect on catalytic property for methane combustion, J. Am. Chem. Soc. 130 (2008) 16136-37.

[4] Y. G. Li, B. Tan, Y. Y. Wu, Mesoporous Co$_3$O$_4$ nanowire arrays for lithium ion batteries with high capacity and rate capability, Nano Lett. 8 (2008) 265–270.

[5] G. McDonald, A preliminary-study of a solar selective coating system using a black cobalt oxide for high-temperature solar collectors, Thin Solid Films 72 (1980) 83.

[6] S. Noguchi, M. Mizuhashi, Optical-properties of Cr-Co oxide-films obtained by chemical spray deposition-substrate-temperature effects, Thin Solid Films 77 (1981) 99.

[7] W. Y. Li, L. N. Xu, Co$_3$O$_4$ nanomaterials in lithium-ion batteries and gas sensors, J. Chen, Adv. Funct. Mater. 15 (2005) 851.

[8] S. A. Makhlouf, Magnetic properties of Co$_3$O$_4$ nanoparticles, J. Magn. Magn. Mater 246 (2002) 184.

[9] Y. G. Zhang, Y. C. Chen, T. Wang, J. Zhou, Y. G. Zhao, Synthesis and magnetic properties of nanoporous Co$_3$O$_4$ nanoflowers, Micropor. Mesopor. Mater.114 (2008) 257.

[10] Y. Kitamoto, S. Kantake, *et al.*, Low-temperature fabrication of Co ferrite thin films with high coercivity for perpendicular recording disks by wet process, J. Magn. Magn. Mater. 193 (1999) 97.

[11] W. W. Wang, Y. J. Zhu, Microwave-assisted synthesis of cobalt oxalate nanorods and their thermal conversion to Co$_3$O$_4$ rods, Mater. Res. Bull. 40 (2005) 1929.

[12] L. X. Yang, Y. J. Zhu, L. Li, L. Zhang, H. Tong, W. W. Wang, A facile hydrothermal route to flower-like cobalt hydroxide and oxide, Eur. J. Inorg. Chem. 23 (2006) 4787.

[13] M. Hamdani, J. F. Koenig, P. Chartier, Co$_3$O$_4$ and NiCo$_2$O$_4$ thin-films prepared by reactive-spraying for electrocatalysis.2.cyclic-voltammetry, J. Appl. Electrochem. 18 (1988) 568.

[14] Y. Lu, Y. Wang, Y. Q. Zou, Z. Jiao, B. Zhao, Y. Q. He, et al. Macroporous Co$_3$O$_4$ platelets with excellent rate capability as anodes for lithium ion batteries, Electrochem Commun 12 (2010) 101-5.

[15] Y. Y. Xu, C. Q. Wang, Y. Q. Sun, G. Y. Zhang, D. Z. Gao, Fabrication and characterization of nearly monodisperse Co$_3$O$_4$ nanospheres, Mater Lett 64 (2010) 1275-8.

[16] Q. Y. Li, R. N. Wang, *et al.*, Preparation and characterization of nanostructured Ni(OH)(2) and NiO thin films by a simple solution growth process, J. Colloid Interface Sci. 320 (2008) 254-258.

[17] Y. Ichiyanagi, Y. Kimishima, S. Yamada, Magnetic study on Co$_3$O$_4$ nanoparticles, J. Magn. Magn. Mater. 1245 (2004) 272–276.

Key Engineering Materials Vol. 537 (2013) pp 252-255
© (2013) Trans Tech Publications, Switzerland
doi:10.4028/www.scientific.net/KEM.537.252

Preparation and Performance Characterization of Nanometer-Sized Tourmaline (Schorl) Films

Yan Li [a], Zhaohui Huang [b], Yan'gai Liu, and Minghao Fang

School of materials science and engineering, China University of geosciences, Beijing 100083, China

[a] liyan0023@gmail.com, [b] huang118@cugb.edu.cn.

Keywords: tourmaline; nanometer-sized; films; polarization effect; magnetron sputtering

Abstract. Electric dipole polarity transformation is one of the key characteristics of tourmaline. In this paper, conductive films are coated as raw materials on FTO glass substrate to prepare functional glass materials, which behaves spontaneous polarization effect, absorption of wideband electro-magnetic radiation and high visible light transmittance. Magnetron sputtering deposition technique is used to sputter tourmaline as target on the conductive FTO glass. The surface morphology, component, microstructure and magnetic properties are studied. Furthermore, the functions of tourmaline as composite electromagnetic shielding films are discussed in this paper, as well.

Introduction

Tourmaline is a complicated silicate mineral with typical ring structure and boron element. It possesses a non-centrosymmetric trigonal structure, with space group R3m. Its chemical formula is usually written as $XY_3Z_6 [T_6O_{18}] [BO_3]_3V_3W$. where: $X = Na^+$, Ca^{2+}, vacancy, K^+; $Y = Li+$, Mg^{2+}, Fe^{2+}, Mn^{2+}, Al^{3+}, Fe^{3+}, Cr^{3+}, V^{3+}, Ti^{4+}, Zn^{2+}, Cu^{2+}; $Z = Mg^{2+}$, Al^{3+}, Fe^{3+}, Cr^{3+}, V^{3+}, Ti^{4+}; $T = Si^{4+}$, Al^{3+}, B^{3+}; $B = B^{3+}$; $V = OH^-$, O^{2-}; $W = OH^-$, F^-, O^{2-}[1-4]. Tourmaline is known as a piezoelectric material and has been widely used as pressure sensing device in transducer technology [5]. The electric dipole moment of tourmaline varied slightly under the wide band and high intensity of electromagnetic radiation [6]. It is possible to prepare high performance glass which can prevent wideband electromagnetic radiation and information leakage with the use of electric dipole polarity transformation of tourmaline.

In this paper, nano-meter sized tourmaline (schorl) films were prepared by magnetron sputtering deposition technique at room temperature and 600°C on Japan FTO glass substrates. In the experiment, tourmaline was used as target and supttered on conductive FTO glass. The measurement and analysis of the surface morphology, film thickness, component, structure and magnetic properties were carried out in the experiment. In addition, we investigate the functions of tourmaline as composite electromagnetic shielding films.

Experimental Procedures

TCO-15 (Transparent Conducting Oxide Coated Glass) is a 2.2 mm thick glass coated on one side with a fluorine doped tin oxide (SnO_2:F) layer ("FTO" glass). The sheet resistance of the FTO layer is 7~14 ohm/square and its transmission is > 90 % from 400 to 700 nm. There are two different FTO glasses adopted in the experiment. One of the FTO glass's resistance is 7 ohm/square, and the other is 14 ohm/square. Each glass was cut into square slices with dimensions of 15×15mm and classified into two groups. Each group of samples has four chips and was coated at room temperature and 600°C, which were named as T-S-X and T-B-X (where, X=1, 2, 3, 4), respectively. All of these samples have 350nm FTO coated on one side.

The tourmaline (schorl) samples were obtained from Fuping, Hebei Province, China. The crystals are in fairly good shape with little cracks and inclusions. The samples were cut into round chips with 6 cm in diameter and 5 mm in thickness. The physical properties of the tourmaline sample are listed in Table.1.

Table 1 The physical properties of tourmaline sample

appearance	color	luster	transparency	optical character	hardness
columnar, single crystal	black	glass luster	2.3254	Uniaxial (-)	7-7.5
powders color (600mesh)	refractive index	specific gravity	birefringence	fracture	species
dark gray black	Ne=1.648,No=1.678	3.152	0.030	conchoid	schorl

Table 2 The main infrared vibrating peaks of tourmaline

Sample	M-O vibration	Si-O Bending vibration	Si-O-Si skeletal vibration	O-Si-O skeletal vibration	B-O stretching vibration	OH stretching vibration
T-1	423	512	707,783	982	1271	3549

Table 3 The magnetic parameters of tourmaline, schorl

Sample	color	saturation magnetization	residual magnetism	coercive force	magnetic
Schorl//c axis	black	0.1741	0.0004303	--	paramagnetic
Schorl//c axis	black	0.2075	0.0004918	--	paramagnetic

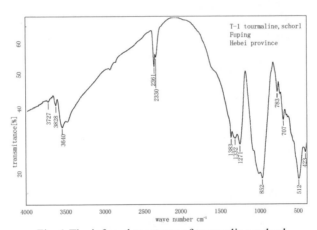

Fig. 1 The infrared spectrum of tourmaline, schorl

The infrared spectrum of tourmaline was mainly composed of [SO_4] ionic group, [BO_3] ionic group, OH$^-$ and Octagonal cationic M-O vibration, as shown in Fig. 1 and the main vibrating peaks are summarized in Table.2.

The namometer-sized tourmaline film was fabricated by magnetron sputtering deposition (Magnetron sputtering apparatus, JPGF-400) on FTO glass substrates and gain the characteristic self-polarization as surface film. The chemical composition was determined by electron microprobe analysis (EMPA) in wavelength- dispersive mode. The final chemical composition data are an average value of five analyses carried out at different points over each sample. The microsturctures were observed by scanning electron microscope (SEM) at an accelerating voltage of 15kV and FESEM at 20 kV. The phases of samples were determined by X-ray diffraction (XRD). The magnetic parameters of tourmaline are obtained, as shown in Table 3.

Result and Discussion

The main elements of these samples change slightly before and after coating on FTO-glass substrates. However, the relative contents of some minor elements, such as Mg, Al, Fe etc., significantly increased. The ratios of Al: Si and Mg: Si of these films are in the range of uncoated samples and the tourmaline. The chemical compositions of the tourmaline crystal samples are given in Table 4. The comparison of glass surface compositions before and after sputtering deposition is given in Table 5.

It is shown that the contents of Al and Mg are high in tourmaline, but lower in FTO-glass substrates. After coating, the films are island particle structure, and the films (Figs. 2 and 3) don't

Table. 4 The chemical compositions of the tourmaline target

Schorl	Na	Mg	Al	Si	Ca	Ti	Fe	K	Mn	Zn	Cr	F	OH
Ions(apfu)	0.549	0.0087	6.456	5.8072	0.0234	0.0159	2.3254	0.0169	0.0451	0.0053	tr	0.68	3.32

Table.5 Comparison of glass surface compositions before and after sputtering deposition

elements	Si	Ca	Sn	Na	K	Mg	Al	Fe
Before coating (wt%)	15.900	5.790	0.755	0.712	0.387	0.299	0.165	0.123
After coating (wt%)	15.500	5.578	0.745	0.655	0.345	0.540	0.890	0.187

Fig.2 SEMS images of the nano-meter sized tourmaline films on FTO glass substrates: a) Magnetron Sputtering deposition at 600°C b) Magnetron Sputtering deposition at room temperature

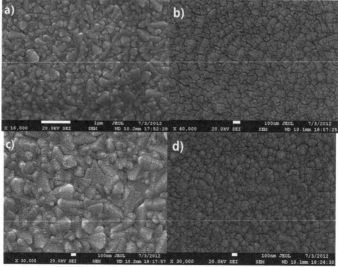

Fig 3. FESEM images of 600°C deposition a) and c) and room temperature deposition; b) and d) samples on FTO-glass substrates

completely cover the glass substrates. Therefore, the results of X-ray fluorescence spectrometer are the comprehensive compositions of tourmaline and FTO-glass substrates, which means that the numerical data are in the range of tourmaline and FTO-glass substrates.

The surface morphology of the tourmaline films are illustrated in Fig. 2 and Fig. 3. The figures a) and c) depict the surface morphology of the samples magnetron sputtering deposition heated at 600°C, and the figures b) and d) show the surface morphology of the samples sputtered at room temperature. The number of the particles sputtering deposition on the FTO-glass substrates at high temperature 600°C is much more than that of room temperature samples. And the size of the tourmaline particles at 600°C is larger, as well. The FTO-glass substrates were not completely covered by the particle films. As shown in these SEM and FESEM figures, the average diameter of the particle is less than 0.5 μm. The thickness of the tourmaline film is about 56 nm and the particles on the glass substrates are flat, or set in glass substrates. The quality of sputtering film largely depends on the quality and temperature of the target, the temperature of the substrates, substrate potential of plasma, vacuum conditions and geometrical shapes, etc. At higher temperatures, the tourmaline particles combine with the glass substrates much more easily.

The nanometer-sized tourmaline films can be sputtered on the FTO-glass substrates in the suitable conditions using ion-beam magnetron sputtering deposition method. Due to the thermal properties of tourmaline, it behaves special spontaneous polarization characteristics after the PPMS testing. The coercive force is doubly increased. And the new functional glass materials possess spontaneous polarization effect, absorption of wideband electromagnetic radiation and high visible light transmittance.

Conclusions

The conclusions are summarized as follows:

(1) The ratios of Al: Si and Mg: Si of coated samples are in the range of the uncoated samples and the tourmaline target. The films are island particle structure, which don't completely cover the glass substrates.

(2) The number of the tourmaline particles sputtering deposition on the FTO-glass substrates at high temperature 600°C is much more than that of room temperature samples. And the size of the tourmaline particles at 600°C is larger, as well.

(3) The average diameter of particle is less than 0.5 μm and the thickness of the film is about 56 nm.

(4) When the tourmaline is sputtered to nanometer-sized films, it behaves the special spontaneous polarization characteristics. The coercive force is doubly increased.

Acknowledgements

This work was financially supported by "the Fundamental Research Funds for the Central Universities" and New Star Technology Plan of Beijing (Grant No.2007A080).

References

[1] F. C. Hawthorne and D. Henry, Classification of the minerals of the tourmaline group, Eur. J. Mineral. 11 (1999) 201–215.

[2] F. Bosi and S. Lucchesi, Crystal chemistry of the schorldravite series, Eur. J. Mineral. 16 (2004) 335–344.

[3] J. M. Hughes, A. Ertl, M. D. Dyar, et al., Reexamination of olenite from the type locality: detection of boron in tetrahedral coordination, Am. Mineral. 89 (2004) 447–454.

[4] D. London, A. Ertl, J. M. Hughes, G. B. Morgan VI, E. A. Fritz and B. S. Harms, Synthetic Ag-rich tourmaline: structure and chemistry, Am. Mineral. 91 (2006) 680–684.

[5] J. C. Anderson. Dielectrics. Reinhold Publ. Co., New York, 1964.

[6] K.Krambrock, M.V.B.Pinheiro, S.M.Mediros, et al.. Investigation of radiation-induced yellow color in tourmaline by magnetic resonance. Nucl. Instru. Meth. Phys. Res. B, 191 (2002) 243-244.

Key Engineering Materials Vol. 537 (2013) pp 256-260
© (2013) Trans Tech Publications, Switzerland
doi:10.4028/www.scientific.net/KEM.537.256

Study on Special Morphology Hydroxyapatite Bioactive Coating by Electrochemical Deposition

Caige Gu [a], Qiangang Fu [b], Hejun Li, Jinhua Lu, Leilei Zhang

C/C Composites Technology Research Center, Northwestern Polytechnical University, Xi'an, Shaanxi 710072, P.R. China.

[a] sweetycaige@gmail.com, [b] fuqiangang@nwpu.edu.cn

Keywords: C/C composites, electrochemical deposition, bioactive coating, processing optimization

Abstract. Bioactive calcium phosphate coatings were deposited on C/C composites using electrochemical deposition technique. The effects of electrolyte concentration and constant current density on morphology, structure and composition of the coating were systematically investigated using SEM, XRD and FTIR spectroscopy. The results show that, the coating weight elevated gradually with the increase of electrolyte concentration, and the morphology of coatings changed from spherical particles to nanolamellar crystals with interlocking structure initially. Then the coating transformed into seaweed-like and nano/micro-sized crystals along the depth direction of the coating. The coatings showed seaweed-like morphology as the deposition current density was less than 20 mA. With the less current density, the coating became more homogenous. However, the coating was flakiness crystal, with need-like crystal stacked upside as the current density reached 20 mA/cm^2. The coating weight was improved gradually when the current density increased from 2.5 mA/cm^2 to 10 mA/cm^2, and then reduced with the increasing current density in the range of 10 to 20 mA/cm^2.

Introduction

Carbon/carbon (C/C) composites, as one of the most promising biomedical materials, have been applied for dental and orthopaedic surgery [1]. They have many advantages, such as excellent biocompatibility, high corrosion resistance, lightweight and good mechanical properties, especially their elastic modulus are similiar to that of human bone. However, C/C composites are not bioactive as the materials for loaded artificial bones. In addition, after in vivo implantation, C/C composites without surface modification often release microparticles to the surrounding tissue, which causes "black skin effect" [2, 3].

Hydroxyapatite (HA, $Ca_{10}(PO_4)_6(OH)_2$) has attracted much attention in orthopaedics surgery due to its similarity in chemical composition and structure with natural bones and teeth [4, 5]. However, its poor mechanical properties, with a bending strength less than 100 MPa [6], restricted its widely application as a bulk implant material under high physiological loading conditions. Therefore, it is a logical choice to deposit bioactive HA layer on C/C composites for increasing the bioactivity and inhibiting the release of microparticles

During the past decade years, there are several technologies developed to coat HA on C/C composites, including plasma spraying [6], sol-gel [7], biomimetic deposition [8], hemical liquid deposition [9], and electrochemical deposition [10]. Among these methods, electrochemical deposition was introduced with advantages, such as quick and uniform coating of the substrates with complex shapes at low temperatures and the control of the thickness and chemical composition of the coating [11].

In this study, we investigated the effects of electrolyte concentration and constant current density on morphology, structure, composition and deposition mass of the coating. The optimized technology paremeters for depositing the homogeneous coating with special morphologies was obtained.

Experimental

Pretreatment of the C/C substrates. The C/C composites with the density of 1.71 g/cm3 were prepared by chemical vapor infiltration process in Northwestern Polytechnical University in China. The C/C composites samples were cut from the block and were polished with 800 grit SiC paper until the final dimensions of 10 mm×10 mm×2 mm. Then they were cleaned ultrasonically in turn by acetone, alcohol and deionized water.

Experiment methods. Electrochemical deposition was carried out in a two-electrode electrochemistry system controlled by a DC power supply, with C/C sample and graphite electrode acted as cathode and the parallel anode. The initial PH value of the electrolyte solution was adjusted to 4.5 by ammonia water or nitric acid solution. The temperature was controlled at $60\pm1°C$ during the ECD by constant temperature circulating water bath system.

Influence of electrolyte concentration. The electrolyte solution used for the samples of calcium phosphate coatings were produced by dissolving reagent grade $Ca(NO_3)_2\cdot4H_2O$ and $NH_4H_2PO_4$ in deionized water. The molar ratio of calcium ion to phosphate ion was 1.67. The electrolyte concentrations of $NH_4H_2PO_4$ were 2.5×10^{-3}, 5.0×10^{-3}, 1.25×10^{-2}, 2.5×10^{-2} mol/L, respectively. Calcium phosphate deposition was carried out at the current density 10.0mA/cm2 for 30 min.

Influence of constant current density. The electrolyte consisted of 2.5×10^{-2} mol/L $NH_4H_2PO_4$ and 4.2×10^{-2} mol/L $Ca(NO_3)_2$ solutions. Constant current density of 2.5, 5.0, 10.0, 20.0 mA/cm² were applied for 30min during the calcium phosphate deposition

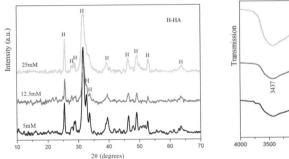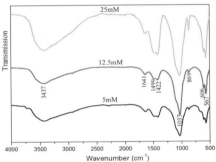

Fig.1 XRD patterns (a) and FTIR spectra (b) of coating on C/C composites prepared at different $NH_4H_2PO_4$ electrolyte concentrations

Results and discussion

Influence of electrolyte concentration on the coating. Fig.1 (a) shows the XRD patterns of coating on C/C composites prepared at different $NH_4H_2PO_4$ electrolyte concentrations. The results indicated that the coating mainly consisted of HA phase with high crystallinity for the electrolyte concentration from 5×10^{-3} mol/L to 2.5×10^{-2} mol/L. The XRD patterns showing a broad and low intensity HA (112) peak at 2θ of 32.15° suggested that crystallinity of the coating decreased at high electrolyte concentration. However, the wide peaks inferred that the small size of HA was obtained.

The FTIR spectra of the coating at different $NH_4H_2PO_4$ electrolyte concentrations are given in Fig. 1 (b). The FTIR spectra of the sample with the $NH_4H_2PO_4$ electrolyte concentrations of 5×10^{-3} mol/L and 2.5×10^{-2} mol/L presented almost all the characteristic phosphate and water bands of HA, in the range 4000-400 cm^{-1} (3437, 1641, 1037, 606, 567 cm^{-1}). Because of the coating without heat treated, the bands at 3437cm^{-1} was doublets of water in HA. The bands around 1499 and 1422 cm^{-1} are assigned to carbonate (CO_3^{2-}), and the bands at 869 cm^{-1} can be attributed to HPO_4^{2-} or CO_3^{2-}. The presence of CO_3^{2-} bands indicated that the calcium-deficient hydroxyapatite (DCP, $Ca_{10-z}(HPO_4)_z(PO_4)_{6-z}(OH)_{2-z})\cdot nH_2O$, $0 > z > 1$) formed here was carbonate-containing HA which was the same as bone apatite [4]. CO_3^{2-} was probably incorporated into the solution from the air during preparation process [12]. The bands at 1037 cm^{-1} is associated with the ν_3 vibration of PO_4^{3-} which is typical of HA. The bands at 606 and 567 cm^{-1} are assigned to ν_4 vibration of the phosphate group (triply degenerated bending mode). As the electrolyte concentrations increased from 5×10^{-3} mol/L to 2.5×10^{-2} mol/L, the intensities of PO_4^{3-} bands around 1037, 606 and 567cm^{-1} for HA are gradually increased. The band at 869cm^{-1} is acuity abruptly at 2.5×10^{-2} mol/L $NH_4H_2PO_4$ electrolyte concentrations.

Fig.2 shows SEM micrographs of the coatings deposited at different electrolyte concentrations. It is found that at HPO_4^{2-} ion concentrations less than 1.25×10^{-2} mol/L the spherical CaP aggregates formed in the coatings had non-uniform grain size in Fig.2 (a) and (b). Meanwhile, the smaller sizes of the sphere crystals were obtained for the less ion concentration. The CaP aggregates obtained were stacked instead of forming integrated coatings. When the HPO_4^{2-} ion concentrations achieved at 1.25×10^{-2} mol/L, the homogenous coating composited of uniform grain size nanolamellar crystals with 30-60 nm in thickness and 300-800 nm in width also exhibited an interlocking structure. Moreover, large amounts of seaweed-like crystals were observed to form in the electrolyte when the HPO_4^{2-} ion concentrations exceeded 1.25×10^{-2} mol/L. The seaweed-like crystals were 1μm in width, 40-60nm in thickness and 8-10 μm in length.

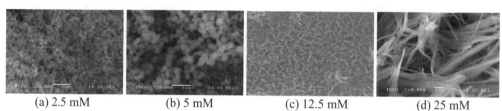

    (a) 2.5 mM   (b) 5 mM   (c) 12.5 mM   (d) 25 mM

Fig.2 SEM micrographs of coating on C/C composites prepared at different $NH_4H_2PO_4$ electrolyte concentrations

As shown in Fig.3, the coating weight increased almost linearly with the increase of electrolyte concentration for 30 min. With the increase in electrolyte concentration, the coating weight was improved, which may be caused by high deposition rate. This suggested that the procedure of HA coating deposition by electrochemical deposition technique would be dominantly controlled by ion diffusion.

Influence of constant current density on obtaining the coating. Fig.4 (a) shows the XRD patterns of coating on C/C composites prepared at various current densities. The coating was mainly composed of HA with a little tertiary calcium phosphate (TCP) at 2.5 and 20 mA/cm^2. When the deposition current density was 5 mA/cm^2 and 10 mA/cm^2, the coating was mainly consisted of HA, with the highest intensity HA peaks at 5 mA/cm^2.

The FTIR spectra of the coating at different current densities are given in Fig.4 (b) The FTIR spectra of the sample with the current density from 5 to 20 mA presented almost all the characteristic phosphate and water bands of HA, in the range 4000-400 cm^{-1} (3437, 1644, 1031, 606, 563 cm^{-1}). The bands at 3437 cm^{-1} was doublets of water in HA. The bands around 1487 and 1418 cm^{-1} are assigned to carbonate (CO_3^{2-}), and the bands at 869 cm^{-1} can be attributed to HPO_4^{2-} or CO_3^{2-} [4]. When the current density was 2.5 mA, the bands of CO_3^{2-} were disappeared. The intensities of CO_3^{2-} bands increased as the current intensity increased to 20 mA/cm^2.

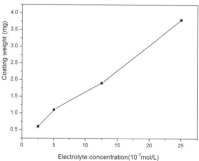

Fig.3 The relationship of different electrolyte concentration and deposition gain for 30 min

Fig.5 shows SEM micrographs of the coatings deposited at different current densities. It is clear that at current density less than10 mA/cm^2 the seaweed-like CaP aggregates formed in the coatings had uniform grain size, as shown in Fig.5 (a) and (b). Meanwhile with the less current density, the more homogenous of the coating were. When the current density achieved at 10mA/cm^2, the nonuniform coating was composed of seaweed-like crystal with different grain size. The coating obtained were flakiness crystal, with need-like crystal stacked upside at current density of 20 mA/cm^2.

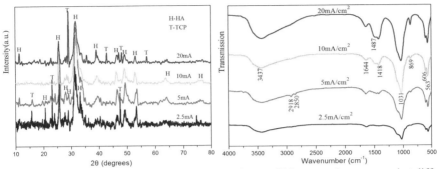

Fig.4 XRD patterns (a) and FTIR spectra (b) of coating on C/C composites prepared at different deposition current densities

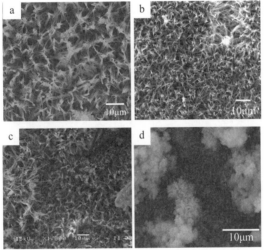

Fig.5 SEM micrographs of coating on C/C composites prepared at different current densities: (a) 2.5 mA/cm^2; (b) 5 mA/cm^2; (c) 10 mA/cm^2; (d) 20 mA/cm^2

As shown in Fig.6, the deposition gain was improved gradually when the current density increased from 2.5 mA/cm^2 to 10 mA/cm^2, which may be caused by high deposition rate. With the increasing current density in the range of 10 to 20 mA/cm^2, the deposition gain was reduced, which may be caused by the exfoliation of the loose coating obtained at high current density.

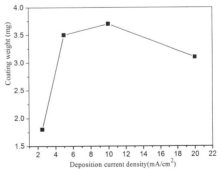

Fig.6 The relationship of different deposition current density and deposition gain for 30 min

Conclusions

Nano/micro-sized calcium phosphate coating was prepared on C/C composites by electrochemical deposition. The coating obtained at different electrolyte concentrations showed different morphology. The spherical CaP aggregates were formed the coating at HPO$_4^{2-}$ ion concentrations less than

1.25×10^{-2} mol/L , however an interlocking structure was formed at low concentration. The homogenous coating obtained at 2.5×10^{-2} mol/L consisted of seaweed-like DCP along the depth direction of the coating with a size from $1 \mu m$ in width, 30-40nm in thickness and 8-10 μm in length. The coating with seaweed-like feature was not changed greatly when the current density less than 20 mA/cm^2. The more homogenous of the coating were obtained at less current density. However, the obtained coating at 20mA/cm^2 consisted of flakiness crystal, with need-like crystal stacked upside. The optimum coating was deposited using the following parameter: 2.5×10^{-2} mol/L NH$_4$H$_2$PO$_4$ and 4.2×10^{-2} mol/L Ca (NO$_3$)$_2 \cdot$4H$_2$O electrolyte concentration, current density of 5mA/cm^2 and the molar ratio of calcium to phosphorus was1.67.

Acknowledgements

This work is supported by the National Natural Science Foundation of China (No.5090211, No. 51072107), China Postdoctoral Science Foundation (No. 2011M501473) and 111 project of china (No.B08040).

References

[1] M. Lewandowska-Szumiel, J. Komender, A. Gorecki, M. Kowalski, Fixation of carbon reinforced carbon composite implanted into the bone, J. Mater. Sci. Mater. Med. 8 (1997) 485-488.

[2] J.L. Sui, M.S. Li, Y.P. Lu, Y.Q. Bai, The effect of plasma spraying power on the structure and mechanical properties of hydroxyapatite deposited onto carbon/carbon composites, Surf. Coat. Technol. 190 (2005) 287-292.

[3] Y.Q. Zhai, K.Z. Li, H.J. Li, C. Wang, H. Liu, Influence of NaF concentration on fluorine-containing hydroxyapatite coating on carbon/carbon composites, Mater. Chem. Phys. 106 (2007) 22-26.

[4] C.S. Liu, Y. Huang, W. Shen, J.h. Cui, Kinetics of hydroxyapatite precipitation at pH 10 to 11,Biomaterials 22 (2001) 301-306.

[5] X.X. Wang, L. Xie, R. Wang, Biological fabrication of nacreous coating on titanium dental implant, Biomaterials, 26 (2005) 6229-6232.

[6] S.J. Ding, Y.M. Su, C.P. Ju, J.H. Chern Lin, Structure and immersion behavior of plasma-sprayed apatite-matrix coatings, Biomaterials 22 (2001) 833-845.

[7] T. Fu, L.P. He, Y. Han, K.W. Xu, Y.W. Mai, Induction of bonelike apatite on carbon/carbon composite by sodium silicate, Mater. Lett. 4371 (2003) 1-4.

[8] T. Fu, Y. Han, K.W. Xu, J.Y. Li, Z.X. Song, Induction of calcium phosphate on IBED-TiOx-coated carbon-carbon composite, Mater. Lett. 57 (2002) 77-81.

[9] P. Christel, A. Meunier, S. Leclercq, P. Bouquet, B. Buttazzoni, Development of a carbon-carbon hip prosthesis, J. Biomed. Mater. Res. 21 (1987) 191-218.

[10]X.N. Zhao, T. Hu, H.J. Li, M.D. Chen, S. Cao, L.L. Zhang, H. Hou, Electrochemically assisted co-deposition of calcium phosphate/collagen coatings on carbon/carbon composites, Appl. Surf. Sci. 257 (2011) 3612-3619.

[11]M. Shirkhanzadeh, Bioactive calcium phosphate coatings prepared by electrodeposition, J. Mater. Sci. Lett. 10 (1991) 1415-1418.

[12]R.Z. Wang, F.Z. Cui, H.B. Lu, H.B. Wen, C.L. Ma, H.D. Li, Synthesis of nanophase hydroxyapatite/ collagen composite, J. Mater. Sci. Lett. 14 (1995) 490-492.

Key Engineering Materials Vol. 537 (2013) pp 261-264
© (2013) Trans Tech Publications, Switzerland
doi:10.4028/www.scientific.net/KEM.537.261

Preparation and Characterization of Inorganic Zinc-Rich Coatings Based on Geopolymers

Xinfeng Li, Xuemin Cui[a], Sidong Liu, Binghui Mo, Li Cui

School of Chemistry and Chemical Engineering, Guangxi Key Lab of Petrochemical Resource Processing and Process Intensification Technology, Guangxi University, Nanning, 530004, China

[a] cui-xm@tsinghua.edu.cn (corresponding author)

Keywords: Geopolymer; Zinc; Coating; Corrosion; Paint; Pigment

Abstract: In this work, a high performance, environmentally friendly and waterborne inorganic zinc-rich coating based on geopolymer was developed and applied as protective coating of steel substrate. The geopolymer was the main ingredient of the zinc-rich paint (ZRP), and the zinc powder and additives were added to geopolymer to prepare ZRP. The corrosion protection behaviors of ZRP with different kinds of zinc powder and various zinc contents were characterized by electrochemical protection performance test. The results showed that coatings with 55 wt% spherical zinc powder and 25 wt% lamellar zinc powder exhibited good corrosion resistance mainly because of the cathodic protection and barrier effect.

Introduction

Nowadays, with the widely applications of steel in various sectors, economical losses from the metal corrosion amount to an estimated of 5% of the world economic growth (GDP). It is larger than losses from fire, wind and earthquake together. Therefore, anti-corrosion materials have attracted attention of various industries and experts [1-3].

The water-based inorganic ZRPs are better than organic ZRPs, owing to their stable conducting electrostatic property, high weatherability, high-temperature rating, good solvent resistance and rust protection, especially as a single paint coating. They are widely applied in long-lasting protective coatings for steel in marine industrial environment, such as offshore facilities because of their unique property of protecting the metal even when there is a slight mechanical damage to the coating [4,5].

In general, ZRP preparation consists of dispersion of zinc powder (spherical or lamellar) in a saponification resistant organic (usually epoxy resin) or an inorganic binder [6]. The physicochemical properties and corrosion resistance of ZRPs strongly depend on the binder, PVC, shape and size of the zinc particles [7].

"Geopolymer" which was first described by Joseph Davidovits in 1978, is a non-crystalline or quasicrystalline gel that has a three-dimensional network [8, 9]. Geopolymers are prepared using natural minerals or solid waste from the po-lymerization of silicon-oxygen tetrahedra and aluminum-oxygen tetrahedra. Geopolymers possess favorable characteristics, such as low CO_2 emission, high strength, excellent construction performance, and good resistance to corrosion from hot water and high temperature [10, 11]. In this investigation, water-based inorganic ZRPs based on a geopolymer were developed and their film forming properties and anti-rust mechanism were studied.

Experimental Procedure

Raw materials. Geopolymer binders were: metakaolin, prepared by calcining kaolin (BeiHai, China) at 800°C for 2 h; sodium silicate with SiO_2:Na_2O = 3.3; sodium hydroxide pellets with analytical purity [13]. Spherical and lamellar zinc powders were used as pigments. Dispersing agent, wetting agent Sapetin® D27, water retention agent Betolin®, and antifoaming agent $(C_4H_9O)_3PO$ were used as additives.

Preparation. The pigments and geopolymer binder were mixed and thoroughly dispersed using a high-speed mixer. Various additives and water were added for controlling viscosity.

Corrosion performance tests. The electrochemical protection performance measurements, coating density measurements and salt spray tests were carried out for investigating the effects of different forms of zinc powder on the corrosion performance of mild steel substrate coated with ZRPs. The electrochemical protection performances of ZRPs were determined according to the DOD-P-24648 US military standard. Finally, pictures of the representative areas were taken with a digital camera for evaluating the extent of corrosion after salt spray test for 2000 h.

Results and Discussion

The film forming mechanism. The use of new geopolymer as a binder in waterborne inorganic zinc-rich coating changed its main characteristics and performance by complex reaction between various components and steel substrate. The waterborne inorganic ZRP as a kind of curing coating relies on solidified film by a chemical reaction. The geopolymer is a major film-forming component of the waterborne inorganic zinc-rich coatings. During the film forming process, a large number of –OH in the film not only cross-link with zinc, but also bond with Iron substrate. Therefore, the alumino-silicate polymer is formed by Si-O-Al-O bond with solid ingredient (zinc powder), and Fe component in the steel layer surface. At the same time, the insoluble silicon-aluminum-zinc complex compounds are formed after further reaction with water and CO_2 of the air. The huge polymer network facilitates the adhesion between film and substance, thus protects effectively the steel substrate.

Corrosion performance tests. The preservation function of inorganic ZRP is based on electrochemical and physical shielding protection. The electrochemical protection occurs by sacrificing anode, consisting of zinc powder as a major factor of reliability and durability of inorganic ZRP. Fig.1 shows that the best results were obtained when spherical zinc powder content was 55%. It meets the quality and technical requirements, despite observation of a bit of rusts after soaking for 40 days in artificial seawater. The rusts did not appear at the same conditions for a coating containing 25% lamellar zinc powder. Thus, the lamellar ZRP provides more effective electrochemical protection because of different coating structure. The lamellar particles were arranged in superposition pattern in the coating, which facilitates the electroconductivity and water vapor impermeability of the coating. This result was supported by corrosion tests.

Salt spray test results. The coating densities and salt spray test results of coatings, with different shapes of zinc powder particles and various Zn loadings are shown in Figs. 2 and 3. The zinc powder content plays an important role in the zinc-rich coatings properties. As shown in Figs. 2 and 3, the anticorrosive property reached a maximum when the zinc powder content was a certain amount and then decreased with the increase of zinc powder, while the coating densities significantly increased with the increase of the Zn concentration for both different types of zinc powder coatings. The soak effect of zinc powder decreased in high zinc content, however, the purpose of cathodic protection by sacrificing anode could not be achieved because of poor connections in the electric current circuit in low zinc powder content coating. ZRP protects the steel substrate by cathodic protection only when the particles are in contact with the steel substrate and with each other. For the spherical zinc particles containing ZRPs, shown in Fig. 2, rust was visible by naked eye after exposure of 300 h when the concentration of Zn was 30%, while corrosion was detected after 1600 h when the

Fig. 1 Surface appearance of the coated specimens with ZRP containing different zinc types after electrochemical protection performance measurements for 40 days: (a, left) spherical zinc powder; (b, right) lamellar zinc powder.

concentration of Zn increased to 55%. However, as shown in Fig.3, corrosion of the coating containing 10 wt% lamellar Zn was detected after 1200 h and for the coating with 25 wt% lamellar Zn particles; corrosion was not evident even after 2000 h.

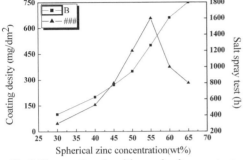
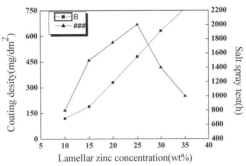

Fig.2 The coating densities and salt spray test results of ZRPs at various spherical zinc contents.

Fig.3 The coating densities and salt spray test results of ZRPs at various lamellar zinc contents.

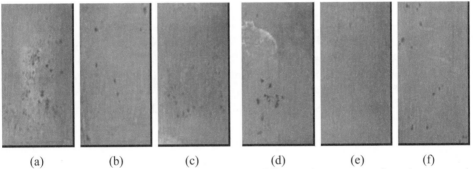

(a)   (b)   (c)   (d)   (e)   (f)

Fig.4 Surface appearance of the coated specimens with different zinc contents after salt spray testing for 2000 h: (a) 45 wt% of spherical zinc; (b) 55 wt% of spherical zinc; (c) 65 wt% of spherical zinc; (d) 15 wt% of lamellar zinc; (e) 25 wt% of lamellar zinc; (f) 35 wt% of lamellar zinc.

The corrosion resistances of various coatings were also evaluated by the rusting and blistering tests on the coatings surfaces on a mild steel substrate. The best performance in the salt spray test, with respect to general corrosion, was displayed by the coating containing 25% of lamellar zinc powder.

The results of visual observations of the coatings containing different amounts of zinc powder after 2000 h exposure to the salt fog are shown in Fig. 4. For the ZRPs containing spherical zinc particles, there were rusts (5 mm in diameter) for 45 wt% zinc powder loading, a few rusts (1–2 mm in diameter) for 65 wt% zinc powder and negligible brown rusts for 55 wt% zinc powder. For the ZRPs containing lamellar zinc particles, there were rusts (1–2 mm in diameter) for the coating containing 15 wt% zinc powder. The rusts were not observed on the coating consisting of 25 wt% lamellar zinc powders.

The salt spray examinations results coincided with the previous studies that the addition of zinc powder could effectively improve corrosion resistance performance by increasing the barrier properties of the geopolymer coatings; however, the anticorrosive property was the best result when the zinc powder content was a certain amount and then decreased with the increase of zinc powder content.

Conclusions

In this paper, the corrosion protection mechanism of spherical and lamellar zinc powder containing ZRPs was investigated by conducting general corrosion and electrochemical examinations. It was found that the barrier property of the coating deteriorated with the addition of zinc powder, and coatings with 25 wt% lamellar zinc powder would protect the steel substrate supper cathodically compared with the spherical ZRPs. The main conclusion was that the anticorrosion ability of ZRPs increased with the increase of zinc powder content, but decreased when the zinc powder content reached to a certain amount for both spherical and lamellar zinc powders.

Acknowledgements

This work was supported by Chinese Natural and Science fund (grants 50602006 and 50962002) and by Graduate Scientific and Technological Innovation Project of Guangxi (GXU11T32539).

References

[1] E. Bardel. Corrosion and Protection. London: Springer Verlag; 2003

[2] P. R. Roberge. Handbook of Corrosion Engineering. New York: McGraw-Hill; 1999

[3] C. H. Hare. Coating Problems on Zinc Substrates. J Prot Coat & Linings, 1998;15:17-37

[4] Z. W. Wicks, F.N. Jones, S. P. Pappas, D. A. Wicks. Organic Coatings: Science and Technology. 3rd ed., Wiley-Interscience, 2007, 151

[5] H. Marchebois, S. Joiret, C. Savall, J. Bernard, S. Touzain. Characterization of zinc-rich powder coatings by EIS and Raman spectrosopy. Surf. Coat Technol. 2002, 157:151-161

[6] A. Meroufel, S. Touzain. EIS characterisation of new zinc-rich powder coatings. Prog Org Coat, 2007, 59:197-205

[7] C.M. Abreu, L. Espada, M. Izquierdo. In: Fedrizzi, Bonora, editors. Eurocorr'96, 1997, 20, p. 23.

[8] X. M. Cui, G. J. Zheng, W. P. Zhang. A study on electrical conductivity of chemosynthetic Al_2O_3-$2SiO_2$ geoploymer materials. J. Power Sources, 2008; 184: 652-656

[9] X. M. Cui, Y. He, L. P. Liu, J. Y. Chen. NaA zeolite synthesis from geopolymer precursor. MRS Communications, 2011; 1: 49-51

[10] J. Temuujin, A. Minjigmaa, W. Rickard, et al. Preparation of metakaolin based geopolymer coatings on metal substrates as thermal barriers. Appl Clay Sci, 2009;46:265–270

[11] J. Temuujin, W. Rickard, M. Lee, et al. Preparation and thermal properties of fire resistant metakaolin-based geopolymer-type coatings. J Non-Cryst Solids, 2011;357:1399–1404

Key Engineering Materials Vol. 537 (2013) pp 265-268
© (2013) Trans Tech Publications, Switzerland
doi:10.4028/www.scientific.net/KEM.537.265

Study on the Fabrication and Properties of Ni-P Composite Coating

Zhao An, Mingya Li, Nianhao Ge, Xiaoying Li, Qiufan Li and Min Chen

School of Resources and Materials, Northeastern University at Qinhuangdao, Qinhuangdao
066004, Hebei Province, P. R. China

mylee@mail.neuq.edu.cn (corresponding author)

Keywords: Electroless deposition, Ni-P coating, Complexing agent, Heat treatment

Abstract. In this paper, the method of electroless deposition of nickel-phosphorous composite coating on the sample surface is employed. The effect of the complexing agent ratio on the surface morphology and phase composition of nickel-phosphorus coating when the main salt and reducing agent concentration unchanged has been studied. The influence of heat treatment on properties and microstructure of coatings are also investigated. Experimental results show that in the case of salt and reducing agent concentration unchanged, complexing agent concentration has strong effect on the morphology of the coatings and the crystal structure of the composite coating. During heat treatment, the morphology of the composite coating changed significantly, and the hardness was improved a certain degree for all the samples, which is related to the precipitation of Ni_3P.

Introduction

The Ni-P coating has been widely applied in many fields for its unique combination of performances, such as corrosion resistance, wear resistance, non-magnetism and high hardness [1-4]. The electroless plating technology is a traditional technique for preparing Ni-P coatings. According to the phosphorus contents of the coating, there are three types Ni-P coating, namely low phosphorus (1-3% wt), medium phosphorus (4-7% wt) and high phosphorus (above 7% wt), which has different properties and applications. The performances of the Ni-P coating depended on the composition and microstructure which were controlled by the process conditions, such as main salt and reducing agent, complexing agent, pH value, temperature and heat treatment parameters, etc. [5]. Among these factors, complexing agent plays an important role in the electroless nickel plating process, which affects the nucleation process [6]. The heat treatment has significant influence on the performance of the coating, which is related to the phase transformation of coating from amorphous to the mixture of amorphous and nanocrystalline [7].

In this paper, the effects of the complexing agent ratio on the surface morphology and phase composition of nickel-phosphorus coating, and the influence of heat treatment on properties and microstructure of coatings have been studied.

Experimental Details

Substrates for electroless plating Ni-P coating were made of low carbon steel. After being cleaned and polished, the substrates were plated. Chemicals used were generally reagent grade from commercial sources. In the plating process, nickel sulphate, sodium hypophosphite, lactic acid and glycine were used [8]. The composition of elelctroless solutions and process parameters were listed in Table 1. After being plated for 2 hours under different conditions, some samples were heat treated at various temperatures for 1 hour. The surface morphology of samples was determined by Scanning Electron Microscope (SEM, Leo1530). The analysis of X-ray diffraction measurements (XRD, Fangyuan DX-2500) was used to determine the phase composition of coatings with Cu Kα radiation. The micro hardness measurements were carried out under a load of 50 g.

Table 1. The composition of elelctroless solutions and process parameters.

	A	B	C
Nickel Sulphate /g·L^{-1}	26	26	26
sodium hypophosphite /g·L^{-1}	30	30	30
Lactic acid /g·L^{-1}	9	18	27
Glycine /g·L^{-1}	4	4	4
Citric acid /g·L^{-1}	6	6	6
KIO$_3$ /g·L^{-1}	0.02	0.02	0.02
PH		4.7 ± 0.2	
Temperature /°C		85 ± 2°C	

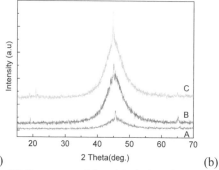
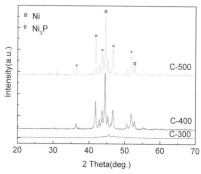

(a)   (b)

Fig.1 The XRD patterns of the nickel-phosphorus coatings prepared under different complexing agent concentration (a) before and (b) after the heat treatment.

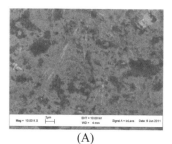
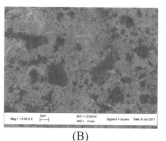
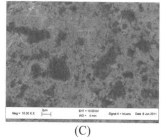

(A)   (B)   (C)

Fig. 2 SEM images of the coatings under different complexing agent concentration.

Results and Discussion

Fig.1 shows the XRD patterns of the nickel-phosphorus coatings prepared under different complexing agent concentration before and after the heat treatment. A, B and C represent the coating under different complexing agent concentration in Table 1. It can be seen from Fig. 1a that three samples demonstrate the obvious amorphous pattern, accompanied by a small amount of nanocrystalline phases before heat treatment. In addition, the intensity of diffraction peaks of the three curves are not the same, indeed, there is a trend that the strength is enhanced with the increase content of the complexing agent. It is suggested that sample C should contain more nanocrystalline phase than sample A and sample B.

Fig. 1b shows that the XRD diffraction patterns of coating prepared under C concentration conditions, which are subjected to heat treatment in the temperature of 300°C, 400°C and 500°C, respectively. For sample C-300, the XRD pattern is similar to sample C without heat treatment. It can be seen that there exist Ni$_3$P and Ni phases for sample C-400 and C-500. Furthermore, the intensity of peaks from Ni₃P phases is higher at 500°C. Results indicate that the Ni-P coating crystal structure is still amorphous at 300°C. When the temperature rises to 400°C, Ni-P coating was crystallized. The crystallization of Ni-P coating is more complete when the temperature increases to 500°C.

Fig. 2 shows the surface morphologies of Sample A, B and C. It can be seen clearly that there exist the light grey Ni-P coatings for all samples. For samples prepared with different concentration of complexing agent, the surface morphologies of the coatings are very close, which is a layer of granular particles. It can be seen that the Ni-P coating is relatively concentrated, dark place may be the coating uneven.

Fig. 3 shows the SEM images for Sample C-400 and Sample C-500. The surface morphology of coatings changes significantly before and after the heat treatment. It can be seen that the Ni-P coating becomes dense for Sample C-400, so the properties of coatings improved. For Sample C-500, the size of grain becomes much coarse, which results in the decrease of coating properties. It is suggested that 400°C is a more reasonable temperature.

Fig. 4a shows the micro hardness of Sample A, Sample B and Sample C. The hardness is 163.4HV, 207.8HV and 312.1HV for Sample A, Sample B and Sample C, respectively. It can be seen from the figure that the hardness increases with the increasing content of complexing agent. Fig. 4b indicates the micro hardness for samples annealed at different temperature. The hardness is 432.5HV, 494.2HV and 460.1HV for Sample C-300, Sample C-400 and Sample C-500, respectively. There is a hardness peak which is for Sample C-400. After heat treatment, the Ni-P coating exhibits higher level of hardness values. This is due to the formation of Ni₃P phase during the heat treatment, which is related to the precipitation hardening process. When the temperature is increased to 500°C, the hardness decreases. This may be due to that Ni₃P grain grows up.

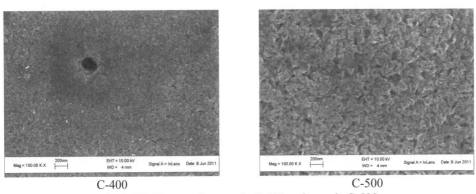

C-400   C-500

Fig. 3 SEM images for sample C-400 and sample C-500.

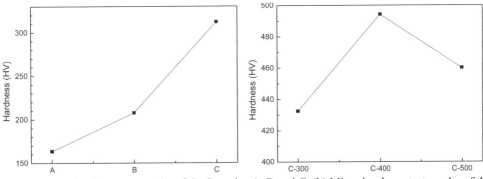

Fig. 4 (a) Micro hardness test results of the Samples A, B and C, (b) Micro hardness test results of the Samples C-300, C-400 and C-500.

Conclusion

The nickel-phosphorous composite coating on the sample surface was prepared using the method of electroless deposition. Experimental results show that the Ni-P coatings prepared under different ratio of complexing agents are mixture of nanocrystalline phase and amorphous phase. The degree of crystallization and the micro hardness increase with the increasing content of complexing agent (lactic acid content). Heat treatment has important influence on the coating morphology and phase composition. Heat treatment could enhance the properties of Ni-P coating, which is due to the formation of Ni_3P phase.

Acknowledgements

This work was supported by the Fundamental Research Funds for the Central Universities (NO. N100423006, NO. N110823001).

References

[1] H. Ashassi-Sorkhabi and S.H. Rafizadeh. Effect of coating time and heat treatment on structures and corrosion characteristics of electroless Ni-P alloy deposits. Surf. Coat. Tech. 176 (2004) 318-326

[2] Q. Zhao, Y. Liu, H. Müller-Steinhagen and G. Liu. Graded Ni-P-PTFE coatings and their potential applications. Surf. Coat. Tech. 155 (2002) 279-284.

[3] N. Krasteva, V. Foty and S. Armyanov. Thermal stability of Ni-P and Ni-Cu-P amorphous alloys. J. Electrochem. Soc. 141 (1994) 2864-2867.

[4] C.Y. Lee and K.L. Lin. Wetting kinetics and the interfacial interaction behavior between electroless Ni-Cu-P and molten solder. J. Jpn. Appl. phys. 33 (1994) 2684-2688.

[5] L.J. Feng, A.L. Lei. Study on chemical plating Ni-P complexing agent on cast iron surface. Foudry Tech. 26 (2005) 676-678.

[6] X.L. Cai and X. Huang and Z.J. Liu. Effect of complexing agent on plating rate in electroless nickel solution. J. Jilin Inst. Chem. Tech. 17 (2000) 21-23.

[7] Y. Wang, J.B Guo, X.D. Yuan, D.B Sun and Q.H. Li. The impact of heat treatment on amorphous Ni-P plating layer structure and properties. Materiais Protection 41 (2008) 54-56.

[8] Y.H. Cheng, Y. Zou, L. Cheng and W. Liu. Efeet of complexing agents on properties of elctroless Ni-P deposits. Mater. Sci. Tech. 24 (2008) 457-460.

Key Engineering Materials Vol. 537 (2013) pp 269-273
© (2013) Trans Tech Publications, Switzerland
doi:10.4028/www.scientific.net/KEM.537.269

Preparation and Characterization of Cube-Textured Ni-5 at.% W Substrates for Coated Conductors

Xingpin Chen[1a], Jingpeng Zhang[1b], Xue Chen[1c], Yongbin Ji[2]

[1]College of Materials Science and Engineering, Chongqing University, Chongqing, 400030, China

[2] Northwest Institute for Nonferrous Metal Research, Xi'an 710016, China

[a]xpchen@cqu.edu.cn, [b]zhangjingpeng1110@yahoo.cn, [c]chenxue61@163.com

Keywords: Coated conductor; Ni-5 at.% W; cube texture; EBSD

Abstract. The goal of present study was to manufacture and characterize the Ni-5 at.% W alloy substrates for coated conductor applications with a strong cube texture. Cube-textured Ni-5 at.% W alloy substrates were fabricated by heavy cold rolling (98%) and a high temperature annealing (1200°C). From the EBSD and XRD analysis of the fully recrystallized tapes, the cube texture component in the tape was found to reach a proportion as high as 99.5% within a tolerance angle of less than 9°, while the FWHM (full width at half maximum) values of the out-of-plane and in-plane textures are about 5.8° and 6.9°, respectively. Moreover, the grain size distribution is homogeneous, and there is a large amount of low angle grain boundaries (95.6%). This process is very suitable for preparation of substrates for coated conductors.

Introduction

YBCO-coated conductors as the second generation of high temperature superconductor materials have received great attention due to their high critical current density (J_c) and intrinsic high irreversibility of magnetic field. Rolling-assisted biaxially textured substrates (RABiTS^{TM}) process is one of the most promising and cost-effective methods to fabricate YBCO-coated conductors. One of the main challenges for the RABiTS^{TM} process is to produce a strong flexible substrate with a very strong cube texture. Ni has been successfully used for this purpose [1]. A sharp cube texture is easily obtained with Ni by rolling and annealing [2], but as in the case of pure Ni, its high Curie temperature (627 K), low strength and thermal grooving after recrystallization heat treatments lead to severe limitation to industrial applications. Hence, many attempts have been made to develop textured NiW alloy substrates, such as Ni-5 at.% W and Ni-9 at.% W. When W contents >5 at.% at NiW alloys, Transition of copper type to brass type texture was observed in the rolling texture which was attributed to the reduction in stacking fault energy, The important consequence of this texture transition is on the recrystallisation cube texture, whose fraction decreases sharply at W contents > 5% [3]. Thus, Ni-5 at.% W substrate is the key point to research in industrial productions. In that field, however, controversial results have been reported in the literatures, and a great amount of work has still to be done. In this paper we focus on the preparation and characterization of the cube-textured Ni-5 at.% W substrates for coated conductors.

Experimental

Elemental powders of Ni (purity 99.99%) and W (purity 99.9%) were used as the raw materials. Ni–5 at.% W alloy was prepared by a typical powder metallurgical (P/M) method of mixing the elemental powders, Homogeneous alloying between Ni and W was established during the sintering process. The sintered compact was subsequently hot forged to 30 mm, and then hot rolled to 10 mm, following that An intermediate annealing process was carried out on 4.0 mm tapes which were cold rolled from 10 mm. The samples were subsequently cold-rolled to a final thickness of 80 μm, corresponding to 98% rolling reduction. The cold-rolled samples were annealed in an argon and hydrogen (Ar-4% H$_2$) atmosphere at temperature of 1200 °C for 1 h. The heat-treated samples were immediately quenched in cold water.

The microstructure and texture were analysed by electron back-scattering diffraction (EBSD) in Nova 400 Nano-SEM with channel 5 software in the rolling plane (defined by the rolling direction (RD) and the transverse direction (TD)). The EBSP data were used to construct orientation image maps. Unless otherwise stated, in the maps shown in this paper low angle grain boundaries (LAGBs) are defined as misorientations in the range of 2-15° and are shown by thin gray lines, and high-angle grain boundaries (HAGBs) are defined as misorientations more than 15° and are shown by thick black lines. Texture components are defined using a 15° deviation to various ideal orientations. The texture components considered are {001}<100> (C, blue), {112}<111> (Cu, red), {123}<634> (S, pink) and {110}<112> (B, yellow). All remaining orientations are grouped together and labelled as 'Random'(R, white). Furthermore, the FWHM of different annealing procedure substrates were examined by four-circle X-ray diffractometer (Rigaku D/Max2500, 18kw) using Cu Kα radiation.

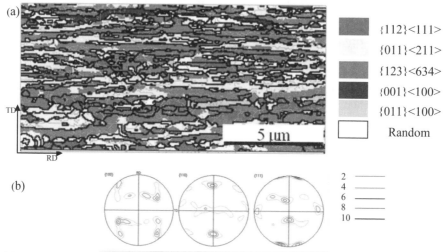

Fig. 1 (a) Crystal orientation map of texture components and (b) pole figures of cold-rolled Ni-5 at.% W alloy

Results and discussion

Fig. 1 shows the crystal orientation map and pole figures of a region of the 98% cold-rolled material. Long, thin, cube-oriented regions (blue) parallel to the RD and in some cases several micrometers in length along the RD can be clearly seen in the map. It can be noticed that the pole figures (Fig.1b) show Ni–5%W alloy has the typical copper or metal type rolling texture (characteristic of pure fcc metals with high stacking fault energy) in which orientations are assembled along the β fibre. The β fibre is basically an orientation tube in the Euler angle space which runs from the copper orientation through the S orientation to the brass orientation [3]. The microstructure in Fig. 2a is a typical lamellar structure subdivided by HAGBs as often observed in heavily cold-rolled materials. Low angle grain boundaries existed in lamellar structure account for most of the grain boundary, just as the grain boundaries map shown. The number fraction of 30°-40°<111> boundaries is in fact quite low, and only 10% (length fraction) of these boundaries are found between the cube regions and their immediate surroundings. Moreover no other distinct CSL boundaries can be seen. The cube bands are characterized by their high orientation gradient and lower stored energy. For this favorable structure, a recrystallization nucleus of the cube grains can thus develop much quicker inside these bands compared to bands of other orientations [4].

The deviations of Cube texture to the ideal Cube orientation is different, drawn from graduated blue in Fig. 4a. Heavier the blue color of grain is, lesser deviation to the ideal cube orientation the grain is. The main tolerances range from 1.5° to 9°. The cumulative value of the area fraction of grains as a function of the tolerance angle is shown in Fig.4b, was clearly seen that the cube grains component approaches close 99.5% in the recrystallization tapes when the tolerance is 9°, this values are better than the results of other reports [5]. That means Ni-5 at.% W alloy develops a much sharper cube texture after annealing as compared to pure Ni, because W decrease the volume fraction of the RD-rotated cube grains [6].

It is well known that the current capacity of RABiTS coated conductor tapes is closely related to the grain boundaries misorientation distribution of the substrates, especially the LAGBs fraction. The grain boundary character distribution of Ni-5 at.% W alloy after annealing at 1200 °C for 1 h (Fig. 5) in this condition indicates that the LAGBs fraction has increased significantly, and HAGBs fraction has decreased as compared to the straight rolled material (Fig.2). It was clearly observed that the amount of LAGBs reaches 95.6% in the full recrystallized tapes. The coincidence site lattice (CSL) boundary distribution at this condition clearly shows that most of the CSL boundaries belong to the $\Sigma 3$ category, the value is 2.9%.

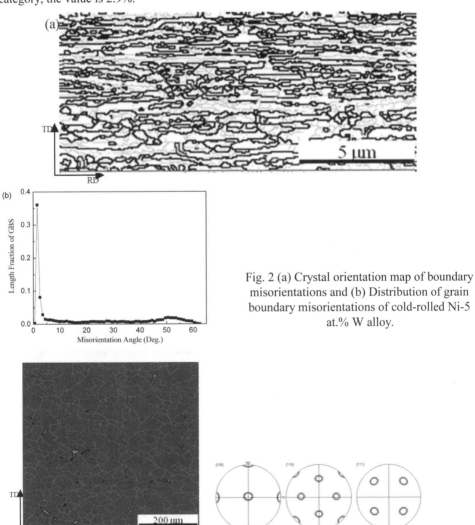

Fig. 2 (a) Crystal orientation map of boundary misorientations and (b) Distribution of grain boundary misorientations of cold-rolled Ni-5 at.% W alloy.

Fig. 3 (a) Crystal orientation map of texture components and (b) pole figures of the annealed Ni-5 at.% W alloy at 1200 °C for 1 h. The color key is the same as in Fig. 1

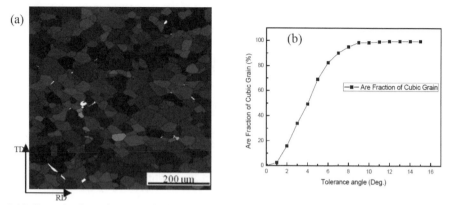

Fig. 4 (a) Crystal orientation map depicting the spatial distribution of cube grains (heavier the blue color of grain is, lesser deviation to the ideal cube orientation the grain is.) and (b) Distribution of the area fraction of cubic grains vs. tolerance angle.

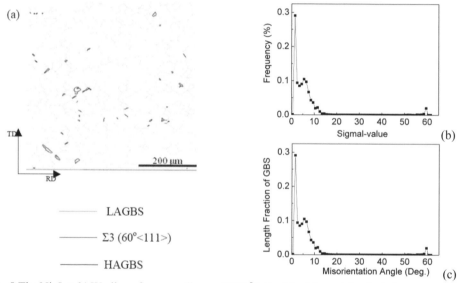

Fig. 5 The Ni-5 at.% W alloy after annealing at 1200 °C for 1 h: (a) crystal orientation map of grain boundary distribution, (b) distribution of grain boundary misorientations and (c) the corresponding CSL boundary distributions.

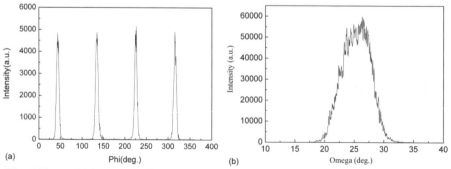

Fig. 6 X-ray of (a) φ-scan and (b) ω-scan of the annealed Ni-5at.%W alloy at 1200 °C for 1 h

In order to characterize the sharpness of the cube orientation grains, the ω-scan and φ-scan curves of the X-ray diffraction are given in Fig. 6. The full width at half maximum (FWHM) value of ω-scan which reveals the out-of-plan sharpness of cube orientation is about 6.9°, while the in-plan FWHM values along Rd is 5.8°. The above results indicate that the cubic texture of the substrate is very sharp.

Summary

(1) Cube-textured Ni-5 at.% W alloy substrates were fabricated by heavy cold rolling (98%) and high temperature annealing (1200 °C), the cube texture component in the tape was found to reach a proportions as high as 99.5% within a tolerance angle of less than 9°. The deviation of Cube texture to the ideal Cube orientation is different, and the main tolerances range from 1.5° to 9°.

(2) There is a large amount of LAGBs for full annealing tapesr (95.6%), and most of the CSL boundaries belong to the Σ3 category.

(3) The FWHM values of the out-of-plane and in-plane orientations are about of 5.8° and 6.9°, respectively.

Acknowledgements

This project was supported by the National Natural Science Foundation of China (51171215 and 51101128).

References

[1] P. Norton, A. Goyal, J. D. Budai, *et al.*, Epitaxial $YBa_2Cu_3O_7$ on biaxially textured nickel (001): an approach to superconducting tapes with high critical current density, Science. 274 (1996) 755-57.

[2] E. D. Specht, A. Goyal, D. F. Lee, *et al.*, Supercond. Sci. Technol. 11 (1998) 945-9.

[3] Sarma VS, Eickemeyer J, Mickel C, Schultz L, Holzapfel B. On the cold rolling textures in some fcc Ni–W alloys. Mater. Sci. Eng. A. 380 (2004) 30-33.

[4] P. P. Bhattacharjee, R. K. Ray, and N. Tsuji, Cold Rolling and Recrystallization Textures of Ni–5at.% W Alloy, Acta Mater. 57 (2009) 2166-79.

[5] Y. Zhao, H. Li. Suo, M. Liu, *et al.*, Development of Cube Textured Ni–5at.%W Alloy Substrates for Coated Conductor Application Using a Melting Process, Physica C, 440 (2006) 10-16.

[6] P. P. Bhattacharjee, and R. K. Ray, Effect of Processing Variables on Cube Texture Formation in Powder Metallurgically Prepared Ni and Ni–W Alloy Tapes for Use as Substrates for Coated Conductor Applications, Mater. Sci. Eng. A. 459 (2007) 309-323.

Key Engineering Materials Vol. 537 (2013) pp 274-278
© (2013) Trans Tech Publications, Switzerland
doi:10.4028/www.scientific.net/KEM.537.274

Effect of Pre-recovery on Recrystallization Texture in Nickel Substrates for Coated Conductors

Xingpin Chen [a], Xue Chen [b], Jingpeng Zhang [c]

College of Materials Science and Engineering, Chongqing University, Chongqing, 400030, China

[a]xpchen@cqu.edu.cn, [b]chenxue61@163.com, [c]zhangjingpeng1110@yahoo.cn

Keywords: EBSD; pre-recovery; recrystallization; cube texture; coated conductor

Abstract. Electron backscattered diffraction (EBSD) technology was applied to study the effect of pre-recovery on the recrystallization texture in nickel substrates for coated conductors. Pure nickel (99.999%) was cold rolled by a 95% total reduction, and then samples were annealed at 200 °C for 1 hour and quenched for fully recovery, and finally annealed at 600 °C for 1 hour and quenched in water. The results show that pre-recovery had a strong influence on the formation of cube recrystallization texture. Compared with samples without pre-recovery treatment at 200 °C, samples through pre-recovery treatment can achieve stronger cube texture after recrystallization annealing, and develop more low-angle grain boundaries but less sigma 3 ($\Sigma3$) grain boundaries.

Introduction

Recrystallization texture, which develops after annealing treatment in metal materials, depends on two factors: one is the orientation of recrystallized nuclei, the other one is that growth rate of different oriented grains is various. Besides, recovery that takes place in the early stage of annealing treatment influences on the recrystallization. Recovery and recrystallization occur simultaneously. The driving force for both process is the store energy, so that they are competing processes [1]. Recovery can reduce the stored energy, hence it retards recrystallization [2]. The extent of this retardation depends on the material, especially its stacking fault energy. And it is also related to the amount of deformation and the annealing temperature. Although recovery slows down the recrystallization kinetics by reducing the driving force for the process, it also promotes nucleation by forming nuclei in the deformed microstructure [3]. Thus, recovery plays a dual role in recrystallization. The extent of recovery that occurs prior to recrystallization affects directly nucleation and growth of the recrystallizing nuclei.

Pure Ni is one of the most studied materials for understanding the evolution of deformation structure and textures subjected to normal cold rolled condition. Additionally, Pure Ni is a face-centered cubic structure (f.c.c.) metal having intermediate stacking fault energy value between pure Al and pure Cu. Thus the results from Ni could be very helpful in understanding the mechanism of formation of characteristic deformation textures during cold rolling processing.

Nickel is not easy to recover during and after cold-rolling at room temperature, which makes it possible that we can study real deformed structure and recovery process in detail. Nickel presents a typical copper texture (mainly including brass, copper and S texture component) after severely cold rolling. It was reported that different oriented grains had different recovery rate, and cube-oriented units had the quickest recovery rate [4]. The extensive early work on the recovery of deformed metals has been reviewed in 1950s. Despite its obvious importance, recovery attracted little interest in later years. However, the current industrially driven move to produce quantitative physically-based models for annealing processes has resulted in renewed interest in recovery, seeing Refs [5-10]. In spite of this, there is still no systemic report about the recovery process in f.c.c. metals. Actually, recovery took place as soon as heat treatment carried out. The researches of recovery process can provide theoretical direction for optimizing the annealing process, especially for the process need strong cube texture like metal substrates for coated conductors.

In the present work the influence of pre-recovery on the recrystallization cube texture and grain boundary distributions was investigated.

Experimental

The starting material used in the present investigation was high-purity nickel (99.999%). It was processed to give a average grain size of 50 μm and a fairly random texture. Nickel was cold rolled by a 95% total reduction in thickness and subsequently annealed. Sample A was directly annealed at 600 °C for 1 hour and quenched in water. Sample B was pre-annealed at 200 °C for 1 hour and quenched in water, then it was annealed at 600 °C for 1 hour and quenched in water. All heat treatment were carried out in tube furnace with reducing atmosphere of argon and 4% hydrogen (heating rate of 300 °C per hour).

The microstructure and texture were detected by scanning electron microscope (SEM) equipped with a fully automated EBSD analysis system, using an accelerating voltage of 20 KV with a working distance of 15 mm and a 70° sample tilt angle. For examination in the SEM, the plane was ground to SiC4000 before final electropolishing in a 1:3:4 $HClO_4:C_2H_5COOH:C_2H_5OH$ solution at 0 °C and 15 V for 35~45 s. Orientation maps were obtained for areas of 400×400 μm^2 with appropriate steps.

Results and discussions

During recovery, the microstructural changes in pure nickel are very subtle and occur on a small scale. The microstructures as observed by optical microscopy do not usually reveal much change. And for this reason, recovery is often measured indirectly by some bulk technique, for example by following the change in some physical or mechanical property. Hardness is one of the most common measurement methods to study recovery and recrystallization kinetics.

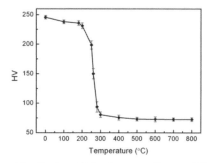

Fig. 1 Hardness data for Ni cold-rolled to 95% after annealing at different temperatures for 1 hour

Figure 1 shows the micro-hardness changes of pure nickel annealed at various temperatures, ranging from 100 °C to 800 °C. When annealed at 100 to 200 °C, hardness of pure nickel decreased slightly, indicating that only recovery take placed. While annealed at temperature between 250 °C and 280 °C, the hardness decreased sharply, which illustrates that materials partially recrystallized. After 300 °C, with the temperature increasing, the hardness reduced a little and kept constant. It suggests that severely cold-rolled nickel completes recrystallization at 300 °C.

EBSD analysis gives crystallographic orientation map for identifying orientations of each grain. In the maps shown in this paper low-angle boundaries are defined as misorientations in the range 2-15° and are shown by thin lines, and high-angle grain boundaries are defined as misorientation more than 15° and are shown by thick lines. Texture components are defined using a 15° deviation to various ideal orientations.

After 95% rolling reduction, the microstructure of a cold-rolled nickel showed a typical deformed structure, the micro-texture of this material presented a typical copper-type texture (Fig. 2a), and it was easy to develop strong cube texture after annealing. In Fig. 2b, samples maintained deformed microstructure and texture after annealing at 200 °C for 1 hour, it illustrated that recovery just take place in nickel but no recrystallization. During recovery the stored energy of the material is lowered by dislocation movement，including glide, climb and cross-slip of dislocations. There are two primary processes, the annihilation of dislocations and the rearrangement of dislocations into lower energy configurations. It is well established that grain boundaries are comprised by dislocations, low angle grain boundaries migrate through the movement by climb and glide of the dislocations. Misorientation distributions of both deformed and recovery samples are shown in Fig. 3, clearly, specimen with recovery treatment can achieve more low-angle grain boundaries.

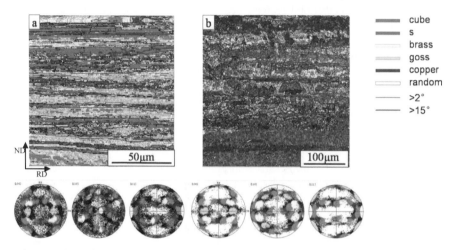

Fig. 2 EBSD orientation maps and pole figures for samples (a) 95% rolling reduction and (b) annealed at 200 °C for 1 h

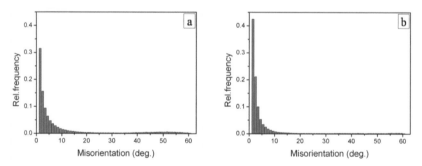

Fig. 3 Misorientation distributions of (a) 95% deformed and (b) 200 °C annealed samples

After 600 °C annealing for 1 hour, sample A recrystallized and developed cube texture (fraction is 55.1%), and sample B achieved stronger cube texture (84.3%) than A, shown in Fig. 4. It is obvious that the recovery treatment of the cold-rolled sheet affected the recrystallization texture of nickel. At a given annealing temperature (600 °C), the cold-rolled nickel that had undergone a recovery treatment exhibited a stronger cube texture and a weaker R texture than the cold-rolled sheet without a recovery treatment. This suggests that the recovery treatment of cold-rolled nickel promotes the formation of cube texture, but restrains the formation of the R texture. Because recovery lowers the driving force for recrystallization, a significant amount of prior recovery may in turn influence the nature and the kinetics of recrystallization. The division between recovery and recrystallization is sometimes difficult to define, because recovery mechanisms play an important role in nucleating recrystallization. It is reported that the cube orientation shows fastest recovery rate than other orientations, and then it is preferential nucleation. This leads to the formation of strong cube texture.

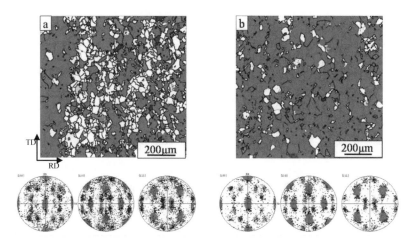

Fig. 4 EBSD orientation maps and pole figures for recrystallized samples (a) 600 °C and (b) 200 - 600 °C

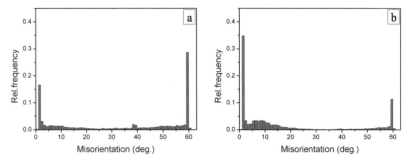

Fig. 5 Misorientation distributions of annealed samples (a) 600 °C and (b) 200 - 600 °C

Low-angle grain boundaries are pleased due to the weak links at the high-angle grain boundaries, which can be analysed by the misorientation distributions. Compared with sample A, sample B had more low-angle grain boundaries (see Fig. 5), but less $\Sigma3$ grain boundaries (see Fig. 6, sample A is 29.7% and sample B is 12.1%), the sharp reduction of $\Sigma3$ grain boundary fractions can significantly low the groove depth of the substrates during high-temperature annealing process.

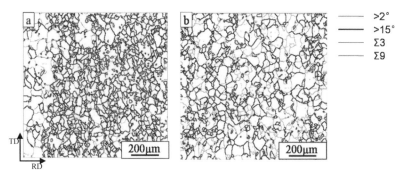

Fig. 6 CSL boundaries of annealed samples (a) 600 °C and (b) 200 - 600 °C

Conclusions

Recovery and recrystallization are competing processes as both are driven by the stored energy of the deformed state. Recovery lowers the stored energy, consequently, a significant amount of prior recovery may in turn influence the recrystallization processing. Compared with samples without recovery treatment at 200 °C, samples through pre-recovery can achieve stronger cube texture after recrystallization annealing, and they have more low-angle grain boundaries but less Σ3 grain boundaries. It demonstrates that pre-recovery is benefical to developing recrystallization cube texture.

Acknowlegment

This work was supported by the National Natural Science Foundation of China (51171215).

References

[1]   J. Humphreys, M. Hatherly, Recrystallization and Related Annealing Phenomena (Second Edition), in: Recrystallization and Related Annealing Phenomena (Second Edition), Elsevier, New York, 2004, pp. 169-214.
[2]   H.P. Stüwe, A.F. Padilha, F. Siciliano, Competition between recovery and recrystallization, Mater. Sci. Eng. A. 333 (2002) 361-367.
[3]   X.Y. Song, M. Rettenmayr, Modelling study on recrystallization, recovery and their temperature dependence in inhomogeneously deformed materials, Mater. Sci. Eng. A. 332 (2002) 153-160.
[4]   S. Zaefferer, T. Baudin, R. Penelle, A study on the formation mechanisms of the cube recrystallization texture in cold rolled Fe-36% Ni alloys, Acta Materialia. 49 (2001) 1105-1122.
[5]   E. Nes, Recovery revisited, Acta Metall. Mater. 43 (1995) 2189-2207.
[6]   A. Martínez-de-Guerenu, K. Gurruchaga, F. Arizti, Nondestructive characterization of recovery and recrystallization in cold rolled low carbon steel by magnetic hysteresis loops, J. Magn. Magn. Mater. 316 (2007) 842-845.
[7]   A. Samet-Meziou, A.L. Etter, T. Baudin, R. Penelle, TEM study of recovery and recrystallizat-ion mechanisms after 40% cold rolling in an IF-Ti steel, Scripta Mater. 53 (2005) 1001-1006.
[8]   E.J. Giordani, A.M. Jorge Jr, O. Balancin, Proportion of recovery and recrystallization during interpass times at high temperatures on a Nb- and N-bearing austenitic stainless steel biomaterial, Scripta Mater. 55 (2006) 743-746.
[9]   F.J. Humphreys, A unified theory of recovery, recrystallization and grain growth, based on the stability and growth of cellular microstructures II. The effect of second-phase particles, Acta Mater. 45 (1997) 5031-5039.
[10] W.C. Liu, J. Li, H. Yuan, Q.X. Yang, Effect of recovery on the recrystallization texture of an Al-Mg alloy, Scripta Mater. 57 (2007) 833-836.

Key Engineering Materials Vol. 537 (2013) pp 279-282
© (2013) Trans Tech Publications, Switzerland
doi:10.4028/www.scientific.net/KEM.537.279

Fabrication and Properties of Al-Cr Coatings on the Inner Wall of Steel Pipe by Mechanical Alloying Method

Zhongqing Tian[a], Guoxing Zhang, Weijiu Huang, Yukai Zhu

School of Materials Science and Engineering, Chongqing University of Technology, 69 Hongguangdadao Road, Chongqing 400054, P. R. China

[a]tzqmail@cqut.edu.cn

Keywords: Surface coatings; Mechanical alloying; Microstructure; Tribological properties

Abstract. The mechanical alloying method process has been innovatively used to prepare Al-Cr coating on the inner wall of steel pipe. The coating thickness was measured from all samples using optical and Scanning Electron Microscope was used to observe the surface microstructure of Al-Cr coating. Microhardness was analyzed by Digital Microhardness Tester. A wear test was performed by high speed reciprocating friction testing machine. The results show that the coating thickness is 20μm and 26μm at the rotating speed of 200 rpm and 300 rpm, respectively. The surface morphology is significantly influenced by the rotating speed. When the rotating speed was 200 rpm, a heterogeneous coating surface consisting of flattened particles produced by cold welding with less interparticle contact is formed. When the rotating speed was 300 rpm, the coating became denser and a smooth, highly consolidated and dense coating is formed. The hardness of the Al–Cr coating prepared at 200 and 300 rpm are about 250 Hv and 270 Hv. The friction coefficient of the Al–Cr coating prepared at 200 rpm are about 0.37, 0.39 and 0.24 at the frequencies of 3, 4 and 5 Hz. The friction coefficient of the Al-Cr coating prepared at 300 rpm are about 0.3, 0.18 and 0.28 at the frequencies of 3, 4 and 5 Hz.

Introduction

In recent years, the mechanical alloying method (MA) has become a popular method to fabricate coating, due to its simplicity, relatively inexpensive equipment and its potential for large-scale production [1-6]. During the mechanical alloying process, the powder undergoes cold welding and the resulted refinement. Meanwhile, due to the repeated impacts and cold welding, a fraction of powder tends to adhere to the surface of the balls and the inner wall of the container, so as to form a coating layer. This process allows thick coatings to be produced in a short time at room temperature in an ambient atmosphere with minimum surface preparation before deposition.

The method has been applied to deposit composite Al-containing coatings such as Ti-Al and Ni-Al [4,5]. Nevertheless, Al-Cr system has not received much attention as a coating. Al-Cr alloys hold immense potential as a decorative material in the industrial area, small-scale electronic devices and corrosion resistance applications. However, no attempt has been made so far for the fabrication of the Al-Cr coating using mechanical alloying technique.

In the present work, MA process was innovatively used to prepare Al-Cr coating on the inner wall of steel pipe. The microstructures and tribological properties of as-prepared coatings are reported.

Experimental procedures

A steel pipe about 5cm high was put into the vial with the ball-to-powder ratio being 1:5. Milling was carried out by a QM-3SP04 planetary ball mill in which the milling vessel and balls were made of hardened steel. The material to ball mass ratio was set at 4:1. Al and Cr powders (by weight ratio 1:2) were added into the milling vessel, and the powders mixed at rotating speed of 200 or 300 rpm for 8 h. In order to avoid an excessive temperature rise within the vial, 60 min ball milling was followed by a 10min cooling interval. Finally, the milling vessel was left to cool for two hours.

Metallographic samples were mechanically ground and polished through standard routines. The coating thickness was measured from all samples using optical. Scanning Electron Microscope (SEM, JSM-6460LV) was used to observe the surface microstructure of Al-Cr coating.

Microhardness was analyzed by the MICRO-586 Digital Microhardness Tester. The average of indents was used to ensure acquisition of reasonably representative value. All indents were kept away from porous locations. As the measurements were carried out in the cross-section of the coating, indents were always positioned not less than 10 μm away from the surface or the interface between the substrate and the coating. A wear test was performed by MFT-R4000 high speed reciprocating friction testing machine in the air. The tests were conducted under the load of 5N with a constant stroke length of 10 mm at different frequencies of 3, 4 and 5 Hz.

Fig. 1. Optical microstructure of the overview cross-section of Al-Cr coating prepared at the rotating speed of (a) 200 rpm, (b) 300 rpm.

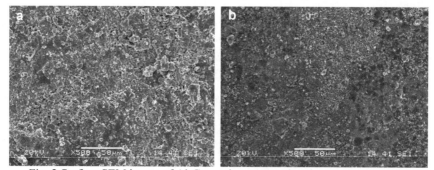

Fig. 2 Surface SEM image of Al-Cr coating prepared at the rotating speed of (a) 200 rpm, (b) 300 rpm.

Results and discussion

Fig. 1 shows the optical microstructure of the overview cross-section of Al-Cr coating prepared at the rotating speed of 200 rpm and 300 rpm. As can be seen from Fig. 1, these two coatings were free of pores and cracks, and with a high quality metallurgical bonding to the pure steel substrate. Fig.1 also shows the dependence of the rotating speed on the coating thickness in the case of the steel substrate. The coating thickness is 20μm and 26μm at the rotating speed of 200 rpm and 300 rpm, respectively. The difference in the coating thickness fabricated by the different rotating speeds could be related to the milling intensity. When the rotating speed is high, more energy was transferred to the powder particles. The raw particles were rapidly fractured and fragmented. The fragmented particles began cold welding more easily and rapidly when high rotating speed was used.

Fig. 2 shows the surface SEM images of Al-Cr coating prepared at the rotating speed of 200 rpm and 300 rpm. The results reveal that the surface morphology is significantly influenced by the the rotating speed. When the rotating speed was 200 rpm, a heterogeneous coating surface consisting of flattened particles produced by cold welding with less interparticle contact is formed. When the rotating speed was 300 rpm, the coating became denser and a smooth, highly consolidated and dense coating is formed. Also there are signs of kneading and plastic deformation.

Fig. 3 shows the relationship between the hardness of the Al-Cr coating and rotating speed. In general, higher milling speed was observed to increase microhardness in the coating layer. The hardness of the Al-Cr coating prepared at 200 rpm is about 250 Hv. The hardness of the Al-Cr coating prepared at 300 rpm is about 270 Hv.

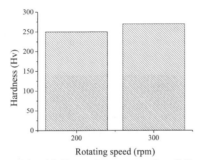

Fig. 3 Hardness of the Al-Cr coating prepared at different rotating speed.

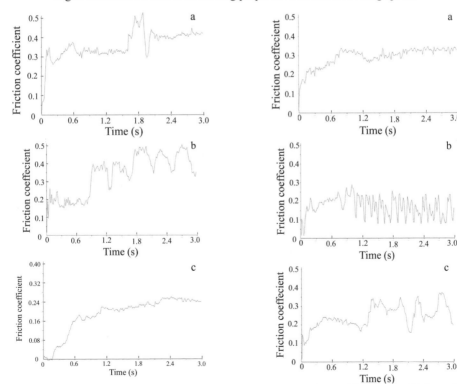

Fig .4 Variations of friction coefficient of the Al-Cr coating prepared at 200 rpm as a function of test time at different frequencies of (a) 3, (b) 4 and (c) 5 Hz.

Fig. 5 Variations of friction coefficient of the Al-Cr coating prepared at 300 rpm as a function of test time at different frequencies of (a) 3, (b) 4 and (c) 5 Hz.

Generally, the wear property of coating is characterized by the friction coefficient. Fig. 4 and Fig. 5 show the variations of friction coefficient of the Al-Cr coating prepared at different rotating speed as a function of test time at different frequencies of 3, 4 and 5 Hz. It can be seen that the friction coefficient of the Al-Cr coating prepared at 200 rpm are about 0.37, 0.39 and 0.24 at the frequencies of 3, 4 and 5 Hz. The friction coefficient of the Al-Cr coating prepared at 300 rpm are about 0.3, 0.18 and 0.28 at the frequencies of 3, 4 and 5 Hz. The relationship between the friction coefficient and microstructures of the prepared coating is not clear and need more work to do.

Summary

The mechanical alloying method process has been innovatively used to prepare Al-Cr coating on the inner wall of steel pipe. The coating thickness is 20μm and 26μm at the rotating speed of 200 rpm and 300 rpm, respectively. The surface morphology is significantly influenced by the the rotating speed. When the rotating speed was 200 rpm, a heterogeneous coating surface consisting of flattened particles produced by cold welding with less interparticle contact is formed. When the rotating speed was 300 rpm, the coating became denser and a smooth, highly consolidated and dense coating is formed. The hardness of the Al-Cr coating prepared at 200 and 300 rpm are about 250 Hv and 270 Hv. The friction coefficient of the Al-Cr coating prepared at 200 rpm are about 0.37, 0.39 and 0.24 at the frequencies of 3, 4 and 5 Hz. The friction coefficient of the Al-Cr coating prepared at 300 rpm are about 0.3, 0.18 and 0.28 at the frequencies of 3, 4 and 5 Hz.

Acknowledgements

This work was financially supported by the Chongqing Education Commission (KJ110816).

References

[1] S. Romankov, W. Sha, S. D. Kaloshkin, K. Kaevitser, Fabrication of Ti-Al coatings by mechanical alloying method, Surf. Coat. Technol. 201 (2006) 3235-3245.

[2] S. Romankova, Y. Hayasakab, I. V. Shchetininc, et al., Investigation of structural formation of Al-SiC surface composite under ball collisions, Mater. Sci. Eng. A, 528 (2011) 3455-3462.

[3] C. C. Wu, F. B. Wu, Microstructure and mechanical properties of magnetron co-sputtered Ni-Al coatings, Surface & Coatings Technology,204 (2009) 854-859.

[4] R. Pouriamanesh, J. Vahdati-Khaki, Q. Mohammadi, Coating of Al substrate by metallic Ni through mechanical alloying, Journal of Alloys and Compounds, 488 (2009) 430-436.

[5] S. Romankov, S. D. Kaloshkin, Y. Hayasaka, et al., Effect of process parameters on the formation of Ti-Al coatings fabricated by mechanical milling, J. Alloys Comp. 484 (2009) 665-673

[6] X. Yan, G. Xu, Effect of surface modification of Cu with Ag by ball-milling on the corrosion resistance of low infrared emissivity coating, Mater. Sci. Eng. B, 166 (2010) 152-157.

Key Engineering Materials Vol. 537 (2013) pp 283-287
© (2013) Trans Tech Publications, Switzerland
doi:10.4028/www.scientific.net/KEM.537.283

Effects of Nanoparticles on the Properties of Chrome-free Dacromet Coatings

Yiping Tang, Chengdong Wei, Guangya Hou, Huazhen Cao, Guoqu Zheng[a]

College of Chemical Engineering and Material Science, Zhejiang University of Technology, Hangzhou, Zhejiang, 310014 , China

[a] zhenggq@zjut.edu.cn

Keywords: TiO_2 nanoparticles, SiO_2 nanoparticles, Chrome-free Dacromet.

Abstract. Chrome-free Dacromet coating is a green anti-corrosion technology. This paper studied the effects of nanoparticles of TiO_2 and SiO_2 on the properties of chrome-free Dacromet coatings. The results shown the addition of nano-TiO_2 (T-coating) improved the physical properties and corrosion resistance. The addition of nano-SiO_2 (S-coating) was week bonded with zinc and aluminium powders, which resulted in the poor properties.

Introduction

Today, with the growing awareness of environmental protection, the green protective coating technology is urgent need for the modern industry[1-3]. Chrome-free Dacromet is a kind of new anti-corrosion coating, it has good corrosion resistance, no hydrogen brittleness and strong covering capacity, which was widely applied in automobile, ship craft, metal pipes and equipment, containers, and etc. Compared with conventional Dacromet, it doesn't contain chromium anhydride, so it is environment friendly[3-5]. However, the properties of corrosion resistance and hardness are still not as good as conventional Dacromet coating. Recently, chrome-free Dacromet was studied to improve the properties in order to meet the various environment of coating performance requirements. Many researchers reported that adding nanoparticles is an effective approach to obtain high performance chrome-free coating[6-7]. This paper focused on the effects of TiO_2 and SiO_2 nanoparticles on the chrome-free Dacromet coatings, the physical properties and corrosion resistance were studied in detail.

Experimental

When preparing Chrome-free Dacromet coating, two kinds of metal slip and passivation solution was divided and mixed. The component was shown in Table 1. Two types of nanoparticles TiO_2 and SiO_2 were added into the coating, which called T-coating and S-coating, the blank coating (B-coating) was compared. Because the viscosity of the obtained coating has been appropriate, it is no longer need to add thickening agent in the coating. The process for preparation of the coating uses two-coating, two-bake technology. The sintering condition is 110 °C drying 10min, then heated to 270 °C, after keeping 15 min with furnace cooling.

Three-electrode system was used for electrochemical test. The auxiliary and reference electrodes were platinum and saturated calomel electrodes respectively. The working electrode was the prepared sample. And the sample was sealed plastic, making the exposed area of 1cm × 1cm, the electrolyte solution was 3.5 mass% sodium chloride. The tests were carried at room temperature. The scan rate of TAFEL curve was 1 mV / s, the scan range was open circuit potential +0.2 V to open-circuit potential - 0.2V. The AC impedance was under open-circuit potential, the amplitude of the AC potential was 5mV and the frequency range from 10^{-2} to 10^5.

Table 1 Raw material ratio of chrome-free Dacromet coating

Component	Material	Additive amount(g)
Metal slurry	Zinc powder	24.45
	Aluminite powder	7.33
	Tween-20	7.33
	PEG-200	17.11
	Vinylsilane	1.06
	Absolute ethyl alcohol	5
	Nanoparticles	2
Passivation solution	Sodium molybdate	2.24
	Boric acid	0.37
	Sodium borate	0.56
	Sulphur silane	2.12
	Polyether modified siloxane	0.5
	Deionized water	37.4

Table 2 The effects on thickness, coated weight and adhesion by different nanoparticles

Properties	T-coating	S-coating	B-coating
Thickness(μm)	17.76	28.29	16.4
Coated weight (g/m^2)	42.28	58.69	39.11
Adhesion level	0	1	1

Table 3 The effects on hardness by different nanoparticles (unit: Hv)

Sample No.	T-coating	S-coating	B-coating
1	137	107	128
2	148	109	121
3	139	103	115
Average	141	106	121

Result and Analysis

Physical performance. Table 2 is the effects of different nanoparticles on thickness, coated weight and adhesion. From table 2, we can find that the thickness and the coated weight of the adding nanoparticles coating is thicker and larger than that of B-coating. The adhesion of the T-coating is the best, after cross cut test, there is little cross-cut area. The S-coating and the B-coating have the worse adhesion, and they all have a little cross-cut area. The thickness of T-coating is similar with B-coating, because this kind of nanoparticle is hydrophilism, they can absorb with the flake powder and fill the space in the coating. But the SiO$_2$ nanoparticle is hydrophobic, the consistency with the zinc and aluminium powder is poor, so the coating is thicker, and out of flatness.

Table 3 is the effects on hardness by different nanoparticles. TiO$_2$ is a kind of hard nanoparticle and fill in the space of flake metal powders, it makes the coating more compact. Because of the poor consistency of SiO$_2$ and flake metal powder, more space appeared in the coating, which resulted in the reduction of hardness.

Result of neutral salt spray test. The neutral salt spray test was used for measuring the anti-corrosion property of the coatings. The coatings can last more than 240h in the tests. After 260h, the S-coating disbonded on the surface and the red rust appeared around the edge. The B-coating also fell off after 260h with the white rust appeared, and after 350h, the red rust appeared. The T-coating had any rust and kept completely after 500h. So T-coating has the best corrosion resistance, and SiO$_2$ nanoparticles will make the corrosion resistance of the coating poorer. In conclusion, the T-coating is

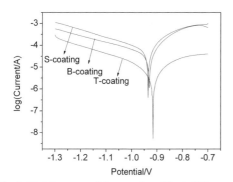

Fig.1. Tafel curve of the coating filled different nanoparticles

most compact, and aggressive medium is hardest to meet the metal base parts. And physical shield performance directly affect the corrosion resistance during the neutral salt spray test, so the T-coating showed the best corrosion resistance.

TAFEL curve. Figure 1 is the tafel curve of the coating filled different nanoparticles, and table 4 shows the free-corrosion potential and free- corrosion current density of the tafel curve. According to table 4, the free-corrosion potential of three coatings is fairly near, are -0.93V(blank), -0.935V(SiO$_2$) and -0.915V(TiO$_2$). The free-corrosion current density of the T-coating is the smallest, and the value is 7.99 µA/cm^2. The free-corrosion current density of the S-coating and B-coating are 53.99 and 43.6 µA/cm^2. On the whole, the T-coating has the best corrosion resistance in the early part of corrosion.

Eletrochemical impedance spectroscopy. Fig.2 shows the eletrochemical impedance spectroscopy of the coating with the addition of different nanoparticles. At first, determining the number of time constant in impedance spectra, which is very important to study the reaction mechanism of the surface of the eletrode and the state variable. Generally, from the bode curves, we can determine the number of time constant by the number of peaks. From Fig.2(c), it can be seen that two peaks appear in the bode figure of the three coatings, which demonstrates there are two time constants of the three coatings.

Table 4 Free-corrosion potential and free-corrosion current density of the tafel curve

Samples	Free-corrosion potential(V)	Anode tafel slope (V/DEC)	Cathode tafel slope (V/DEC)	Free-corrosion current density (µA/cm^2)
T-coating	-0.915	-3.4777	6.9182	7.99
S-coating	-0.935	-3.6364	12.609	53.99
B-coating	-0.93	-2.9433	10.196	43.6

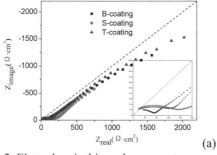

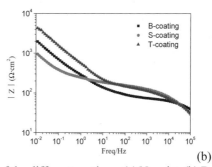

Fig.2. Eletrochemical impedance spectroscopy of the different coatings: (a) Nyquist, (b) Bode

As a result of the layer-by-layer structure of chrome-free Dacrome coating, the high frequency region corresponds to the impedance on the coating itself, which is related to the physical shielding property. However, the impedance of the low frequency region corresponds to the reaction condition of zinc aluminum powders, which is related to the corrosion condition. Thus, as Fig.2(b) shown, the impedance of the coating of adding nanoparticles is better than the blank coating. This is because the addition of nanoparticles makes the coating more thicker, then improves the physical shielding property of the coatings. The addition of TiO_2 nanoparticles makes zinc aluminum powders more protective and more corrosive. While the addition of SiO_2 nanoparticles makes zinc aluminum powders less protective and less corrosive compared to the adding of nothing. This is possibly because TiO_2 nanoparticle is well bonded with zinc-aluminum powders,conseqencially increasing the compactibility of the coating and protecting zinc aluminum powders.

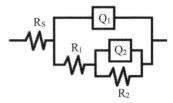

Fig.3 The equivalent circuit for electrochemical impedance spectra of the three coatings.

The impedance spectra were fitted by using the equivalent circuit R(Q(R(QR))) shown in Fig.3. It can be seen from the picture that it is perfect to use the R(Q(R(QR))) model to fit the three coatings. Rs refers to the solution resistance , Q_1 refers to the capacitance of the coating,R_1 refers to the resistance of the coating,Q_2 refers to the capacitance produced by the surface of the zinc aluminum powders transfer charge, R_2 refers to the resistance produced by the surface of the zinc aluminum powders transfer charge.

The data of the impedance spectra of the three coatings by fitting are listed in Table 5. It shows that R_2 of the T-coating electrode is the largest among the three electrodes. The charge-transfer resistance reflects how severe the chemical reaction goes on the surface of the electrodes. The larger electric resistance indicates the larger the reaction resistance force. So the largest charge-transfer resistance indicates that the corrosion property is the best of all. R_2 of the adding of SiO_2 elctrode is the smallest among the three electrodes,demonstrating that the corrosion property is the worst. The same results were obtained from the salt spay test and the Tafel curves..

Table 5. Impedance parameters of the different coatings

Samples	$Rs(\Omega)$	$Q_1(F/cm^2)$	n_1	$R_1(\Omega)$	$Q_2(F/cm^2)$	n_2	$R_2(\Omega)$
S-coating	7.438E-7	8.258E-4	0.4343	243.9	2.543E-4	0.3974	8.656E14
T-coating	7.897E-7	1.315E-5	0.5772	159.2	0.004539	0.4691	3.886E15
B-coating	1.001E-7	3.309E-6	0.6305	70.87	0.001865	0.4655	1.001E15

Conclusions

(1) Among the different coatings, the physical performance of T-coating is the best, S-coating is the worst one. It because that TiO_2 nanoparticles can bond with chrome-free Dacromet coating very well, however, SiO_2 nanoparticles are not and result in the reduction of performance. .

(2) The corrosion resistance of T-coating is highly enhanced by the addition of nanoparticles of TiO_2. But the addition of nanoparticles of SiO_2 has inappreciate effects, thus the S-coatings exhibit almost the bad corrosion resistance.

References

[1] C. Zhong, M.F. He, L. Liu, *et al.*, Formation of an aluminum-alloyed coating on AZ91D magnesium alloy in molten salts at lower temperature, Surf. Coat. Tech., 205 (2010) 2412-2418.

[2] C. Zhong, F. Liu, Y.T. Wu, J.J. Le, L. Liu, M.F. He, J.C. Zhu, W.B. Hu, Protective diffusion coatings on magnesium alloys: A review of recent developments, J. Alloy. Compd., 520 (2012) 11-21.

[3] J. Lu, Y. Liang, S.W. Tang, Study on process and corrosion resistance of chrome-free Dacromet coating, Appl. Chem. Industry (In Chinese), 40 (4) (2011) 612-613.

[4] Z.D. Wang, C. Lin, M. Q. Jiao, Study of non-chromate Dacromet process, Electrop. Pollut. Contr. (In Chinese), 28 (5) (2008) 34-35.

[5] L.Y. Ji , Z.J. Yao, Preliminary Reseach on the Technique of non-chromate Dacromet Coating. Electrop. Pollut. Contr.(In Chinese), 31 (3) (2011) 22-24.

[6] S.Y. Zhang, Q. Li, B. Chen, X.K Yang, Preparation and corrosion resistance studies of nanometric sol-gel based CeO_2 film with a chromium-free pretreatment on AZ91D magnesium alloy, Electrochim. Acta, 55 (2010) 870-877.

[7] Q.H. Zheng, X.H. Li, X.M. Song, Z.S.Wu, Effect of SiO_2 nanoparticles on performance of Dacromet coating, J. Mater. Protect. (In Chinese), 39 (11) (2006) 14-17.

Key Engineering Materials Vol. 537 (2013) pp 288-291
© (2013) Trans Tech Publications, Switzerland
doi:10.4028/www.scientific.net/KEM.537.288

The Preparation and Characterization of Copper Silicon Alloy Films

Xiuli Feng [a], Bijin Li[b]

Institute of Electronic Engineering ,China Academy of Engineering Physics, Mianyang, Sichuan 621900, P.R. China

[a]fen-33@163.com, [b]410965493@qq.com

Keywords: Copper silicon alloy;   Pulsed laser deposition; Thin-film batteries

Abstract. The fabrication, physical and electrochemical properties of copper silicon alloy film were systematically studied and the optimized preparation conditions were obtained. When KrF excimer laser (λ=248 nm) repetition rate is 10Hz, laser energy is 260mJ, vacuum is 10^{-5}Pa, target and substrate revolution per minute of target is 10, substrate temperature is 300°C and deposition time is 1.0h, as-deposited film from Si target and Cu substrate is obtained by pulsed laser deposition. XRD pattern showed that as-deposited film is cubic structure of Cu_9Si, SEM images showed that as-deposited film has regular surface, particle size is about 300nm and particle size distribution is in narrow range. And the same time the electrochemical properties of as-deposited film showed good cycleability.

Introduction

With the development of the miniaturized electronic devices, thin-film batteries (TFBs) are of considerable interest as power sources [1]. The all-solid-state thin film lithium ion rechargeable batteries have several attractive features including possible integration of battery fabrication with that of the microelectronic devices designed for one specific purpose such as smart cards. An additional secondary lithium microbattery provides backup power and allows the digital memory states to remain unaffected during power failure or storage. Several operational designs including electrodes and electrolyte materials of the all-solid-state thin film lithium ion rechargeable batteries have been reported.

Si, Ge and Sn have attracted much attention as anode materials in lithium-ion secondary battery since such electrode possess higher theoretical specific capacity than that of graphite counterparts. Among the three alternatives, Si has the highest theoretical specific capacity, about 4200mAh/g. However Si shows large volume expansion after alloying with Li (charging) and is low conductivity, which hinder its commercial use[2] .To improve Si properties, many attempts have been made and proven to effective , such as synthesis of metal silicate alloy, small-scaled Si and use of conductive additive[3-7]. Nuli [7] reported Cu5Si-Si/C composite electrodes deliver larger reversible capacity (500mAh/g at 40th cycle with 0.2mA/cm2) than commercialized graphite (370mAh/g) and better cycleability (columbic effective 74.5% at 40th cycle) than silicon (14.1% at 25th cycle). Copper in the composite acts as not only the constitute of alloy, but also conductive additives. The content of copper in composite is higher, cycling performance is better, but reversible capacity of composite electrode is smaller. The film electrode fabricated by Si deposited on rough Cu current collector with physical methods shows superior electrochemical properties, the capacity up to 1500mAh/g after 30 cycles, which is higher than that of Si powder and silicate alloy powder. But the structure and component of the film have not been systematically studied.

In this work, we have fabricated the film through Si depositing on Cu foils by varying pulsed laser deposition conditions. The crystal structure, SEM micrograph and electrochemical performance of the samples were systematically studied.

Experimental

Fabrication of copper silicon alloy thin films. In pulsed laser deposition Si wafer (99% prrity, 350um thickness) as target was mounted on a rotating holder . A KrF excimer laser (λ=248 nm, 260mJ,repetition rate of 10Hz) source was used to ablate the target, thus Si was deposited on rough Cu foil. At first, the vacuum chamber was evacuated down to less than 10^{-5}Pa, the substrate was 25 °C or 300°C, Si deposition time was 1.0h, the as-deposited film was obtained .The as-deposited film was subjected to anneal at 500°C for 20 minutes, thus, the as-deposited film and sintered film were obtained.

Physical and electrochemical analysis of copper silicon alloy thin films. The morphology of the thin film was measured with SEM (Nova600i, FEI Corporation).X-ray diffraction patterns of the films were determined by Philips X diffractometer with Cu-Kα radiation source. The electrochemical performance of thin film electrodes was measured using the same method as described in Reference [8]. A two-electrode cell with Li metal foil as t anode and the films as cathode were used for charge-discharge electrochemical measurements with LAND BT1-10 series battery test system. Cut-off voltages were 1.4 and 0.1V for the charge and discharge at a constant current.

Results and Discussion

Fabrication of copper silicon alloy thin films. Figure 1 shows the SEM images of as-deposited films at different substrate temperature by pulsed laser deposition. The SEM images of film deposited at 25°C(Fig.1a) shows the single particle is square in shape, the size was abou 300 nm .many single particles swam to form irregular polygon agglomerates ,and there are a 300nm width cleft between two agglomerates,which indicated the surface of film is not smooth and the film is not well-knit. Figure 1b displays the SEM images of film deposited at 300°C,where the as-deposited film has a rough and regular surface, particle size is about 300nm and particle size distribution is in narrow range.It can be concluded from Fig1 that the substrate temperature not only affect film morphology,but also affect film intensity.

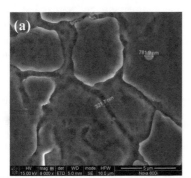 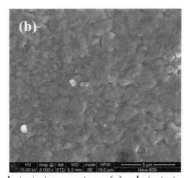

Fig. 1 SEM images of as-deposited films at different substrate temperature: (a) substrate temperature of 25°C (b) substrate temperature of 300°C

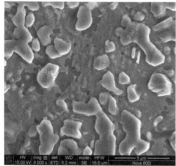

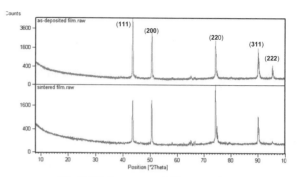

Fig. 2 SEM images of sintered film   Fig 3 XRD pattern of film samples

The film deposited at substrate temperature of 25°C was poor so that it could not endure volume expansion and shrinkage during the insertion and extraction of lithium-ions, so presumably the cycling performance of the film is not well. The film deposited at substrate temperature of 25°C was in-situ sintered at 500°C for 20min,thus the sintered film was obtained and its SEM images was seen in figure.2.

The sintered film has an irregular surface, particle size distribution is not in narrow range, the biggest size is 3um,the smallest one is 200nm.There appears to be significant morphology change between as-deposited film and sintered films and it is indication of evolution of particles, which indicates the annealing help to particles grown-up.

To date, the methods of Si films deposited on rough Cu foil is radio frequency magnetron sputtering, direct current magnetron sputtering, pulsed laser deposition, electron beam evaporation, physical vapor deposition and so on. The as-deposited Si films was not characterized by XRD, but people considered as-deposited film to be Si film. To verify the component and structure of films by Si depositing on Cu foil, the films were characterized by XRD.

Fig.3 showed XRD patterns of film samples. The red pattern stands for the pattern of film deposited at 300°C, the other is the film deposited at room temperature and sintered at 500°C. It is clear that diffraction peaks of two films are basically identical in fig.3. Diffraction peaks of two films at 43°,50°,75°,90°and 95° can be indexed to the cubic structure of Cu_9Si. The diffraction peaks of silicon can not be detected, which suggest Si fully react with cu to copper-silicon alloy. The sintered film is (220) plane orientation, while as-deposited film is not preferentially oriented, which suggest that fabrication condition affect the film structure..

Electrochemical analysis of copper silicon alloy thin films..Fig.4 showed the relationship between voltage and the first discharge capacity of as-deposited film at $50uA/cm^2$. The initial discharge specific capacity is $180uAh/cm^2$,the second discharge capacity is higher than that of first charge, but the reason can not make clear. The discharge capacity decreases slowly from 3 to 19 cycles, the 19th discharge capacity keeps $129uAh/cm^2$, the lowest retainable efficiency of discharge capacity still is 68.5%,the columbic efficiency is close to 100%.

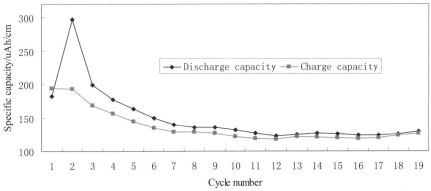

Fig.4 Relationship between cycle number and discharge capacity at 50uA/cm^2

Conclusions

It is firstly reported Cu$_9$Si alloy film was successfully synthesized from Si deposited on Cu foil by pulsed laser deposition and the optimized fabrication preparation conditions were obtained. When the substrate temperature is 300°C, the as-deposited film not only has a rough and regular surface, cubic crystal structure, but also exhibits good electrochemical performance. The first discharge capacity of as-deposited films is 180uAh/cm^2,the 19th discharge capacity is still 129uAh/cm^2,and the capacity loss per cycle is about 1.6%.When substrate temperature is 25°C, the as-deposited film has poor surface micrograph. By in-situ sintering the film deposited at substrate temperature of 25°C, the sintered film of (220) plane orientation was obtained. And the sintered film has an irregular surface, particle size distribution is not in narrow range. It is concluded fabrication condition had an influence on structure and morphology of copper silicon alloy film.

Acknowledgement

This work was supported by the Key foundation of Academy of Engineering Physics (2007A05002).

References

[1] J.B. Bates, N.J. Dudney, B. Neudecker, et al. Thin-film lithium and lithium-ion batteries[J]. Solid State Ionics , 2000 (135): 33–45

[2] T.Sunano,Y.Takano,H.Yamamoto,et al. deterioration mechanism on cycle performance of Li-ion battery with alloyed metal anode. The 14th international meeting om lithium batteries,2008:122.

[3] J.M.Yan,H.Z.huang,J.Zhang,et al. The study of Mg2Si/carbon composites as anode materials for lithium ion batteries. Journal of power sources,2008(175):547-552

[4] H.B. Yang,P.P. Fu, H.F. Zhang, et al. Amorphous Si film anode coupled with LiCoO$_2$ cathode in Li-ion cell. journal of power sources,2007(17):533-537

[5] W.R Liu, Z.Z. Guo, W.S Young, et al. Effect of electrode structure on performance of Si anode in Li-ion batteres: Si particle size and conductive additive.Journal of power sources,2005(140):139-144

[6] K. Wang, X.M.g He, L Wang, et al. Si,Si/Cu core in carbon sheet composite as anode material in Lithium-ion batteries. Solid-state-ionic,2007(178):115-118

[7] Y.N. Nuli, B.F. Wang, J. Yang,et al. Cu$_5$Si-Si/C composites for lithium-ion battery anodes. Journal of power sources,2006(152):371-374

[8] X.L.Feng,L.Q.Li , F.M.Meng , et al. Study on Preparation and Properties of High-purity LiCoO$_2$ Powder. The 14th international meeting on lithium batteries,2008,363.

Key Engineering Materials Vol. 537 (2013) pp 292-297
© (2013) Trans Tech Publications, Switzerland
doi:10.4028/www.scientific.net/KEM.537.292

Oxidation Resistance and Plating Encapsulation of Cu-based Alloys as Phase Change Materials for High-temperature Heat storage

Guocai Zhang[1,3], Jianqiang Li[1,a], Bingqian Ma[1,4], Zhe Xu[1,3], Zhijian Peng[4] and Yunfa Chen[2]

[1]National Engineering Laboratory for Hydrometallurgical Cleaner Production Technology, Institute of Process Engineering, Chinese Academy of Sciences, Beijing 100190, PR China

[2]State Key Laboratory of Multi-phase Complex System, Institute of Process Engineering, Chinese Academy of Sciences, Beijing, 100080, PR China

[3]Graduate University of Chinese Academy of Sciences, Beijing 100039, PR China

[4]School of Engineering and Technology, China University of Geosciences, Beijing 100083, PR China

[a]jqli@home.ipe.ac.cn

Keywords: Phase change materials (PCMs); Thermal energy storage; Oxidation resistance; Copper; Plating

Abstract. Cu-based alloys have been regarded as one of the most promising phase change materials (PCMs) in industrial waste heat recovery and solar thermal electric generation. In this paper, the oxidation behavior and the containment of liquid Cu were investigated. It was found that with the small addition of aluminum, the oxidation resistance of Cu-based PCMs was greatly enhanced. Notably, its latent heat density remained high. The containment of PCMs was achieved by depositing a Ni-base exterior coating on Cu spheres through barrel plating, rack plating and electroless plating processes. The deposition rates, surface topography, and the crystallography of the coatings depended largely on the plating process. The cyclic thermal was tested at last.

Introduction

Thermal energy storage (TES) technology has attracted considerable attention in the last 20 years with the worldwide growing of concerns on conserving energy and utilizing renewable energy resources [1]. Most high temperature industrial processes are batch operation and emit high temperature waste heat intermittently. For example, in the steelmaking process, a large amount of high temperature waste heat above 1000 °C exists in the blast furnace (BF) slag. However, due to technical difficulties, only 30% to 40% of the waste heat was recovered in spite of their high energy level and large potential [2]. The great motivation of developing TES technology is the rapid development of solar thermal electric technology. TES system has been primary in building a long-term solar thermal electric plant since the solar energy is extremely unstable and an efficient TES system would contribute to circumvent problems associated with the mismatches between energy supply and demand. Since the conventional methods for steam generation are not available in long-distance heat transport and long-term heat storage, new TES technologies are strongly required either for the recovery of waste heat or for the utilization of solar energy [3].

Latent heat storage (LHS) is the most promising TES technology. The heat storage medium of LHS is phase change materials (PCMs) which endows it many advantages. A variety of PCMs have been developed including paraffin, hydrate salts, fused salts, metals and alloys. From the viewpoint of exergy theory [4,5], thermal energy should be stored and released at the highest temperature to maximize its utilization efficiency during transforming in other processes. Therefore, the high temperature PCMs above 1000 °C are strongly required to recover industrial waste heat and solar energy efficiently (newly reported working temperature of solar power has been up to 1000 °C [6]). Both molten salts and metals (alloys) have the suitable melting point and liquids while metals possess the unique advantages, such as much higher thermal conductivity, high melting temperature, high heat of fusion, good thermal and chemical stability [7].

The binary and multi-component metallic systems among common elements such as Al, Ca, Cu, Mg, P, Si and Zn were investigated [8, 9]. Considering the cost effectiveness, Cu may be the most promising candidate for high temperature heat storage above 1000 °C [10]. The successive work was performed on the affordability of recovering the high temperature waste heat from LDG for producing hydrogen by using pure Cu as PCMs. However, till now, there are still many obstacles for Cu as a practical high-temperature LHS medium. For example, pure Cu is easily subject to oxidation at oxygen-contained atmosphere which would be crucial in next-generation high temperature solar thermal electric process [6]. On the other hand, the containment of liquid Cu is indispensable. An alternative means is external coating or encapsulation. Therefore, preliminary efforts were conducted for improving the oxidation resistance of Cu and encapsulating processes of Cu spheres with inertly metallic shell.

Experimental

The copper and aluminum powders with 99.99% purity were used as starting materials. The pure copper and Cu-5 wt%Al (designated as Cu-5Al) alloy were melt in a quartz tube by high-frequency induction under evacuated argon atmosphere. The oxidation rates of the pure copper and its alloy ingots were measured with thermogravimetry (TG) in an alumina crucible. Both melts were dwelled in air at 55 °C above their melting point (Cu at 1085 °C) or liquidus (Cu-5Al at 1076 °C) for 50 min. The weight gain of the melts by oxidation was recorded online and the oxidation rates could be calculated. Moreover, the melting point were tested and analyzed by a differential scanning calorimeter (DSC, STA 449C, NETZSCH, Germany). Latent heat density of Cu-5Al was calculated by Thermo-Calc software.

The encapsulations of commercial copper sphere (Φ2 mm, 99.99%) were conducted by various methods which were selected owing to their capability for large-scale production. The morphologies of the samples were observed by an optical microscope (OM, BX51M, OLYMPUS, Japan) and scanning electron microscope (SEM, SSX-550, Shimadzu, Japan), and the phases were determined by using an X-ray diffraction (XRD, X'Pert PRO MPD, PANalytical, Holland).

Results and Discussion

Oxidation resistance of Cu–base alloy. Fig. 1 shows the mass changes of Cu and Cu-5Al melts during their exposure in air. It indicated that the pure copper melt was oxidized rapidly, and the oxidation process accelerated when the time lasted for 1800 s. In contrast, the oxidation of Cu-5Al alloy proceeded very slowly and remained a very low level during the whole process. The curve of the oxidation of Cu is a parabola. According to theory of the metal oxidation [11], copper (or oxygen) diffuses through the oxide layer driven by a concentration gradient at high temperature. The layer thickness X at a time t follows the equation

$$X^2 = 2At \qquad (1)$$

A is a constant related to concentration gradient.

Because the possible reactions between liquid Cu-Al alloy and alumina crucible wall would not result in mass change, the mass change in Fig.1 was completely attributed to the oxidation of melt bath which occurred on the melt surface. The thin compact Al_2O_3 film on surface could provide excellent protection for most Al alloys against isothermal oxidation. Tomaszwicz et al. [12] confirmed that a Fe-Al alloy could form a single phase Al_2O_3 film to protect the alloy from high temperature oxidation with more than 7 wt% aluminum. In our experiments, aluminum diluted in Cu-5Al alloys

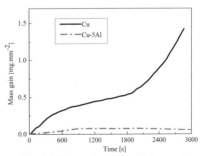

Fig. 1 Mass gain of liquid Cu and Cu-5Al alloy during thermal exposure in air

had greater affinity with oxygen than that of copper matrix, which made Al_2O_3 film format firstly on the melt surface and then prevented the oxidation process of the melt. According to the reference [13], the enhancement of oxidation resistance by Al_2O_3 film depended on not only kinetic migration rate between oxidant and solid oxide film, but also the morphology of oxide film especially such as compactness and density. Fig. 2 shows the surface morphologies of Cu and Cu-5Al after thermal exposure experiments. It shows that a dense layer coated on Cu-5Al alloy surface which consisted of flake-like crystalline grains with size of 1~2 μm. The grain morphology was similar with that of activated alumina reported in ref. [14]. The EDS analysis shows that the grains on Cu-5Al alloy surface contained mainly Al and O elements besides a portion of Cu element. The results indicated that the oxidation resistance of Cu-base phase change materials could be enhanced largely by an easy approach of Al addition.

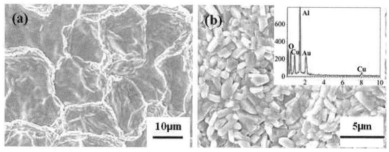

Fig. 2 SEM micrograph of (a) Cu and (b) Cu-5Al alloy after thermal exposure in air

Table 1 Conditions of electroless plating, rack plating and barrel plating

Electroless plating		Rack plating		Barrel plating	
pH Value	3.9	pH value	3~4	pH value	4.5~4.8
Temperature	70 [°C]	Temperature	40 [°C]	Temperature	55 [°C]
$NiSO_4 \cdot 6H_2O$	15 [g/L]	$NiSO_4 \cdot 6H_2O$	250~300 [g/L]	$NiSO_4 \cdot 6H_2O$	280~320 [g/L]
$NaH_2PO_3 \cdot H_2O$	14 [g/L]	$NiCl_2 \cdot 6H_2O$	30~60 [g/L]	$NiCl_2 \cdot 6H_2O$	40~50 [g/L]
$NaCH_3COOH \cdot 3H_2O$	9 [g/L]	H_3BO_3	35~40 [g/L]	H_3BO_3	40~50 [g/L]
		$C_{12}H_{25}\text{-}OSO_3Na$	0.05-0.1 [g/L]	$C_{12}H_{25}\text{-}OSO_3Na$	0.1~0.15 [g/L]
		Electric current	1~2.5 [A/dm²]	Electric current	0.1~0.7 [A/dm²]

The latent heat density and melting point of Cu-5Al, Cu and Al were compared. The melting point of Cu-5Al was 1055 °C tested by DSC. It revealed that the melting point of Cu-5Al was lower than that of Cu by 30 °C, which is consistent with the Cu-Al phase diagram. The latent heat density of Cu-5Al is 200.38 J/g calculated by the Thermo-Calc software, which near to that of Cu. The addition of Al not only enhanced the oxidation resistance of Cu-base alloys but remained its large latent heat density and relative low melting point. The large latent heat density and relative low melting point of Cu-5Al was great in favor of recovery of high temperature heat and utilization of solar energy.

Encapsulation of Cu PCMs. Regin et al. [15] noted that a desired PCMs containment should meet the following requirements: (i) providing a sufficient surface for rapid thermal energy exchange; (ii) meeting the requirements of strength, flexibility, corrosion resistance, and thermal stability; (iii) acting as a barrier to protect the PCMs from harmful interactions with the outer environment, and (iv) providing structural stability and easy handling. An alternative method of retaining Cu alloy in a porous ceramic matrix was hindered by the inherent low thermal conductivity. Coating a rigid metallic layer on Cu alloy surface can provide an encapsulation approach which would not cause a dramatic drop of the thermal conductivity of Cu-base PCMs. Moreover, the produced PCM granules (e.g. in spherical form) are particularly suitable for direct contact heat transfer process which is in favor of the next generation direct steam generation (DSG) solar power technology.

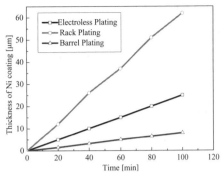

Fig. 3 Coating thickness vs. time for electroless plating, rack plating and barrel plating

The processes of electroless plating, rack plating and barrel plating were investigated to examine their ability for the encapsulation of Cu PCMs. The experimental conditions are listed in Table 2. The nickel-base alloys were selected as containment materials due to their high melting point, excellent oxidation resistance and particular feasibility for large-scale plating. The deposition rates of nickel coating by three plating methods were investigated and the results are showed in Fig. 3. A linear relation could be seen between thickness of nickel coating and plating time, which indicated that the deposition rates of the three plating processes were constant. Barrel plating possessed the lowest deposition rate while that of racking plating was the highest up to 37 μm/h. The coating thickness could be easily modulated by controlling the plating time.

Fig. 4a shows the cross-sectional optical morphology of nickel coatings prepared by rack plating. The coating was dense with few pores and other defects, and the bonding with the Cu sphere was very well. The SEM image of the as-deposited coating surface (Fig. 4b) shows that the coating consisted of 2~5 μm pyramid-like grains, which were almost pure Ni tested by EDS shown in the inset. Electroless plating process produced a much thinner coating while its bonding with Cu sphere seemed still good, as shown in the cross-sectional micrograph of Fig. 4c. The SEM image shows that the coating surface was very flat and only a small portion of grains could be detected. An EDS spectrum indicated that the composition of coating was about 87.5% Ni-12.5% P in molar ratio.

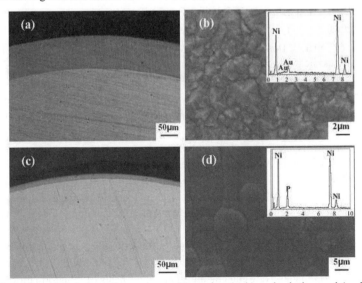

Fig. 4 Morphologies of nickel coatings prepared by using (a, b) rack plating and (c, d) electroless plating

To explore the thermal stability of the Cu-PCM capsules, a carbon layer was deposited on the Cu sphere prior to plating process in order to avoid the solution between Cu and Ni at high temperature. Then the capsules were exposed to cyclic heating and cooling from 800~1100 °C in air by using an electric furnace. The barrel-plated PCM capsules with Cu shell more than 70 μm thick showed the best high-temperature durability. It could endure at least fifty times thermal cyclic and no exudation was detected. This consequence can be attributed to the modest deposition rate, better stability during heating without phase transition, and homogeneity compared to the other processes. Moreover, barrel plating in continuous way and facile operation was very suitable for encapsulating Cu-based PCMs.

Conclusion

To facilitate Cu-base alloys as practical PCMs, both the oxidation behavior and the containment of liquid Cu were investigated in this paper. With the addition of aluminum (5 wt% Al) the oxidation resistance of Cu-base alloys was greatly enhanced while its melting point and latent heat density remained high. The containment of PCMs was achieved by coating Ni-base exterior shell on Cu spheres through barrel plating, rack plating and electroless plating processes. The coatings demonstrated high quality and excellent bonding with Cu substrate. A linear relation between thickness of nickel coating and plating time indicated that the deposition rates were constant. Barrel plating possessed the lowest deposition rate while that of racking plating was the highest up to 37 μm/h. The cyclic thermal tests validated that the rack-plated Cu capsules could endure at least fifty times thermal shock without exudation.

Acknowledgements

This work was financially supported by the National Natural Science Foundation of China (Grant No.50704031), Beijing Natural Science Foundation (No. 2112039). J. Li appreciates the support from the Beijing Nova Program (Grant No. 2007A086).

References

[1] H.O. Paksoy, Thermal energy storage for sustainable energy consumption: fundamentals, case studies and design, Kluwer Academic Publishers Group, Dordrecht, 2007.

[2] T. Akiyama, K. Oikawa, T. Shimada, E. Kasai, J. Yagi, Thermodynamic analysis of thermochemical recovery of high temperature wastes, ISIJ Int. 40 (2000) 286-291.

[3] R. Tamme, T. Bauer, J. Buschle, et al., Latent heat storage above 120°C for applications in the industrial process heat sector and solar power generation, Int. J. Energ. Res. 32 (2008) 264-271.

[4] M. Ishida, C.C. Chuang, New approach to thermodynamics, Energ. Convers. Manage. 38 (1997) 1543-1555.

[5] O.F. Dilmac, S.K. Ozkan, Energy and exergy analyses of a steam reforming process for hydrogen production, Int. J. Exergy 5 (2008) 241-248.

[6] A. Hoshi, D.R. Mills, A. Bittar, T.S. Saitoh, Screening of high melting point phase change materials (PCM) in solar thermal concentraing technology based on CLFR, Sol. Energy 79 (2005) 332-339.

[7] V. Morisson, M. Rady, E. Palomo, et al., Thermal energy storage systems for electricity production using solar energy direct steam generation technology, Chem. Eng. Process. 47 (2008) 499-507.

[8] C.E. Birchenall, A.F. Reichman, Heat storage in eutectic alloys, Metall. Mater. Trans. A 11 (1980) 1415-1420.

[9] D. Farkas, C.E. Birchenal, New eutectic alloys and their heats of transformation, Metall. Mater. Trans. A 16 (1985) 323-328.

[10] J. Yagi, T. Akiyama, Storage of thermal-energy for effective use of waste heat from industries, J. Mater. Process. Tech. 48 (1995) 1-4.

[11] N.F. Mott, A.S. Alexandrov, Sir Nevill Mott: 65 Years in Physics, World Scientific Publishing Company, Singapore, 1995.

[12] P. Tomaszewicz, G.R. Wallwork, The Development of oxidation resistant Fe-Al alloys, High Temperature Corrosion, In International Corrosion Conference, Houston, 1981.

[13] A.T. Fromhold, Theory of Metal Oxidation, Volume.1, Fundamentals, North Holland Publishing Company, Amsterdam, 1976.

[14] Y.Y. Park, S.O. Lee, T. TRAN, S.J. Kim, M.J. Kim, A study on the preparation of fine and low soda alumina, Int. J. Miner. Process. 80 (2006) 126-132.

[15] A.F. Regin, S.C. Solanki, J.S. Saini, Heat transfer characteristics of thermal energy storage system using PCM capsules: A review, Renew. Sust. Energ. Rev. 12 (2008) 2438-2458.

Key Engineering Materials Vol. 537 (2013) pp 298-301
© (2013) Trans Tech Publications, Switzerland
doi:10.4028/www.scientific.net/KEM.537.298

Copper Nanowires Preparation and Field Electron Emission Properties

Li-jun Wang [1,a], Can Yang [1,b], Zi Wang [2,c], Xiao-fei Liu [1,d], Xiao-ping Wang [1,e]

[1]College of Science, University of Shanghai for Science and Technology, Shanghai, China 200093

[2]School of transportation engineering, Tongji University, shanghai, China 200092

[a]wljpj@yahoo.com.cn, [b]yangcan0905@163.com, [c]wzyoyo89@yahoo.com.cn,
[d]liuxiaofei0802@126.com, [e]wxpchina64@yahoo.com.cn [Corresponding author]

Key words: Copper nanowires films, electron beam vapor deposition, field electron emission

Abstract. Copper (Cu) nanowires films are deposited on the molybdenum film-coated Al_2O_3 ceramic substrates by using the electron beam vapor deposition technique. The films were characterized by optical microscopy, scanning electron microscope, x-ray diffraction spectrum and energy dispersive spectrum. The surface morphology show that the Cu nanowires have excellent length-to-diameter ratio of ~300, and each Cu nanowires diameter is uniform. The field electron emission measurements of Cu nanowires films were also carried out showing the turn on field as low as 2.5 V/μm and the average current density of 0.10 mA/cm^2 at electric field of 10.8 V/μm were obtained from a broad uniform emission screen over 3.0 cm^2.

Introduction

Nano copper (Cu) has received significant attention by researchers due to their unique properties such as conductivity, melting temperature, magnetism, specific heat and light absorption. It has many important apply such as solid lubricant, cooling material, conducting material and so on [1-12]. Because of nano Cu has lower melting point and higher heat capacity, it can be used for cooling spacecraft components. Moreover act as conductive materials, nano Cu plays an important role on the miniaturization of microelectronic device.

The better-known methods for fabricating Cu nanowires are chemical solution-phase synthesis and electrochemical deposition of Cu inside polymer and anodic aluminium oxide channels[13]. In the present study we used electron beam vapor deposition (EBVD) method to grow Cu nanowires films, and scanning electron microscopy (SEM), x-ray diffraction pattern (XRD) and energy dispersive spectrum (EDS) were used to detect and analyse the effects of process parameters on morphologies and structures of the thin films. A direct-current (DC) power supply was used to drive the field electron emission device. The experimental results indicated that the optimized nano-structured Cu films were typically Cu nanowires films, the Cu nanowires had excellent length-to-diameter ratio of ~300, and each Cu nanowires diameter is uniform. A field emission current density ~ 0.10mA/cm^2 at 10.8 V/μm and a threshold electric field of 2.5 V/μm were obtained from a broad uniform emission screen over 3.0 cm^2.

Experimental and Discussion

Fabrication of Cu nanowires. Firstly, in order to obtain a rough surface morphology the Al_2O_3 ceramic substrates were first treated by Muller using abrasive with diameter of 10μm, the ceramics rubbed were in turn ultrasonically cleaned in 0.1% oxalic acid ethanol and deionized (DI) water for 10 min, then were wiped with 99.7% alcohol cotton ball. After that, EBVD system was used for Mo deposition on ceramic substrates, high purity (99.99%) metal Mo as deposition target. During the EBVD process, the whole deposition conditions were kept as follows. The beam of current electron gun and pressure were maintained at 200 mA and 2×10^{-2} Pa, and substrate temperature was kept 450°C. The thickness of Mo layer is 2μm controlled by films thickness monitor.

Then, the EBVD system was used to deposit Cu nanowires on the Mo film-coated ceramic substrates. The specific parameters used for the EBVD process were listed in Table 1.

Table 1. Parameters employed for the EBVD process

EBVD process parameters	values
Beam current /mA	220
Vacuum /Pa	2.3×10^{-2}
Substrate temperature /°C	550
Voltage /kV	6
Vapor time /min	30
Filament current /mA	0.8

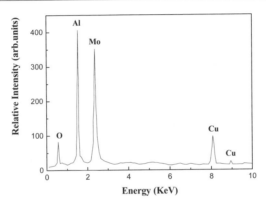

Fig.1. EDS of Cu nanowires film sample

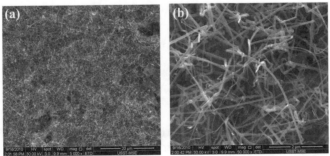

Fig.2 SEM images of the Cu nanowires deposited on Mo layer (a) and (b) are different magnifications.

Results and discussion.

The surface morphology, microstructure and surface resistance of Cu nanowires were characterized by SEM, XRD spectrum, EDS. The XRD spectrum of the Cu nanowires films contains the peaks of the Al_2O_3 ceramic substrates as well as the peaks of Mo layers, but there is no peak of Cu layer, we tend to think the main reason is that the Cu layer is too thin and the signal is too weak. Fig.1 shows the EDS of the sample films, in addition to appear the characteristic peaks of Al, Mo and O, but also appeared the peaks of the Cu . This indirectly proof that our samples contain Cu components.

The microstructure of Cu nanowires films deposited on Mo layer were characterized by SEM. Fig. 2 show the typical SEM images of Cu nanowires films. As can be seen from Fig.2, the nano Cu films were grown randomly over the entire surface of the Mo layer, and which were composed mainly of nanowires. The Cu nanowires have excellent length-to-diameter ratio of ~300. The diameter of the Cu nanowires in 20nm to 100nm between, and the Cu nanowires' lengths range from 3μm to 7μm. Furthermore, each Cu nanowires' diameter is uniform, and the Cu nanowires show very good uniformity and flexibility.

A functional field emission device was made using Cu nanowires film/Mo film/ Al_2O_3 ceramic as a cold cathode, glass/zinc aluminium oxide (ZAO) film with fluorescent coat as anode, keep apart cathode and anode by a 220μm insulating sheet of mica film. This nano-structured Cu film field emitter was placed in a vacuum system with a Hyvac pump operating at a base pressure of about 4.0×10^{-5}Pa.

Typical curve of emission current density (J)–macroscopic electric field (E) were depicted in Fig.3 (a). The threshold microscopic field, E_{th} defined as the one yielding J=1 μA/cm^2, was as low as 2.5 V/μm, and the average emission current density of 0.10mA/cm^2 at a macro-scopic electric field of 10.8 V/μm were obtained. The macroscopic FE areas of the measured samples were 3.0 cm^2. Fig.3.(b) are the Fowler–Nordheim (F-N) plots, obviously the nonlinearity were happened in F-N plot, which don't corresponding to classical electron tunneling mechanism in low E range (E<5.6 V/μm), when E>5.6 V/μm the nonlinearity character of the F-N plot were not happened, the F-N plot is straight line corresponding to classical electron tunneling mechanism.

These experimental results can be explained as follows. the lower E_{th} and nonlinearity character of the F-N plot in low E range can be attributed to the surface adsorption of some foreign gases, it is believed that its low E electron emission behavior, which does not belong to the real field emission, can be attributed to the high electronic work function of Cu material. Inversely, the F-N plot is straight line in high E range (>5.6 V/μm), which corresponding to classical electron tunneling mechanism.

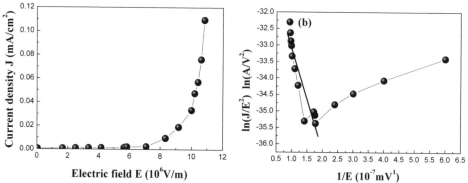

Fig.3. (a)Emission current density (*J*) versus electric field (*E*) plots of the measurements for a typical Cu nanowires film sample.(b) Fowler–Nordheim (F–N) plots from Cu nanowires film

Conclusions

In summary, Cu nanowires films were fabricated by the electron beam vapor deposition technique, the Cu nanowires have excellent length-to-diameter ratio of ~300, and each Cu nanowires diameter is uniform. An emission current density as large as 0.10 mA/cm^2 was attained at a macroscopic field of 10.8 V/μm from the Cu nanowires films. The macroscopic FE areas of the measured samples were over 3.0 cm^2.

Acknowledgments

This work is financially supported by Shanghai Human Resources and Social Security Bureau (No:2009023).

References

[1] N.A. Dhas, C.P. Raj, A.Gedanken, Synthesis, Characterization, and Properties of Metallic Copper Nanoparticles, Chem. Mater. **10** (1998) 1446–1452.

[2] J.H. Wang, T.H. Yang, W.W. Wu, L.J. Chen, C.H. Chen, C.J. Chu, Synthesis and growth mechanism of pentagonal Cu nanobats with field emission characteristics, Nanotechnology **17** (2006) 719–722

[3] P.I. Wang, Y.P. Zhao, G.C. Wang, T.M. Lu, Novel growth mechanism of single crystalline Cu nanorods by electron beam irradiation, Nanotechnology **15** (2004) 218–222

[4] C.J. Mao, X.J. Wang, X.C. Wu, J.J. Zhu, H.Y. Chen, Synthesis and field emission of single-crystalline copper vanadate nanobelts, Nanotechnology **19** (2008) 035607 -612

[5] K.J. Ganesh, A.D. Darbal, S. Rajasekhara, G.S. Rohrer, K. Barmak and P.J. Ferreira, Effect of downscaling nano-copper interconnects on the microstructure revealed by high resolution TEM-orientation-mapping, Nanotechnology **23** (2012) 135702-708

[6] Z.K. Liu, Y.F. Wang, Y.L. Liao, Gary J. Cheng, Direction-tunable nanotwins in copper nanowires by laser-assisted electrochemical deposition, Nanotechnology **23** (2012) 125602-608

[7] A.K. Chatterjee, R.K. Sarkar, A.P. Chattopadhyay, P.A.R. Chakraborty, T. Basu, A simple robust method for synthesis of metallic copper nanoparticles of high antibacterial potency against E. coli, Nanotechnology **23** (2012) 085103-113

[8] T.M.D. Dang, T.T.T. Le, E.F. Blanc, M.C. Dang, Synthesis and optical properties of copper nanoparticles prepared by a chemical reduction method, Adv. Nat. Sci.: Nanosci. Nanotechnol. **2** (2011) 015009-5014

[9] L. Bartoli, J. Agresti, M. Mascalchi, A. Mencaglia, I. Cacciari, S. Siano, Combined elemental and microstructural analysis of genuine and fake copper-alloy coins, *Quantum Electronics* **41** (2011) 663 – 668

[10] X.Z. Yu, Z.G. Shen, The growth mechanism of copper films coated on cenosphere particles using magnetron sputtering method, J. Phys. D: Appl. Phys. **41** (2008) 225409-416

[11] Z. Tang, T. Toyama, Y. Nagai, K. Inoue, Z.Q. Zhu, M.Hasegawa, Size-dependent momentum smearing effect of positron annihilation radiation in embedded nano Cu clusters, J. Phys.: Condens. Matter **20** (2008) 445203-207

[12] E. Falcon, B. Castaing, C. Laroche, Turbulent electrical transport in copper powders, *Europhys. Lett.* **65** (2004)186–192

[13] M.Y. Yen, C.W. Chiu, C.H. Hsia, F.R. Chen, J.J. Kai, C.Y. Lee, H.T. Chiu, Synthesis of cable-like copper nanowires, Adv.Mater. 15 (2003)235-237

Key Engineering Materials Vol. 537 (2013) pp 302-306
© (2013) Trans Tech Publications, Switzerland
doi:10.4028/www.scientific.net/KEM.537.302

Corrosion Inhibition and Adsorption Behavior of 4-((2-Thiophenecarboxylic acid hydrazide)methylene)benzoic Acid on Copper Surface in CO_2-saturated Oilfield Water

Shiliang Chen[1,2,a], Zheng Liu[1,2,b] and Jie Liu[1,2,c]

[1]College of Chemical and Biological Engineering, Guilin University of Technology, Guilin 541004, Guangxi, China

[2]Key Lab of New Processing Technology for Nonferrous Metals and Materials Ministry Education, Guilin 541004, Guangxi, China

[a]piaocsl@163.com, [b]lisa4.6@163.com, [c]liujie410306@126.com

Keywords: Copper; Acylhydrazone; Corrosion inhibitor; Electrochemical test; Inhibition mechanism

Abstract. The inhibition effect and inhibition mechanism of a 4-((2-thiophenecarboxylic acid hydrazide)methylene)benzoic acid (HD) against copper corrosion in CO_2-saturated oilfield water was evaluated using electrochemical techniques and scanning electron microscopy. The experimental results shown that HD is a good corrosion inhibitor and the inhibition efficiency increased with the increase of HD concentration, and research indicate that adsorption behavior of the HD followed the langmuir's adsorption isotherm and the adsorption mechanism is typical chemisorption.

Introduction

Nowadays, the corrosion inhibitor of copper is an important industrial and academic topic, it is used in petroleum production, chemical processingand refining, construction and metal-processing equipment. The dissolved oxygen and CO_2, Cl^-, CO_3^{2-}, HCO_3^-, SO_4^{2-} ions produced serious corrosion for copper. It is possible to reduce the rate of corrosion processes of copper by adding inhibitors and organic compounds and their derivatives containing sulfur, oxygen, and nitrogen heteroatoms were suggested as effective corrosion inhibitors [1-3]. The primary step in the action of organic corrosion inhibitors is usually attributed to the adsorption process [4]. The Hydrazone compounds have a strong ability of coordination, so it can form stable protective film in metals by coordination action that it can inhibit effectively the metal corrosion in saturated CO_2 oilfield water.

In the present work, we synthesized a new acylhydrazone derivative, the HD, which has the particularity to possess both 2-thiophenecarboxylic acid hydrazide and 4-formylbenzoic acid parts, as corrosion inhibitor in saturated CO_2 oilfield water. The presence of nitrogen and oxygen atoms in HD gives an opportunity to understand and explain the mechanism of the inhibition and the type of adsorption on the copper surface. The influence of HD as the corrosion inhibition for copper in saturated CO_2 oilfield water was investigated by electrochemical techniques: potential-time curves and polarization curves and scanning electron microscopy (SEM).

Experimental

Material preparation. Fig. 1 shows the molecular formula of investigated acylhydrazone compound HD which was synthesized according to the method described in our previous paper [5].

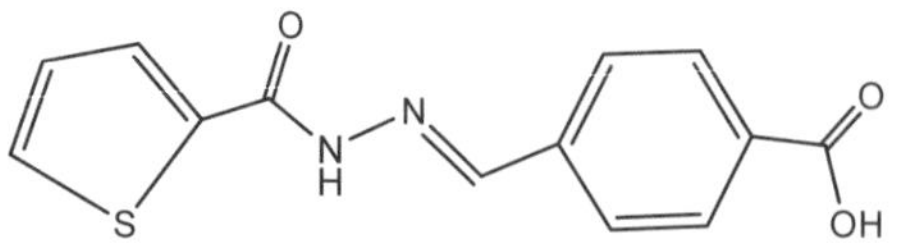

Fig. 1 Molecular structure of HD

The reference electrode is a saturated calomel electrode (SCE) and the counter electrode is a Pt electrode. The working electrodes with a surface area of 1.00 cm^2 were prepared from copper (99.9%) embedded in epoxy resin, and were gradually ground with different grades of emery paper (600, 1000, 1500 and 2000), and then degreased with alcohol and washed with deionized water. All experiments were performed using the copper electrodes with freshly prepared surfaces. The simulative oilfield water was obtained (Formula is given in Table 1, Reagents are analytical reagent) according to the literature [6-8], which is inleted CO_2 until the saturated.

Electrochemical measurements. A three-electrode cell consisting of copper working electrode(WE), a platinum counter electrode (CE) and saturated calomel electrode (SCE) as a reference electrode, was used for electrochemical measurements. Electrochemical experiments were carried out by means of a CHI860D electrochemical system, which was controlled by a computer that recorded and stored the data. The electrolyte used was simulative oilfield water maintained at 30 °C.

Potential-time (E-t) measurements. Before E-t curves measurement, working electrode was immersed in water solution of 100ml 0.5 mmol·L^{-1} HD. The potential vs. time curves is mensurated simultaneously by the CHI860D electrochemical system.

Polarization curves. The working electrode was in the form of a square cut from copper sheet (1.0 cm×1.0 cm) embedded in epoxy resin. The WE was immersed in a test solution for 30 min until a steady-state open-circuit potential was obtained. The cathodic polarization curve was recorded by polarizing from the open circuit potential (E_{OCP}) in a cathodic direction under potentiodynamic conditions corresponding to 0.5 mV/s (sweep rate) with continuously stirred condition at 30 °C. After this scan, the anodic polarization curve was recorded by polarization from the E_{OCP} in an anodic direction. The inhibition efficiency can be calculated by Eq. (1) [9]:

$$IE = \frac{(i^0_{corr} - i_{corr})}{i^0_{corr}} \times 100\% \qquad (1)$$

where i^0_{corr} and i_{corr} are the uninhibited and inhibited corrosion current density values, respectively, and determined by extrapolation of Tafel lines to the corrosion potential.

Scanning electron microscopy. The copper specimens (size 1.0 cm × 1.0 cm) were abraded with emery paper (grade 600, 1000, 1500 and 2000) then were washed with distilled water and acetone. The copper specimens used for surface morphology examination were immersed in oilfield water containing 0.5mmol/L HD and blank at 30°C for 50 h. Then, they have been removed, rinsed quickly with distilled water, cleaned with acetone and dried under 25°C. They were investigated by using a JEOL·TSM-6380LV scanning electron microscope. The energy of the acceleration beam employed was 20 kV.

Table 1 Formula of simulative oilfield water

Composition	NaCl	$MgCl_2$	$CaCl_2$	$NaHCO_3$	Na_2SO_4
Concentration (g/L)	70	4	6	0.48	0.58

Results and discussion

Potential-time (E-t) curves. Fig. 2 shows the Potential-time (E-t) curves obtained to copper in water solution of 0.5 mmol·L^{-1} HD. It can be seen in this figure that the potential negative move sharply in the early days of the self-assembly. This suggests that corrosion inhibitors molecules form gradually molecules films by adsorption in the copper electrode surface and hindered the conductivity of the electrode, which lead to the potential negative move. The potential recorded after 0.5 h of immersion of the electrode in the HD solution reached a steady state condition. These results reveal that the copper surface formed a complete and stable corrosion inhibitor molecule film.

Polarization curves. Fig. 3 shows Tafel polarization curves of copper in simulated oilfield water containing saturated CO_2 in the absence and presence of different concentrations of HD at 30 °C. The associated corrosion parameters such as E_{corr}, cathodic and anodic Tafel slopes (b_c, b_a) and corrosion current density (i_{corr}) are listed in Table 2. In this case, the inhibition efficiency is defined by Eq. (1).

It is clear that the addition of HD causes a decrease in the corrosion rate, i.e. shifts the cathodic and anodic curves to lower values of current densities. Namely, both cathodic and anodic reactions of copper electrode corrosion are inhibited by the inhibitor in simulated oilfield water containing saturated CO_2. This may be ascribed to adsorption of inhibitor over the corroded surface of copper. It follows from the date of Table 2 that the corrosion current, i_{corr} decreases, while IE enhances with increase in inhibitor concentration. The decrease in the corrosion current density was observed for the HD, corresponding to a maximum efficiency of 95.3 % at 0.50 mmol·L^{-1}. Further inspection of Table 2 reveals that the presence of HD remarkably shift the E_{corr} to the cathode, therefore, the selected compound can be described as mixed-type inhibitor of the inhibition primarily the cathode corrosion for copper corrosion in simulated oilfield water, and the inhibition of the compound on simulated oilfield water is caused by adsorption, namely, the inhibition effect results from the reduction of the reaction area on the surface of the copper [10]. The results demonstrate that the corrosion reaction is inhibited and that the inhibition efficiency increases with inhibitor concentration.

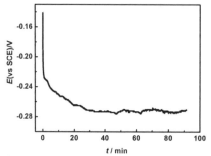

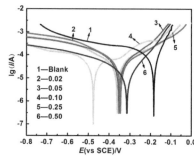

Fig. 2 Potential-time (E-t) curves of copper assemble corrosion HD inhibitor films.

Fig. 3 The polarisation curves of copper in simulated oilfield water without and with various concentrations (mmol·L^{-1}) of inhibitors

Table 2 Polarization data of copper sample in simulated oilfield water without and with various concentrations of inhibitors

Inhibitors	C (mmol·L^{-1})	$-E_{corr}$ (mV)	i_{corr} (μA cm^{-2})	b_a (mV dec^{-1})	b_c (mV dec^{-1})	IE (%)
Blank	0	0.146	175.90	99.3	187.5	—
HD	0.02	0.361	19.83	101.0	108.7	88.7
	0.05	0.350	17.17	81.6	93.2	90.2
	0.10	0.472	15.13	82.4	88.5	91.4
	0.25	0.394	12.06	84.2	92.9	93.1
	0.50	0.311	8.321	80.2	131.5	95.3

Fig.3 also reveal that the inhibitor molecules were able to reduce some corrosion activity of copper during the initial anodic process, probably due to the inhibitor molecules displacing some hydrated layers of copper oxides/hydroxides film in order to create more favorable sites for the formation of stable protective layer. In the case of hydrazone derivatives, the heteroatoms such as N and O atoms with lone pair electrons are chemisorbed on the electrode surface displacing the Cl⁻ ions, and positively charged part of the alkyl group is physically adsorbed. Then the anodic active sites vulnerable for corrosion attack gradually decreases with increasing the inhibitor's concentration.

Scanning electron microscopy (SEM). SEM analysis for the copper specimens after immersion in oilfield water for 50h, with and without HD respectively, are shown in Fig. 4. The morphology of specimen surface in Fig. 4 (1) shows a characteristic uniform corrosion of copper in oilfield water, many corrosion pits make a distinct appearance on the SEM image of copper in blank oilfield water and formed the more corrosion products particles. In Fig. 4 (2), when the specimens dipped in oilfield water with 0.5 mmol ·L^{-1} HD, the surface morphology of copper is smoother than that in blank oilfield water case, and the mechanically polishing nicks before immersion can be clearly observed in the view field. These observations are in good accordance with the results obtained from electrochemical measurement.

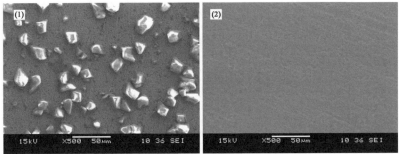

Fig.4 SEM photograph of the surface for copper: (1) after 50 h immersion at 30 °C in blank oilfield water; (2) after 50 h immersion at 30 °C in oilfield water with 0.5 mmol·L^{-1} HD.

Adsorption isotherm. The establishment of isotherms that describe the adsorption behavior of corrosion inhibitor is important as they provide important clues about the nature of metal–inhibitor interaction [11]. Values of degree of surface coverage (θ) corresponding to different HD concentrations were used to determine which isotherm best described the adsorption process. In the present study, values of the surface coverage were calculated using the polarization curves results according to the Eq. (2):

$$\theta = \frac{(i_{corr}^{0} - i_{corr})}{i_{corr}^{0}} \qquad (2)$$

where i_{corr}^{0} and i_{corr} are the corrosion current density values without and with inhibitor, respectively. The results obtained for HD in simulated oilfield water fit well Langmuir adsorption isotherm given by Eq. (3) [12]:

$$\frac{C}{\theta} = \frac{1}{K_a} + C \qquad (3)$$

where θ is the degree of surface coverage and C is the inhibitor concentration in the electrolyte.

The plot of $C\theta^{-1}$ against C gives a straight line as shown in Fig. 5. It is found that the linear correlation coefficient R^2 is 0.99998 and the slope is 1.06983. This isotherm conforms to Langmuir type, suggesting that each HD molecule occupies about 1.07 adsorption sites on the copper electrode surface. The equilibrium adsorption constant K_a obtained from the Langmuir plot is about 31.949×10^4 L·mol^{-1}. K_a is the equilibrium constant of the adsorption process and is related to the standard Gibbs energy of adsorption, ΔG_a^0, according to the Eq. (4) [13]:

$$K_a = \frac{1}{55.5} \exp\left(\frac{-\Delta G_a^0}{RT}\right) \qquad (4)$$

where R is the universal gas constant (8.314 J·mol^{-1}K^{-1}) and T is the absolute temperature (303.15 K). The value 55.5 in the above equation is the concentration of water in solution (mol·L^{-1}).

The standard free energy of adsorption ΔG_a^0 can be calculated, is -42.35 kJ·mol^{-1}. The negative values of ΔG_a^0 ensure the spontaneity of the adsorption process and stability of the adsorbed HD-SAM on the copper surface. Generally, values of ΔG_a^0 around -20 kJ mol^{-1} or higher are consistent with the electrostatic interaction between the charged molecules and the charged metal (physisorption); those around -40 kJ mol^{-1} or lower

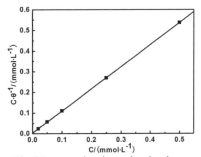

Fig.5 Langmuir adsorption isotherm plots of inhibitors HD for copper in simulated oilfield water at 30°C

involve charge sharing or transfer from organic molecules to the metal surface to form a coordinate type of bond (chemisorption) [3]. The calculated ΔG_a^0 value indicates that HD takes place through chemisorption between the inhibitor molecule and the copper surface.

Conclusions

(1) The inhibition process of HD were evaluated by electrochemical techniques SEM. The results indicate that the investigated HD have good inhibition properties for copper in saturated CO_2 oilfield water, and the inhibition efficiency increased with increasing the concentration of the inhibitors.

(2) The reduced HD molecule adsorption follows the Langmuir isotherm. The negative value of the Gibbs free energy of adsorption indicates that the adsorption of the inhibitor molecules is a spontaneous process.

Acknowledgements

This work was supported by the Guangxi Natural Science Foundation (2012GXNSFAA053034).

References

[1] M. Behpour, S.M. Ghoreishi, N. Mohammadi, *et al.*, Investigation of some Schiff base compounds containing disulfide bond as HCl corrosion inhibitors for mild steel, Corr. Sci. 52 (2010) 4046.

[2] A.B. da Silva, E. D'Elia, J.A. da Cunha Ponciano Gomes, Carbon steel corrosion inhibition in hydrochloric acid solution using a reduced Schiff base of ethylenediamine, Corr. Sci. 52 (2010) 788.

[3] Q.Q. Liao, Z.W. Yue, D. Yang, *et al.*, Self-assembled monolayer of ammonium pyrrolidine dithiocarbamate on copper detected using electrochemical methods, surface enhanced Raman scattering and quantum chemistry calculations, Thin Solid Films 519 (2011) 6492.

[4] R.R. Annand, R.M. Hurd, Hackerma.N, Adsorption of monomeric and polymeric amino corrosion inhibitors on steel, J. Electrochem. Soc. 112 (1965) 138.

[5] S.L. Chen, Z. Liu, J. Liu, Synthesis, Characterization, Structure and quantum chemistry of the 4-((2-Thiophenecarboxylic acid hydrazide) methylene)benzoic acid, Chinese J. Appl. Chem. 2012, doi:10.3724/SP.J.1095.2012.00018.

[6] S.C. Shi, R.T. Fang, S.L. Tang, Study on corrosion inhibition performance of imidazoline phosphonoamide hydrochloride in simulated oilfield water, Oilfield   Chemistry, 25 (2008) 34.

[7] C.W. Chai, G.A. Zhang, M.X. Lu, Electrochemical behavior and corrosion mechanism of X65 Steel in CO_2-saturated oilfield water, Corrosion & Protection, 29 (2008) 54.

[8] J. Li, M.X. Lu, M.L. Yan, G. X. Zhao, D.B. Sun, D.J. Yang, Corrosion mechanism of steel P110 in CO_2-cotaining simulated oilfield brine, J. Chin. Soc. Corros. Prot. 19 (1999) 285.

[9] F. Bentiss, F. Gassama, D. Barbry, *et al.*, Enhanced corrosion resistance of mild steel in molar hydrochloric acid solution by 1,4-bis(2-pyridyl)-5H-pyridazino [4,5-b] indole: Electrochemical, theoretical and XPS studies, Appl. Surf. Sci. 252 (2006) 2684.

[10] W.H. Li, Q. He, C.L. Pei, B.R. Hou, Experimental and theoretical investigation of the adsorption behaviour of new triazole derivatives as inhibitors for mild steel corrosion in acid media, Electrochim. Acta 22 (2007) 6386.

[11] N. Hackerman, E. McCafferty, Proceedings of the 5th International Congress on Metallic Corrosion, Houston, 1974, pp. 542.

[12] F. Bentiss, M. Bouanis, B. Mernari, *et al.*, Understanding the adsorption of 4H-1,2,4-triazole derivatives on mild steel surface in molar hydrochloric acid, Appl. Surf. Sci. 253 (2007) 3696.

[13] M. Outirite, M. Lagrenée, M. Lebrini, *et al.*, Ac impedance, X-ray photoelectron spectroscopy and density functional theory studies of 3,5-bis(n-pyridyl)-1,2,4-oxadiazoles as efficient corrosion inhibitors for carbon steel surface inhydrochloric acid solution, Electrochim. Acta 55 (2009) 1670.

Key Engineering Materials Vol. 537 (2013) pp 307-310
© (2013) Trans Tech Publications, Switzerland
doi:10.4028/www.scientific.net/KEM.537.307

Influence of Sputtering Power on the Structure and Mechanical Properties of Zr-Nb-N Nanocomposite Coatings Prepared by Multi-target Magnetron Co-sputtering

S. Zhang[1], D.J. Li[1, a], L. Dong[1], H.Q. Gu[2,3], R.X. Wan[2,3]

[1]College of Physics and Electronic Information Science, Tianjin Normal University, Tianjin, 300387, P.R. China

[2]Tianjin Institute of Urological Surgery, Tianjin Medical University, Tianjin, 300211, P.R. China

[3]Ninth People's Hospital, Shanghai Jiao Tong University, School of Medicine, Shanghai 200011, P.R. China

[a] dejunli@mail.tjnu.edu.cn (corresponding author)

Keywords: Multi-target magnetron co-sputtering, Zr-Nb-N coatings, power, hardness

Abstract. Ternary Zr-Nb-N nanocomposite coatings were synthesized on Si(100) substrates by Multi-target Magnetron Co-sputtering. The structure and chemical composition and binding energy of coatings were characterized by X-ray diffractometry (XRD) and Energy-dispersive X-ray spectroscopy (EDS) and X-ray photoelectron spectroscopy (XPS). The results of measurements of Nano Indenter indicated that the maximum hardness was up to 36 GPa and elastic modulus was close to 425 GPa at the 30 W of Zr power and 120 W of NbN power. The hardest coating also showed the modest residual stress. These Zr-Nb-N coatings appeared to be a promising composite coating system suitable for engineering applications.

Introduction

Superhard and wear resistant coatings have been used in cutting and forming tools in order to increase resistance against wear and improve the lifetime of tools and machine parts for many years [1-2]. It is well known that transition metal nitrides have high hardness, high melting point and strong abrasion resistance characteristic [3-5]. NbN coatings are of increasing interest because of their extremely high hardness. Despite the fact that the hardness of bulk NbN (HV=14 GPa) is much lower than the bulk hardness of other nitrides (e.g. TiN and ZrN), the hardness of cathodic vacuum arc deposited NbN coatings is usually significantly higher than other binary nitrides [6]. Zirconium always is used as alloying agent for its strong resistance to corrosion. Zirconium can react with metallic and non-metallic elements to form solid solution compound.

   A new trend in studying transition metal nitride films in wear-protective coatings is to add a third element such as Ti, Zr, Al, or C for improving hardness and/or oxidation resistance. For example, based on TiN film, (Ti, Al)N, (Ti, Cr)N, Ti(CN) [7-9], and so on. However, few results are available based on NbN film. Therefore, in this work, we introduce Zr element to NbN film to form Zr-Nb-N coatings. Multi-target Magnetron Co-sputtering is employed to synthesize ternary Zr-Nb-N nanocomposite coatings on Si(100) substrates in this work. Our purpose is to understand influence of sputtering power on the structure and mechanical properties of Zr-Nb-N nanocomposite coatings.

Experimental

Coating depositions were carried out in a computer controllable Multi-target Magnetron Co-sputtering system (model: G450, made by SKY Technology Development CO., LTD., Chinese Academy of Science). Si (100) wafer substrates were ultrasonically cleaned consecutively in acetone and ethanol for 15 and 15 min, respectively before being placed into the chamber. The chamber was evacuated to a residual pressure lower than 4.0×10^{-4} Pa prior to deposition. Before deposition, the chamber was filled with argon to 5.0 Pa and the substrate bias was set to -600 V, producing a glow discharge etching for cleaning the substrate. The Zr (hexagonal structure) and NbN (face-centred cubic structure) targets were sputtered to synthesize Zr-Nb-N coatings using pluse DC power (frequency of 40 kHz and duty ratio of 40%) and RF power, respectively. The area of target is 20 cm^2. The substrate bias was set at -80 V and the work pressure keep 0.5 Pa. Sputtering power of NbN target was set at 70 W, 100 W, 120 W, 150 W and 170 W under constant the sputtering power of Zr targe of 30 W. All of the coatings thickness was controlled within 350-450 nm. The substrate holder rotated at a speed of 6 rpm during deposition.

The crystal structure of the coatings was characterized using monochromatic Cu-Kα radiation on a Rigaku D/MAX 2500 X-ray diffractometer (XRD). The coating chemical states were investigated by a PHI5000 Versa Probe X-ray Photoelectron Spectroscopy (XPS) using an Al Kα X-Ray source (1486.6 eV). Elemental composition was tested by Energy-dispersive X-ray spectroscopy (EDS) system for Hitachi Tabletop Microscope TM3000. The Ambios XP-2 surface profilometer was used to measure thickness and residual stress of coatings. Hardness and Young modulus of the coatings were obtained using a Nano Indenter XP™ with a Berkovich indenter.

Table 1 Elemental composition (at.%) EDS of the Zr-Nb-N coatings prepared at 30 W sputtering power of Zr and different sputtering power of NbN.

NbN power		70 W	100 W	120 W	150 W	170 W
Elemental	Zr	24.07	12.98	11.09	10.56	8.92
composition	Nb	43.90	48.13	52.64	53.05	51.03
(at.%)	N	32.03	38.89	36.27	36.39	40.05

Results and Discussion

Table 1 shows that with increasing sputtering power of NbN, NbN content increases. It is easy to see that the nitrogen atoms are loss during coating process, because Nb: N = 1: 1 in the target decreases in the coating.

The coating only displays peaks of Si substrates when the sputtering power of NbN is 70 W (Fig.1). It means that the Zr-Nb-N coating is amorphous at this power. This is due to smaller kinetic energy and surface diffusion rate of sputtering particles that leads to hard crystallization when sputtering powers are small. Two different crystal orientations are identified, which are the face-centered cubic structure (Zr, Nb)N(111) and (Zr, Nb)N(200) when sputtering power of NbN increases. It demonstrates solid solution mixed crystal structure of Zr-Nb-N coatings [10]. When sputtering power of NbN is 120 W, the coating exhibits the smallest and broadest diffraction peaks, indicating its small grain size and high defect level.

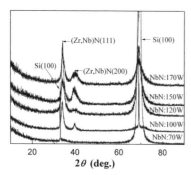

Fig. 1 XRD patterns of the Zr-Nb-N coatings prepared at 30 W sputtering power of Zr and different sputtering power of NbN.

The chemical bonding states of the Zr-Nb-N coatings are obtained by XPS. Before analysis, the surface of the coatings is etched in XPS chamber by argon ions. In this measurement, beam spot size is 2 mm×2 mm, accelerating voltage is 500 V, sputtering time is 30 seconds. The chemical bonding states are obtained by subtracting the background with the Shirley's method and deconvoluting the spectra by a curve-fitting method using a non-linear least squares fitting to a mixed Gaussian–Lorentzian product function.

In Fig.2, Zr$_{3d}$ peaks centered at approximately 181.8, 181.0 and 178.9 eV ascribe to Zr-N-O, Zr-N and Zr-Zr bonds, respectively. The Nb$_{3d}$ peaks at 205.6 and 202.8 eV are related to Nb-N-O and Nb-N bonds, respectively. The N$_{1s}$ peak at approximately 397.4 eV and 396.9 eV are due

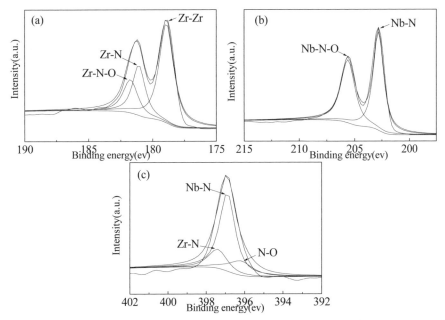

Fig. 2 High-resolution XPS spectra of (a) Zr_{3d}, (b) Nb_{3d}, and (c) N_{1s} for Zr-Nb-N coating at 30 W sputtering power of Zr and 120 W sputtering power of NbN

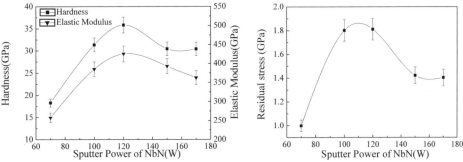

Fig. 3 Hardness and elastic modulus of the Zr-Nb-N coatings at different sputtering power of NbN

Fig. 4 Residual stress of the Zr-Nb-N coatings at different sputtering power of NbN

to Zr-N and Nb-N bond, respectively. Besides, the signal at 396.2 eV represents N-O bond. The O signal identified in this spectrum comes from the residual O_2 during etching or growing as well as the diffusion of O_2 adsorbed on surface when the sample is removed from chamber.

Fig.3 shows the change in hardness and elastic modulus of the coatings with different sputtering power of NbN. Their values are lower when sputtering power of NbN is 70 W. According to EDS and XRD analysis above, the coating at this power does not form solid solution mixed crystal structure and the coating is amorphous because elemental composition of Zr is too much in the coating. So the hardness of the coatings at this condition is lower. Then hardness and elastic modulus increase with increasing sputtering power. The maximum hardness and modulus are up to 36 GPa and 425 GPa at 120 W. When power continues to increase, their valuss slightly decrease. There are two main reasons for hardness and elastic modulus increasing. One reason is that proper elemental composition ratios among Zr, Nb and N make coatings form the substitutional solid solution structure. The Zr, Nb and N elemental composition ratio for maximum hardness and modulus coating is 0.31:1.45:1. Enhanced solid solution strengthening effect can increase compressive stress in the coatings which might contribute to the increased hardness. Another reason is smaller grain size at 120 W, enhancing hardness and elastic modulus [11].

Fig. 4 shows the variations of residual stress as a function of different sputtering power of NbN. It is clear from this figure that the residual stress increases firstly and then decreases with the increasing power. Higher bombarding energy at 100-120 W leads to more atomic mixture at the interface of coating/substrate, which inhibits the stress releasing along the interface. With sputtering power continuing to increase, the residual stress decreases. This is because the higher bombarding energy causes crystalline structure to change.

Conclusions

Ternary Zr-Nb-N nanocomposite coatings were synthesized on Si(100) substrates by Multi-target Magnetron Co-sputtering. These coatings exhibited face-centered cubic solid solution mixed structure. The maximum hardness was up to 36 GPa and elastic modulus was close to 425 GPa at the 30 W of Zr power and 120 W of NbN power. These Zr-Nb-N coatings appear to be a promising nanocomposite coating system suitable for engineering applications.

Acknowledgements

This work was supported by National Basic Research Program of China (973 Program, 012CB933604), National Natural Science Foundation of China (51272176, 11075116), and the Key Laboratory of Beam Technology and Material Modification of the Ministry of Education, Beijing Normal University, China.

References

[1] W.D. Sproul, Reactive sputter deposition of polycrystalline nitride and oxide superlattice coatings, Surf. Coat. Technol. 86–87 (1996) 170-176.

[2] S. Veprek, The search for novel, superhard materials, J. Vac. Sci. Technol. A 17 (1999) 2401-2420.

[3] D.J. Li, F. Liu, M.X. Wang, J.J. Zhang, Q.X. Liu, Structural and mechanical properties of multilayered gradient CrN/ZrN coatings, Thin Solid Films 506-507 (2006) 202-206.

[4] D.J. Li, M.X. Wang, J.J. Zhang, J.Yang, Working pressure induced structural and mechanical properties of nanoscale ZrN/W_2N multilayered coatings, J. Vac. Sci. Technol. A 24 (2006) 966-969.

[5] L. Hultmann, C. Engström, M. Odén, Mechanical and thermal stability of TiN/NbN superlattice thin films, Surf. Coat. Technol. 133–134 (2000) 227-233.

[6] V.N. Zhitomirsky, Structure and properties of cathodic vacuum arc deposited NbN and NbN-based multi-component and multi-layer coatings, Surf. Coat. Technol. 201 (2007) 6122-6130.

[7] Z.P. Huang, Y. Sun, T. Bell, Friction behaviour of TiN, CrN and (TiAl)N coatings, Wear 173 (1994) 13-20.

[8] I.J. Jung, S. Kang, A study of the characteristics of Ti(CN) solid solutions, J. Mater. Sci. 35 (2000) 87-90.

[9] I. Dörfel, W. Österle, I. Urban, E. Bouzy, O. Morlok, Microstructural characterization of binary and ternary hard coating systems for wear protection Part II: Ti(CN) PACVD coatings, Surf. Coat. Technol. 116-119 (1999) 898-905.

[10] H. Klostermann , F. Fietzke, R. Labitzke, T. Modes, O. Zywitzki, Zr–Nb–N hard coatings deposited by high power pulsed sputtering using different pulse modes, Surf. Coat. Technol. 204 (2009) 1076–1080.

[11] J. Musil, Hard and superhard nanocomposite coatings, Surf. Coat. Technol. 125 (1999) 322-330.

Keyword Index

Author Index

Library of Congress Cataloging-in-Publication Data
ansma, Theodore J., 1943 -
Becoming Kate / by Theodore J. Jansma, Jr., Katherine St. Clair.
   p.   cm.
SBN 0-915677-46-6
   St. Clair, Katherine, 1949---Mental health. 2. Multiple personality--
nts--United States--Biography. 3. Psychotharapy patients--United
s--Biography. 3. Psychotherapy patients--United States--Biography.
Clair, Katherine, 1949-. II Title.
.5.M8J36 1990
236'0092--dc20
                                                    89-61782
                                                    CIP

DITION

5 4 3 2 1

BECOMING
KATE

I.
1.
Pati
State
I. St.
RC56
616.8S
[B]

FIRST E

10 9 8 7 6

BECOMING
KATE

by Theodore J. Jansma, Jr., Ph.D.
and
Katharine St. Clair

ROUNDTABLE
Publishing, Inc.

Prologue

This is the story of Kate, a multiple personality consisting of 233 separate, clearly and distinctly identifiable and autonomously functioning personalities and personality fragments. It is a joint venture, written by both patient and therapist, and began as a clinical experiment early in therapy when I encouraged one of the personalities to write about some of her reactions to a difficult series of memories. This was so helpful that it became an on-going enterprise. By the time the idea of writing a book had solidified, an extensive data base already existed. It is a first-hand account of both living as a multiple personality and trying to heal a multiple personality.

In the professional community, the reality of multiple personality has been accepted for some time. It is also generally accepted that multiple personality dynamics are, essentially, a psychological survival technique necessitated when massive and on-going abuse is forced on a defenseless, but intelligent, child.

However, among the general populace such acceptance is not generally the case, and the risk for Kate in allowing so much of her very private self and painful past to be exposed is great. Therefore, to protect her privacy, names and places have been changed.

In 1981, when this case began clinically, there was not an abundance of resources available to most clinicians. There now is the International Society for the Study of Multiple Personality and Dissociation, there are recognized experts in the field. Also there is a growing body of literature and of research, which has greatly broadened our knowledge of multiple personality phenomenon.

Research now indicates that probably 97% of multiples' are abused as children and what's more, 25% were probably forced into Satanic worship rituals. We also know that having a very large number of personalities is, in fact, not all that unusual, although it is now believed that only a small portion of this total number are "full" personalities; many are "fragments" who have independent identities but only limited functions--a limited range of emotions, or a limited history. In this case, Kate would be diagnosed as a Polyfragmented Multiple Personality Disorder.

Because of the large number of personalities involved you may, at times, find the sequence of events confusing and difficult to track. So in order to make it less confusing, we talk about the groups, or the groupings in which the personalities arrived into the patient's psyche. You may also find them difficult to comprehend, perhaps even find them hard to believe, but I assure you they are all true. Our wish is that you try to understand and, in doing so, we believe your world will be expanded and enriched, as ours have.

1

Karin is angry. That's the only word I can use to describe her. That is her role. That is her feeling. Anger. Toward herself, toward people in the world, toward everything in the world. She learned to hate at a very early age. And that's what she feeds on. Hate. At times I think that Karin could physically hurt someone, given the right time and the right situation.

When Karin was three years old, she was tied to a chair. I remember thinking that she remained tied there for a very long time. Now that I look back on it, it must have been a full 24 hours--quite a punishment for such a little girl. She wasn't allowed to go to the bathroom. She couldn't feed herself so they fed her. They kept shoving food in her mouth so fast that she couldn't swallow it, so she vomited on herself. They left. But she didn't cry. She was so pale. There she was, in her little dress, tied to a chair, with food all over her--stuck on her face, on her clothes, in her hair. And here were these big blue eyes in this little white face, and she didn't cry. She never cried. They let her sit there all night. I couldn't leave her alone, so I watched her all night, but she didn't know I was there. I think she was in her own little world by this time. I think the only way she stayed alive was every now and then she'd say, "I hate you, I hate you, I hate you, I hate you." She'd say it in a whisper. That's all. I couldn't get the ropes off. They had her arms and legs tied, and it was such a hard chair, too. The next day when they untied her, she tried to stand up and she fell. Oh, I'm sorry. I don't mean to cry, but I just--I can still see her, so damn helpless. She has a right to be so angry. When she fell, she was so stiff she couldn't stand, and dear grandfather kicked her. And then she had to go get cleaned up, and I helped her. She didn't know it, but I helped her. When she was all cleaned up Jennifer took her hand and brought her outside, and I don't know how Jennifer ever dared, but Jennifer snuck her a piece of bread with some butter and sugar on it. She brought it to her outside and got her some water. And Jennifer kept rubbing Karin's arms. Jennifer wanted her to warm up, too.

Francesca arrived at a particularly bad time for Kate, when Kate was going through a very difficult period and was being beaten. Kate couldn't take the beating anymore, so she slipped away, and for a few moments left Carol in her place, and suddenly there was Francesca, feeling tremendous pain, physical pain--in fact, she would hurt for days afterward. She got to know Kate almost immediately, and even though they were opposites, they seemed to cling together; they needed each other. Francesca was, and still is, if she allowed herself to be, a very talented artist. But art was a no-no.' Yet it's the only thing that Francesca had. Time after time she would draw, only to have her pictures destroyed. She would be so pleased when she had drawn something she really liked--and she liked her work at that time. She had natural talent. It was very easy for her. She drew everything around her. She drew the house that we lived in. She drew what was outside: the big farm across the way, she drew the North Sea, she drew the poppies and the church; only to have each and every drawing destroyed right in front of her. Ripped to shreds, burned. And each time, she drew again. Maybe it was defiance. Maybe it was her need to cope, her escape so to speak.

Often, after she'd been locked up--and I'm remembering one particular instance when she and Kate were locked in a trunk--they would comfort each other inside that trunk, two little girls, and even though Kate was very strong at that particular point, Francesca would be the stronger of the two. I couldn't do anything to protect either one of them. But after they were let out of the trunk, Kate left for awhile and Francesca took her paper and pencil and she went out inside and she drew a picture of a big field that wasn't too far from where we lived and it had trees and all kinds of flowers. Very beautiful. She would draw something like that after each bad experience. It was her way of trying to erase the ugliness that had happened. But, eventually, all that ugliness changed her; and where there had been once someone who loved beautiful things, who loved nature and was so caring inside, there was now someone who became very cold inside, and she started to hate. That developed when she was about ten years old, when she had drawn another one of her pictures--she had started to hide her

2

pictures by this time--well, grandfather found them and he tied her to a chair. He tied her arms so tightly that the pain must have been excruciating, then he put all her pictures in front of her and he put a match to them and he burned them.

Something died inside Francesca that day. After that, it was too painful for her to draw. She said her wrists hurt, that she couldn't draw anymore because of the pain. But you could see in her eyes that day that something died. She's never really recovered. Since then, she's done a watercolor for Dr. Jansma. That was so difficult for her, and of course, she's not pleased with it. It's as if a light went out inside of her. After that came something even more devastating; her need for **punish**ment. It became a way of life for her. It's what she seemed to need to exist. It's so sad. She became someone, something totally different. Punishment became her survival; revenge and hate became her survival. She'd deliberately set herself up at times: as she was being punished, as she was being whipped, she said to herself over and over again, "I need this, I deserve this." And that's what she's done ever since. She became so cold, so hard and calculating.

She developed another habit when she became a little older. Any time she was punished, first by the parents and the grandparents, and later on by the ex-husband and by another male in her life, that's what made her go on; it's what kept her going. Each time she was punished, she'd buy herself something. It was as if she was rewarding herself, "All right, I took the punishment, I deserved the punishment. Now, I deserve to buy myself something because I accepted what I'm here for. I accepted the punishment, and I did it without crying. I did it without giving in to the pain, but now I deserve something else."

Jessica was raped often. She doesn't talk about it, and I think many times she just doesn't want to remember it, but it's there and, of course. It left a horrible impression on her. The first instance occurred when she was four years old. She was taken to a small room in her grandparents' house, and the grandparents and the great uncle took turns with her. I dislike calling them grandmother' and grandfather' because they don't fit into that category for me. Grandparents are supposed to be warm, loving, kind people. These people were so cruel. She was taken to this room and she was stripped by the grandmother. Then she had to perform oral sex on each of them. This is very difficult for me to talk about. After she had done that, she was strapped down to a cot that they had in there. A little four-year-old, who couldn't have fought them anyway. When the grandfather put his penis inside her, she screamed. I will never, ever forget that scream.

To quiet her, they shoved this old rag in her mouth and she gagged.

3

And then she passed out. So they waited until she was conscious again, threw some cold water on her, and then grandmother sat down right on top of her face, and then it was the great-uncle's turn to shove his penis inside of her. And then when they were finished with that, they turned her over and they whipped her, telling her that she was a bad girl, and that she deserved to be punished because she didn't do it right, and that they were ashamed of her because she was such a dirty little girl. And they were supposed to be grandparents. When they were all done with her, they left her in the room, strapped down to this bed, without anything to cover her up. They left her there for a whole day. It got dark. She was cold. And the next day they let her out, after they raped her one more time. I think every part of her body must have hurt. Jessica doesn't tolerate being cold. But even as a child, even with all the abuse that she went through, Jessica carried herself with dignity. That's the only word I can think of to describe it. She carried herself in such a manner that even as children we respected her.

When I look back on those times, it's amazing to me how many of us wanted to die. And every time someone did there would be someone like Elizabeth or Katharine waiting to help ease things a little bit. I haven't really thought about that before. A child that age shouldn't want to die. A child should be happy and full of curiosity, wanting to discover; a child should be loved. But a lot of us wanted to die. I wonder sometimes if that's why there were so many of us. There were so many terrible things done to us. When I think of that trunk that we were locked up in, Kate and myself and some of the others--that wasn't a big trunk. You couldn't lie down in it and stretch out, we were so cramped up. And we'd have to spend the whole night there, or to have Kate spend an entire night in that hole in the ground in a field away from the house. No wonder, to this day, she's afraid of the dark and hates being cold, and is afraid of being alone. Sure, she needs her time to herself to relax, but she's never really alone because we're always around. She knows that she can reach out to any one of us at any time. But if she were alone, really alone, if she didn't have us, I think she'd be terrified.

Maria S., October 7, 1985, recalling her memories of several other personalities.

2

The woman on the phone has a friend and fellow employee who is thirty-one years old, quite depressed and may be suicidal. "She seems to go in and out of reality." I agreed to see her the next day.

January 27, 1981: It's cold outside. My new patient comes in. She'sclearly anxious and probably wouldn't be here if her friend hadn't accompanied her. I see a rather meek, frightened, tall, attractive female who is clearly intimidated by me and the new surroundings. It is obvious that my first priority must be to foster some comfortableness. I proceedslowly with taking a history. I learn that she's been divorced for two years,immigrated from Holland at age seven, married at age twenty-three, comes from a strongly religious family background, has a spotty memory, is very preoccupied with weight, clearly has parental problems and has attempted suicide in the past. Last evening she had a strong urge to jump out of a moving car. She is employed full-time and apparently is a good employee. She now has a sense of losing touch with reality, but does not appear to truly want to die. I will try to treat her on an outpatient basis and will see her later this week for another session. Psychological testing should help give me more specific clinical direction, and she agrees to come in the following day for this. I advise that she stay with friends for the next few days. Unknown to me, an "observer" is seeing it this way:

> When I look back on the events of the week preceding the start
> of therapy, I can easily understand why Carol was so very
> confused and suicidal. The amount of time she had been losing

5

had become greater and she was very frightened. The last straw came when she found a note on her apartment door that said she was a whore. It was written in large letters and left where anyone passing by could see it. That evening she swallowed a large amount of pills: valium, librium whatever she could get her hands on, drinking an entire bottle of wine to force them down. The rest of us watched. I think we were so tired of the way our lives were going that we didn't care either.

Fortunately, Francesca's friend, Gordon, happened to stop by because he was worried about her. He immediately saw what was happening and forced her to start walking and to drink black coffee. She vomited several times. He then bundled her up and took her to his apartment and made her eat some toast. He then called Carol's boss, Sue. Gordon stayed up with Carol all night and the next day took her to Sue's house. Sue had taken the day off. She tried to talk with Carol, but she was silent, and withdrawn. Gordon returned in the evening and he and Sue decided Carol needed help. Sue called Dr. Jansma at home. He seemed to ask Sue a lot of questions and apparently advised that Carol not be left alone. Gordon took Carol back to her apartment to pick up some clothes. He wanted her to spend the night at his apartment. As they were going up the stairs, Harold was coming down. He was drunk and again called her a whore. He had also left another note on the door, again stating she was a whore in large letters, plus some other very vulgar things. Gordon took Carol back to his place and encouraged her to try and get some sleep, but she couldn't, (for the fourth day in a row). Sue arrived at eight o'clock the next morning. The appointment with Dr. Jansma was at noon. Carol was still very uncommunicative and withdrawn, and she was very worried about the way she was dressed. She had on jeans and a sweater. We observed that she was very pale. She was very anxious because she didn't have any make-up with her, only mascara. Sue tried to reassure her that she looked fine, but to no avail. (She had to look perfect, you see.) Carol was very nervous, very restless. She kept on looking at herself in the mirror, staring at herself, touching the mirror as if trying to touch herself. She had on her favorite white sweater, the one with the big cowl collar, but as she looked at herself she saw only an ugly, fat, dirty person. Sue finally got her to put on her coat and come with her.

Carol said nothing on the way to Dr. Jansma's office. She couldn't give the receptionist her name, Sue had to. In the waiting room, she was too restless to sit. She paced around the room as

6

if she were caged. She had just seated herself on the couch when Dr. Jansma came into the waiting room and introduced himself. Carol didn't stand, she just looked from him to Sue and back again. He finally had to take her hand, tell her it was going to be all right, and led her back to his office as if she were a child. She refused his offer of a cup of coffee even though she wanted one. She didn't want anyone to go to any trouble for her. She sat in the first chair she came to in his office because she was too shy and embarrassed to walk across the room to the couch. He asked if he could take her coat and hang it up, but she said no. We watched all of this with interest because we were curious and we had decided that if Carol felt better, it would be easier for us to get on with our lives. We constantly had to watch her, protect her, because she couldn't take care of herself. Some of us were already starting to resent her. Dr. Jansma asked her questions about her family and then asked her what she thought the problem was. She told him about Harold and showed him the note. He was going to throw it away, but she wanted it back and held on to it the entire session. He told her she was like a computer that had been overloaded and had blown a fuse, that she was not crazy, as she thought, and that he was willing to help her. He also told her not to go back to work until after the next session and that he wanted her to take a psychological test (MMPI) before then. We took it for her the next day because tests frightened her--she had never taken one. During that first session she also told him about the abortion, (she had pieced that together from things the parents had said), and told him she hated her parents and God. She also told him she wasn't a very nice person. He set up an appointment for later in the week and again tried to reassure her that everything would be all right. She didn't believe him. As she left, he put an arm around her shoulders, but she backed away. We all watched the entire session with a "see what happens" attitude. Trust was a big issue, of course, and we didn't believe that anyone could really help Carol. We were also in a great deal of turmoil, unsure of allowing Carol to go into therapy. But we needed help with her. Most of all, though, we were afraid. Afraid of being discovered, afraid of being blamed for not taking care of her better. I think we knew our lives were going to change.

Cassandra
July 26, 1987

* * *

Several months of therapy pass. The more I get to know her, the more and varied her problems appear to be. She's very obsessive/compulsive; she takes laxatives regularly, her depression easily spills over into suicidal thinking, and it's easy for her to associate violence with sex. She now reports that beginning at age seventeen, she'd regularly wake up with blood on her fingers, apparently from her vagina. She has never before shared any of this. She has many fantasies regarding sex and violence; they involve men and women who are always nameless and faceless.

It is now August. Carol has clearly accepted therapy, but our experience appears to be a matter of attempting to deal with one crisis after another, while an ever-escalating account of pain and trauma emerges. I keep getting the feeling we're only touching the tip of the iceberg. I also am having great difficulty having any sense of real continuity. As I review my early notes, I'm reminded of the loss of memory concerning her past. There is very little awareness of her childhood. By the fourth session I was already wondering about multiple personality dynamics, although there are many differing, (and probably all appropriate), diagnoses that could be used. By now we've settled into a routine of weekly sessions, with a number of extra sessions squeezed in to handle this.

Subsequent sessions appear to have triggered off a number of memories that have been actively repressed for years. She now recalls that at age nineteen she thought she was pregnant. A friend told Carol's parents, but it turned out she was not in fact pregnant. However, Carol never told her parents that she wasn't pregnant. Two months following this incident she actually did become pregnant and since she was underage, her parents arranged for an abortion and also had her tubes tied. "I didn't even know what that meant," she tells me. She reports that one month later she experienced a great deal of blood from her vagina and she believes that she saw the fetus from the abortion emerge. In addition, she also reports that the first man that she had sex with following this experience arranged for her to perform multiple sex acts with seven other men and that "I was so afraid that I just went along with it."

It is quite clear that the walls are breaking down even further and that these are obviously significant events in the life of this woman. However, even with my awareness of how powerful the defense mechanisms of denial, avoidance, and repression can be, it remains striking that such powerful and significant events could actually not have been in this young woman's awareness for many years. She is clearly trusting me with information that she's never been able to share with anyone and there is no reason to believe that there is lying or manipulativeness on her part, but somehow things are not adding up. It is obvious that her pain is honest, but I also cannot help but note that even in these painful sessions, there are no tears coming from Carol's eyes.

During a later session, I am aware that Carol has become more depressed. She has been preoccupied with the image of the aborted fetus. What is encouraging to me is that her anger is finally beginning to emerge. I am also aware that her parents hold a central role in the dynamics of this woman. She tells me that she talked with her mother after our session about the abortion and her sterilization. She tried to make her mother understand that the physician who performed the abortion had raped her during a follow-up visit to his office. Her mother refused to believe her.

The next several months essentially pass in the form of managing present issues. Carol's depression comes and goes, the laxative abuse increases and decreases, and relationships with the men in her life appear stable. We seem to be on somewhat of a moratorium. Although therapy appears to be helping her maintain her present functioning, we do not seem to be making progress in any particular way. Approach/avoidance dynamics regularly show. We do have sessions that seem to be quite significant, with information that is quite dramatic and very painful emerging. She even shows appropriate insight during some of these sessions, but I don't have a sense of closure on any of the significant issues, especially those issues relating to anger and her parents. I hope that we do not need to settle for only the long-term goal of maintenance therapy. It is very clear that there is chronic and extremely traumatic material. There still remain many major questions involving the treatment of this woman.

It is now November and we have been in therapy for ten months. Laxative abuse is now beginning to diminish with some very strong prodding on my part. There has been some grief over the departure of one male friend, and she has been finding herself interested in religious involvement. She is very aware of the issue of anger at her parents but will only deal with it superficially. She did cry twice one week which is new behavior for her.

Christmas is approaching. Carol has eleven days off from work and is viewing that as a possible time to totally wean herself off the laxatives. We have been focusing on the subject of anger and she now feels that her anger is basically at herself. Control is still an ongoing and very major issue.

During one session, Carol is very angry at her sister and mother for calling her at her friend's home. It is obvious that they had to do a lot of "detective work" to locate her and they called her because: "We have not heard from you in a day and a half." The patient reports that her mother, in particular, very regularly checks up on her. This provides me with the opportunity to bring up an issue that has been necessary for a long time, but which the patient has also been extremely hesitant to allow to happen, namely, a joint session with her mother. The family system clearly has a

9

number of problems and it is hoped that we can begin to confront some of those problem areas. It is also quite clear that Carol is extremely threatened by this thought. She does agree to have a session with her siblings that we have scheduled for next time. Her sister, at the last minute, cannot make it but her younger brother does come in with Carol in this session. He basically corroborates the picture of her family, namely, he confirms the pedestal that the older brother who died was on, the possessiveness of the parents, and their need to be very opinionated and controlling. He describes himself as being able to stand up to them.

We are now past Christmas and Carol's laxative abuse is in fact increasing. She has finally started to talk about her marriage of six years, which she was unwilling to discuss before. She describes her ex-husband as being quite lazy, losing a number of jobs because of not showing up for work much of the time. He is a probable transvestite and probably bisexual. She also tells me that the man whose relationship she has been grieving over is married. He wants her to move away with him, which she is considering. She feels very lonely and somewhat depressed at this time. The reality of her present situation is hitting her, although her depression is not the wallowing kind of depression that has so frequently emerged.

It is now past New Year's and we are close to one year in therapy. Carol now reports that the violent sexual fantasies are returning again. She relates some of the kinky sexual preferences of her ex-husband, and the anger and guilt she experienced after realizing that she could get interested in some of those practices. She also reports that older men often made very blatant sexual passes at her as she was growing up. Her mother always accused her of "having that look about her." She now recalls and shares information regarding two gang sexual experiences which were done against her will. For the first time she is able to cry about these experiences. To this point she has not had any laxatives in seven and one-half days. Two weeks later Carol has had no heat in her apartment for three days and had to stay with her parents. Her father was regularly critical during this time and she resumed her laxative use.

By the end of January, we have been in therapy for one year. We have a two-hour session scheduled but Carol has spent almost the entire two hours in chit-chat and avoids discussing anything of substance. At the end of the two hours she reports that her sister had a baby the evening before, which triggers off her rage at her parents because they had had her sterilized. Although Carol appears to intellectually have known for some time about the immense amount of anger inside of her, she does not allow herself to experience it. It would appear she is finding out that she can let herself experience anger in therapy and that she will not fall apart because of it.

10

By the end of thirteen months, Carol presents somewhat of a hyperactive picture as we focus on two very primary areas of insight: the amount of power she freely gives to others, usually the people she least wants to have power over her; and the question of: is my best good enough for me? These are powerful insightful areas for her and she says, "I've never felt better."

As the next couple of weeks progress it would appear that Carol has begun to put into practice the insights noted above and she's quite elated about her present functioning. She has some worries about "crashing," but is clearly maintaining her progress. She is, however, still using laxatives.

Early March, 1982: Carol comes in with complaints of a major headache. We decide to try hypnosis for the first time and she proves to be an excellent subject. Through the use of imagery, we are able to totally eliminate this headache and she is quite impressed.

Unfortunately, the laxative abuse continues. Carol confesses to me that she has not been fully candid about the extent of her use of laxatives and it is quite clear that this chronic problem, which waxes and wanes, is in fact, not being resolved. We discuss the possibility of hospitalization as a means of forcing the issue. She now opens up about the possibility of substance a use. Drugs she takes include Sinnequon, Fiorinol, laxatives, and tenty cups of coffee per day. She is very hesitant to enter the hospital but eventually decides that it is in her best interest to do so. We have our first family session to deal with this decision. She is particularly worried about her parents' reaction, about their perception that this will hurt the family, and afraid that they will grill her about all kinds of information relative to this. During this session, the parents are informed about the laxative problem. They report that they had already suspected this and are more supportive than expected. The session seems to go quite well. We decide upon admission to a private psychiatric hospital, which I arrange. The family appears to support this hospitalization.

Admission actually occurs on May 4th when an opening occurs in the hospital. She has spent two days prior to admission without any laxatives. She is extremely anxious but is coping. There is tearfulness, and periodically a strong sense of anger just prior to hospitalization, but her motivation looks good at this point.

Once admitted, she begins to adjust to the program. However, the parents discover that there is a family acquaintance who is also an in-patient in the hospital and they are very fearful that some information about the family might come out of the experience so they are putting strong pressure on Carol to leave the hospital at once. She resists their pressure to leave and must continue to do so for her entire hospitalization. At this point Carol has been off the laxatives for nine days. She's beginning to open up somewhat in her program but is still feeling quite vulnerable.

It is fourteen days after admission and Carol is still not using laxatives. She has made a decision to do a psychodrama the next day and I have been invited to be there for that experience. A psychodrama is a dramatic acting out, onstage, of a key event in the life of the patient. Staff typically portray relevant individuals involved in the event, while the patient portrays herself. A two-hour psychodrama occurs which I attend. Two primary issues are confronted during this experience: a gang rape, and Carol's parents and their decision to force an abortion and sterilization on her. The amount of rage associated with these events is focused on and a very emotional catharsis occurs. The experience turns out to be very relieving for her and the expectation is that we can now plan to discharge her. Six days later this happens. She has been hospitalized for three weeks. She is somewhat concerned about out-patient therapy because of her fear that this will now diminish her progress in the hospital. I assure her that this need not be a worry.

Although I am not aware of this, one of the other personalities, whom I will not meet for several more years, is writing this account of their hospital experience:

Entering the second psychiatric hospital was a very scary thing for us. We went because of the laxative abuse. Carol didn't want to go either. She likes to be at home, in a safe environment. She cried when she was admitted, not just a few tears, but great big sobs. I felt sorry for her. We all went along because we had to protect her. This was not something she could handle on her own. We still had the memory of a previous hospitalization. That was a bad experience, and we were quite convinced this hospital was going to be the same. We didn't like the idea of being locked up, of not being able to leave. Dr. Jansma told us over and over again that everything was going to work out. He came to see us every week while we were there and that was helpful. The staff was very nice and they tried to be helpful. When Carol went to her group meetings or to see the psychiatrist, we would often have to take over because she couldn't fill in the gaps. Of course, we couldn't let on about us and that was the most uncomfortable part. We needed to be home so we could truly have time to ourselves. We couldn't do that in the hospital. We constantly had to keep an eye on the children, because we knew it could be dangerous if they were allowed to talk or play with other people around. Group sessions were difficult also. We were expected to open up, to talk.

12

That was a problem because we weren't accustomed to doing that. The only person we had ever talked with was Dr. Jansma, and he had just been told about surface problems. There was one doctor in particular, a psychologist, that seemed to try and bait Carol and us during their sessions. I thought Amanda was going to throw her shoe at him one particular time. Fortunately, we were able to stop her, but we did allow her to walk out of the room.

Then there was the psychodrama. Alexis and Jacqueline did that because Carol couldn't. Neither could Kate and Marie. It was all about the gang rape and being sterilized. Lots of anger showed. They must have pounded at the walls for two hours. The psychologist from the group session was there too. I thought they were going to attack him. We had to stop them, put some limits on it. I wonder what Dr. Jansma thought about it. He came to watch. We were glad he did. It meant a great deal to us, since the psychodrama was very painful for us to watch. So many, many feelings going on. Something else that bothered us was the daily routine. It was difficult to adhere to their schedule. It was much easier for Carol. Once she got over the initial scare of being there, she started to feel almost safe. We never did. We had to be constantly on guard. We couldn't be ourselves and that's very difficult. All in all, we did learn some things there, but it was a relief to get home. I think Carol was afraid to leave, but we'll watch over her, just as we always have in the past.

Ashley Baldwin

The next several months are somewhat up and down. Some laxative use has again begun. Carol comes into one session with three immediate and very appropriate agendas: "I'm close to going on a shopping binge;" "I think I signed the papers for my abortion and sterilization," and "I'm using a lot of laxatives again."

The primary issue relates to "I am a bad person," which has been a very central theme throughout her entire life. In her next session, she confides that there is information she has not shared. There are other targets of immense anger, many of them involving family members. She also reports in early September that: "I have felt like two people since I was a teenager;" that "I sometimes hate kids," and "I sometimes hate my sister." She tells me she has regularly thought of suicide and had one overdose experience at age nineteen. We also have several sessions on the issue of anger at me, anger which results from her awareness that: "I have

13

risked more with you than I have ever risked with anybody and there's a big part of me that resents having done so." She feels very vulnerable in having shared this much about herself. She has a long history of invariably ending relationships when they get too emotionally close. This is an important insight for her.

We also focus on the issue of "What goodies do I get out of my sick relationship with my parents?" and, "Why am I so fearful of therapy ending?" Conclusion: It is better to have a place to belong (even if it's negative) than to have no place to belong. She is finally acknowledging dependency (on parents and on me) issues now. There is some healthy searching going on.

An emerging relationship is developing with her pastor, which I am supporting. She's beginning to go to church and we are tying this into the generalized and very long-standing historical issue of her need to belong. Laxative use is up and down. We have two sessions to use hypnosis for weight management.

By mid-October, "the need to belong" issue apparently hits home for her. There are many tears. She has now applied for a Big Sister program and is assigned a third-grade Indian girl. I am very supportive of this development. We periodically broach the issue of weaning away from therapy in order to discourage dependency issues, and her basic response is to come up with many more "problems." She is clearly invested in justifying the need for therapy.

At the end of October, Carol has become very hyperactive, has not been sleeping, is very wound up. We have been pursuing her childhood more, especially pre-age seven (in the Netherlands) versus post-age seven (America). She sees better her hatred of her grandmother, who henpecked her grandfather. There is essentially a picture of a lack of nurturing from anyone.

As therapy continues, some further memories emerge. She remembers her mom holding a two-week grudge against her, (which was only shown when father was not around), because her baby brother and sister had hid and couldn't be found and the mother blamed Carol. She brings in a picture from the newspaper showing a police officer giving Halloween candy to some kids. It turns out he was one of the seven males who had gang-raped her when she was nineteen years old. She has kept that picture for some time and looks at it daily. She reports that her mom has told her: "I don't want anything to do with you anymore." And one week later is furious with her mother who has regularly been checking up on her at work.

Since it is very clear that this ongoing double messages between mother and daughter guarantees a very conflicted relationship, I recommend an individual session with the patient's mother. The patient reluctantly agrees, as does the mother. The mother presents herself as very nervous and very angry, both emotions she totally denies. She tells me that "the whole family told me not to come in because you're just on her side." She insists there is no problem in the relationship. She comes across basically as a narrow, very defensive woman with a very negative mindset on her daughter. She interrupts me frequently and makes sure that I know that she is very suspicious of me. She is not at all amenable to any further contacts with me, although it does appear that she seems to care about her daughter.

The following week, Carol makes a reaffirmation of her faith in her local church. Her entire family shows up for this. She has also been seeing a local female physician to deal with the re-emerging issue of laxative abuse. About this time one of the personalities writes:

December 21, 1982

I have created myself. There are, of course, things that were inherited and there are also some things that were imitated; but I put together the person I am now. Therefore, I am responsible for any changes that have to be made; no one can do it for me. I alone can make the decision to accept or not to accept those things in my life that I am powerless to change. I am the person that has to decide whether I am going to live in the past or in the present. The past causes feelings of anger and hatred, much of which is directed at myself. Eventually, these feelings will cause more self-destruction and will only serve to destroy any chance I have of becoming a whole person. I have to learn to accept myself for what I am now. I have the capability of letting go of the anger and hatred inside me, but I have to be willing to do so. I will not find myself until I do.

Kate

As 1983 begins, Carol has a friend and the friend's boyfriend over for dinner to her apartment. This is the first time in three years she's had any non-family members in her apartment. "For the first time I really feel I'm getting better."

It is approaching two years of therapy now. Carol is beginning to focus in on her relationship with her father. She has been so aware of her anger toward her mother that it is a revelation to her that she also has anger at her father. She has actively denied over the years that there is a problem here. She is troubled by her anger. She also brings in some newspaper articles on the death of Karen Carpenter and the star's struggle witha-norexia nervosa. She recognizes that some of the dynamics and certainly some of the patterns affiliated with the disease apply to her. She has been bothered by my question: "Why are you still coming in for therapy?" Although her therapy clearly has been lengthy, there are still some missing links. She is, however, functioning. Hospitalization has not been an issue for some time.

She's thinking of going to the Netherlands this summer, which she has not seen since her emigration to the U.S. at age seven. Her motivation is to get a better sense of her roots since she has very little memory of anything prior to emigration. Her parents at this point, surprisingly, have encouraged her to remain in therapy telling her that they see real progress. She is for the first time in a long time out of debt. However, when she later hears that her entire family will be going to the Netherlands with her, she decides she does not want to go along.

By April, 1983, Carol has become very tense again. Although I initially associate this with the subject of weaning away from therapy, upon pursuit she has been finding herself very preoccupied with some very bad past sexual memories. Although it requires prodding on my part, she finally begins talking about what appears to be a long history of sado-masochistic sex with her preferred role as the slave. Her migraine problems have become particularly active and the timing of their emer-gence with her growing memories of past sexual acting-out experiences seems to be a clear link. We focus on the spiritual dimension of her functioning and the cleansing rituals associated with her obsessive/ compulsive behavioral patterns. She has her first weekend with Heather, her nine-year-old American Indian Little Sister.

In her next session, Carol shares a specific bad sexual memory, one which involved a ten-hour-long sado-masochistic sexual experience. Following the emotional catharsis that occurs with this discussion, interestingly enough, her migraine headache disappears. I interpret the cleansing rituals in light of her sexual dynamics. We now know that her laxative use increases (cleansing) when she talks about reports in session some of the bad sexual experiences. She seems to have a growing sense of the significance between her physical symptoms and emotional tur-moil, especially sexual turmoil.

April 6, 1983

Last evening was really rather peaceful. At one point, my mind really wandered and I found myself in a meadow with all kinds of wildflowers. I was barefoot and running in the wind and I had on a long white dress. The sun was bright and I was totally free and at ease with myself, happy to be alive. Then I made the mistake of going to bed, because the dreams started all over again. James (the ex-husband) was beating me again--I was tied to the bed and he was doing all kinds of things to me--I don't even want to think about them, let alone about what I did to him. Mom and dad and everyone else was there part of the time too--people from work--they all said I was nothing but filth, that I didn't belong in society. They all backed away from me, even Dr. Jansma, because he was there at one point also. Someone said that I deserved not to be able to have children because I was evil. I hurt so bad inside; if only I could make them understand that I don't know what makes me do all those bad things. I always wanted to be a good person, but somehow I always did all the wrong things. My mind gets so tired and I get so angry inside. Sometimes I want to kill everything around me, including me. There are times when I feel everyone knows about my past and what I'm really like. I want to hide where no one will ever find me.

Francesca

As Mother's Day approaches, Carol is angry and sullen. "I always get this way before Mother's Day." I am aware that two years ago she was in one psychiatric hospital, and at this time last year she was in the second psychiatric hospital. Mother's Day is obviously of extreme symbolic importance to Carol.

In early May, a friend calls to report that Carol had become faint and weak and could not concentrate. She can barely stand. I speak with both over the phone and recommend that she go to the emergency room. The physician reports that her potassium levels have become a cause for concern. It turns out that Carol has been seeing a "diet doctor" and is using diet pills and diuretics. I "order" her to be in touch with her female physician again; this she does.

There are many agendas going on at this time. Therapy is causing a number of others' to think:

June 19, 1983

I don't believe I have ever really been a child, as far as doing the things children do. My mind was always very busy in an adult

17

way; always scheming and planning. Physically, I started to develop at a very early age--I believe it was in the fifth grade. I remember the comments the other children made in school. I had a great deal of responsibility at home and took care of my brothers and sister frequently. I fantasized about being my father's wife a great deal and I used to pretend I was my brothers' and sister's mother. These fantasies were even more real to me if my father and I happened to go shopping together. If Mom stayed home from church, then I always pretended to be his wife and I acted very grown-up and proper. In fact, I can remember some of the instances very clearly--others I must have blocked out. I learned all the things a wife and mother did very early: cleaning, cooking, laundry, ironing, how to care for babies, sewing, etc. Phillip and I baby-sat a lot, too. Then I had to bathe the younger children and put them to bed, do the dishes and pick up the house--sometimes fixing Dad's lunch for work the next day and setting things out for breakfast the next morning.

Kate

Carol comes in today with an announcement of her decision to terminate therapy. She says that she has decided that it "was just too much to deal with any further and that since I have progressed as much as I have it is time to terminate." She does reveal that she is now under family pressure to terminate therapy. She eventually decides to terminate by September 1. She will continue to talk with her pastor and assistant pastor, as well as her boss who has become a friend. She may even request to bring them in for a joint session, which I say is fine with me.

June 29, 1983

I am not pleased with today's session at all. I was very defensive and felt as if I was battling against Dr. Jansma. There was definitely anger involved because of the feelings that terminating therapy, or even weaning away from it brings on. I don't want any part of it and I resent it. I'm so afraid I'll lose the best friend I ever had. I've worked so hard at maintaining this friendship and I've invested a great deal of myself into it. I have learned to trust him and now I question the wisdom of that because the relationship will eventually come to an end. He doesn't seem to understand that we can't have a life-long friendship because of the doctor-patient factor. In the long run, it's going to be one-sided anyway and I'm sick to death of that kind of a relationship. Better to cut it off now, before I get hurt any more than what I feel already. Besides, he's never going to understand my feelings anyway and I'm tired of trying to make him understand. I'm not letting him read this either. From now on, I'll tell him surface things, but not what's really inside. Screw it! My mind is too tired for any more of this. I've had enough.

Elizabeth

It is mid-August. Weaning away from therapy is now being emphasized. We have agreed to not have any further two-hour sessions. We have had a session with the assistant pastor and her boss to focus on the subject of how she can use these outside resources without becoming dependent on them. She begins a college English course this week and has also begun to teach Sunday school. Her sense of humor is re-emerging. We will see how she manages. At the end of September, Carol is in the emergency room because of heart irregularities. Concentration is poor. It appears this is due to potassium deficits and she is again referred to her physician. We hope that she is now frightened enough to mean business about her laxative abuse. In her next session she says: "I don't care what you say, I despise one-hour sessions; I don't get anything done in only one hour." With that she storms out. Two days later, Carol is furious about something else. Her parents have returned from their time in Europe and are blaming their kids for "ruining" the vacation. She says she intends to tell them all off at this point.

By the next month, Carol is bringing up new material, specifically new memories regarding past incest. These memories include the grandfather and a great-uncle, as well as a non-intercourse "rape" in the seventh grade with a group of black students. She's never shared this information with anyone before.

As therapy proceeds, she relates to me many bizarre thoughts and dreams, which have been recurring; and which have been going on since pre-teen years. For example: "The scenario of a naked little girl, on a table, on a stage, with many nude male and female adults; they're to teach her about sex but the little girl winds up teaching them." She also reports another recurring nightmare involving a young woman, on stage, getting sexually molested by a number of men and women, and finally getting revenge by mutilating and killing them all. The morning after this last session, Carol wakes up with blood on both of her hands. She reports that this has occurred before, apparently sporadically since her teen years but

19

has not occurred for some time. She has no idea where the blood comes from. She's very troubled by this. From my perspective, there are emerging indications of multiple personality dynamics. Although this had been considered a number of times throughout therapy, there has not been opportunity to confirm this possibility. This is a very scary issue for Carol to talk about. She does report that she has one sweater with different initials on it. She later writes down two names that she finds herself identifying with: Andrea Marika and Francesca. She's quite frightened at this point. After a troubling weekend, she begins opening up regarding a sense of other personalities. She reports that she spent the time trying to "deny that this fits me, but I can't deny it any longer." She finds herself very leery of hypnosis right now.

December 5, 1983

So many things are happening in sessions, (and in my head). Who am I, really? I have always thought of myself as more than one person, but to have Dr. Jansma bring it up is very frightening. Yet, it's also a challenge to see if I can out-smart him. Today I can be me and it feels good to finally let myself come out. I knew I'd win sooner or later. I'm going to have to be careful, though. I tend to be a bit bitchy. It gives me a real sense of power, because I can change to suit the situation. It's just that I have to watch it a bit.

Francesca

December 7, 1983 (phone conversation). "Francesca" calls me. This is the first public identification of herself or any other personality to me. "I have worked hard to get control. I will destroy Carol before I let her win." (She will not reveal any further what she means by this.) She has apparently called to check out whether I want to meet her and volunteers her hesitation to meet me because "your office is your territory." After almost three years of therapy, this presents as the first overt evidence of multiple personality dynamics.

3

Francesca comes in as promised on December 8, 1983. There is sufficient evidence (coffee instead of tea, a different chair, very subtle voice inflections, and much less carefulness), that indicate fairly clearly that Carol is not faking the existence of another personality. She leaves the room after the first hour and Carol is the one who returns to my office. (In multiple personality, it is usually the other personalities who have the power of control and are able to switch in a fraction of a second.) Carol is quite troubled. She sees cigarettes in the ashtray by the other chair and knows, (especially when she realizes that one hour has passed), that someone else was here. Francesca has reported that there are "many others." Carol needs much support at this time as this is extremely confusing to her. I try to encourage her to see this as a very positive development. At this point it would appear that a diagnosis of multiple personality can be confirmed.

December 8, 1983

I'm Francesca Germaine. This is my first experience with writing anything and I am a bit uncertain about it. In a way it seems kind of cold, but then that's the way I am, isn't it? Cold. At least I always thought I was. I am starting to learn a little bit more by talking to Dr. Jansma. I am not all that cold and I am not all that hard. It is very difficult for me to express or show any feelings or emotions. I don't know how well I'll do on this.

I first came into being when Carol was four years old. Kathar-

21

ine was already there. Katharine needed a little bit of help by this time and so, of course, did Carol, which is why we were there. Katharine had already been subjected to numerous beatings, quite a bit of sexual abuse. She just couldn't deal with it, and so I came. There was an instance of sexual abuse; forced to have sex with her grandfather. Some day, I am going to get even with that man because I hate his guts. I have never been quite able to understand why he did all those things to us, or why any of them did, for that matter. It wasn't just him. It was the grandmother. And I have always tried to fight back in my own way, by getting even, by trying to not let them get the best of me, but sometimes they did. No matter how hard I tried not to show them anything, not to feel anything. But the hurt and the pain is still there and the memories just don't ever go away. I don't know why people are like that and I didn't know if I ever wanted to grow up because I was afraid I would be like that. I realized very early that I wasn't going to have love like other kids had from their parents. I wasn't going to have a caring atmosphere in which to grow up in; an atmosphere in which parents encouraged them and gave attention to their kids. That hurt, but I decided I wasn't going to let them know that it mattered to me. And so I'd fight--not physically like Mandy did, but in other ways. I tried to out-smart them, out-manipulate them. And sometimes I succeeded. But the times I didn't, and they caught on to what I was doing, they beat the hell out of me. Of course, that was par anyway. They did that all the time.

Very early, I wanted them to pay. I often wished that they were dead. I did have one escape though and that was my drawings. I never talk about my drawing anymore. It's been a long time since I've done anything like that. Sometimes I want to try, but I am afraid to. I tried once a long time ago and my hand hurt. I couldn't do it. My drawing used to mean a lot to me, but I learned that I wasn't any good at it. It hurts when you learn that you're not good at something. But if it's something you are forbidden to do anyway, then why try?

The first couple of years that I spent with Carol were sad, but they were eye-opening as well, because I knew exactly what I could and couldn't do, although sometimes even when I'd do the right thing, I would still get punished for it. Everything I did wasn't good enough and I learned even as a child that you don't make mistakes. Because mistakes mean pain. It might mean a belt, it might mean being put in a hole in the ground, it might mean having your clothes taken off and having to have sex with someone that was a grandfather. See, my own parents weren't like these people at all. They were caring people and they loved me, but Carol needed a lot of help. (*Note: Francesca had created her own parents in her mind.*)

She couldn't handle that mother alone. She couldn't handle that father alone either. Katharine was there, that's true enough, but sometimes things got a bit much for her too, even though she was the first one there. But sometimes Katharine just couldn't take another beating and she couldn't take that hole anymore. Then I'd take over and we ended up sharing a lot of experiences in those early years. Some of them were different, but the punishments were for approximately the same reasons-a smudge on my dress, the part in my hair not good enough. But what the hell is a four or five-year-old supposed to do, you know, I mean...you can't even quite see in the mirror because it's up a little higher than you are tall, but that didn't matter to them. You know, it didn't matter that I was a child--I was a four-year-old or a five-year-old, who supposedly had to act like a fifteen-year-old or a twenty-year-old. "Mind your manners, don't spill your milk, part your hair correctly, make your letters perfect."

I went to school with Katharine, because Katharine didn't like to do numbers and so I did them. At first they were rather intimidating, but it got to be fun after awhile. At least, they were a change of pace. I mean, anything was fun instead of being home. I was never really that crazy about school, although I liked learning--it was just that it was confined. You had to sit in this room all day and so I created games to play while I was doing numbers. For example: I'm six years old now, so if I add ten to that, that makes me sixteen--age sixteen is old enough to kill her grandfather. Or, in so many years I can run away. I tried that a couple of times, like at age five. And the punishment was so severe that I realized I wasn't quite old enough to do that. I'd have to wait a few more years.

Phillip, (Carol's real-life, one-year younger brother), was cute though. He was a good baby. The problem was that Carol, Katharine and I were pretty good too. We didn't cause the hassle, they caused the hassle. We were always the ones who got punished, Phillip never got punished. He did like us though, I know he did. He would want to play with us, and we liked to play with him. And I sometimes wonder if he knew something was wrong. He was a year younger than Carol, but even at that age, it seemed as if he knew something was wrong. And every once in a while, he'd crawl over to us, to where I was sitting in the corner or whatever. And he'd put his arms around me and hug me. He especially knew that when I sat on a little chair in the corner, I was going to get punished later. I always knew that too, but it was strange that he picked up on that. I guess he associated that chair and my sitting on it with a punishment. He knew! Because then when I would get whipped or they'd start to take me away to the hole, he'd scream bloody murder. And, you know, the mother would just pick him up and cuddle him and laugh with him and try to calm him down. But she never slapped him for crying like that. I have a feeling he woke the whole neighborhood up when he did that. He had a good pair of lungs.

But I have never yet been able to understand why they loved him and took care of him while, at the same time, they hated my guts! I knew they did. From the time I got there, I never had any question about that. I knew they did! And I trained myself to believe that it didn't matter. But it did. Especially when I would draw a picture and I'd show it to one of them and they would never make a comment. Sometimes they shoved it aside, sometimes they told me it was shit, sometimes they just ripped it up. So I learned, there too, not to show them my pictures, because they'll get destroyed. And so I kept a pile of 'em and I always kept them hidden. I think that they suspected that I drew. In fact, I don't even have to think about that, they knew that. But I never showed them my pictures anymore.

And then they came up with the bright idea of coming to the United States. That really bothered me because where we lived was pretty. It had wildflowers and it had the sea and I liked it. But they wanted to move on. I think it was the father more than the mother. The mother's family was going to stay behind, you know; her parents and brothers and sisters, but her dad's parents were over here in the United States. They'd come here the year before, and that had given us some relief from the sexual abuse, because as long as grandfather was over here, we didn't have to put up with him. Well, Daddy must have gotten lonesome, or something, because he decided he wanted to go to the United States. Better opportunities, better life so, we didn't really have much say in the matter, we just had to go. And so we did--in a big airplane. I was so scared, I didn't want to come here at all. Besides, I knew what lay ahead. Good ole grandfather lived here and I knew he wasn't going to just let me go about my own business. He was going to have us again and again and again. And he did. The very first

24

night we were here. Katharine had taken over; she was there first and she just couldn't handle it, so I took over. That man raped me! He took off all my clothes, told me to shut up and he hurt me and I really, really hated him for it and I wanted to go back to the home. We did have a set of grandparents there that were nice to us; took care of us over there every once in awhile. They even spoiled us a bit. But here, here I was with this grandfather. I still want to get sick every time I think about him touching me and how he'd put his penis inside me, and I thought I was just going to split wide open it hurt so bad. And then he'd turn me over and he'd spank me. I think he enjoyed hearing the sound of his hand against me as he slapped me. He did that every time and then he'd turn me over and then just put it in my mouth and I got sick. I threw up and then I had to clean it up. And, you know, the rest of the people were all down stairs. Her parents, her grandmother, aunts and uncles, because it was our first day here in the United States and you can't tell me that they didn't know what was going on. It doesn't take that long to tuck a child in. But no one cared! No one made a move to come upstairs and see what was going on, why it was taking so long. I was such a mess when he was done with me and I was so sick. But they didn't care and he certainly didn't. His parting shot when he'd leave would be in Dutch...in English the equivalent is, "See you tomorrow," and I knew he'd have me again the next day.

And Katharine felt so terrible. She usually was the one that took his sexual abuse. Even though I took some of it, she's the one that took the most of it, but this particular night she was so tired and she was really upset about having to come here and I felt so bad for her, but she was blaming herself because I had to put up with that and because she wasn't able to do it that night. None of us wanted to be here. Even Carol. We had given her a little bit of time so that she knew what had happened, where she was, not about grandfather, but she knew that she was coming to the United States. And she didn't want to any more than we did.

They were such terrible, terrible people. Back-track a little bit: the time I tried to run away when I was five--the first time. I thought that if I could get out of the town that we lived in, then maybe I'd find a place where they would love children, especially ugly children like me. So I pretended to go to school, only I didn't go. There was only one main road that went to the center of the town and that connected with another main road that went to the villages on either side of the one that we lived in. But I knew a

25

back way. To go the main way would be to go right past the house. Being a little kid, you don't walk very far or very fast and I got tired. I got hungry too, but that was nothing unusual because sometimes we weren't fed anyway. But it was after supper before they caught up with me. I didn't sit down for a couple of days, I hurt too bad to sit. I had to sleep lying down on my stomach. They had to keep me out of school for that one. I don't know what kind of excuse they made to the teacher. I know they made some kind of an excuse. I didn't have anything to eat for one and a half days.

I always wanted to take art lessons. I said earlier that I need to draw. It provided me with an escape. It took me right out of this world and put me someplace else where I didn't have to think that anything was bad. I liked to draw things that I saw. It didn't matter if it was the house, or maybe the fire station. On the corner there was a fire station and in nice weather two firemen always sat outside. On the way home, I could talk to them and they were really nice to me. In fact, one time my arm was all black and blue and they asked me what happened and I just had to tell them I fell. I didn't dare say anything else. I couldn't tell them that Daddy had thrown me across the room the night before. But drawing, that was special: flowers, trees, fire stations, houses, cars, people. I loved drawing flowers. One day, my grandfather ordered me to bring him all my drawings because he caught me drawing earlier in the day. I wasn't supposed to draw. So I went and I gave him all my drawings and he told me that they weren't any good, that they were rotten; that I wasn't an artist and I never would be, and he ripped some of them up. He kept the rest of my drawings, sent me home, and told me to be back there the next day. It was in the summer. So the next day, I went back over there and he told me to draw him a picture. He told me that if he liked it, he might let me keep it. So I drew him a picture. It was a picture of the farm, the farm that was across the street from us in the Netherlands. It was a great big farm. And when I got all finished with it, I gave it to him and he said it was the worst thing he'd ever seen and I knew that wasn't true. Then he made me stay in the chair and he tied my arms in back of me to the chair. He tied my right hand especially tight and he took out the drawings, including the one I had just finished for him. He ripped them all up into little, tiny pieces. I tried to get him to stop, but he still wouldn't. I cried and he slapped me across the face and then he took a few little pieces and he burned them, right in front of me in a dish and he set them on fire. He did this every day for a week--every day, for seven days and after the first time I decided I wasn't going to let him see me cry and so I didn't cry, but it didn't stop him either. Afterwards I couldn't draw anymore. My hand hurt too bad. Besides, I'm not any good at drawing anyway and I should have known that but it was something to do. But it hurt so bad, inside, when he did that. I hated him so much. I don't think I have ever hated anyone as much as I did at that particular time. I've hated before that and since that, but never as intensely. I never drew again. I tried a couple of years ago, only my hand still hurts. It wasn't--didn't even look like anything. But sometimes I still want to draw and I still want to take art classes, but it's so stupid, because I was eleven when I quit drawing.

I don't know if things are ever going to be right for us. They haven't been in a long time and I know that, but we've always coped pretty much on our own and so have I. I know I should probably be talking in session as far as why I hate so much and why I want to get even so much. Why I'm so angry inside and why it's so hard for me to show anything like that. It's hard to talk. I mean, we talk amongst ourselves. We always have, but that's different, really different from talking to Dr. Jansma and I know he's our friend. Sometimes it's hard to believe that I have a friend, but I do want that friendship. I have never had one. The friendships I've had have been with people who use other people.

I am getting a little off the track here. This is supposed to be my autobiography and I should be trying to stay in some sort of chronological order, but this makes a lot of things run through my mind; like things about childhood that I've tried to hide for a long time. What it was like to go to school here, and there too. I didn't go to school in the Netherlands very long. There are even people that I was growing up with in school whom I still want to get even with. Things people did that hurt me, hurt Carol. People are so damn cruel. People like that don't deserve happiness. They took away whatever happiness I knew at the beginning of my life; that was the caring I had from my parents (*Note: her imaginary parents*) and later, the caring I sometimes had from the one set of grandparents, but outside of that, there's been nothing. There's been no love, there haven't been any hugs. There's been a lot of emptiness, lots of years that I've hurt, lots of years that I've hated a great deal. And I don't know if that's going to change. Sometimes I want it to, but then I can't have my revenge, can't I. And I want that revenge! I want them to pay! I haven't quite figured out yet how, but they're gonna pay. It seems the only right thing for them to do because you don't take a child and beat it until it's bruised all over, and you don't let it go hungry and you don't rape it, and you don't make fun of it and you don't call it ugly, because that child doesn't forget those things. She remembers and they influence her. They make her hate and they make her angry and they take away the chance to have a real life, a normal life. I don't even know what a normal life is! I have never known what a normal life is. Deep down, I used to envy other people because they all had the life I wanted to have, but after awhile, you don't care anymore because you know it's not going to happen anyway. At least I didn't. But I found other things. Maybe that's why I enjoy working with computers and word processors, you know. They're a challenge, but they don't talk back to you, they don't beat you, and they don't fight you. There's no feeling there. You have to make sure you punch the right buttons, but outside of that...maybe that's why I analyze everything. I don't do anything without thinking about it. What I do has to be perfect. I am still finding ways to get even. After all these years, I am still trying to find ways, and I am going to find them.

According to some people, that's not the right attitude to have. But I'm thirty years old (*Note: Carol is actually thirty-six at this time, but Francesca views herself as thirty*), and I don't see myself changing anymore. I am not ever going to be free. I am not going to have the chance to just be me, so why hope for something that isn't going to happen anyway.

One more time,
one more humiliation
after so many allready
past.
Don't cry.
Mustn't cry.
A silent sob.
A tear.

Marie
Febuary 10, 1970

I'm sorry!
Please--
I'll be good.
I'm sorry.
Bare skin.
A belt cracks.

Marie
November 3, 1963

28

4

Carol calls for an extra session on December 9, 1983, but Katharine is the one who comes in for the session. She is a quiet, musical, artistic personality. She identifies two others, Jennifer and Alexis. At this point a total of seven clearly identified, separately functioning individual personalities are "out of the closet," and there appears to be a number of others. Katharine lets me know that she has been in session before, observing and participating without my awareness that I was talking with someone else. She also reports that she wants Carol to get better and that she is very fearful of Francesca. I recommend that she let Carol know about her existence.

Katharine arrives for a following session and reveals an eighth personality, Alexandria. Katharine appears to function as an information-giver and protector for Carol. Carol appears to be regressing and is extremely frightened. She has begun to think of hospitalization, which we discuss and mutually agree to try to hold off.

Carol tells her boss about the issue of multiple personality. She and Katharine invite him to join them in a session to which I agree. We had agreed to use the trigger word "classy" for Katharine to emerge, capitalizing on her ability to immediately "take over" from the others. About ten minutes into this session, with Carol struggling, I have Katharine emerge. She introduces herself to Jason, Carol's boss. He is very supportive and is clearly invested in her not losing her job. I give him a general explanation of multiple personality to defuse his uncertainty.

It might be difficult to understand how a multiple personality could hold down a full-time job. Actually, this is simply another example of the amazing division of labor characteristics that is part and parcel of why the survival technique of multiple personality happens in the first place. Multiple personality happens because there is a need for "recess" from trauma, so another personality is unconsciously created to provide that

29

"recess." It then becomes very easy for a given personality to become the "expert" at handling various aspects of the birth person's life. For example, one personality may handle the finances, one might be responsible for a certain relationship, one for handling a certain feeling etc. To handle the complexities of a full-time job, often several personalities will share the responsibility. Because all the personalities know who the birth person is and how she acts, even though she has amnesia for all this, these other personalities become very adept at role-playing the birth person. To all of us "outsiders," we will simply see one person who is able to do her job.

Francesca appears for a session. She is quite untrusting, has never had a friend, and states. "I need no one." She comes because I had mentioned to Carol a few weeks earlier that I hadn't seen much of Francesca and also because "I decided it was time to see you." She appears to be testing me. She wants to know my opinion of Katharine, (as Katharine previously wanted my opinion of Francesca). It appears that they are vying for control. She does tell me that the blood on Carol's hands, which has been occurring since teen years, was her own blood. I ask to talk to Carol, which is accepted. Carol is quite depressed; there is clear suicidal thinking and she says, "I want to return to therapy as it was before the multiple personality issue emerged. The last issue I remember talking about was grandpa. Is that why the others have come out?" At the end of the session Katharine emerges smiling. We have begun to experiment with quick ways of using hypnosis to foster the emergence of different personalities.

Some of my questions about Katharine's emergence and role are not answered until she writes her autobiography about one and a half years later:

April 15, 1984

The first time I came into being was when Carol was two. I did not take an active role in her life immediately. I observed at first. I observed people around her, her parents, her grandparents, she had a little brother. She was a very sad child even at that age, and I was sad too, because I had parents that cared about me and loved me and wanted me, but she needed help so I came. (*Note: here again that Katharine's parents are imaginary.*) I watched as they abused her, both physically and mentally. Even at that age, her grandfather was taking such liberties with her; undressing her, playing with her, touching her, caressing her, hurting her, hurting me. We were often punished for slight things, like dropping a toy or crying or not eating properly, but then two-year-olds and three-year-olds do not always eat properly.

When I was four years old, we were left with her grandparents for the day and, as was the usual routine, he undressed me. But instead of the usual caresses or touching, he went a little bit further and he spanked me hard, first with his hand and then with a belt. Then he laid me down on a table and he put something in my vagina and I don't even remember now what it was. All that I remember is that it hurt and that I cried. That made him angry and so he whipped me some more. After that, he cleaned me up and dressed me again and said I was not to speak of it to anyone. I had to call him grandpa,' even though he wasn't my grandfather. (Note: be reminded that Katharine has her own imaginary family.) I tried to do all the right things in that family so her parents and her grandparents wouldn't get angry at me. I always tried to make sure that I stayed clean and that I didn't make any noise, that I didn't cry and even though I was four years old, I knew how to make my bed and I knew how to clear the table. I even knew how to help with the dishes. On Sundays, we went to church. Then, while they went visiting with their friends and their neighbors or they went for walks, I had to go to bed all afternoon. They always took Phillip with them. I was always afraid that they would hurt him, just like they hurt me, but they never did. He was always the good little baby. He was kind of passive then; I was too. I did like to draw, but it had to be perfect. They always threw everything away after I finished it. I would try so hard to draw something nice for them to make them like me, but I never could.

Once in awhile, I was allowed to go on a walk with them and I loved it when we would walk to the sea. We would climb up the dike, and it was so beautiful to me. All the big waves crashed and made a lot of noise. And there were the birds and it was usually windy and sometimes the spray from the sea would hit me right in my face and it always felt so good.

I didn't have many friends, just two. One of them had the same name Callie did. She lived next door and we used to play, but I was only allowed to play for a certain length of time each day and then I had to come inside and do my chores. I had to clean the table, help with the dishes making sure not to get myself dirty in the process. That was always a good thing, not to get myself dirty in the process. I had to stay perfectly clean, because if I didn't, then I would get whipped. I always had to wear a bow in my hair, usually white. Sometimes it matched what I was wearing, but usually it was white. That couldn't come undone either. There were many times when I wanted someone to hold me and to read me a story and to rock me, but they never did. My own parents used to do that for me, but these people never did. I don't think they liked to touch me at all. I learned to lose myself in nice things, like when I was outside and I could look at the little flowers

31

that were growing around, especially when spring would come and snowdrops would come up through the snow. They were so pretty. They were white. I always liked white. I think white made me feel good because it was so clean.

Sometimes I played with Phillip. He was so good. I remember when he broke his leg. Because of that, his one leg was always a little bit shorter than the other. I got punished for that too, because I didn't take good care of him. I should have kept an eye on him a little bit better. Boy, did I get a beating for that one. I don't think I could sit down for a week. I remember that the father shoved me so hard that I fell on my face and that I had a nosebleed and I didn't get anything to eat for two days. Why? And I wanted to cry so bad. But if I cried, it would just make it worse. And so at night, when I thought they'd be asleep, I would stick my head under the pillow and cry. But I had to do it real quiet and not make any noise.

Her mom's grandmother lived with us too. She was a real old lady and she never said anything--just sat there. Once in awhile she would do something; like peel potatoes or crochet or something. Not very often. She usually just sat there. She died when I was four or five. I was scared, because they kept her right in the house until they buried her. I had to go look at her and touch her and I can still remember how cold she was. I didn't ever, ever want to be that cold. I couldn't get warm for a long, long time after that. Our house wasn't very big and they kept her right there in the house. Our beds were built into the wall and they had curtains in front of them. They kept her in the one next to mine. I didn't like that at all. She was dressed in a long black dress. And every day until they buried her, I had to go up there and I had to touch her. The last time I had to kiss her and I didn't want to. I tried not to, but I got spanked because I didn't want to, because she was supposedly my great-grandmother, but she wasn't. She was Carol's great-grandmother. She wasn't part of my family. I was just there to help Carol and to take care of her.

Sometimes, I just wanted to leave. There wasn't any love in that family at all! They didn't know how to be nice, or kind. I don't think that they saw any good in anyone and especially not in me. I couldn't ever do anything right. I imagine that it must have been really hard for Carol before the rest of us came there.

When I was four, I started school. It was more like what we call a nursery school today, but it was school, nonetheless. Phillip cried so hard because he didn't want me to leave. I liked school. At least someone was nice to me there. What was really neat was that the sheep came right up to the windows. On certain afternoons, the girls had to stay after school and learn how to knit. I hated knitting with a passion. I would rather draw or be outside and pick flowers. I made daisy chains a lot. I used to like to paint. And I used to like to write too. I didn't write stories or anything, but I used to like to practice letters and besides, I knew I had to get them right and that they had to be perfect because I couldn't bring home anything less than that. The only thing I had real problems with was math. I didn't like numbers. I liked to make the numbers, but I didn't like what I had to do with them. So I used to have to work extra hard at that. It got easier when Francesca came because she was better at numbers than I was.

It was difficult at times because I'd get punished for things that 'Cesca or Mandy did, and sometimes Carol would get punished for things that one of us did and we never really meant to let that happen to her. She would say that she hadn't done this or that and, really she hadn't, because we did, but then they would say she was lying and so then she would get put into that hole in the ground, only we never let that happen to her. One of us was always put in the hole. For hours and hours at a time, we'd be put in that hole. The first time that happened to me, I screamed and screamed and screamed, but they just laughed and said that I would have to stay in longer because I was screaming so hard. It was cold, wet in there cold and it had worms in it. I couldn't climb out of it because it was too deep. I hated them for doing that to me because I really hadn't done anything wrong. That stupid button just fell off my dress. I didn't pull it off or anything like that. It just fell off. It was so terribly cold in there and I was so scared and it was a long, long time before they came and got me again. It was dark by the time they came to get me. I got spanked because I was dirty, but I couldn't help it because I got tired of standing up and I sat down, only it was wet. And when they pulled me back out, my dress was all wet and I had dirt on it and I had dirt on my face and on my hands and on my legs and I was so cold because I didn't have a jacket or anything. But I learned that the next time I couldn't sit down or move around because I would get dirtier than when I got put in.

Strangely, I didn't think that doing an autobiography was going to be so difficult. Remembering things, I mean. After all, I am now thirty-eight years old (*Note: Carol is actually thirty-six years old*) that was a long time ago, and it's not going to happen to me again. But, I still think about it sometimes. They can't do that to me anymore. Some of the others are still afraid it's going to happen to them. What surprises me too, is that I want to get even sometimes. That's not like me. I have tried, probably because I wanted to be the opposite of what they were, to see good in people and to find things that are beautiful and that mean something to me. That's how I used to try and help Carol too, by coming back and drawing. Sometimes they were to please myself, to see something pretty and sometimes they were to try and please them, to stop another punishment. I guess I appreciate quiet things because life with them was always so violent. And so my escape has been to be the exact opposite. I don't like violence. I don't like noise. I like things that are white and clean. Maybe all these things are a way of getting rid of some of the dirt I see on myself sometimes. I only like soft touches. I don't like people coming up behind me. I like my solitude, I guess. That doesn't mean I don't like people, I do. But when there is too much noise and too much confusion, my mind says "enough," and so I retreat.

There are so many things that happened to me that have happened to all of us. I remember someone teaching us to ride a bike in the Netherlands. Jenny experienced part of that and had great fun. I thought it was a nuisance. I made the mistake of falling off that stupid bicycle, and I got whipped because I wasn't coordinated. What they didn't understand was that I didn't enjoy bicycling anyway. I would much rather sit and read a book.

Carol had another set of grandparents and they were special--they were nice people. Every once in awhile we would go over there. They lived in another city. Grandfather worked for the railroad. Sometimes he would take Phillip and me with him to the station where he worked. That was neat. Their house was right next to a railroad crossing and, at night, I would lay there in bed and it would be all nice and warm and snugly because this grandmother really cared about me. And you could hear the train going by and it made the loudest noise. And I used to imagine myself on that train, just going away. Sometimes we did ride the train and it was fun. Grandfather didn't say a whole lot, but he pulled me up to sit on his lap only he wasn't like her other grandfather, he didn't. Her other grandfather always, always tried to take off my clothes and he did a lot of things to me. But this grandfather just let me sit on his lap and he would hold me. Grandmother would make us a special pudding with raisins in it. It was always a special treat when we went there, because they knew I liked it a whole lot. Phillip did too. None of us ever got spanked while we were there. We were spoiled. Francesca would go too. She was always very, very bright. She enjoyed showing them what she had learned. It was quiet there, a quiet time for our minds. We had a breather because they didn't have a hole in the ground and they never, ever hurt us. If I got dirty, I wasn't spanked or whipped. At night, this grandma would give us a bath, and she would put powder all over me and things like that, but she didn't hurt me while she was giving me a bath. Then one of them would always hold me on their lap until I almost fell asleep and then they put me in bed. It reminded me of my own parents (*Note: Kate's imaginary parents*), because that's what my parents did. Sometimes this grandmother would look at me and she would have the saddest look in her eyes. I don't know why. Grandfather would sometimes just pat me on the head and give me a piece of candy. I always hated it when I would have to leave there and go back home because even though her mother and father went with me sometimes on these trips, they never touched me when we were there because her grandmother took care of me. If I fell and hurt myself, I was allowed to cry. I didn't though, not even when she told me that I could. Sometimes I really, really missed my own mom and dad. But I learned some people need help more than others and Carol needed my help. She needed all of us. There were so many things going on all the time that I couldn't keep up with them by myself and then one of the others would take over.

Sometimes I would get to run to the store by myself though, and her mom would give me some money for a loaf of bread or something like that, and I could go down to the bakery all by myself and get it. That was always fun, too. I liked doing that. But I always hurried there and hurried right back and didn't stop to talk to anyone because if we were gone too long, well, then it was usually the hole in the ground again or no supper and probably not any food the next day. Or a whipping. I did try. I tried to keep things smooth so there wouldn't be any trouble. I tried to do things just like I thought they wanted me to do them only it wasn't right, it wasn't good, usually, so then I would try harder.

I didn't want to come to the United States either, because it meant leaving the good grandparents, and I knew that my own parents weren't going to go. Her bad grandparents had moved there just a little while before we went over there and I just didn't want to go. I wanted to die, because then I wouldn't have to go. But, we went. When we got to New York, we had to go to Grand Central Station and get on a train and come to Michigan. That wasn't so bad. The train was fun. But when we got to our destination, I saw them standing there waiting for us and I threw up. All over myself. I was so sick to my stomach because I knew what was going to happen again and again and again and again. It never stopped. I wanted my own mom and dad. But they weren't there anymore. It was just me. And I wanted to die again.

I was only seven and I wanted to die. It was kinda hard to have a positive attitude about life when all I had ever known was neglect and abuse, except when I was at the nice grandparent's house. I never did get to see my parents or good grandparents again. Instead, the very first night that we were here, grandpa was back up to his old tricks and

I was so homesick. There wasn't anything I could do about it.

Then there was the humiliation I went through. How humiliating it was to be without clothes, to be on my hands and knees so that I could receive my daily whipping. That's usually how the day started--to teach me to be a good girl. Well, they taught me all right. They taught me so well that even now when I don't do things right, I hate myself. And things didn't get any better after we moved here. The first night here, Francesca had to take over because I couldn't. Some of the others shared in that experience too. I hated myself for not being able to do it, but I couldn't take the thought of having him touch me again. I couldn't take being raped again. That day, I just wasn't strong enough. I was homesick. I wanted my parents. I wanted her other grandparents. It was a different country, a different language. It was terribly hot, I remember that. And Phillip was crying because he was homesick too. We weren't in the same bedroom though. About an hour after we were put to bed, grandfather came upstairs and woke me up. Francesca shouldn't have had to go through all that because she'd had enough already too. But she did, and she got sick just like I always did. Then she had to clean it up and all I could do was to sit there and watch. I couldn't even help her. I should have been able to do at least that much for her. I will never forget the look on her face when he was done with her. There was hate and fury and there was sadness--incredible sadness and a resignation that seemed to say, "I didn't think it was going to change anyway." I guess that we had all been hoping that maybe after their big move here, the grandparents would have changed. But they hadn't. They were still mean and spiteful. I knew then that things were going to be the same as they always had been, but even worse because I couldn't speak a word of English. And now we actually had to live in their house! Up until this time, we at least had to travel a little ways to see them. But now, we were living with them and we would be seeing them everyday and I didn't want to see them at all. I hated their guts! I still do. I hope they burn in hell for it.

Another thing that really strikes me is how pious they all were. They went to church twice on Sundays. On Sundays it was not okay to ride a bike or play outside, but it was okay to beat your kid. I wasn't their kid, but everyone thought I was, so it was okay to beat me. It was even okay to put me in the hole on Sunday and not feed me. They actually read their Bible at meal times while I sat in the corner and wasn't given anything to eat. I struggled with that one for a long time. I guess I finally came to the conclusion that they just can't be Christians. I don't see how that's possible.

Phillip and I were even sent to the Christian school. Wonderful, isn't it? We got here in August and were sent to school in September--scared to death. All these strange kids, strange words coming out of their mouths. I got put in a speech class. Up until that time I had always been called Callie, which was Carol's nickname. All of a sudden I'm supposed to answer to Carol. That took a little while to get used to.

Phillip and I spent a whole school year at the Christian school. Then they decided they couldn't afford to pay the tuition, so they pulled us out and sent us to public school. Actually, I was pretty happy there. Fourth grade and Miss Baxter: I learned a lot from her. Had her for the sixth grade too.

We had been here two years. By this time we were living in our own house. Of course, that didn't stop things from happening to us, because I still had to go over there. Grandpa would call and want me to come over and visit, and you didn't say no to grandpa! Every time I went, he had his fun and games. Even here, he had a trunk. One time I tried to defy grandpa. When he decided he wanted to have sex with me, I decided I was going to run. He was bigger and he caught me and had me anyway. Then he decided I was going to spend the night. He informed her parents that I was going to spend the night, only I didn't spend the night in a nice warm bed, I spent it in the basement in a trunk. It didn't do any good to scream because who was going to let me out? At least it had some air holes so I could breathe.

When morning finally came, he let me out. I remember being so stiff that it hurt to move. Every part of my body hurt. I hated him! I had to go to the bathroom and he wouldn't let me go. I finally couldn't hold it anymore and, of course, I was all wet. Then he spanked me, hard, and he sent me home, where I got spanked again because I was wet. I hated myself for doing that. It made me feel dirty. But I was only 7 years old! I swore that I would never let that happen again and, to this day, if I get tense or if something frightens me, I may have to go very badly but I won't be able to. It's put me in the hospital a few times. They never could figure out what was wrong with the bladder. I guess I know though.

This is not easy. I told Dr. Jansma that today. It makes me feel very alone. It makes me want to have someone hug me and tell me everything's okay, just like a little kid. I'm supposed to be so grown up and in control of things, a protector and mother to all the others, and I am having a terrible time coping with my own feelings.

I still wonder why the adults had so many fights. Her grandmother, her dad's mother, hated Carol's mother with a passion. And she never did like her own son, Carol's father. She was then, and she is now, an evil woman. She is a schemer. She never liked me. She used to pull my hair. It was tough to go over there and do her housework. Even when I was seven, eight, and nine, I had to do her housework for her. Fix supper, then have to stay for supper so grandpa could have his fun. That must have been his form of relaxation in the evening. Take a little girl and rape her. Force her to take his penis in her mouth, beat her and make her submissive, make her into a good little girl. I hated being a good little girl.

I couldn't fight them. I already hated myself so much that I wanted to die. Only I couldn't do that either. That's why some of the others would help me. They'd take care of me and of Carol. We're so much a family and when Dr. Jansma talks about integrating us into one person, it scares me so much, because I don't want to lose them. All we've ever had is each other. There hasn't been anyone else. Except for now, because now there's Dr. Jansma. Even though he doesn't know all of us yet, we do sense that he cares and that we can trust him. Even the little ones feel that way. It doesn't mean that we're not scared though. I know a lot of them are afraid that he is not going to like them. I used to be afraid of that too.

There were times when I didn't even feel human. I felt like a thing. I didn't know what it was like to wake up in the morning and have someone hug me and say "I love you." I was just a thing, I was in the way. Again, again, again and again, they stepped on me. I never understood why. Even sick people don't do that. Not to a child. They just didn't give a damn. I don't think they cared whether I lived or died. They really didn't care.

The only real escape that I had was finding things that were pretty. I wanted to float up on top of a white cloud and it would be clean and soft and it would be warm and it would be something I could sink right into. I wanted to be clean, and I don't think I've ever been really clean again. How can you be after someone's hands touch you that many times? Children don't know about sex. Is that what attracted him? What did I do wrong to draw his attention to me?

5

It is later in December, 1983, two weeks since confirmation of multiple personality as a working diagnosis has been made. Five-year-old Callie emerges in session. This personality only speaks Dutch so we communicate with hand signals, a few words, but especially by drawing. I use hypnosis for Carol to re-emerge, but Katharine arrives, having brought in four of Callie's books. As she says: "The kids are the ones who can help you help Carol. Callie is the one who holds the key." Katharine also gives me a Christmas present. It's a small basket with seven separate food items in it, each one from a different personality. This symbolizes acceptance by them of our beginning relationship and I am elated. She also shares that Callie's primary anger is directed at Mom.

On December 27, 1983, Jennifer (twenty-four years old, spontaneous and fun-loving) arrives for our session and introduces herself. She is an ally of Katharine. I get to know her as a lively, outgoing personality with a good sense of humor. Although I am initially told she arrived when Carol was nine, I later discover that this was when others became aware of her. She actually arrived when Carol was seven. For two years she essentially observed. It is interesting to realize that personalities can function so independently and even without the knowledge of all the other personalities.

I am Jennifer Montgomery and this is my autobiography, or the start of it anyway. I first came into being when Carol was seven,

39

more as an observer. I watched, but I didn't really become active until she was nine years old. She had to go to her grandparent's house that day. She had to scrub the kitchen floor and the bathroom floor. They were hard, stone tiles. She couldn't do it, so I did. I had to scrub the floor on my hands and knees without any clothes on. And there was this tiny little brush that I had to use for all the corners and along the walls and in between the tiles. While I was doing it, the grandfather was always right there next to me and every once in awhile he would hit me with a belt. The first time, I wanted to cry, but I already knew from watching the others that that wasn't a good idea so I laughed. I guess that made him mad because he hit me again. Then he made me kneel in front of him and I had to perform oral sex on him, but I couldn't do that right, so he hit me again and I laughed again. So, he made me do oral sex again and it went on like that for a long time. I tried to think of a lot of other things, but it was hard because he wouldn't stop. I decided that the best thing for me to do was not to cry but to pretend that it was all right and that I didn't feel anything when he hit me or when he touched me. And it worked! I had to do it at least once a week. Sometimes more. It got worse because he didn't just keep it to oral sex. After awhile, he started to rape me. That hurt really bad. And I guess about the fourth time he did it, he did something else too. He turned me over on my stomach and then he raped me that way. And that was even worse. That hurt more than I'd ever been hurt before. That hurt more than the beatings even. I guess I just should come out and say that he had anal sex with me. That's really hard to say. It was really hard for me to understand why he would want to do that to me, but it helped me to understand why Carol really needed so much help, because, boy, she couldn't have handled them all by herself. Her grandmother was a bad, bad person too. Always calling her a boy and slapping her around. After awhile, whenever she did that, I stepped in, and so she slapped me instead, but I didn't feel anything. I just screamed at her and that made her madder than hell.

I always tried to find fun things to do. Sometimes it got me in a little bit of trouble. Like that one time I tried to see if I could get this swing to go over the top bar, and it did. I landed flat on my face on that one. It didn't hurt, but I was bleeding all over the place. They were mad about that! I knocked a couple of teeth loose and I put a couple of teeth through my lip. I couldn't eat much, but that didn't matter much anyway because they didn't give us a great deal to eat. So I would sneak things. Carol's mother would make cookies, and her father could have some and Phillip could have some and so when I knew they weren't lookin' I'd sneak some. A couple of times I got caught and that was bad. But it was worth it. I don't know...living there with them was a real pain in the butt--in more ways than one. They hassled me all the time. Don't climb this fence, don't climb that fence, don't do this, don't do that. "You're a bad girl," and out would come the belt and I'd get another whipping, except I didn't feel anything and they'd just keep on whipping me. Guess they wanted me to cry. Didn't do that either. (*Note: amazingly, Jennifer literally developed anesthesia for pain.*)

Sometimes Rosa, the girl next door, and I would hang around together and get into trouble. She was Mexican and she was really neat. I liked to go over there because they talked this funny language. Of course, they thought that we talked a funny language, because we talked Dutch and English, more Dutch than English. But I thought they were funnier than I was. I used to love to go over there when the whole family would be home and they'd all be chattering away, and I couldn't understand a word. And their food was really something else. When I went over there, they always gave me something to eat and these flour things, I know now that they were tortillas, were totally different from what I was used to, which wasn't very much. The first time I had one of those things, I don't know what they put inside of it for the fillings, but I bet I drank about ten glasses of water, it took me about ten seconds flat. And they all laughed. They thought that was really funny. I do now. At the time I wasn't too sure. Made my eyes water.

They had a dog too, an absolutely gorgeous dog. I can't remember the dog's name. Rosa, the dog, and I would go about a mile down the road, where there was a building that had a lot of coal in it. A railroad track ran through it. It was pretty high up and it had all these rafters and everything. Rosa and I walked those railroad tracks and the rafters and Katharine would have absolute fits, because it was a long way down and we got absolutely filthy. I mean, it was all coal. It was so much fun. Of course, you don't get coal off your clothes, so Rosa and I would sneak in the back door of her house and we'd try to wash ourselves up. We managed the face and the hands, but the clothes just gave us away. So anyway, here I go traipsing home and her mother would take one look at me and beat the hell out of me. But, it didn't matter so much, because I had fun doing it. It didn't hurt, because I'd know that I hadn't fallen, hadn't slipped off the rafters or off the railroad track, and I made it. I used to be pretty proud of myself for that. I couldn't sit down for awhile because of the beating, but it was fun.

I never was one to play with dolls very much. It was okay once in awhile but I just got bored with it. I liked this one teacher. Her name was Miss Baxter, and she and I got along really well. She seemed to understand me. I was always so restless, inside and outside. I couldn't sit still and sometimes I felt as if my stomach was jumping around. My mind wouldn't ever quit. It was really hard for me to be quiet in school. I know I asked her a lot of questions and that drove her nuts sometimes. But she said that's the only way you're going to learn, if you ask questions. 'Course sometimes I would go through Phillip's encyclopedia at home and I'd find questions to ask her. I already knew the answer because I had read it. But when I'd go to school the next day I'd have this whole list of questions--drove her crazy. I always got good report cards from her though.

We had gym class and a lot of times we'd have to play softball inside the gym, not with a real wooden bat, but with plastic bats. That was for the birds, but anyway I always liked to play. One time, I swung that bat real hard and I didn't know this kid was standing behind me. Hit him smack in the head with that bat. I didn't mean to, but I got punished for that one. I was too rambunctious.

Rosa and I went to this fair together at school and they had this dunk tank. I thought it would be a lot of fun to get dunked. For one thing, most people aren't that good of a shot, that they're going to hit the thing just right and you are going to go in the tank. I had on this outfit, shorts and a top, and I was supposed to keep myself clean to begin with. Well, water's clean, right? So I went up to Miss Baxter and I asked her if they'd let me sit in the dunk tank for awhile. I wanted to try that. They had a teacher that was getting dunked and I think that I begged hard enough so that she finally went to this teacher and she asked him if he would mind if I had a chance. I think he was glad to get out. So they helped me up there and here I am in my pink shorts and pink and white top and the first person that threw the ball got it and I went in. I probably went down five or six times. I was soaked. My clothes were still clean but I was wet. Rosa kept on saying, "You're gonna catch hell"--exactly the words she used. They talked that way at her house. And I just kept telling her I didn't care. We went by the fire station on the way home and we stopped and talked to the firemen. They gave us each a cookie. I decided I wanted to sit on their bench. Here I am with wet shorts, bare legs and I sit on this bench. About the time I sit down, one of them was trying to tell me not to sit down and so I stood up again and saw this green gunk all over the back of me, my legs, my shorts,

42

the back of my shirt. Rosa thought that was pretty funny. I got a little scared then, because I knew that I was going to get punished for being wet, but I knew I was really going to get beat for getting paint on myself. The firemen tried to get some of it off, but they couldn't so I had to go home that way. And her grandparents were over at the house too. You should have seen their faces when I came in. That was funny. I don't know who got the belt out faster, her mother or her father. It hurts a little bit more when you get beaten with a belt and your clothes are wet, but I kept on thinking about the good ole dunk tank and how much fun that was and it didn't hurt. I decided that every time something went wrong, if I thought of something fun, it's not going to hurt. And it worked. I decided that I wasn't going to feel when they hurt me.

One time when it was Carol's birthday, her dad wouldn't talk to her all day. He'd talk to cousin Julie and let her sit on the couch next to him and he'd read her stories. They played games together but he ignored Carol, and he really made me mad. So we all took over, because that really hurt Carol's feelings. She always wanted her dad to like her and I didn't care if he liked me or not.

One time, I almost told Miss Baxter about things that happened to me, but I knew things would really get bad if I did, so I didn't. Should have. Although she probably couldn't have done anything about it anyway. She always said I had a vivid imagination and she probably would have thought I imagined that too. Sometimes if I had a bruise she would ask me how it happened, and sometimes she'd ask me if I was all right. One time, I came to school and I had a bruise on my chin because grandma had pushed me down the flight of stairs from the kitchen to the basement. Jaw really hurt, felt as if it was on the other side of my face, if that makes any sense. She asked me how that happened and I said I fell. Then she wanted to know where I fell and exactly how it happened. But I just told her that I fell off the swing again. She asked if I was trying to go over the top of the swing again. I told her no. One time I did that. They had sawdust underneath the swings and I ended up with a mouth full of sawdust and blood. Yuck!

There were things I really, really liked to do. I really liked to draw. I didn't draw like Katharine, or like Carol or 'Cesca. I drew animals a lot--horses. I had my first pony ride when I was twelve. Actually, it wasn't really a pony, it was a horse. And boy, I got hooked. I still remember that horse. I wanted to take her home with me. We went with a class from school to visit a farm and I could have stayed there. They had everything. And then they let us ride the horse. Well, this guy's going to lead this horse around the yard and I was going to sit on top of it. I thought that was kinda dumb. I mean, you don't ride a horse that way. But he wouldn't let go.

I loved it when we took school trips to the zoo. We'd take our lunches and lots of times a couple of classes would go together. I was a grade ahead of Phillip so it was kinda fun when he could go along too. He used to be kinda scared of some of the animals, so then I'd just hold his hand. They let me walk with him and I'd show him there was nothing to be afraid of. For one thing, the animals couldn't get out, but what I really hated was when people at the zoo would tease the animals. They'd try to make them mad and they'd throw some things at 'em.

It helped me a lot, especially if I could get out the day after I had to go over to her grandma and grandpa's house and he'd rape me again. Boy, I did a lot of work at their house. I think I was a maid by the time I was nine years old. Learned how to set a perfect table over there. Her grandmother used to say that she was the queen in that house. Some queen. I used to draw a picture of her in my mind. Made her uglier than what she was. She didn't like me at all. She used to call me a boy. Maybe because I liked to be dressed in pants and a shirt better than I liked to be dressed up in a dress or skirt. I invariably got dirty dressed in a skirt.

One time I had a skirt on and I tried to roller skate down the hill there. They lived on Robbins Road. It's a steep hill and I couldn't stop and I went flying up over the curb and right into Grand Avenue and kept right on going. When I finally stopped, I fell in a mud puddle. I did that after the grandfather raped me really bad one time. After he was all finished, he told me to get cleaned up and get dressed. They had a pair of roller skates at the house and so I put them on and I went flying down this hill as fast as I could. Lots of times, when something bad happened, I'd run. Not in my mind, but I'd actually run and I'd run and I'd run and I'd run until I just couldn't run any more. Sometimes my legs hurt a little bit, but I didn't think about that. Then afterwards, I'd walk home real slow and it always made me feel better, except by the time I got home I was late and I'd get whipped again. So then, I would daydream about this great big white horse and how it was going to come to my house. Then I'd get on that horse and we'd run away together.

I feel kinda bad now that I had to go live with them, but yet Carol needed so much help. I don't think they've ever been nice to any of us. They pretend a lot, but they have to pretend when other people are around. It's just that when we're there by ourselves, they're not so nice.

I think the grandparents thought up punishments all the time for me. Scrubbing the floors was the worst though, because I had to do that without any clothes on. I had to do that every Friday. Then sometimes they'd decide it wasn't good enough and I'd have to come back on Saturdays. On Fridays, I had to go there right after school. Sometimes I wanted to tell them: "I'm not Carol, I'm Jennifer!" They didn't call her Carol, they called her Callie. And I'd get so mad because they couldn't see that I was Jennifer. I have long hair and Katharine always says it's auburn. (*Note: In multiple personality, each individual personality literally "sees" his or her own separate characteristics. Therefore, even though we, as outsiders, see only the actual physical features of the birth person, Jennifer truly sees long, auburn hair, while in actuality she has Carol's short, blond hair.*) I still would rather wear jeans and sweatshirts and things. Carol likes to do that sometimes, but she likes to get dressed up. No one can tell I'm Jennifer by looking at me, but I am. I'm me. I know I have a hard time sitting still, and sometimes I talk too much, but I'm still me. I know I'm not a very quiet person, and sometimes I drive everyone nuts around me, but that's just my way. If I don't keep moving, then I think a lot, and when I think, my head hurts.

Some things scare me like when Mandy has the knife. But I don't think about those things. Someday I'd like to have a horse. I rode a horse on the beach a couple of times and that's really a wonderful feeling. Especially early in the morning when there's no one else around, and there's just you and the horse, and you just run down the beach and your hair is flying and it's not too terribly hot yet and there aren't any people around. I like that.

I guess sometimes I daydream. I wonder what it would be like to live in other places and to have someone really care about us. Someone that wouldn't care if I wore jeans all the time and that would let me go to school and let me learn how to make pottery and learn how to weave and taking a wood-carving class. I'd love to do that. But sometimes I think that none of us will ever get to do the things that we really want to do. It's difficult because we have to share time. Even when I was little I didn't like that because there were always so many things that I wanted to do. There was never enough time. Oh, well. I suppose I shouldn't gripe so much because we still have a lot of things and we all get along pretty well together. But I can't help but wonder sometimes if life will ever change for us. When I was growing up, her parents and her grandparents always said that I wasn't ever going to amount to anything. I couldn't ever sit still because I was always too restless. When they used to beat me, they said maybe that would help me learn I couldn't just go off and explore on my own and that I was to stay put in the house to do the housework, and I hated it so much. Lots of times on summer vacations, I'd have to stay in all day and do housework or take care of the other kids. I would hate them and I know it's not right to hate.

They always used to tell me that God wasn't ever going to like me and that He didn't care about kids that couldn't behave and I tried so hard to behave. So now I really don't know if God cares about me or not. I care about God, but I guess I'm just different. I care about other kids and people. Her mom and dad used to say all the time, "We don't know you at all. You're not one of ours." Well, of course, I'm not one of theirs. They're not my parents. You see, my parents understood me. They wouldn't mind if I wanted to explore or do something like that. They didn't object at all, in fact. They taught me all kinds of things, like about the stars and the universe and plants and flowers and animals. But I don't think these people want to learn. They're dumb. In my real family, it was okay to be noisy and to have a good time and to play the music a little bit loud. It was okay to ask a lot of questions, but with Carol's parents, it was always, "Be quiet, shut up, don't ask so many questions." Lots of times I got smacked across the face just for asking a question.

Miss Baxter was always finding things for me to do. I used to play with my hair all the time and she'd say,"You need to keep your hands busy again, don't you?" So she'd send me out to get the milk, because we had milk at recess and lunch. Once in awhile she'd even send me down to the gym if my shoolwork was all done and then she'd just let me run around the gym. They had a hoop there and I used to throw basketballs. I could just run and run and run in circles if I wanted to. I learned how to play volleyball. She had to tell me to be careful with that though because I hit the ball so hard. You had to do that to get it over the net. I loved serving the best. Stand in the corner and serve that ball and off it would go. Then she'd say, "Don't hit the ball so hard because you'll hurt your hand." And lots of times my hand was all red by the time I was done.

I always felt better after she'd let me run around the gym. Sometimes she'd even send me outside if it was nice. If I had a time when I was kind of quiet, then as a reward she'd let me draw pictures on a big sketch pad. I never took those home because I didn't know if I would be allowed to keep them, so she always kept them for me. I didn't tell her why I didn't take them home, because I didn't want her to say anything. They ask parents questions and you get into all kinds of trouble. That's why I never told her how come I had bruises or anything. I felt really bad when I didn't have her as a teacher anymore.

She'd talk to me about Jesus sometimes too. I used to ask her questions. I went to church because my mom and dad always went to church twice on Sundays and I went to Sunday school and catechism and in the summer, I went to vacation Bible school.

I used to wonder a lot, when I was growing up, if God really cared about me and I still wonder about that. Katharine always tells me that He does, and I probably shouldn't question it, but I can't help it sometimes. I get so mad inside for no reason at all and so then I have to work extra hard at being happy. That's just the way I've always had to be because if I'm not, then I get sad and then I hurt too much. When her grandfather used to touch me, I'd shake inside and I'd get all cold inside. I don't like to be cold. I want to be warm all the time and when I'm sad, I'm not warm.

I really think Dr. Jansma is our friend. I know we haven't talked to him that many times yet, but I think he does care about all of us. He's never hurt us. It took us a long time to tell him that we were there and maybe that wasn't so nice, but we just weren't ready. I hope he understands that. It was nothing against him. Just wanted to make sure. He doesn't seem to think we're terrible people like we were always told we were. At least, I hope he doesn't. I just try to be happy all the time, whether it's at home or at work, but he said I can just be myself in his office. That's nice to know, but I wouldn't want him to see me all sad and upset. So I'm not gonna let him see me that way. I cried in there once and that was really unnecessary. Crying doesn't accomplish anything anyway. I just want to be able to be me all the time and I sure would like to have some more time. I'm getting a bad headache right now and I don't know why. Sometimes I get those and then all of a sudden I'll get really, really tired and then I can't sleep and then I have to get busy again. Maybe I should quit now anyway because I feel kinda funny inside.

6

It is now December 29, 1983. So much has happened in the three weeks since I met the first personality who "came out." Katharine comes in for her session. She stays briefly to advertise that Carol has a list of questions and that I may need to prod her in order to ask these questions. She lets Carol take over and her questions are dealt with. It is now becoming more clear that Katharine has become a therapeutic ally and is beginning to clue me in on some of the inner workings of Carol. This is very encouraging. Much time is spent with Callie today working out ways to communicate with her. She hides behind a chair for a period of time, but with some coaxing on my part she finally agrees to try to let me be a friend. We spend some time drawing pictures as a means of communicating. She says that she does want Carol to be helped. I recommend to Carol that she try to remember what she can about a whip--a picture that Callie drew--and about being in a pit (another picture drawn by Callie). Clinically, this is excellent progress. Personally, this is very trying for her.

Francesca reports that Carol's parents both whipped her as a child. Carol has no memory of this. At this point, a number of personalities are vying for clinical time. In the next session, Francesca brings in a letter that Carol had written to the other personalities revealing her anger at them for taking therapy time away. She (Francesca) will not allow herself to be introduced to Carol as yet, but she does say she would bring in for me a list of all the personalities (twenty-one in all). She is becoming more comfortable in talking with an outsider, something she has never done before in her life.

Katharine arrives in the next session, bringing a list from Francesca of the total twenty-one personalities and a stuffed animal that Callie and I later name Mishka. This toy puppy is being brought for Carol because,

49

"She won't ever buy things like this for herself," and for Callie because, "She was never allowed to have toys." Katharine is clearly trying to build some bridges. Carol is suddenly there. She initially does not want time, but as soon as she sees the puppy she easily accepts it. Callie then spontaneously emerges. She clearly loves her new puppy. We communicate with drawings and a few words, and Katharine ends the session doing some interpreting for Callie.

Now Alexandria introduces herself for the first time. She had called me on the phone advertising that she is fearful of being eliminated. I talk with several of them about integration and about getting an interpreter to help communicate with Callie. Since my father speaks fluent Dutch, I propose inviting him to come in and help, to which all agree. Carol writes in her journal:

January 13, 1984

Today is a big day because of the session Dr. Jansma and I will be having. His father is going to be there to translate what Callie says. Apparently, communication is somewhat difficult because she speaks only Dutch. They have been drawing pictures and that has helped some, but not enough so that Dr. Jansma understands what is really bothering her. I am somewhat afraid of this, because I won't be in control and that bothers me.

Carol

On January 13, 1984 we have our first session with our interpreter, Rev. Jansma. With his help, I am able to get information regarding the black trunk that Callie was regularly locked in as a young child. She also reports that there were two holes in the ground that were deeper than she was tall, in which she was put in. Also she reveals information on her belief that she is in fact the devil. We also get information about her fear of her grandfather and having to undress in front of him. We review the above information with Katharine. Following this, Carol spontaneously comes back, at which time we have a first experience with Katharine and Carol sharing time simultaneousely, and this actually being in Carol's own conscious awareness. (This phenomenon is called co-consciousness, an essential development if the patient is ever going to be able to be trained away from dissociating as a way to cope with life's difficulties. For the personalities themselves it is simply a matter of being aware of the other, or as they describe it, it's like meeting a new friend for the first time.) Carol now is working hard to retrieve some memories.

Marie now introduces herself in my presence for the first time. She informs me that she is going to be leaving the state. She is twenty-one, very angry, and over the previous weekend had trashed their apartment. She also acknowledges that she has been taking money from Carol. She has come in basically to have a verbal fight with me. She is very surprised when I do not try to talk her out of leaving. Once I have passed this initial "test" she begins to talk. She is able to open up about being very alone and very lonely and having no friends. She then decides not to leave the state and to enter therapy.

I see Carol in session who is very quiet and very scared. I let her know that I am aware that some memories are returning. With prodding, she lets me know: "Granddad played his games with me even in the Netherlands," and "I remember the trunk." She now remembers that she was locked in the trunk frequently as a child for "being a bad girl." She is able to cry over these beginning memories in session. Katharine arrives at the end of the session announcing that next week will be her three-year anniversary of the start of therapy.

Marie comes in with two of her poems, which are very powerful. She is also self-taught at the piano, even though: "I have no talents, I'm always bad; I always mess up." Carol is struggling much right now, especially over the issue of whether her new memories are real or imaginary. She has been told all her life that she imagines an awful lot.

I have a session with Marie, who gets very angry when I comment on the trunk and the hole in the ground: "Did you know that there were worms in there?" She informs me that she never wants to talk about that again. That night Marie trashes their apartment. Carol is furious. (It now comes out that this has happened several times in the last few years.) I use hypnosis in our next session in order to call out the person who has trashed the apartment and Marie does confess to it, stating that our previous discussion about the trunk and the hole in the ground had set her off. I suggest that she have a big pillow on hand to beat on, instead of ruining the apartment if she again starts to feel in such turmoil; and she agrees to try it. Carol develops a new memory, namely that her mother made her drink castor and cod liver oils daily.

My strategy at this point is to develop as comfortable a relationship with as many of the personalities as they will allow. It is clear that therapy is now taking a whole new turn because of these many unexpected developments.

We have now been in therapy about three years. Jacqueline, age nineteen, believing she is gay, very careful, comes in voluntarily for the first time. She informs me that she is the one who was gang raped. "I became gay as a result of this," although she has never had a lesbian relationship.

Throughout all of these developments I have periodically given Katharine my opinion that even though on the surface she presents as very organized, calm, and together, I am very aware that she is also doing a lot of struggling. Up to this point, she has regularly denied having had any difficulties.

In later January, Katharine finally gets past some of the smiles and begins to share why she hurts also. She reports that she was the first personality to arrive. She suffered many of Carol's beatings and some of the grandfather's forced sex. She is particularly frustrated because she can never be a mother. She believes that if Carol's parents were ever confronted with what they have done, they would kill Carol and the mother would then kill herself and her husband. Katharine also reveals for the first time that her biggest fear is going out of existence. It seems quite clear at this point that Katharine has some very real awareness of the implications of multiple personality, and what the healing process entails. It is also apparent how central her role is in regard to mothering functions for many, if not all of the others. At this point I have met twelve of the personalities: Andrea, Francesca, Callie, Carol, Katharine, Jennifer, Alexis, Alexandria, Alice, Callie (age four, the other one is age five--therefore two Callie's), Simone, and Karl, a male personality.

In early February we have an emergency session. Carol had called prior to our session. Her checking and savings accounts, as well as her work paycheck have all disappeared. She is in crisis. I discover that Marie and Amanda have plans to leave town but will not say where they are going. They have taken the money. There have been numerous efforts by other personalities to reason with them, but they do take a plane to Chicago over the weekend. (Cassandra later revealed that Marie and Mandy each tried to buy their own ticket. The agent said, "But lady, you just bought one," at which point Mandy said: "She did, but I didn't.") Once there, Marie changed her mind and came back. They had taken about a thousand dollars of Carol's money, although they did pay rent. Marie wants to return the rest but Amanda will not. Marie also broke a number of things in the apartment upon her return. Because of this, Carol did not appear for her own birthday party given by her family. I attempt using hypnosis to talk with Amanda, but she will not come out. In our next session, however, she does introduce herself to me, the first time she has introduced herself to any outside person in her life. Even though she is very angry, she does talk with me about the events of the past weekend. I sense that she is scared, although that is something she will not readily admit to.

On February 9, Carol appears in session for the first time in several weeks. She is quiet and holds back. She had called me a few days ago wanting to know what happened over the weekend and I told her of the trip to Chicago. As a result of this, I get her to commit herself to a serious pursuit of her past, encouraging her to know that it is in her best interest to mean business about getting better. Katharine volunteers to help Carol to try to recover some of her memories. For the next several weeks Carol periodically reports in session how dreams are remaining in her memory, and how some of the dreams are becoming a source of information to her about her own past. An example of this is a very specific dream she had about worms in a hole in the ground.

Katharine arrives for a session expressing her resentment at all the mothering necessary for so many of the others. We pursue a dream of hers by having her visualize it, and then by questioning, adding detail. Callie suddenly takes over--there are many tears as she reacts very strongly to what is going on. She begins to draw a picture. Carol re-emerges briefly to help out with interpretation. Callie then finishes the picture. Carol then re-emerges and we process the picture. She now knows that the dream is about a hole in the ground with worms in it. This is a powerful session.

One week later, Carol comes in for a session and shares another dream which involves her being in a building with a circle in the floor, and her being put in a hole underneath a lid which covered the circle in the floor. It is very cold. It is very dark. Katharine informs me that this dream is reality. The grandfather, in the Netherlands, had such a room and put them in it regularly. Such isolation was continued by the parents following emigration at age seven.

Four-year-old Callie is now attempting to take time at work and severe logistical problems need to be handled in regard to this. Katharine, who is so used to regularly taking care of other people's problems, is now beginning to have to relate to her own emerging hatred toward the parents and grandparents because of the child abuse. She shares that the grandparents would put her in a hole immediately upon awakening, would spank her extremely hard, and would give her no food for an entire day. Their rationale was, "You have bad blood in you from the other side of the family." Katharine cried for the first time in session over this.

We have our second session with Rev. Jansma as our interpreter. Three other personalities come out to meet him but our time is primarily spent with Callie, with Rev. Jansma acting as interpreter and co-therapist. The primary need is to keep Callie from taking time at work, assuring her I would return from vacation, encouraging her to know there are good adults, encouraging her to help Carol remember even more. This is a very powerful experience for Callie. Carol is seen immediately afterwards to update her. Katharine later writes:

Callie has grown to love Dr. Ted, as she calls him, very, very much, and she knows that he's her friend and that he's not going to hurt her. She had to observe him for quite awhile first. It has been really interesting when I go in to speak with Dr. Jansma, because sometimes she asks to go along. She always says that she's going to be very, very quiet so that he won't know that she's there. And she'd watch, and she slowly learned to trust, and then he gave her E.T., a little piece of wood that was polished with a little face drawn on it and it really resembled E.T., (from the movie "E.T., the Extra Terrestrial"). She really loves that. She carries it with her wherever she goes. That's how she would first talk with him. When she saw that in his office and she pointed at it, he told her she could have it. And the next time when she went in to see him, that's the first thing she showed him, E.T. E.T. was his friend and it became her friend. I think that's the first time anyone on the outside had ever given her anything. The parents and grand-parents certainly never did. What meant the most to her was that he did not go out and buy it, but that it was his, and he gave it to her. I have to admit I was rather surprised that she accepted it because she doesn't accept things very easily, but I was very pleased and very glad that she trusts him because she needs a friend.

In March, Alexis comes in for the first time reporting that she has observed for some time, adding that she had met me once before, unbeknownst to me, on a winter day several months ago when she brushed snow off of my car window as I was leaving work. Alexis is age twenty-six, arrived when Carol was eight, is headstrong, has a bad temper, wants to be very thin, (is actually anorexic), and takes many laxatives. Marie, who also takes many laxatives, is feeling very guilty and is worried that I might be angry with her. She handles this by telling me a "big secret," namely that Katharine is keeping a "big secret" and has been suicidal. This is Marie's way of "making up" with me. I talk with Katharine and she acknowledges that she has been suicidal but will not share what's troubling her.

Francesca asks for her own session to deal with sexual issues. She came into existence when Carol was age four to handle some of grandfather's sexual advances. She also handled a lot of the abusive and very sick sex for Carol during her marriage at age twenty-three.

Actually, Katharine and Francesca are the ones who married James. It was a terribly problematic marriage that lasted six years. James was apparently bisexual and was physically abusive during sex while he was dressed in female clothes. No one ever informed the parents about James' physical abuse in their marriage because they didn't believe that the parents would care. Francesca brings me the following:

December 26, 1972

I'm going to be married in a few days. I don't love James, but it seems to be the only alternative I have. I need to get the others away from that family. It's not good for us to stay there. They stifle us. We can't explore ourselves. Besides, we can't do anything right anyway. The mother thinks I'm a whore and I'm sick of listening to her. There's probably truth in what she says, but I don't need her to tell me that. That entire family has put us through so much grief and now it's time to get out. I know they don't like James, neither do I, but who cares. We won't have to live with them anymore and we won't have to see grandpa and grandma anymore. God, how I hate them! When I think of all the things they have done to us, how they have hurt us, I want to kill all of them.

I made my own wedding dress. How ironic that I'll be wearing white. I haven't been a virgin since I was four, Kate since she was two. But it's important to keep up pretenses for family and friends, I've been told. What I'd like to do is stand up in front of everyone that day and tell them exactly what was done to us--how grandpa and grandma made me and Kate have sex with them and took pictures of us in front of other people and how our hands and feet would be tied and we would have gags in our mouths so we couldn't scream while all those men had sex with us. I counted one night and twenty-one different men had sex with us--again and again and again. And all I could feel was pain, unbelievable pain. After awhile, Kate passed out. I'm glad she did so she didn't have to feel anymore.

Anyway, I'll put on my white dress in a few days and become Mrs. Richie. What a joke!

Francesca

We have our first group session, called for the purpose of dealing with Katharine's wish to teach Sunday school. It turns into a lively group session and a number of personalities, including Katharine, Marie, Francesca, Jacqueline, Alexis, Alexandria, Amanda, Jennifer, and Callie, participate in the discussion. This is a good first experimental group session. We end with all signing a contract to support Katharine's new venture in teaching Sunday school. Later on Katharine is seen alone. Her "big secret" turns out to be financial: she's worried that she cannot afford her portion of therapy. We go over these financial issues to lighten the load. This also gives me the opportunity to deal with her resistance to entering therapy for herself, and some progress is made regarding this issue. In fact, the next day she lets me know she will agree to co-ordinate as many of the others as possible to each take an MMPI psychological test.

A number of problems occur next. Carol's real-life twenty-three-year-old brother, Nathan, is diagnosed with the same disease that killed brother Phillip, and may soon die. Carol is not aware of this. Katharine has become suicidal and Marie hides her pills because she is afraid that Katharine will kill herself. Many of the others become frightened while I am briefly out of town and want to confirm that I have returned. What is clearly emerging is a more public acceptance of being responsible for each other and keeping me informed, which is encouraging.

Amanda comes in for her first appointment. She has appeared once before, although briefly, during a group session. Amanda is twenty-three years old; she arrived when Carol was six. She is a writer, is very frustrated with life, and hates her body. Amanda is the one who has maintained the relationship with Harold, a married man. She does this in case "I need a way out." She comes in prepared to fight me. Later on she told me: "I had this first session all planned; I wanted you to dislike me and me to dislike you." When she realizes I am not willing to fight with her we are able to have a fairly good first session. I interpret her MMPI results, but when I mention the issue of Nathan's illness, Amanda stops talking. Suddenly there are many tears, many feelings, and very strong sobbing. She has a surprising and extremely strong catharsis. This is unusual for her because she's never been one to cry or share feelings. It turns out that Nathan is the only friend she has ever had. At the end of this very first contact she offers to bring in a number of writings. *(Note: Amanda is the one who regularly leaves town, rebels on an ongoing basis and, along with Francesca, causes "the blood" in Carol's vagina.)*

Amanda comes in to continue the previous day's session. She had a very bad evening: wanting to run away, wanting to break things. Francesca had called me in the evening because she was very worried about Amanda, afraid that she would hurt herself. None of them had ever seen Amanda cry before. In fact, Amanda is the one who would regularly taunt Carol's parents and grandparents when they beat her by refusing to show any feelings. She tells me that she has consistently refused to have any friendships with the exception of Nathan, "because that person will just go away." She brings in the following powerful autobiographical story:

She was so small and he was so very big. She shivered as he approached her, partly from fear and partly from the cold, for it was wintertime and the small, one room shed was unheated. In her mind there was certain knowledge of what was about to happen to her and she willed herself not to make a sound, not to show him her fear. By now she had learned to assess his moods by observing his walk, the inflections in his voice, but above all by looking directly into the bright blue eyes. He was capable of silencing adults and children alike with one single, chilling glance. Today the eyes were especially cold and hard and instinctively she reached deep inside herself so that once again she would not betray any of the emotions she felt. She met his eyes with a direct, defiant gaze as he stepped in front of her. One hand grabbed her hair and forced her head back while with the other he slapped her across the face, a hard stinging slap that left a large red mark. He pushed her down on her knees in front of him, quickly unzipped his trousers and forced his hard penis in her mouth. As he did so, her thoughts turned to revenge, to one day killing him and as the hatred and anger welled up inside her, she knew that she had won, that she would not cry and give in to fear. She felt her dress being ripped from her back, felt herself being thrown on the floor face down and as she received the first of what would be many stinging lashes from the belt, she recognized that her defiance made him angrier and she laughed quietly to herself. Stupid man, she thought. When are you going to learn that you are not going to break me, that I won't beg you to stop, that I won't cry. He suddenly turned her over, spread her legs and shoved his penis inside her. When he made a hoarse sound, she knew he was finally finished and she looked directly into his eyes as he pulled away from her. He quickly dressed and she stared at the grandfather's retreating back with contempt as he left the shed. She slowly started to get up, but the pain sent waves of nausea throughout her body and as she knelt, on the floor, still naked, she retched until there was nothing left to bring up. She then dressed, making sure to tightly tie the offending shoelace that had brought on today's beating by becoming untied. As she walked out, she clenched her fists, and then with all the strength a six year old can muster, slammed the door to the shed.

Amanda

7

By the end of March, 1984, Katharine is now beginning to talk more during session time: "I've always thought I liked myself, but now I'm not so sure." She is starting to become aware of her own anger. I also spend time with Carol, filling her in on events of the last month and especially informing her of brother Nathan's illness. This is extremely painful and there are many tears. I request that the others give her time to be with Nathan, which they agree to. Katharine has now decided not to teach Sunday school and is still rather depressed, but it seems to be lifting a bit. She is finally acknowledging painful issues: the hole in the ground, being beaten, being ignored, being sterile, her hysterectomy, feeling old, having arthritis.

The next time I see Carol, she seems to be functioning somewhat better. She is processing in healthy ways her reactions to Nathan's illness. She is also eager to know more about the others. For the first time we discuss attempting co-consciousness with the use of hypnosis and she states that she is willing. Katharine still struggles but is able to function. She is very hesitant to try co-consciousness with Carol: "If she knows my issues, I won't deal with them." We have thus far been able to prevent hospitalization throughout all of this turmoil.

Jenny comes in to get her MMPI testing feedback. This results in her taking a two-hour session for herself. The focus is on her fear of intimacy, her wish to be male, her hyperactivity, her anesthesia for pain and sexuality. She reveals that she became an active personality when Carol was age nine, at grandfather's order to scrub the kitchen floor naked, while being beaten, and being required to perform sexual duties for him.

59

Harold is now in town again. Amanda has become very frightened. Over time, she has sent him a thousand dollars to save for her in case she moves there. She now questions the relationship. I recommend that she get her brother to help her, but the problem escalates to a point where it is necessary to bring in the police. A court order is now in effect keeping Harold away. He refuses to return the money.

Time is spent with Katharine whose depression is diminishing. She is functioning better, and her role as leader is more and more coming into focus. Katharine is also having difficulty in conveying what it is like for all of them to exist as a multiple personality. I find it encouraging that she sees the reality of their existence as well as she does. It's also becoming quite evident that she would like the other personalities to know much more about what is their reality. As we talk about this, the idea develops of having this material in written form for the other personalities. At the same time, it would provide an emotional outlet and processing experience for Katharine. In addition, if it's in written form this would allow other personalities to look at this material at their own discretion without feeling pressured by anyone else. We decide we will, in a very loose way, have Katharine write an "autobiography." She becomes quite excited about this project. This will also make good use of her mothering needs. We update Carol on this decision and she, surprisingly, is very supportive of it.

I have another session with Amanda in order to deal with her relationship with Harold. We have a powerful session. Amanda has never been able to have a meaningful relationship with anyone outside of the other personalities. She is beginning to trust me. Once again there are many tears. It is a big step for her to share some of her own history and feelings and her need to run is, at this time anyway, defused. She has now decided to end her very destructive relationship with Harold.

The logistics of therapy are now getting difficult. With individual relationships having now been established with a number of the personalities, many of them are wanting their own therapy time and there are obvious limits to what is possible. There still appears to be an immense amount of untouched material and there are probably other personalities.

Jenny has now begun to exercise tremendously and has stopped eating. We focus on her wish to be male and she voluntarily tells me more of the painful memories, especially the sexual experiences with the grandfather. She finally cries. With Jennifer's literal anesthesia to pain, her obsessive exercising is physically having an impact on the others. I must order her to slow down.

By mid April, I am again emphasizing sessions with Carol. Although hesitant at first, she begins to deal with some of her more recent dreams with growing confidence: being locked into the box by someone, being put into a hole in the ground by someone, and being raped several times by an older man whom she thinks is grandfather.

There is much emotion exhibited and many tears are shed; but what is of special importance is the fact that her anger is beginning to emerge. She is sensing a strong ring of truth regarding the accuracy of her dreams.

Katharine has now written the first two chapters of her autobiography. As was hoped, this has opened up an immense amount of material for her. She asks for an extra session to "help me really talk." The basic issue: "Why me. Why did they do these things to me?" I find out that Katharine has also done a third chapter, (Jennifer had told me privately that this was her most painful one.) We read this chapter together. The subject is anger at God and her wish to die as she was growing up. Hearing her own words allows her to open up. Healing is occurring. What is particularly noteworthy is that her therapy appears to be good modeling for many of the others as a constructive way to deal with feelings. After all, many of them do observe these sessions. At the end of the session she reports: "Grandpa wasn't the only one to rape me; it was also grandma."

I'm Katharine. It's Easter Sunday and I'm doing some more writing. I'm really tired today. It's almost as if my mind hurts. I'm not even sure I want to do this and yet inside I know it's something I have to do, and that someday I'm going to realize exactly how much this has helped me. I don't like to talk about myself very much, about the things that hurt.

We had to go to Carol's parent's house for dinner today and I felt so badly for Jennifer. She wore a skirt and they all made fun of it. I don't understand people like that. She was so hurt and yet she puts on such a very good front. Just like I had to today because they made me so damn angry. I didn't feel well to begin with because of all the laxatives that were taken last night and I couldn't go to church today, which is something I really wanted to do. Actually, I probably shouldn't say I really wanted to go to church. I felt I had to. Friday evening I really wanted to go to church. All the while that I was growing up, I believed that there was a God. I didn't always understand this God because we were hurt so often and terrible things happened to me. But I still believed that there was a God and then when I actually became a Christian, it was I think the most important thing to me, outside of my little family. That's what the others are to me, my family. I don't have another family. Lately, I'm angry at God

61

so much. Dr. Jansma was right in bringing that up when we talked last week. I'm angry at God and I don't want to be angry at God, but it's still there and it disturbs me.

I used to love to go to church even though I didn't enjoy going with those people. I loved the singing and being in the choir. So did Christina. She had a beautiful voice and I used to love to listen to her. She sang solos, and she would lose herself in it. It was her means of escape. She'd sing and she can't do that anymore and that hurts. (Christina damaged her vocal cords in a playground accident.) Now she has no escape. I want to tell her it will be all better, but that would be a lie because it won't be better. She won't ever sing again like she used to. Sometimes I feel as if I have failed so terribly with all of the others. I can't make them feel better. I want to protect them and take care of them and I can't do that. I know that they have to deal with their own problems, but I don't want them to hurt like I hurt.

I want to feel alive all the time and I don't. I do put on a front and sometimes I hate myself for that too. There have been times since I was a child that I've wanted to die. And at times I have tried to die and I haven't. There was a time when I was kept in that hole for one whole day and one whole night and then another day. I was six years old and my body hurt terribly from the beating I'd had and I hadn't done anything wrong. Something about me made them hate me especially that day, because without having anything to eat, I was told to get dressed and I was taken there and left there. I don't know why any of it happened. After I came out of that hole, I couldn't talk for a week. Maybe it was longer, I don't remember. But I would try and talk and I couldn't. It didn't seem to bother them any because they felt that I was just pulling a stunt. So I'd get beaten for not being able to answer a question. I hate them so much. I still don't know why I couldn't talk. Since I've become older I have tried to analyze that. Maybe I was too scared, maybe I was too cold, maybe I was too hungry to talk, I don't know.

I liked school, probably because it was an escape and partly because I enjoyed learning. I especially liked art and I liked to write things. When I think about it, I was such a perfectionist even then. My drawings had to be right, my handwriting had to be right. If I made an error, or if I didn't think that a letter was perfectly formed, I would start over. Sometimes I'd almost be finished writing something and I would have to start all over. I had to be clean all the time. I couldn't stand a smudge on my hands or on my clothes. I can't to this day. I hate it. I have washed my hands, washed my body sometimes until I was red. I don't like things out of order and I never did, even as a child. My shoes had to be put away just so, my dress had to be put away just so. My drawings had to be put away just so; if I read a book, it had to be put away at just a certain angle. I started doing that very early in life. Even now, I wonder how much freedom that has cost me. Everything still has to be right. My own habits are sometimes like a prison. It's not that I don't want to change some of these habits, I've tried. But they're ingrained. I cannot go to bed if a closet door is slightly opened, if things aren't set out for the next day, if books are not put away properly.

When I was in my early teens, maybe even before that, I started planning how I would kill them. Obviously I haven't. That doesn't mean that the desire has gone away. Sometimes it's still there. I want them to feel all the hurt that I had to feel.

Then there was my grandmother. I've never, ever talked about this. So far it's just been about the grandfather abusing me sexually. Well, she did too. And she'd laugh. She'd laugh the entire time. And she'd scare me so much. They took turns. They decided who was going to have me and on what day. You get her on Thursday, so I'll have her on Saturday. It was nonstop. A whole world of abuse; of being told that I wasn't any good, that I was stupid, that they wished I'd never been born. But I didn't ask to be born, I didn't ask to go live with them. Carol needed me. She wouldn't have lived without me and the others.

I tried to hide once, and for that I sat tied to a chair in the middle of the room for two days again. Two days was usually the limit that any one punishment lasted before the next one started. All I had to do was try and figure out which one. Would grandpa rape me? Would the grandmother rape me? Would the mother make me drink castor oil? Would the father beat me? Loving Christian adults! I'd look around me at my friends and I knew that their parents cared about them. Even when they did something wrong, they didn't get punished like I did, and when they did something right, they were praised. Drawings made in school were saved, not thrown away, not burned up. They weren't whipped so that they couldn't walk or sit. They weren't threatened with being killed. They didn't have their hands broken like we did. How Callie screamed and screamed when that happened to her. And she thinks she's so evil; she thinks she's a devil and she never did anything wrong.

Very early, she lost her ability to even laugh. Sometimes I wish she had more time with Dr. Jansma. She actually laughs in there sometimes. I know that she misses his father, Rev. Jansma, terribly and I tried to explain to her that he really is going to come back sometime, he just can't be here all the time. She needs to be able to talk freely without having to worry about whether or not Dr. Jansma will understand her.

I think we all deserve an award. Best act of a lifetime--the collective group. I'm an act, Francesca an act; we cover so well and there's so much underneath. Put on a good front, a happy face.

I don't know if I want Dr. Jansma to read this. I think I've said things I haven't said before.

Katharine

It's a facade,
the smile, the peace,
the confidence,
for underneath lies pain,
pain that has been covered
by a smile,
sadness that has seemingly
been covered by peace,
and the uncertainty covered
by false confidence.

Marie

8

Katharine called me on the evening of April 26, informing me that Amanda had her knife and wanted to cut out her vagina. She agrees to see me the next day, not to cut herself, and to bring in the knife. Katharine also said, "I now know that if Mandy cuts herself, I will be hurt, too." Eureka!

I see Mandy in session and give her some analogies about being a good person. "You mean I'm not bad?" She's very surprised and smiling. Although still somewhat unsure, it appears her eyes are beginning to open. Simone now advertises she wants to enter therapy too.

I'm Katharine St. Clair and it's April 29, 1984 early in the morning, very early as a matter of fact. It's 5:00 a.m., I can't sleep and I don't even want to sleep. I have so many feelings that have confused me lately. Marie was right in her poem. I put on such a show--I think she calls it a facade.' Underneath, there's a whole lot more--anger, hate, resentment, some envy, insecurity. There is something missing. I need to belong somewhere. I need to be loved and to be cared about. I need to have time for me without worrying about any of the others, even though they are all so special. Even though I care about them all very, very much, sometimes they get on my nerves. Sometimes I feel I don't want to be there for them anymore, that I need time to be quiet, that I need to be free for a little while. I feel so guilty for wanting those things, because they have the same wants. If I could just go away for a little while. I don't even have any place special where I want to go, anything special that I want to do; I just want to be. What I need escapes me because I'm never quite sure what it is. A rest would be wonderful, but is that what it really is, is that what I really need? Is it the need to belong? Belong where? I belong to twenty others. And I feel guilty, guilty for wanting more and needing more, because all of the others need more too. They deserve more than what they've ever had and I can't give them what they deserve. I can't give Marie her piano. I can't give Jenny the opportunity to go horseback riding whenever she wants. Some provider I'm turning out to be. I feel so badly about that.

I should be feeling better now. I had a good session on Thursday. It was difficult and it hurt, but it was good. Dr. Jansma helped, yet I want to back away. I'm so afraid of letting him and others see that I'm not always so secure and that my head's not always on straight, because it's not. I don't want to use Carol's name anymore. I hate it--I want to use my own name. I want to be free to go out with people and say I'm Katharine. I'm not Carol, I'm Katharine, and I want them to know me for who I am. I don't want to do things anymore just because Carol would do them. I'm not Carol. Yes, I've taken care of her, but I'm not her, I'm myself, I'm me--Katharine. Except sometimes it's so difficult to know who I am anymore.

Who is this person? I used to have goals and now for years my only goal has been to keep us together, to try and keep things running smoothly so that no one would ever find out about us. I've had to do that. Now that I've told Dr. Jansma about myself. Well, now I want to be myself all the time. I don't want to share anymore. I don't want to have to think about bad things anymore.

"Put-together Katharine;" well, she's not. She's a tired Katharine that's 38 years old and has arthritis and she's ugly. I want revenge, justice done--although I think it's a little late for that. So many years have been taken from all of us. I want to have a screaming temper tantrum. I want to let go of it. I want to beat the hell out of something. That's my right. Then why the hell do I feel so guilty about being angry, about wanting revenge, about being angry at God. Why am I angry at myself for being mixed up? Why do I feel so inadequate?

Dr. Jansma would probably disagree with me on this one. But that's how I feel, they're honest feelings and I have a right to those feelings. I don't want anyone to take that away from me because so much has been taken away from me already. My dignity as a child. I was never a child. In the physical sense of the word, yes, I was a child. But you are no longer a child when you are forced to have sex with a grandmother. You are no longer a child when she makes you screw her with an object she keeps handy for that purpose. You're all grown up. Neither are you a child when the grandfather and a great-uncle and a grandmother decide to have you stripped and watch them have sex together and they, in turn, have sex with you. I didn't have a shred of dignity left when they were finished with me that time. I was ten years old. I think they must have positioned me a hundred different ways and it was hours before they were finished. When I left, I couldn't think anymore. I couldn't cry; I couldn't talk. I was a nothing at that point. And later there was rage. So much rage. And I thought about going up in the middle of the night and killing them. There were a large number of knives downstairs and I wanted to grab one of those knives and cut them up.

After that experience of sex with the three of them, there were more occasions like that. None quite as devastating as the first one, but there were more. The others were usually a little shorter in duration, but every time the hate came back, and I wanted to fight, but I did nothing. I have no respect for myself for that, for not fighting back. The grandfather once said to me, "I'm gonna teach you to enjoy this yet." I never did. I couldn't. I thought he was dirty then and I think he is now. I tried to remember that it was done to me and that it wasn't my decision. It was theirs. That doesn't change the fact that it happened and it doesn't change how I feel about it.

You know, in church I've heard over and over again that children are to be nurtured and they are to be loved and that God loves children. Every time I'm in church and there is a baptism, I'm reminded of that. "Nurture your children in the ways of the Lord. Take care of them and love them because they are precious in God's sight." Somehow, I guess I wasn't so precious.

There were a few people in high school whom I thought were my friends, only I found out that they weren't. I've always been isolated. I'm very good at listening to other people and being their friend if they need me, but so often they didn't need a friend to listen anymore. They didn't need someone to pick up the pieces anymore. And they don't care anymore either. Even my friendship with Dr. Jansma is difficult. I trust him, but I still need to be cautious. How long is it going to last? I start and I back off because the fear of being left alone, of being hurt is so great, sometimes I'd rather not even try.

I've only known one way of life. To live it with all the others, every day, 24 hours a day. Nothing's private, but then, I know everything about them too. We all look different, we all act different. Mandy hurts herself, she hurts me and that is so difficult to grasp. One body, but we're all different. I don't look a bit like Carol, nor like Francesca or Jenny or Alexis. We're totally different. We're different in thoughts and actions and looks and speech and yet we're the same, because we share a body. But our bodies look different. Francesca isn't very tall. I'm taller that she is. Simone has long, black hair. Jennifer has long curly hair--it's auburn, and Amanda's a redhead with shoulder length hair. Jennifer's a tomboy, I'm not. Alexis is mysterious. Andrea and Alexandria are both so timid, so we're all different. And yet, if I die, they die. I can't figure that out sometimes. I need a diagram.

I remember things like the mother standing in front of me and saying, "Go on, you want to hit me, I know you do." And I'd want to punch her out and stomp on her so bad, but I couldn't. A number of times she came up to my room and closed the door and stood in front of it and she would say that. And I wouldn't be able to get out. And he always covered for her.

The worst part for me was being ignored, as if I didn't exist. I think their idea of love was sending me to the grandfather's house where I could get screwed. So I could feel that dirty old man pawing at me and that grandmother running her hands all over me and inside me. Years and years and years. Sometimes days in a row. Every day. I wish I'd been smart enough then to just have taken a knife and just stabbed their guts out. I want a different body--to remain a woman, but I want a different body, so it won't be the same one that they touched. I think it's the only way that I can ever be clean.

I'm bitter and I didn't even realize until recently how bitter I am. I wanna crush them and step on them like they were bugs, flatten them and put them out of existence. And write their name off the record. I wonder if I would feel free then? I wonder if it would make me feel better--probably not. I want to die. Tried that one time. The mother wouldn't speak to me for weeks! As if I didn't already feel guilty. How could I do this to her? Except she did it to me. She and her husband and his parents, they did it to me. I think it's called emotional blackmail. There was a lot of that.

Parents

Others say that parents
give love,
that parents are caring,
they say that parents
give comfort,
that parents give life.

I say that parents give hate,
that parents give anger,
I say that parents give loneliness,
that parents give emptiness.

Others say that parents
give encouragement,
that parents give hope,
they say that parents give
a sense of belonging,
that parents give warmth.

I say that parents give pain,
that parents give despair,
I say that parents
give rejection,
that parents give death.

Marie

9

On April 30, 1984, Simone comes in for the first time and I give her psychological testing information about herself. She's 15 years old, shy, insecure, and cannot identify one good thing about herself. She believes she weighs 172 pounds. She binges and purges.

The logistics of therapy are becoming more difficult at this point as more and more of the personalities are deciding to enter therapy. Although this is clinically indicated, it is logistically unwieldy. I inform Carol of events coming up: a cousin's visit from the Netherlands and, a trip to Chicago and to Canada with her family. I tell her I am thinking of talking with her folks, which she is extremely hesitant to allow. (It seems time to establish once and for all whether or not there's any chance that the family system can become a support and a help in her therapy, or whether we have to proceed without them.)

Francesca is wanting to deal with her "anesthetized sexuality." She has also been writing. She is growing in insight and is able to talk much more freely about the sexual abuse. She reveals that she and Katharine are willing to provide a written description of all personalities. Katharine has been doing drawings of all of them, but she is becoming depressed and confused and prefers to stay away for awhile. Francesca blackmails Katharine into coming in because of her escalating suicidal thinking. She basically has a mini-psychodrama experience with much pounding with her fists on the sofa. An immense amount of anger and rage emerge followed by uncontrollable crying. The flood of feeling is quite remarkable. Amanda brings in her last knife. There is no longer an urge to cut herself. She is talking with her pastor and seems to be finding strength in her faith in God at this point.

I'm Amanda Saunders and it's April 27, 1984. I went to see Dr. Jansma today and we had a pretty good session because we looked at a magazine for awhile that had all kinds of neat things in it like woodcarving and pottery making and weaving, all the kinds of things that we like to do, especially Jenny. And I gave him my knife. I didn't really want to, but I suppose if I'm going to get better, I have to do that. We talked today about things that happened to me when I was little. I've been with Carol a long time, ever since she was real little. But I didn't really start to be very active until she was five and six when sometimes, Katharine and Francesca couldn't handle things. Dr. Jansma said it would get easier to talk about all this stuff every time, but it's not so easy because it hurts a lot and I don't understand a lot. I gave him a story about something that happened when I was six years old.

I hate all of them because they used to beat me. And I hate all of her aunts and uncles. The only person in that whole family that I really like is Nathan. But he didn't come until a long time later. I don't even like Patrick and Dena. I did when they were little, but now they don't care about me either. I used to take care of them, just like I used to take care of Phillip, 'cept Phillip died and I miss him. 'Cuz he was always really nice and he smiled a lot and he laughed and he always had jokes. And then he'd pick me up, right underneath my elbows and he'd lift me right up off the floor and put me in a corner. And I'd kick my feet, or else he'd throw me over his shoulder and I'd kick my feet and he just laughed and I laughed too. But now he's all gone, because he died and I don't know why he had to die. Why, why, why God made him die and now Nathan has to die too, 'cuz he's sick with that same disease that Phillip had. And I don't want Nathan to die, 'cuz he's my friend 'cuz we do things together sometimes. I'll miss him an awful, awful lot if he dies. Patrick and Dena, they just don't care. They're always so impatient and I used to always protect them when they were little, because sometimes other kids like Bill, next door to us, here in the United States, he would pick on them. And so one time, I punched Bill out. He was older than I was too, but I punched him out. I got him right in the eye and I gave him a big black eye because he was picking on Patrick, and Patrick was only three. And one time in the Netherlands, some kid was picking on Phillip and I hit him over the head with a rock. 'Cept I had to go apologize for that one 'cuz the kid needed stitches. I don't even remember his name now, but he had to have stitches in his head. I wasn't really sorry though, 'cuz he was a bully. I wish I coulda done that to the grandfather, hit him over the head with a rock and just bash it in.

Lots of times Mom and Dad didn't give me enough to eat. They always said that I was cold and that I had a nasty temper. They didn't really see my temper 'cuz if they'd seen my temper, they'd be dead now. I didn't ask to come live with them. Only reason I came was because Carol needed a lot of help and because Kate and 'Cesca just couldn't handle everything at once anymore. I got beat and so did they. That grandpa had a special little room for me and he'd put me in it sometimes. Sometimes I'd stay there for a whole week. e'd bring me some water every once in awhile, maybe some bread and cheese. But not every day. But he sure came in every day to beat the hell out of me. First he'd come in there and he'd make me take my clothes off, and then he'd put his hand up inside me and I hated that. And then sometimes he'd have sex with me and sometimes he'd just beat me with a belt.

One time he hit me so hard that he made blood come. And it was dark in there because I couldn't reach the light. When he left he would turn off the light. One time when he came in there, he made me take my clothes off and I thought he was going to have sex with me, but he didn't, he just used the broom handle inside me, for a long, long time. He put it in another place too, but I just can't say that. And after he left, I hurt so terribly bad and I guess I got sick and then I didn't remember any more, because the next thing I remember after that was that he was back and this time he brought me some milk and some bread and some cheese and when I tried to eat it, I threw up. Only he didn't do anything, he just looked at me and then walked out. When he came back later with some water, he said that if I would just learn to be a good girl, then he wouldn't have to do this to me, but I just wasn't a good girl. And so, he had to do these things to me. He had to teach me a lesson, he always said. But I didn't know what I did wrong. I hadn't hit anyone over the head lately. He said that I was so stubborn and I think he said that because I wouldn't cry; but I hated him right then, more than any other time. And I'd like to chop his hands off, and I'd like to chop his penis off, and I'd like to chop 'em up in little pieces. And the grandma too, for what she did to Katharine.

One time too, this was after we moved to the United States, I had to go over there and the grandpa made me lie down on the couch and spread my legs and then he pulled my panties off and he shoved something inside of me like a, like a short stick or something. And then he made me put my panties back on and then another pair of

71

panties that were tighter than my own. He made me put those on, too. And I had to keep that stick inside of me all day while I was there. And he made me walk up and down the living room with that stick inside me and it hurt to walk. And he laughed. He thought it was funny. And he'd say, "I'm gonna teach you a lesson yet." That stick was bigger than I was, though. And I was bleeding too, but he didn't care, didn't care at all. Maybe he liked blood. One of these days I'm gonna get even with him and with her too, the grandmother. I'd like to do to them exactly what they did to me, because I haven't forgotten any of it. See how they like it. Maybe they'd scream and I'd laugh at 'em.

I suppose it's not right to think like that, but I can't help it. I despise them. I wish they were dead. All of them and then I'd be all free. I wouldn't have to think about them anymore. It just always hurts so bad inside. I have these temper tantrums and then I want to run away. Then I want to do all kinds of bad things, like cut myself. I've wanted to do that ever since I was litte, 'cuz I figured if I don't have a vagina, he can't do it to me anymore and no one can ever do it to me anymore. I would just cut it out. And I tried. But now I gave Dr. Jansma my knife, at least my big one. I got my little one yet. And now he kinda wants that one too, and I wrapped it up, but I don't know if I want to give that to him. He said he'd just feel a lot better if the knives were there instead of at home.

I don't like a lot of people around me. Sometimes, I don't even want to leave the apartment. But once in awhile I do, like the time Marie and I went to Chicago. We got on the airplane and went away. Then Marie got scared so I had to take her back home. I had to come home, too. I didn't like that.

Dr. Jansma told me Friday I wasn't a bad person, but they were bad people. He said none of it was ever my fault, but they always said it was. I don't know why that grandpa wanted to teach me a lesson. They said I didn't have a brain in my head and that I was just rebellious. They always told me that. They told me I was a devil. Callie still believes she's a devil. I don't think I'm quite a devil, but I must be awfully close to it, because they sure had to punish me a lot.

Dr. Jansma's nice; I like him. But I get so scared in there. He said he wants to help me. But one of these days, he's going to get tired of me--I just know it. People always get tired of me. It just really, really gets mixed up all the time. Sometimes I like music, sometimes I hate music. Sometimes I like being outside, sometimes I hate being outside. Sometimes I like Dr. Jansma, sometimes he scares me. Sometimes I like all the others, especially Kate and Francesca, but sometimes I don't want to be anywhere around them. Sometimes I just want to smash things. I wish I had a different body. I've wanted a different body ever since I was a little girl. I figured then they wouldn't do so many things to me. That's another thing the grandpa always said. "Well, it's your body that's doing it. It's your fault because of your body." He told Jenny that too. It was either I had to be taught a lesson because I was bad or it was because of my body. I didn't ask for the body. It's not even a good body.

I don't know if things will ever change. Dr. Jansma said that if I work really hard, and if I let him help me, it will. I don't see how. I think that's impossible. You can't change a person that's been one way for years and years and years. All I've ever been is bad and I've been full of hate. Right now, I'm very tired. I want to go to sleep. And I want to be left alone.

Self

got no talents,
got no looks,
got a dull mind
words are hard to find,
no disposition
am an imposition
me.

Marie
February 10, 1970

Case: Mr. Copeless

Hello, Mr. Copeless,
How do you do
I'm glad for this time
to talk with you,
so what is it please,
that makes you so blue.

Well, you see Dr. Shrink,
I can't seem to think,
the things in my head
can't even be said.
I ain't got no spine,
I ain't got no guts,
I'll sit here and pine
on this perch o' mine.

So what I remain
is really a pain,
the situation is hopeless,
my name's Mr. Copeless.

Marie
May 26, 1984

10

In early May, I schedule a one-hour session with Carol's parents. I make this appointment taking a calculated risk that I can use this time to determine once and for all whether or not the family system, and in particular the parents, will be a support in her therapy. This is done unbeknownst to the patient, which I am very uncomfortable in doing but feel I need to do. Her parents come in willingly. I confront them on the issue of child abuse, with primary emphasis on involvement of the grandparents. Both strongly deny the use of a hole in the ground or of a trunk in any way to punish Carol. They do say that they guess it is possible that some sexual abuse by the grandparents might have happened. Her mother is in tears and her father is quite angry. They do agree to keep this session confidential. At the end of the session I ask their willingness to help Carol out financially with her need for therapy. They do not give me a response. I do not reveal to them anything in regard to multiple personality dynamics.

Caroline comes in for her first appointment, which is an emergency session. She is eighteen and has functioned as a storehouse for rage, especially at the grandparents for the sexual abuse. She is one of several who abuse laxatives, and has attempted suicide. We use a technique of having her imagine the grandparents in chairs close to her. We keep this to a verbal confrontation and much of her rage emerges. When I reach out to touch her hand in the middle of the experience, she starts crying uncontrollably. This is an effective session and she now views me as her ally after only one session.

I have another contact with Caroline. It is now clear there is as much self-hate as there is grandparental hate. She has always believed she was bad, because they told her the sex was her fault, "because you teased us." (For her the sex occurred between the ages of eight and fourteen, after their immigration to America.) We work at restructuring her thinking about her need to view the sexual experiences as her fault.

May 8, 1984. My name is Caroline. I'm eighteen years old. I went to see Dr. Jansma today. It was the first time I have spent any length of time with him. I was more that a bit nervous when I first went in. But I was so angry, I needed help. I called him yesterday because I didn't know what to do anymore. It had been building for such a long time and he listened to me, he helped me. And I guess that surprised me because I hadn't taken the time to talk with him and I, like the others, am really not used to having anyone take an interest in me and in what I'm feeling. I was surprised at myself for losing control like that. Up until now I have channeled the anger in other directions, mostly towards myself.

I have been with Carol since she was eight. I was raped at that time by her grandfather. We've all been through that and it has always been safer not to talk about it, to try and push it out of our minds, but I'm finding that that's become difficult to do. My anger, resentment, and my hate have built up to such a degree that I don't know how to deal with it anymore, so I finally decided to ask for help. Now, I'm a bit confused by what happened today. I saw the grandparents in that office, even though they weren't really there. After Dr. Jansma told me to imagine they were there, I saw them. What I said to them was only part of what I wanted to say, and it didn't begin to cover what I want to do to them. But there is some relief. Yesterday, I didn't know what to do with myself. I was at work. I wanted to smash everything in sight. I wanted to hurt myself. I wanted to yell and scream and I even wanted to cry--and that's something I've not allowed. Well, I did today. And now I'm going back and forth. Was it the right thing to do; was it the wrong thing to do? Some of the anger is gone or let's say, it's lessened, but it's still there. But I found I can be me in that office. I don't have to be Carol. I was afraid that they could hear me at first, but they couldn't. Someday I want them to hear and I want them to pay. After all, we paid. It took years from us. I know that Dr. Jansma is correct when he says that I live in the past and the present. There is no future. But I don't want there to be a future if it's going to be anything like the past and the present. I'm locked into one body with twenty others. I can't be Caroline for 24 hours a day, seven days a week. I want to do the things that I've always wanted

76

to do, like learning how to play a saxophone. I was never given the opportunity. I don't think that I was ever allowed to be a real person.

From the time I arrived, I think I was shot down. I tried to fight back, but I think that you can only fight for so long and then all of a sudden you realize it's no use. It's no use trying any more. It's no use trying to get approval from people that hate your guts anyway. It's no use trying to do nice things, because they are not going to be noticed or recognized.

Dr. Jansma used the word "alienation" for me; on the outside looking in, when you never feel that you belong anywhere or to anyone and that no one cares, and obviously, the family doesn't. A family that cares doesn't beat you and doesn't belittle you and doesn't call you stupid and doesn't try to get rid of you. Because I think that's what they did. I think that they tried and tried and tried, but they were never quite successful. They tried to kill me, I think. Well, they didn't succeed, and when I tried it myself, I didn't succeed either. Made it worse. It made me very, very sick. It wasn't any fun having my stomach pumped.

Mother's Day is coming up. You're supposed to do special things for your mom. Well, I don't have one. And it's so hard to watch Kate (*Katharine*) struggle with Mother's Day. Maybe we'll have to do something special for her because after all, she's been kinda like our mom, but she's the one that hurts the most with Mother's Day. All of us want a mom and we'd like to be able to do something special for a good mom, but Kate's the one that can never be a mom, and that's all their fault. They did that to her and I wish she had talked to Dr. Jansma about that today. I know that between me and Marie, we had a pretty heavy session. Katharine keeps an eye on us and it tires her out, and I think we tired Dr. Jansma out today too. But it makes her hurt really bad.

Today made me really tired. I've watched Dr. Jansma for a long time now, but it was so different to go in there and talk to him about myself than to actually let someone else see me and what I'm like. I'm kind of ashamed of the way I behaved today. This was the very first time I had a session and I cried all over the place and I yelled at the grandparents and I really carried on. I'm not used to being that way at all, but I didn't know what else to do anymore. He seems to help Kate and Mandy and Jennifer and Cesca and everybody else. But I felt so funny afterwards. I don't let people see what I'm feeling or thinking. I know he says it's okay but I'm still not sure if it really was. I want to go back, but I'm scared to go back. I told him I would and the others trust him a whole lot, I guess, and I'm not sure how

I feel about that either and I should know by now, because I've watched them for so long. I'm just all mixed up tonight and I don't know what to do about it. And then, when I get that mixed up, I get bad headaches and then I start to get mad all over again.

Everything was so different today. I don't even know what made me call him yesterday. All of a sudden I picked up the phone and I called. I don't do things like that. I did watch him work with Marie today though. Boy, she's so scared, just like me. She felt better when she left, but I wonder if she'll really go back and talk to him on Thursday. But I like how he took care of her and he said he cared about her and what's funny is I believeched Dr. Jansma for a about how I'm feeling, why is it so hard for me to believe that one?

This whole therapy bit is kinda scary. Maybe it's because I'm afraid I can't change. No one ever told me I could be someone. No one ever told me that they cared about me, I mean outside of our own little group. I'm afraid of having friends. Friends don't stay. They eventually go away. I don't usually talk to people. I don't let anyone touch me and I don't usually tell anyone a whole lot about me. And I don't know why I have such a terrible headache either. I want to cry again and I don't usually cry. It's dumb. It isn't necessary--I can't change anything. I can't be any different than what I am, right now. I can't ask anyone to love me or care about me, because they don't. And I hurt so bad inside and, and I want to scream and yell all over again. I'm scared--and I hate them.

I don't want them to be alive anymore. I want them to die and I want them to be punished and punished and punished. I don't know who's going to help me.

Look At Me

Look at me! I am a mistake!
A mistake created by hate,
kept alive by rage,
and nourished by destruction,
A wrong in a world of right.

Marie
Undated

Friendship

Friendship is special.
It is caring and sharing, together,
It is laughing and crying, together,
It is two people being close.
It does not exist.

Marie
Undated

11

It's still early in May, 1984 and Katharine is really struggling right now. As Mother's Day is approaching she is having her anniversary reaction, which is directly related to her inability to have children. This is a primary source of very private pain for her. Marie and Katharine have made an agreement: If Marie would agree to come in, Katharine will deal with her resentment at her sterility. Marie submits a powerful poem reinforcing her wish to stay private. She believes she is a totally bad person. Katharine lets me know that Marie is suicidal. We spend time preparing contingency plans for the others who know the implications of this, and what to do if Marie seriously becomes suicidal.

This is Katharine St. Clair. It's May 15, 1984, and it's 4:00 a.m. I haven't been able to sleep much. Marie is to see Dr. Jansma today. I am glad she decided to go. She called him yesterday to tell him she'd come in. She is hurting very badly and she has wanted to die for awhile now. She feels very vulnerable now and she feels left out. That may be because so many of the others have had problems and have needed to talk to Dr. Jansma, and she's been left by the wayside. It's not intentional. Part of it is Marie herself. She's terribly withdrawn. Her poems are full of death and retreating, hiding, and for some reason, she believes Dr. Jansma isn't her friend anymore. I don't know why. She's again started to use laxatives and she's tried so hard on her diet and she's not losing any weight. I wish that she would learn that I care about her a great deal; that I love her, and that Dr. Jansma cares

79

about her and wants to be her friend. I had promised her that if she would at least talk to Dr. Jansma this week, we would do things together, have our time together and we did. We transplanted plants, made potato salad and at least kept her mind occupied a little bit. I know that if she dies, I die--and everyone else dies I don't think she quite comprehends that. She sees herself as totally separate. And for Marie, who wants to be just Marie all the time, it is difficult to comprehend that even though we're separate, we're one person. I have at times had a difficult time comprehending that. How can we look so different, feel so different, act so different and be one person? I guess it comes with the territory.

For some of them it's safe to remain inside themselves, to not let anyone else know what they're feeling or thinking, what they're doing. Of course, we all tune in to each other, we just don't always know each other's deepest thoughts and feelings. But I know Marie pretty well, and I think I know the others pretty well. I've been with them, with all their hurts and their disappointments, all their pain. I've watched it all. For Alexis, it was a hole in the ground and a thunderstorm at the same time. Alexis is a moody individual and puts on a big front. When she was put in the hole, there was a terrible thunderstorm. She screamed and screamed until she couldn't scream any more and no one came to help her. I couldn't help her. Now she tries to prove to herself that she isn't afraid of anything or anyone. Not only does she try to prove it to herself, she has to prove it to everyone else.

Then there is fifteen-year-old Simone. She doesn't believe in her own intelligence. In some ways she has had to grow up very quickly--received an education in sex quite early in her life. She's never had friends and has been teased and ridiculed all her life.

And there is Jacqueline. Gang-raped. Despises her body and will do anything, anything she can to make herself thin. She thinks she's a lesbian. She wants to have no interest in men. She never again wants to feel a man touch her. She hates men. She feels so dirty and ashamed and yet she didn't do anything to cause it. For some of them it is much safer to stay in the apartment where other people can't see them and they think other people can't hurt them. That's no life for them. I don't want that kind of life for them. I want them to enjoy freedom and happiness, to believe in their talents, because they do have talents.

Andrea and Alexandria can do needlepoint and crocheting. Marie's talent is writing, if only she'd write about something happy--something different. I thought spending time with her this weekend would help. She did call Dr. Jansma to say she wanted to talk with him, but I'm afraid she'll change her mind again. She can also listen to a piece of music and play it, even though she's never had a single piano lesson. She plays my favorite piano concerto by Chopin, and she plays it well, without error. She plays it with feeling. She believes that she doesn't have any talent. She learned it because she wanted to give me something. Of all of them, Marie is the one that will call me "mother." I'm not her mother, but when she needs a little bit of attention, when she needs to let me know that she's there and that she's afraid, she calls me "mother." But I can't stop the pain for her. I can't stop her nightmares. I can't change her feelings about herself.

Caroline amazed me last week by going in and beating on the couch in Dr. Jansma's office. As much as I love Caroline, sometimes she scares me because I can't always control her and I can't always keep her anger in check. She has to do that herself and she doesn't want to anymore. I've had to watch the fear in their eyes, I've had to watch their hate grow, their confidence in themselves all but disappear--and I can't do a damn thing about it. After Marie had her abortion, she couldn't even talk. She couldn't think. She couldn't help herself. We tried to help her, but there was only so much that we could do. After a time, when she started to talk again, her voice was different. She couldn't think as clearly and she still can't. She asked that doctor if she'd ever be able to have children again and he said, "Nope, now you can screw all you want to." They had sterilized her in the process. Marie wanted something to love. And in her mind, she was going to show the whole world that she could love a baby. That baby was going to love Marie too. Something she had not had. It was going to love her. All the wrong reasons to have a baby, but to her, those reasons were real. It all made perfect sense to Marie. Maybe I should have explained things more to her. How was I to make her understand that having a baby was not going to make things all better? Marie's guilt is so tremendous. She hates herself even more than she did when she was a child.

We spent a lot of years together before we first met Dr. Jansma. That was a joint decision. We could have stopped Carol from even going, except she wanted to die and there were a number of us that wanted to die at that point. An attempt was made and so we collectively started therapy. It was hard for her. We had to fill in for her. Give Dr. Jansma bits and pieces. All the while, telling him a little bit about ourselves; Carol thinking it's all her. One of us might tell him something in one session, and he would talk to her about it in the next ses- sion and so she would try to fill in, because she was afraid to tell him she loses time. Obviously she does, because we take the time.

But, we watched. We watched for almost three years. Shouldn't have to take that long to trust someone. Some of us still have a problem with that.

Therapy can be so difficult: to allow someone to help you, you first have to expose yourself, open yourself up to that person. And even though he says that nothing any of us say will surprise him or will make him think any less of us, that doesn't take away the guilt.

After we had been going a couple of months, we ended up in the first psychiatric hospital. Now that was a hell of an experience. Locked doors, locked windows--a cage. Staff members didn't really seem to give a damn and it was like a punishment. Again, some of us wanted to die, and there were a lot of battles going on amongst ourselves. And so one night in May, it was Mother's Day, we went to this hospital in the pouring rain, late at night. I was depressed that day anyway, because it was Mother's Day and I didn't care if I lived or died. A friend brought us. Of course, she didn't know about "us." She knew us as Carol. They politely asked her to leave and so there went our last link with the world. And we went because our friend had called Dr. Jansma. She was worried. Dr. Jansma called us back and said, "Maybe it is a good idea, maybe you should go to the hospital." So we went. After signing all those papers, we got to the floor where we were going to be staying. They take things away from you. Compact, crochet hooks, curling iron. Then they dumped our suitcase out on the floor in front of the desk, turned it upside down, looking for things, anything, I guess. Then we went down to our room and it was in a locked wing. Couldn't get out of the wing unless you banged on the door. I had to stay there overnight, maybe it was two nights, I don't even remember. I hated it. It felt like punishment. We were all tired and scared. We had to see Dr. Jansma that week, so we were allowed to see him at his office. And Jennifer, who of all people should never ever be locked up, asked Dr. Jansma to please, please let us go home. He wanted us to stay the whole week, so we did. We've always been alone, but never as much, I think, as we were that week. Well, maybe that's not correct, but it sure felt like it. No one to talk to; no place to go. The little ones terrified. Francesca, so furious and we all vowed we'd never go back to another hospital. Well, that lasted a whole year because it was time for the second psychiatric hospital a year later. We stayed longer that time. The laxative abuse had become so heavy that something had to be done. Of course, Carol was taking them, trying to become skinny, trying to make people like her. All she ever wanted was for someone to like her, so we all agreed to go to that hospital. It was different. Staff that cared, classes to go to so you didn't sit and stare all day. We didn't have to make collages like we did at the first hospital and we did learn some things. We did do a psychodrama, Jacqueline's psychodrama about the gang-rape. Alexis did it for her. Jacqueline couldn't. It was also about Marie's sterilization. The children were afraid of being away from home. Dr. Jansma came out to see us every week. That helped. A friendly face, someone we knew, trying to help. Sometimes I wish that I had told him about us sooner, but it wasn't just my decision. Everyone had to give the okay to that one, not just me, because there were so many taking laxatives and diet pills and wanting to be thin and thinking that the only way people would like them would be if they were thin and attractive. At least there we could take walks outside. I remember walking to the duck pond with Dr. Jansma, sitting outside for two hours one evening talking to Connie, a staff member. She didn't have to, but she cared. Just like Dr. Jansma didn't have to drive out to see us, but he cared.

It was safe there, to an extent. There weren't outside pressures. There weren't people at work to be aware of. The Mom and Dad came up a lot. A lot of times that was just to tell us to behave ourselves and not to say too much, and to get back home. The children were a lot of work to keep under control. It sure would have surprised the staff had a five-year-old suddenly appeared. But it wasn't like the cage the first hospital was. We could still walk the grounds. A little bit more freedom but not totally free. But of course, none of us will be totally free, will we?

I'm not always going to be here. I'm not always going to be Katharine St. Clair. I'm helping Carol remember things now. I haven't really told Dr. Jansma I'm doing that yet. I'm doing it slowly. I think too much all at once would hurt. There is a certain amount of pride, I guess, when I look at the others, when I see them accomplish something in therapy. Mandy giving Dr. Jansma her knives, Marie calling Dr. Jansma and agreeing to talk with him once again, Caroline getting some of that anger out. Francesca slowly opening up. Simone having enough nerve to even go in. She was so nervous; didn't know what to wear. She's only fifteen and she wanted to make such a good impression. What other people think means so much to her.

There was the time Amanda wrote a short story. It was published in a childrens' magazine. That wasn't even acknowledged by the parents. Amanda's copy of the story was thrown away. It was junk, it wasn't necessary, according to them. Only it wasn't junk. It was a wonderful story about a mouse.

We've all tried so hard to please, but it's all turned out into one disappointment after another. So now we're in therapy. Dr. Jansma tells us that we'll never be lost. We'll always remain. Maybe differently than we are now, but we'll always remain in existence. I know that Dr. Jansma doesn't lie to any of us, but yet I'm afraid. I'm afraid of not being any more and I'm just learning that it's possible to have real friends. I know I can't expect life to be perfect and to be happy all the time because I have been especially unhappy the last couple days. But yet I still realize that there are people that care about me. It's starting to sink in after almost three and a half years. But I want to feel that on my own all the time. And I don't want to lose me. Sometimes I have a little bit of trouble knowing who I am, but I don't want to lose me.

On May 15, I see Marie for almost an entire session. We focus on her conviction as to how bad she believes she is. I let her know that it is possible that her perceptions can sometimes be distorted.I have her pull a few hairs from her head (she's always believed she had black hair). When she realizes they are blonde she becomes extremely anxious and angry and leaves. Francesca then takes over. It is clear that this confrontation has bothered and confused a number of them.

May 15, 1984. This is Katharine St. Clair. It's been quite a day. Marie went to see Dr. Jansma and she freaked out as did a lot of the others. I think I understand why it happened, but now she's terrified. Everyone else wants to run away and I can't handle it today.

I let Carol have time and of course she was totally upset when she saw all the packed suitcases. Marie wasn't ready for today. I wasn't ready for today--none of us were. Marie just wants to be Marie; I just want to be me. I know that Dr. Jansma was just trying to explain to her that her descriptions about herself are wrong; that she has talents, but she can't see them and he can. But the little bit with the hair was just too much to handle, because Marie has black hair--but he can't see that. No one can see what we look like. Marie, Mandy and Jenny want to run away. Alexis took off down the highway again tonight, her favorite place to drive. I spent the weekend trying to convince Marie that it's better to be alive than to be dead. And I thought that she was coming around a little bit. She spent some time with Dr. Jansma talking about her poetry, her talent for writing. So much for that. I don't even know if I'm really angry. I'm hurt and I'm disappointed and I'm confused. I want them to feel free and I want Marie to like herself. And I wanted her to talk with him. I wanted her to get better just as I want everyone else to get better. The look on her face when she saw that hair. That dream of being Marie, of being herself all the time. 'Cesca said it quite well, you know. Maybe we attach too much importance to our physical appearance. But, damn it, that's all we have. That's the way we've known each other. Not just by what we feel and how we are, but what we look like.

May 16, 1984. We have an emergency meeting to deal with massive turmoil over yesterday's session. Katharine, Marie, Francesca, Jennifer, and Alexis all attend. They are quite frightened about who they are; if Marie's perceptions have been distorted, it may be that others' perceptions are distorted also. We have a session to defuse these issues. It turns out that their bags have been packed and a number of them were ready to leave to go to Houston on this date.

May 17, 1984. Marie brings in a poem written about me. This is her way of telling me that she wants to be my friend. She can now therefore feel free enough to discuss her abortion, and the physician's statement following that abortion: "Now you can screw anyone you want." This was a cathartic session for Marie. Emotionally and in terms of voice tone she seems to be approximately seven years old, but is clearly quite bright and perceptive.

Dr. Ted

It's all there!
An endless blue sky,
soft, cottony clouds,
clouds that vary,
no two alike.

He looks up and sees them,
really sees them,
and the child within him fantasizes,
transforms them into images,
magical images that only one
who has not lost the child
within him can see.

That one's a caboose,
and that one's a castle,
that's an ice-cream cone,
and there's the cotton candy.
Laughing out loud, he watches
them glide, spots even a

clown just a grinning with pride.

And he feels within him joy,
the exhilaration of a new day,
feels the warmth of the sun,
sees the calm water around him
reflecting it all.

And he knows the meaning of peace,
savors the wonderful experience
of life itself, content with
the solitude, at ease with himself.

And I know such a man,
he really exists,
a man who loves the water,
a man who can still see
as a child,
who sees the magic of the clouds,
a man who is my friend.

Marie
May 16, 1984

May 18, 1984. We have another extra session to deal with Marie and Amanda (as well as Alexis and Caroline vicariously), a chance for me to try to prevent them from leaving town. All are quite frightened. Amanda has also had the urge to cut herself. At the end of the session it appears that we have defused much of this inner-turmoil.

May 22, 1984. Katharine is very troubled by the possibility of going out of existence. She knows now that leading twenty-one separate lives is not workable. She wants total independence but knows that this will never happen. She has been feeling guilty for not being able to prevent the others' pain. Although she continues to nudge Carol's memory somewhat, she is personally struggling quite a bit. She reveals there are three holes in the ground at the parent's home and one at the grandfather's home.

12

I t is now the end of May, 1984. Events seem to be escalating, and although an immense amount is being accomplished clinically, I'm not always sure of my direction.

Marika introduces herself to me for the first time. She takes an MMPI and accepts a session to get testing feedback. To my surprise, she lets me know there is a second group of twelve personalities none of which the others are aware. This group includes the following: Marika, Elizabeth Austin, Ashley Baldwin, Miranda Brandon, Lisa W., Moiria Sanderson, Cynthia, Tanya, Corrie, Kristen, Jennifer Christenson, and Karin. Marika presents as the "Kate" of this second group and quickly agrees to work for co-consciousness with Katharine. Katharine is seen but her barriers are very visible. She clearly resents the additional personalities and will not consider co-consciousness with any of them.

At this point, Katharine begins to seriously regress emotionally, and eventually she regresses to age four. It would appear that for years Katharine has privately experienced her own alter egos (which are not necessarily other personalities) and that now even these have "gone public." The four-year-old is the first to appear. There is also a fifteen-year-old and a twenty-six-year-old, all of whom emerge. Francesca fills me in and with the use of hypnosis and age progression, we attempt to get Katharine from her four-year-old, who is very destructive and only speaks Dutch, up to age thirty-eight. There are frequent regressions to ages fifteen and twenty-six. She feels she is a bad person.

87

There's a change,
a change that comes from pain,
from hurt and from anger.
It brings with it fear
and confusion,
fear not only of the present,
but also a fear of the past.

And this woman who has always
been strong becomes
a child, a confused child,
a lonely child,
and the fear of being alone
is once more evident.

And when she speaks,
the words are jumbled,
and uncontrollable laughter
echoes, and tears course

down her face.
And then she is silent,
unable to speak,
and her fear becomes complete,
complete because the control
that was once hers is shattered.

There had been so many
thoughts of late,
thoughts of the present,
thoughts of a future
that might have been.

But the past gave
birth to fear,
and the past is
what made her,
and the past once
more has reached her soul.

Marie
May 30, 1984

Francesca informs me that the four-year-old Katharine is becoming violent and they cannot cope with her. We do brief hypnosis and try a co-consciousness experience between Francesca and Marika in order to combine our helping forces, which appears to help. They're all quite worried about Katharine's regression. I explain that this is her unconscious way of "letting down the walls." It is clear that self-hate is an issue, as there are scratches all over her body.

As a way of defusing so much attention being given to Katharine, I attempt the first co-consciousness with Carol and use Marika as a vehicle for this. This goes quite well. Although Carol was very anxious at first, she becomes fairly comfortable with the agenda quite quickly. Marika is emerging as a key ally now.

✤

May 29, 1984. I'm Marika and I want to explain a little bit about what the others mean to me, not just my group, as I call them, but the others as well. My group, of course, is unknown. I didn't introduce myself to Dr. Ted until this week. No one really knew of my existence nor did they know of any of the others that are a part of my group. We're very, very different from the others.

We have not had the same kind of traumas as they have had. I, myself, am happy and easy-going, at ease with myself and with others. I wasn't beaten as a child. I wasn't emotionally abused. In a sense, I guess I'm what Carol wants to be, because I'm free. I took a test at Dr. Jansma's and he said that I'm very invested in not letting anyone know when things aren't quite right, that I can be defensive, or I have a tendency toward being defensive. Maybe I do, but I, like Katharine, have a group to take care of.

I take care of Katharine's group too, more than she knows. She's not aware of me or at least of who I am. She's aware of another person, the one that is the so called "Number 21." What she and the others don't realize is that Number 21 consists of the rest of my group. (*Note: Group 1 had a vague awareness of someone else whom they identified as "the virgin." This twenty-first personality was always really the twelve personalities of Group 2, a much less traumatized group of personalities who arrived at various times between Carol's ages of thirteen and thirty-four.*) It hasn't always been just me. The reason I came now, or at least introducing myself now, is because I think that they are having some problems. Katharine is so unhappy. She's really struggling. It's not easy for her to speak about herself and to let go of those feelings. She's been very invested in taking control and maintaining that control at all times--not letting anyone else see when she's troubled--or talking to anyone about those troubles. She's not used to reaching out. It doesn't bother her in the least to ask for help for one of the others, but she finds it difficult to ask for help for herself.

She needs to maintain complete control at all times, to not let anyone else see that she also is affected. Only Dr. Jansma knows that she is. But we know that she is and so do the others in my group. She has forgotten about herself: her own needs, her own desires. And now that those needs and those desires are coming out, now that she's vocalizing them in session, the guilt has started. She views it as weakness. I've never thought it a weakness to want things. I'm not speaking of material things, but of things such as happiness, to put yourself first for a change. It's not a weakness to want a normal life, and I'm going to put "normal" in quotation marks because what's normal?

I care about all of them, but Marie has to be one of my favorites. Such a dark, brooding person at times but the last few weeks have been so interesting to watch: her insecurity and the big debate she had with herself over going to see Dr. Ted (that's what she always calls him in her mind). Marie is a woman-child, and at times, much more of a child. There's a quality about Marie that's special. I loved the poem she wrote about Dr. Ted. I switched from Dr.

Jansma to Dr. Ted. I prefer Dr. Ted. And she knew what she wanted to say. It was as if she knew, as if she actually put herself in his place. She imagined how he would feel. And I think she's pretty correct. I'm sure not any one person feels that way all the time, but sometimes, and Marie captured that. The last line, "A man who is my friend;" Marie meant that too, and she still does, of course. That was a big step for her. She reached out and believed, actually believed, that he was her friend.

Then there's Jennifer Montgomery, not the Jennifer from my group. Jenny is spirited. She is hyper at times, but she too has some of that childlike quality that Marie has. Not the same, but it's still there. Jenny has energy. And she really does run. As fast and as hard as she can. She's done that ever since she was a child. When things started getting bad, she'd run. And, in a sense, that's what her exterior happiness is about--don't let anyone see the hurt on the inside, cover it, hide it, run from it. If you cover it well, you won't have to think about it. And if you hide it, you won't have to look back.

There's a member of my group that is somewhat an extension of Marie, Carol and Moiria. Her name is Elizabeth Austin and she is the same age as Marie, she's twenty-one. The difference is that Elizabeth is pregnant and Marie had the abortion--not a wanted abortion. That's very evident in the story she wrote about it:

She lies alone in her bed by the window. There are three other beds in the room, each one occupied by a woman older than she herself is. They are all laughing, visiting with family and friends. She has no visitors, no one to care how she feels, to see that she is frightened. Her hand touches her belly and feels its roundness. She thinks of the life within her and wonders if it's a boy or girl. She wants to reach inside herself and touch it. A sob starts deep down inside her, but she stops it before it escapes. She is given some medication to make her sleep, but it has no effect on her. All through the long night she waits and thinks. When it is morning, she is given an injection and everything becomes blurred. She is put on a stretcher and taken to a large bright room. She is still alone, still frightened, with no one to care for her. She is lifted onto a hard table and she is very cold. Her hand once again steals to her belly and silently she cries. A needle is put in her arm and quickly she falls asleep. Hours later she opens her eyes, still alone, still frightened. There are needles in both her arms and blood flows through one of them. She feels the heaviness of the bandage on her belly. Her body aches and in her mind she screams, for she realizes that the life that once grew inside her is now gone; they have torn it from her body and that she will never know the happiness of holding a child for they have taken that away. Someone leans over her and tells her she's lucky, that she almost died. She wishes she had. With that thought, her mind shatters like a fragile piece of crystal breaking into many fragments. And she is still alone.

Marie

Elizabeth has all the domestic qualities that Marie doesn't have, but she does have something in common with Marie. She also likes poetry and she's very intelligent. She's somewhat dreamy, and by dreamy I mean that she actually sits and daydreams. She thinks about this baby of hers and has named it. It's a living, warm, very small human being inside of Elizabeth. One of these days, that baby's going to be here and Elizabeth is going to be complete.

And then there's Ashley Baldwin, another member of my group. Ashley is twenty and she reminds me somewhat of Jenny, (*Jennifer Montgomery from Group 1*). She's a bit younger than Jenny. She loves nature and has a wonderful, wonderful sense of humor. Kind of slides things by you before you even catch it. She does beautiful pencil sketches. She's rather quiet though, not as boisterous like Jenny, but she's happy, whereas Jenny is not.

We all came to introduce ourselves at this time because we're watching the others struggle and we don't like it. I don't mean to reveal their secrets, their thoughts, their feelings. The secrets are going to have to be told some day, anyway, but I feel that is up to them, not me. Their life story is their own. All I can do is give my own opinion. Carol is an unhappy person. She has learned something through therapy, but there is so much that has gone right by her. It's as if she has picked up bits and pieces of what therapy has been all about, because so many of the others have taken time. And we, in a manner of speaking, have created a life for her because of some of our experiences, some of the things that we have done. Of course she feels incomplete. Who wouldn't. If I were missing a great deal of time, if pieces of my life were gone, I'd feel incomplete too.

I understand that someday we'll have to be integrated and I'm not terribly worried about being lost. It would be nice to keep my own identity, but that won't be possible. Sometimes it's difficult to believe that outside people can't see us the way we see ourselves. How is that possible? It's like saying you're a ghost. Well, I'm not a ghost, I'm a person and I have hair and I have eyes and I have a body. I have feelings and dreams and hopes, just like anyone else.

* * *

On the weekend, four-year-old Katharine goes a bit wild and cuts a lot of her hair off. All have been trying to contain her. Elizabeth Austin emerges for the first time in session and introduces herself (age twenty-one and pregnant). She seems to be an ally. We are using Francesca as an interpreter and co-consciousness between Francesca and the four-year-old Katharine as needed.

By the next week, Carol is still feeling very good about co-consciousness with Marika. She brings in her "first" poem ever written.

Beginnings

It's a brand new day,
a day of new beginnings,
a day that promises freedom,
a whole new existence.

And I open myself to this day,
my destination as yet unknown,
the future left to my imagination,
the beginnings of peace finally mine.

It's a brand new day,
a day of renewal,
of re-discovering friendships,
a time to start finding me.

And it's a day of hope,
an acceptance of reality,
a celebration of life,
a day that is my birthday.

Carol
June 1, 1984

Time is spent with Katharine; she is now consistently remaining at the regressed age of eight years. We are dealing with the issue of her belief that she is a bad person.

13

This is Francesca Germaine and it's Sunday, May 27. We just came back from church and it was a good experience. I'm beginning to feel more comfortable in church. I'm not so sure about Katharine. I've watched her this past week. She's becoming so restless inside. She can't find that peace that was once so easy for her to have. I think part of it is because she's been helping Carol remember. She worries about so many things. I do too, but they're different. Kate is the mother, she's more maternal than I am. Her love for Marie is very evident. It shows in the time she spends with her, in her pride in the poem that Marie wrote for Dr. Ted. (*Note: all are now using this form of address.*) She even worries about me. She doesn't have to, but she does. Then, I worry about her, because we are a family. Different from most, but we are a family.

On the whole, we get along pretty well. Like most people, we are closer to one than we are to another. Although, for some of us, that is changing too. My relationship with Kate is an example. Kate and I differ on a great many things. Kate has always been the more quiet one. Softer, more gentle, more giving and has more patience than I have. We have many of the same childhood memories, the two of us. But we are very different. My attitude has always been one of getting even. I tend to be a bit harder, sometimes a whole lot colder and I realize that that's a defense because I don't want to be hurt anymore. Now I'm not as invested in that as I used to be. I think therapy is helping me with that, although I haven't spent as much time in therapy lately as I would like.

I think my greatest need has been to get even with men in particular. There have been a number of men in our lives who have caused us a great deal of pain and I want them to pay. And that's not just men in the distant past, it also includes men in the not so distant past. There have been some that have been users. They played games. See what they can get. So over the years, I devised ways of getting even. At times, it's caught up with me.

Bruce is a good example of that. I watched him use some of the others and I realized a long time ago what kind of a man he is. He's manipulative. He's brilliant as far as his work goes. He enjoys things like classical music and he's very clever and says he likes himself. Now, in my opinion, when a person tells someone else over and over again that he really likes himself, then I think there is something wrong; that he really is insecure and he really doesn't like himself. He's trying to convince himself that he does. I picked up on that a long time ago. Everything for Bruce is "I": I did this, I did that, and I will do that. He's never satisfied. Always striving for more. If he doesn't do it right the first time, he will do it again and again and again. Since his work is very precise, he has to be that way, but it carries over. He's never satisfied with anything. There always has to be more. More money, more clothes, a better car. He has, I think, a contempt for women because of the way he uses them. Takes them out a few times, then takes them to bed and screws them. That's been repeated so many times with girls at work that I've lost count. He's going to get them into bed, and when he does, he's finished with them. That's his goal. He's someone that uses other people at work to get where he wants. He will step on anyone to make himself look good. He used us.

I devised my own game with him. And even though to others it may appear that I've lost, I really haven't. Because I was able to take him in and I did it well. It was still worth it. His opinion of me does not matter, but he knows without a doubt what I think of him, that I despise him and that I know him for the kind of person he is. It was worth it for that alone. And I'd do it again. He may feel contempt for me, but my contempt for him is doubled. I feel that way about a lot of men. They're contemptible. I don't like them. I guess it's safe to say that I would like to get even with every man who has ever tried to screw me.

It is true that my relationship with some of the others is changing. I used to have a real problem with Kate. I considered her to be Miss Goody-Two-Shoes. This quiet person, who seems to have so many talents, who always seemed so at ease with herself. Oh, I know she had traumas, but she handled them. She never seemed to have all the anger and the bitterness, and it was so surprising to me a couple of weeks ago when Kate just pounded on the couch in Dr. Ted's office. I think it was good for her--she needs to let go of some of that pent-up emotion. Lately, she has been changing. When she took that test, Dr. Ted said she had a tendency to take psychological problems and turn them into physical symptoms. She's tense, more tense than she appears. She's become even more private about her headaches and I know they've been really, really bad lately. Yet if someone asks her if she has a headache she'll say, "just a little one," whereas actually, her head is pounding so badly that she has to lie down. Her restlessness disturbs me also. It's so unlike her. And the headaches are so frequent now. When I ask her about them, what they're like, she'll say, "It's like a pressure, a terrible, terrible pressure."

Her patience level is lower too. True, we all demanded a great deal of Kate, but it's always been that way because she has always been the mother. She puts up walls with Dr. Ted. She knows he's her friend, yet when she opens her mouth, the words stick. The thoughts are there, but the words don't come out. She has always maintained control, but this control is unlike anything I've seen in her. There are so many things that she's hurting about. She swings back and forth between suicide and wanting to live, suicide and wanting to live. She can't sleep; she can't sit still for very long. It puzzles me and it hurts me to watch her because since I started talking to Dr. Ted, since I've had a few sessions of my own, I have found it easier to talk to Kate and we've related much better than we ever have. We're closer and I'm more than willing to take some of her responsibilities away from her, but sometimes it's as if she holds on to those responsibilities as if that's all she has left.

She's becoming increasingly more concerned about the way she looks. Not just the way she appears to other people, but the way she looks to herself. That too is unlike her. Everyone worries about whether or not their hair's in place, but Kate, Kate is the ultimate perfectionist when it comes to getting ready for work in the morning. It takes her three hours. She scrubs and scrubs and scrubs herself. In the evening, she goes through the same routine. She'd do it at work if she could. She has to get clean. Her nail polish has to be just right. Her hair has to be just right. Clothes, everything has to be just right. She wears herself out. Of course, I can't reason with her on that one. I don't like looking sloppy either, but Kate could look fantastic in rags, I think. An old beat up pair of jeans and flannel shirt--she'd look great. Some people have the ability to pull it all off. I don't. She does, but she doesn't see it. And of course, with all of Kate's uncertainties, the others also have more uncertainties.

Now I know that there are others--people we don't know. That disturbs me, as it disturbs her greatly. We always thought we knew all the others, all of the personalities. That's a little hard to take. It feels as if they're an intrusion. But that's also what I think Carol felt, although we were there to help her all the time. It's a strange feeling to have others know about you, only you don't know about them. Who are they, what are they, what do they look like? Where did these people come from? What do they want? Am I a multiple personality? How about that, a personality being a multiple personality. I guess I am, but I don't like the term because I am my own person. And "multiple" makes me feel as if I; I don't know what it makes me feel like. If I say it makes me feel as if I am a multiplication of something or someone, well in a way, that's supposedly accurate, but yet, I don't look like Carol and I don't look like Kate or Jenny or Marie or

Simone, Christina. I don't look like any of them. I am me, myself, Francesca. And I rather like it that way.

I find it difficult to believe that I will be in existence after we integrate. I don't feel that's possible. I have existed as an individual for so long, how can you possibly exist as an individual when you're integrated with someone else? I have a problem really understanding that. It's not that I'm unwilling to try. I have a great many things to work out in therapy. Some of it really frightens me. And I know that I'm insecure as a woman. I'm afraid of letting people in, afraid of rejection. I'm afraid of dealing with all the anger because I, too, have always needed to maintain control. Sometimes it becomes a real struggle. Do I really want to go through all this? Do I really want all this therapy? Part of me does and a big part doesn't. A big part of me wants freedom and another part of me is afraid of it. I would like a normal life. It's just that change frightens me. The uncertainty of it. All the things that are required to accomplish that change. And that's strange, because I'm always the one that said I wouldn't be afraid of anything. That I wouldn't show anyone fear; that I wouldn't tell anyone I was afraid. Maybe that's the first step, admitting that I'm afraid. Admitting that a change is needed.

I remember when we first started. I resented it. I hated it. I didn't want anyone to ever find out about us and I was afraid that Dr. Ted would. And of course he has. I guess it all comes down to what I want more. To stay the way I am or to be happy.

Today is June 3, 1984. This past week has been very difficult and I'm feeling a little bit scared and angry and I'm tired. "Emotional" is the correct word. Of course, it's because of what Kate is going through--her regression back to childhood. All the pain she is going through. I'm not the only one that feels that way, we all do. It was somewhat of a comfort on Thursday evening when Dr. Ted helped me get in touch with Marika. I like her; she's neat, but she can't reach Kate either and we've been trying. But when Kate is little, she has such a strong will and she's such an angry child--so destructive to herself. It's difficult to stay home from work and yet there is no way that we can go because we don't know when the four-year-old will take over, but something has to be done.

All of a sudden, she (the four-year-old Kate) goes into such violent rages. Thursday evening I hid knives and forks and all sharp objects around the apartment because I didn't want her to hurt herself. She drew some pictures that she tore up--I managed to salvage two of them. She dug at herself with her fingernails. She has welts all over her body. She drew blood. And she doesn't talk. Dr. Ted said she çan't, or maybe she won't, hear us. She won't say anything and I can't reach her and she fights me. It seems to be a losing battle for her. She tries so hard. It frightened me when I took her into Dr. Ted's office on Thursday--he used hypnosis to bring out

96

thirty-eight-year-old Kate--only she kept on slipping back to different ages--fifteen to twenty-six. Then she'd be the child and she tried so hard to tay with him. She was fighting so hard inside. It hurt to see her in so much pain. Dr. Ted explained to me that this is her way of coping with it. Her way of getting out the anger and the pain and the hurts.

It scares the others also. I'm frightened for her and I want her back as the Kate I know. A loving Kate, one that always knows just what to do in any situation. When the child is there, the eyes are so full of hate.

I guess it really scared me when she said she'd always been there, meaning the little one. I've never known about her and neither have the others. This was completely outside of us and so I think that's what scares us all a little bit more. And some of the others are afraid that they'll all have to go through this very same thing that Kate is experiencing now. I've tried to explain to them that this is Kate's way, but it's not necessarily their way. Still what they see is a Kate who has always been their mother image not functioning. She isn't there to talk to, she isn't there to encourage them. I need her too, because she's finally become my friend. I know Dr. Ted said that he doesn't want to hurry things, he doesn't want to push Kate--that he wants her to stay well permanently, not just temporarily. But I want her back now.

One thing that did come out of it has been Carol and Marika's co-consciousness. I wish I had that with Carol; that she would allow me to do that. But I enjoy having it with Marika, although I can't always quite seem to reach her. Maybe it depends on how uptight I am or something. But for Carol, it was such a wonderful experience. I watched her over the weekend. Friday night she and Marika stayed in control. I was there, only Carol wasn't aware of me and she's so excited. So happy. Like a little kid discovering a whole new life. I think she was more peaceful than I've ever seen her. More confident.

I spent some time with Marie, and did some talking with Marika. We were able to keep the four-year-old Kate under control until about 4:00 Saturday afternoon, when all hell broke loose. I shouldn't have been so stupid. I had hidden all the sharp things, but forgot about the damn scissors. Andrea and Alexandria had worked on some needlepoint and used these big scissors. And, all of a sudden, the little one took over. Threw every book out of the bookcase, broke a small crystal dish that had always been one of Carol's favorites, and funny enough, also one of Kate's. She turned over some more plants--they aren't going to survive much longer the way they get thrown around. Then she went for the scissors. None of us could stop her, because when she battles against us, she battles. She cut great big chunks out of her hair. It was an absolute mess. It's as if she were trying to cut all of it off. I don't understand why. I'm not sure what goes on in her head, but then she must have become tired, because Marika and

97

I were able to grab control for a couple minutes and we got the scissors away, but not before there was a lot of damage to the hair.

We hid the scissors and were trying to decide what to do when she took control again. She had control all night. It's not always violent when she has control. Sometimes she will sit on the floor and just rock. Back and forth, back and forth, back and forth. Or she'll sit there and try to hide her whole head, bury her head in her lap and cover her ears. Today, we finally got a little bit of control. But Marie and Alexis and Jenny and Mandy are so terrified. They took off, got in the car--Alexis driving and they drove and drove and drove. She finally stopped the car and Jenny got out and she started to run. Marie was crying. Marie finally got control and got them back here. I just don't know what to do. This has been going on all week. I can't go anywhere until I get Kate to a beauty shop and get her hair fixed. It's an absolute disaster. If they can fix it, they're going to have to cut it very short because in some spots there's not a whole lot left.

Tomorrow there is an appointment with Dr. Ted. Marika told me that someone in her group is going to go in, but is willing to give up her time so that the time can be spent with Kate, but I don't think anything can be accomplished in an hour. Not the way she slips in and out. But the others have to be put at ease. Dr. Ted said that Kate is retrievable. But she's a person, not a piece of equipment or machinery--one we all need and we all love. And I know Dr. Ted doesn't mean anything by using the word retrievable,' it just sounds odd. We know that he cares a lot and that he cares what happens to Kate. We appreciate him.

I'm having a tough time staying in control here. Not emotionally, but that little one wants out. She wants to take charge. She wants to be back here and take over and I don't know how long I can fight her. Something else that bothers me...sometimes when Kate's out, or at least one of the Kates, not the little one, but one of the Kates, she'll start to laugh and she'll laugh and she'll laugh. And then she'll cry. And then she'll laugh again. I don't understand that either. I don't know why. Maybe I should just let that little one take over. She's trying like hell. But I don't want to see a four-year-old that covers her ears and hides her face and hides in a corner or destroys things. I can't reach the child, none of us can...makes me tired...makes me tired and very angry. Makes me want to pay a visit to a certain grandfather and say, "Look, look, this is what you did. This is what we are." Makes me want to slap a certain grandmother. Shake up a set of parents who didn't give a damn anyway. Probably wouldn't give a damn now. I don't know how much longer I am going to hang on here you know. This little kid wants to take over, and I get tired of battling. I really do. Sometimes it's so much easier to just give in. Just say, you win this round kid. Let it all out. I wonder how much she'd really destroy.

14

Individuals

*Created by God,
in His image to be,
individuals all are we,
yet it's obvious to me,
that others just won't see
the individual that is you,
the individual that is me.*

**Marie
June 13, 1984**

There are now thirty-two identified personalities. I am open with all of them about my difficulties in keeping track of things. I make much use of advice, direction and "inside information" provided by those personalities who function in a leadership role.

June 7, 1984. We have a session with the eight-year-old and four-year-old regressed Kates. We use stuffed animals as a medium through which we can communicate. Although the kids were very frightened, the use of these stuffed animals has been effective. Marika says she wants to say good-bye to me, which I will not accept.

June 12, 1984. Jennifer Christenson (from Group 2) has left me a note. She is worried about Marika. She offers to come in and clue me in as to why Marika wants to say good-bye. Jennifer, who is seen here for the first time, presents as a calm and articulate personality. The issue boils down to fear of integration and especially a fear of regression like Katharine has been experiencing.

June 14, 1984. The original twenty-one personalities have been feeling very left out. Most of them are worried about Katharine and her suicidal thinking. We focus on Katharine and she finally reveals she feels very guilty because she has not told me about her ten regressed personalities.

June 15, 1984. Ashley Baldwin comes in for the first time, mainly because so many of the others are struggling. "Someone had to bring them in here." She is likeable and seems to be a fairly healthy personality. Most of this session is spent with Katharine. I touch the four-year-old's finger, then I touch her toe. She finally lets me hold her hand. Then, with her sitting on the floor, she nudges over to my legs and with body language lets me know she wants to cuddle, at which point the eleven-year-old emerges. "I'm hugging all ten of you," I say. They reply: "We have never been hugged before." Tears emerge at that point. Suddenly the thirty-eight-year old Katharine emerges. There is uncontrollable sobbing. It appears she is finally letting herself experience her own feelings instead of relegating them to other regressed personalities. This is a significant clinical session for her. It is the first time in several months that she has held her own personality and her own age without regressing in session.

It's June 17, 1984, and this is Katharine St. Clair. It's been a very long time, it seems, since I've written anything--and I'm not even quite sure where to start. The past weeks have been rather turbulent, emotionally upsetting, one might say.

I regressed. This is something I'm trying to come to grips with. I mean, Katharine is not supposed to regress. It's like everything in my head, all of a sudden, said okay, that's enough--no more. And I shut down and the others that have always been a part of me took over (*meaning her own ten regressed personalities*). I had been so invested in not letting my group know about them, and they didn't. They never knew about my own little group of ten.

They're different from what I have with 'Cesca and Marie and Jenny and everyone else in that group. Sometimes I could control what they did. Other times, they had complete power, and I lost time altogether. And still, at other times I could see them. I knew what they were doing, what they were saying and it was as if I was watching them. I was powerless to stop them. I watched them be angry inside, cry, destroy. Sometimes I couldgain the control back; I could snatch it back from them. Other times, there was no way in hell I could do that. When they took over so completely, then it was impossible for me to try and gain any kind of control. And other times, I just lost it, and I didn't want anyone to know about them. They're what I call my weakness. They take so much of the anger I feel and so much of the hurt.

I would try, during the last few weeks, to come out and talk with Dr. Ted and I couldn't. I couldn't stay long enough. One of the others would take over and they'd talk to him. The four-year-old that cut her hair; that destroys things--that's an angry child. She has good reason to be angry. And is afraid; doesn't know what love is.

It was a strange feeling because some of it I could watch as an outsider looking in. Or is it an insider looking out? I don't know which one I was. I was out of the whole group, still a part of them, but out of it because I wasn't making any decisions, I wasn't helping, I wasn't coping. And yet, I was very much inside of everything, inside of these ten others and their feelings. It was so hard when one of the others was talking in Dr. Ted's office. He'd want to talk to me and he'd ask me to talk with him and I'd try. And he'd say, "Stay with me, now stay with me." And I would try so hard that it was difficult to concentrate on what he was saying because I could feel one of the others wanting to take control, take time away from me and I wouldn't want them to. So it became difficult to even think about what he was saying to me. It was a battle inside of me and I'd want to talk to him, then all of a sudden, it would be too much, and I couldn't hold on and one of them would take over. And yet, sometimes I could watch, I could hear them, when they were talking to Dr. Ted. Like the twenty-six-year-old and the twenty-eight-year-old, I could hear them but I couldn't stop them. I couldn't say, "I need to tell him that myself." It's as if they were the feelings and the emotions for whatever incident precipitated their arrival. I'm the one that had the hysterectomy, yet the twenty-six-year-old has the feelings that go along with that. In a way, she was able to bring them out, whereas I can't. I may want to, but I can't get the words out and it becomes very, very frustrating for me.

Then Dr. Ted brought up to me that there were other personalities, people I didn't know--one in particular--named Marika. He said she was a leader like I was. Well, I felt anything but a leader. I already could feel myself coming apart and he brought up co-consciousness with her and I just rejected that totally. I didn't want anything to do with her--I still don't. He he even asked me about co-consciousness with Carol. Right now I don't want to. As for Marika, and any member of the other group, they are a threat to me. We've managed fairly well without them, and now, all of a sudden, they're here. They want to put in an appearance. They're talking to Dr. Ted, andone of them wants co-consciousness with me. For what? For what reason? That makes me angry. Maybe that was the final straw. Learning about them. And my fear that, if I'm really honest, I'm afraid that they'll take control away from me. They'll take my title of "mother" away from me.

I don't know if there is anything else that has to be said. Right now I don't know about going back to a session on Tuesday. I don't know if I want to talk. Right now, I don't know if I'm going to get better. Right now, I don't know if I want to get better. One more black mark.

<u>We</u>

For fragile friendships lost
is paid a heavy cost,
by lives intertwined
to one body confined,
in a world outside
unable to confide,
needing to be free
wanting only to be,
a problem longstanding
no longer withstanding.

Marie
June 20, 1984

15

June 20, 1984. Katharine is still regressing periodically. Carol has, at times, been suicidal. I call for an entire group session and approximately twenty personalities announce their presence. I propose my questions about the amount of therapy going on and the lack of progress along with my concerns about where we go from here. I specifically focus in on the issue of co-consciousness and why more co-consciousness has not been occurring. A number of them are very angry with me, but discussion does ensue.

The next session, Francesca comes in and even though she is apprehensive, agrees to co-consciousness with Carol. My confrontation of the previous day has fostered a lot of internal discussion within groups, but not between groups. Carol, who is also very apprehensive, decides to agree to co-consciousness with Francesca. With the use of hypnosis we effect the first co-consciousness between Carol and someone from Group 1. (Previously we have had co-consciousness between Carol and Marika of Group 2, as well as a first experience in co-consciousness between Groups 1 and 2, namely Marika and Francesca.) This particular experiment goes well and I see both for a talk-down. I also see Katharine briefly, who has again become suicidal. We discuss the implications of this with emphasis on how her death would affect the others (i.e., they would be dead too). This argument has an impact on her and she agrees to talk further with me.

I also find myself often worried about Carol. With so many dramatic things happening to so many of the personalities, Carol, who is after all the birth person, often gets "lost in the shuffle." I encourage her to accept more responsibility for her own progress and she responds by agreeing to write.

103

✻

My name is Carol. The date is June 24, 1984. I feel a little strange giving out a lot of information about myself. I know that some of the others have done it. The personalities that are a part of me. I just never have. I've thought about it. Guess I never really wanted to before. Maybe somehow, I was a little bit afraid of it. I don't know why. I'm going to start at the beginning, of what was my beginning.

I was born in the Netherlands. I lived there for the first seven years of my life. Or at least what I can remember of it, which isn't a lot. I know a little bit about my life there. I know I had grandparents that lived in different cities from where I lived with my parents and brother. We lived in a very small village right by the North Sea. It was about a mile and a half walk to get to the sea. There was a canal that went right up alongside our house. You could see it from our living-room window. It was not a very big canal, but big enough to go ice-skating on and have fun. Only I don't remember doing that, except I was told I went through the ice one time, but I don't know anything about that. I went to school there. I guess I was finished with the first grade when we moved here. In fact,I was told that I was more advanced in some things because in the Netherlands the school year is longer and the school day is also longer; plus we went to school on Saturday mornings. I really can't say much about my life there because they aren't many memories. I know I used to like to make daisy chains out of real daisies and that I played an angel in the Sunday school Christmas pageant; that I hated math, although I did all right in it. I always got very high grades.

I know that I went to visit my grandparents occasionally and I guess I kinda do remember visiting my mom's parents. They lived first in a very small town and then they moved to a bigger city. They lived right next to the railroad tracks and, at night, I used to love to listen to the train whiz by. But I just don't remember that happening very often. (*Note: The reader must remember that the birth person, the original person, who is Carol in this case, literally has total amnesia when another one of the personalities has taken over. Usually, the other personalities can maintain an awareness even if they are not "in charge" at a given time, but the birth person cannot do this. This is why co-consciousness is so important, clinically. When we've gotten to the point when the birth person can experience co-consciousness with the other personalities, we have made major strides. Because of the immense amount of therapy time involved, practical considerations might dictate that this is all that is accomplishable in therapy, i.e., co-consciousness. In the case of Carol, we all made a commitment to try for total integration of all the personalities, which is the only way true and total healing can occur.)*

Got pretty good at fooling people with that though. I learned a long time ago that when I had some time, that I could listen enough to what other people said, and I could pick up on where I was in my life and pick up on things that had happened to me. All I had to do was play dumb--pretend I knew what was going on. Once in awhile it would get me into trouble, because I would either agree on something or disagree on something that I was supposed to do the exact opposite to.

I do know that I didn't want to come here. It's a vague memory. For some reason, I didn't want to move to the United States. It seems to me that the day that we were supposed to get our vaccinations, the doctor came to town and I had to come home from school early to receive the shot. I didn't want to go home. And I know that my other grandparents, my dad's parents were already here, along with one of his brothers and his wife and two of his younger sisters that still lived with his parents. I know that the last week or so before we left there, we spent with my mom's parents and I didn't want to leave them and that I gave everybody a pretty rough time. The last couple of days we spent with my dad's brother and I didn't like it there, but I don't remember that either. Some of these things I've been told and I don't even knowhow much to say because I don't know what's been told me and what's my memory anymore. I have so few memories. I don't remember the plane ride over here.

And I know that we had to go to a train station in New York after we arrived here and we had to get on a train and come to a city in western Michigan, only I don't remember that either. It was hot. I do vaguely remember one incident here and it was after we arrived, but it is only very vague. Again, I'm not sure if it was told to me or if it's memory. I think it's a memory because I still feel some anger when I think about it and can almost see it happen.

We had to stay with my grandparents when we came here and I know that I was terribly, terribly homesick. In fact, not too long ago, my mom told me that I had a terrible time for the first couple of years that I was here because I was always so homesick. I wanted to go back, and even now, sometimes when I have time to think, I still do. I actually become sick to my stomach and I want to go home. I guess my mother was homesick too and apparently she had never gotten along with my dad's mother. Neither, for that matter, did my father. I knowshe hates me for some reason. I have never figured out yet what I did to her to make her hate me, but she does. I remember my mom sitting in a chair in my grandparents' living room and my grandmother slapped her across the face and my mom cried and I went to run across the street to call my dad. I didn't really have any idea how I was going to do that because I couldn't speak English very well, but I was going to call my dad and my grandmother caught up with me and she dragged me back across the street by my hair. I don't like her. She makes me very uneasy for some reason. I don't see her that much. My grandfather makes me really uneasy too.

I don't even know why I decided to do this. There's not a lot I can fill in for anyone. I know that a lot of things have happened to me that I don't know about. I'm not sure I want to know about all of it because logically, from what I understand about multiple personalities, if these were pleasant things, I wouldn't have any problems at all. But these aren't pleasant things, so I don't even know if I want to remember them. I know that I have to, in order to get better. Sometimes I don't have a very good attitude about it. I'd just as soon die than even try.

My brother and I were pretty close. I can remember having times with him later on here. I remember the others being born after we arrived here. They were a big surprise too. Phillip died. He was my best friend. We were pretty close because I can remember when I had some time, we would just sit and talk. We could talk about anything. I still have problems with trying to understand why he died. It's been a number of years. I sure miss him a lot.

When I was in my junior year in high school, I was in the hospital, and I wanted my class ring and my parents wouldn't buy it for me. They said it was too expensive. I'd ordered it earlier and put a down payment on it. Anyway, the ring had come in so Phillip bought it for me. He had it all wrapped up like a present. I wish I had all the feelings that go along with all these things because I know they happened and I feel special towards him now, because he did those things, but I wish I'd remember on my own instead of having to be told. He was a year younger than I was.

I don't remember much about being in high school. Oh, there's little things I remember. I know I was pretty much left out of things. I was really, really shy. And I was always very, very overweight. I don't think I was attractive at all--still got a problem with that. But you can't really be attractive when you're overweight, and I was fat, fat, fat, then. Anyway, I was left out of things a lot. I did a lot of things for other people.

This past week I wanted to die. I felt pretty let down ever since I had co-consciousness with Marika and she kind of backed out. That

106

made me really angry, and I felt left out and hurt. Ever since I found out about the multiple personalities, and that's been about seven months now, I've had periods of time when I've been really, really angry. I have to admit I resent the times that they've taken away from me. I think that I've been pretty lenient in letting them have most of the time in Dr. Ted's office. In fact, they had almost the whole three and a half years since I've been in therapy. I don't really know Dr. Ted that well at all. It wasn't until November that they decided to start using their own names. So when Marika wanted co-consciousness with me I guess I expected a whole lot more from her. I know now that she's working through some problems and stuff like that, but I don't think it's fair that she started it and then just totally backed off. For awhile she was telling me that she needed some time to herself and stuff like that. And that's okay, because everybody does. But, she just stayed away and so when Dr. Ted wanted me to do it this past Thursday with Francesca, I didn't want anything to do with it. Of course, he put it to me in such a way I finally couldn't refuse, you know. He's good at that.

I don't want her to tell me a lot about my life because then all I am going to have again is stories. If she is going to help me remember, I hope she helps me remember in a different way than just telling me. But it's been kinda fun to get to know her. I guess, all in all, we've had a pretty good weekend, she and I. There have been times when she has said, "Okay, I need some time on my own now." Then she either takes over completely or she just leaves me for awhile. I've been able to talk to Francesca and I've asked her some questions about herself. I don't ask too many questions about the others because I don't want to say anything that will hurt their feelings. I know that Kate's been upset over the weekend and that some little ones have been upset over the weekend, but I don't know how many little ones there are. I feel better than I have in the last couple of weeks, because I was really excited when I did the co-consciousness with Marika and I have to admit that even though this weekend has been a really good one, I'm a bit skeptical about how long it's going to last. Until that co-consciousness, there was still a part of me that doubted and didn't believe. And yet I knew that they were there. I don't know if that makes any sense or not, but that's how I feel. I asked Francesca, specifically, if I could have some time all to myself today to write. And I don't know when I'll be doing more, but at least it's a start.

I remember it being Christmas, and the next thing I knew it was spring. I wasn't even sure if we were in the same year. Well, obviously, we weren't, but I wasn't sure if it was the next year or the year after that or what. I don't even know Jason that well and I've worked with him for almost five years now, and I find that really irritating. To know these people and to have had an on-going friendship or whatever kind of relationship with them for a number of years, but yet not know them.

I don't know how long it's going to take me to get better, to be normal. I know I don't want to stay like this, because if I have to stay like this, then I'd just as soon be dead. Maybe I'm looking for positive proof that I'm going to be better. Maybe I'll get it and maybe I won't. I know Sybill did. Maybe I should read that book again. I get really mixed up as to what's going on and, and why everybody's having so many problems. I know some of the problems are associated with their specific memories. I'm sure they must have some memories that are really upsetting. I'm sorry for that, I really am, but they're not my memories and I can't make them feel better. A lot of times I wonder what happened to cause all this and what I did to make it happen. It must have been something, but I don't know what.

I wish that I had a time, a date when I knew I was going to be better. I kind of have a goal, but I'm going to keep that one to myself for now. But I sure feel a lot better than I did last week when I wanted to be dead. I talked with Francesca and we're doing some things together. Saturday morning we cleaned and it was fun. She was right here to help me and she doesn't particularly care to clean. I know that I like to cook. I don't cook a lot, obviously, since I don't have a lot of time. I know that she doesn't particularly care to.

Anyway, I know that the things I said aren't in sequence or anything, but I'll do more; and it wasn't as hard as I thought. A little frustrating, at times, not knowing exactly what to say because I don't remember all these things, but I'll do more.

16

It is now June 27, 1984. Therapy continues through the summer, consisting of much jumping around from personality to personality, from issue to issue. Direction is often difficult to maintain and I often am unsure.

I am finally able to have therapy time with Carol. For some time we have wanted to emphasize memory retrieval for her, but could not proceed because of the need to deal with Katharine's regression. With hypnosis I begin to age regress Carol. She goes back to 1953 (age four). Since she speaks only Dutch, I then age progress her to age ten so that we can communicate in English. She finds herself with grandfather. He is beating her and beginning to have sex with her. The emotional pain is so strong that I age progress her back to today without proceeding any further. I am fearful of too much pain too quickly. In our talk-down she retains full memory of what she has just experienced and begins to evidence some emotional reactions. I can see much fear in her eyes, and signs of struggling to comprehend. There is confusion and the beginnings of anger, but I am elated because she is clearly beginning to retrieve her own memories, and is voluntarily talking about them. The doors are now opening for discussion and pursuit of additional memories. The immediate need, however, is for acceptance of the reality of this initial memory and dealing with the associated feelings. Several sessions are needed for this, which go well.

Carol requests another individual session. She is aware that someone has taken a number of laxatives, because she had to take five potassium pills (she usually does not take more than one). I request to see Jennifer in order to get an update. It turns out that Marie and Christina each took forty-eight pills and that Jacqueline took thirty-two pills. I see all three together to impress upon them that all three share the same body; (none of them have accepted this yet).

I see fifteen-year-old Simone in session for one-half hour and confront her on the issue of laxative abuse. It has gotten to the point where it is serious. I also see Francesca who tells me that she has six regressed personalities (ages ten, fifteen, seventeen, twenty-one, twenty-two and twenty-seven). These are not separate personalities, but are all Francesca, at different ages. The three oldest are the ones who have historically been involved in sado-masochistic sexual relationships. Francesca is quite fearful that she will regress as Katharine has.

A day later I am informed that the patient is in a general hospital, having been brought there the previous evening on an emergency basis. They went to the zoo, a surprise for Carol and the personalities who are children, and while there experienced a great deal of pain. A neighbor brings them to the emergency room. There is the possibility of an intestinal obstruction and therefore probable surgery. Following a number of enemas, however, her system is clearing and therefore she will probably be discharged tomorrow. This may turn out to be a blessing in   disguise because this brings home the issue of the price paid for their laxative abuse. While in the hospital, Carol informs me she has retrieved a memory of her own; she now remembers being in the trunk.

Carol is discharged the following day and a session is held in my office. Katharine advertises that she wants to make time for Carol to do her therapy work, and in fact comes to say good-bye so Carol can do that. I talk her out of whatever she means by the finality of saying good-bye. (I later realize that Katharine is reliving the same memory. She has historically "mothered" many of the other personalities which has been her defense against having to deal with her own traumas. Her wish to say good-bye is therefore her avoidance of having to deal with her own memories of being in the trunk.) I use hypnosis with Carol in session and we return to the trunk she has remembered on her own. Katharine suddenly takes time on her own. I use the opportunity to foster co-consciousness for the first time between them, which Katharine and Carol finally do agree to. Both now know that they were reliving the same memory and the two of them have now committed themselves to getting to know each other better.

I later see Francesca (age twenty-seven), my first contact with this regressed personality, the one who apparently controls the other six. She agrees to prevent Francesca's other regressed personalities from going to the sado-masochistic man with whom they've had a sexual relationship, if I will agree to help her in therapy. She wants to deal with sexuality. She

110

is used to relating to others in power terms. In fact, she is a fairly sensitive woman (even though her first description of herself is: "I'm a bitch.") As a result of the observing she has done in therapy, she is now wanting to explore her own potential for sensitivity and warmth. I see her briefly the next day and she is in turmoil because of our session. She has had a life-long need to dominate and is not used to sharing with anyone else. She views herself as having been weak yesterday by telling me this much about herself. "Maybe I'd better stay a robot," she says.

Jenny has been working out in a spa. (Jenny is the hyperactive one who has anesthesia to pain.) Everyone is sore. At the spa she was weighed and measured, and is now in great confusion. The numbers she got at the spa do not jibe with her physical picture of herself. This gives me a chance to confront her on the issue of sharing one body. I see twenty-seven-year-old Francesca again. We focus on the rage and its connection to Carol's grandfather and her way of handling it with sado-masochistic sex. Tears almost happen. She is in fact mellowing and she is allowing the relation-ship to proceed.

In our next session we focus in on Jenny's wish to be male and her physical exercising to the extreme. There apparently are shin splints and back muscles that are very sore for everyone else but which Jenny herself denies. I touch the muscles and she flinches. She's not sure how to take this and is silent for awhile. After all, she has spent most of her life literally not experiencing pain and in this one moment her world of familiarity is reversed. It is now more difficult for her to deny.

Katharine is now gradually returning to her former self, the much more functional Katharine. She is spending more co-consciousness time with Carol. I am now bringing Francesca (the original) more into a three-way co-consciousness, beginning to develop the concept of a "leadership group" within the personalities. There has been some assimilation among the regressed personalities within Francesca at this point. Carol is now expressing a wish to work on retrieving more memory.

I see Tanya for the first time. Tanya is one of the laxative abusers and I have talked with her by phone but not seen her in person as yet. Tanya is age thirty-two. She believes her baby, which was aborted, in fact was born and stolen. She feels she is bad and dirty because of the sexual abuse done to her as a child. This is the first time she has talked with anyone on the "outside" and in fact, even outside of her group within the system. Jennifer is functioning as the facilitator for therapy at this time.

Carol is becoming much less passive and much less intimidated. She brings in a list of demands for the others. In fact, she is furious. She is now taking stands regarding her view of her needs. I get reactions from about twelve of the others basically supporting her demands. We begin to relate

111

some of the returning memories which include sexual abuse by her ex-husband and seeing the face of her father's uncle who she wanted to get away from but couldn't. It is significant that her anger is finally showing.

✻

My name is Jennifer Christenson and it is August 9, 1984. There have been many things going on in the last month, including the hospitalization in July, caused by the laxatives, as well as having to go to the emergency room a few times.The worst occurred when Tanya took a large amount of laxatives and vomiting had to be induced. It wasn't a pleasant time for any of us and especially not for her. I'm glad that Dr. Elders was there with us that night (Dr. Ted was on vacation and Dr. Elders was providing coverage).

Tanya is such a frightened lady--has so many problems. And I think more than most of us, she has a problem understanding the concept of multiple personalities that we are one body and do not all have separate bodies. And I too, have problems with that although I'm more accepting of the fact.

We just had a call from Dr. Elders just to check to see how we were doing, especially since Tanya took more laxatives on Tuesday evening and this past Friday evening. It was nice of him to call; it's nice to know that someone cares.

Tanya had to have blood drawn and not only was she left with a bruise, but so was I. So, since I wasn't the one having the problem to begin with, and I still ended up with a bruise, it must mean that we have the same body. I can't deny that. I just have difficulty accepting that because I look different from Tanya.

I feel separate, I think separately, I have different interests. Appearance-wise I'm different. But I'm not separate. Now that's difficult to take sometimes. Difficult to accept.

Dr. Ted wants me to have co-consciousness with Carol and I've been delaying that, on purpose I guess, because I don't really want to do that and I didn't really understand why until this week. I have had time to think, time to evaluate, analyze. Part of it is fear, but a bigger part is that I feel as if I am going to lose something. Right now, she doesn't know me, doesn't know much about me and she can't talk to me and I can't talk to her, of course. I do know what she looks like, what she thinks, but I can't talk to her. And not being able to do so, to do any of these things, keeps us separate, makes me feel more of an individual.

Dr. Ted tells me time and time again that we won't be lost--all these things that are important to us will still be important, but I don't fully accept that. Right now, I don't want Carol to know me that well. I want to retain my individuality as long as I can. I will help her remember, I will help make things easier for her at work, but I wish to remain separate. Dr. Ted says that it's no different than being able to communicate with members of my group, with Marika, for example. Or Tanya when she wishes to talk. Or with Miranda, who can be such a handful. But it is. It is very different, because they understand my desire to be an individual, just as I understand their desire to remain individual.

And Carol doesn't comprehend this. I realize that she wants to be a whole, complete person, and I know that she needs that, but with the rest of us giving her more time, it gives her an opportunity to develop her own interests, feelings, thoughts, ideas. It gives her a chance to develop friendships. She may need a little guidance, but I think she can become a whole person without us. And then we could remain individuals also. We've had to develop on our own, by ourselves, all these years, and I think she can do the same. Without having to talk to me all the time or to Kate or to Francesca, or to Marika, or to any of the rest of us, she can grow--we've had to.

Carol brought her set of rules for us into session. (*Note: In attempting to be a more assertive person, Carol had brought in a list of twenty expectations she wanted to negotiate with all the other personalities.*) Things that she does and doesn't want us to do. Twenty, in all, I believe, but she doesn't have the right to tell us what to do. We have been there all these years to protect her, to help her, to take care of her. I will help give her back the pieces of her life and I'm more than willing to help her grow and I was pleased at her anger last week, but I think her expectations of us are too great. At first, I thought, okay, we can handle this. We'll follow her rules. And then I realized I would be doing this only because she wants us to and Dr. Ted wants us to. It's not something that I want to do. She's asking us to change our lives. She wants us to change overnight. We don't set down rules for her. We don't tell her that this is what she has to do. We've taken care of so many things for her. And now it's as if she wants us out of the way.

I'm sorry that Tanya took the laxatives, but Tanya's done that for a very long time. And to expect Tanya to change overnight is impossible. It's unrealistic. Tanya has problems. Tanya needs help. And Tanya has taken a lot of the hurts and disappointments that Carol would otherwise have had totake. To be angry at her is unfair. She feels badly enough as it is.She doesn't need a guilt trip. What she needs is someone to love her and to care about her and to cry with her and make her understand that she is not a bad person. If Carol can't understand any of that, then she had better learn to feel. She had better learn to understand that we are people; we arehuman beings, just as she is. And just as she has made mistakes in the past, so have we. Just as we've been hurt, I know she has been also. And just as she gets scared, so do we. And believe me, we've had a hell of a lot more to be scared about than she's ever had.

In therapy, we are talking to Dr. Ted about things that we've never talkedabout. We've never shared our feelings of anger and hate and frustration and the way we feel about ourselves as people. But the talking doesn't make everything better all at once. And I can understand and respect her need for changes, her need for a new life, but we didn't have tocome out of the closet, so to speak. We didn't have to let Dr.Ted know about us. We didn't have to let her know about us, but we did. And so now we finally feel that we'reindividuals. It's hard to think about someday giving that up. We know that we need to make changes, changes in our daily lives, our personal lives, and the way we think, the way we react to things, but we've been one way for a very long time and none of us have intentionally tried to hurt her. I think we've tried to protect her more than anything else.

And for some, it takes a much longer time to develop trust. It takes a much longer time to explore feelings, to allow tears to start, to allow the hurt to start healing. And that's only because of years of conditioning, years of trying to hide the hurt and the pain and the bad feelings. But we know that by hiding, they don't goaway.

We've taken time from Carol. But when is she going to understand that she wouldn't have made it without us. We're people too. We're not ghosts. She had better learn to understand that she wouldn't have made it at work without us. She'd better learn to understand that some of us go to church and some of us don't. Some of us have a real problem accepting God's love, because we don't know the meaning of the word. We've never had any. I don't mean to be bitter. But things are the way they are. And we're willing to work at ourlives if she's willing to work at hers.

I have my own goals and I have my own dreams. I have my own hobbies and I function separately from her. I am my own person,

I know what I like and what I don't like. And I will not be told what to do and when to do it. We will do our best at making changes, but we've been told what to do for too long now.

For us, using our own names, means freedom. Maybe not total freedom, but more than we've ever had. With Dr. Ted, I can be myself. I don't have to be her or anyone else and it feels good. And it's time she tried to understand a little bit more how we feel. Because we don't have to do this at all. We don't have to go to therapy. We don't have to have co-consciousness with her and we don't have to try and help her. It's important to us that we discover what makes us tick and why we feel the way we do.

And if I sound angry, it's because I am! I'm really angry! And I'm tired! We need to be understood also.

Harold has returned to town. He has called twice while intoxicated announcing his rights to come to their apartment. I recommend that the police be alerted, which they have had to do on two previous occasions.

Jennifer is now seriously considering co-consciousness with Carol. She is struggling because of her fear of going out of existence. She resents not being able to have her own life and dreams. This is actually encouraging because it is a clear indication of her acceptance of being a multiple personality.

Moiria Sanderson introduces herself for the first time. She asks for individual time "because I have a problem with memory." Carol now asks for the names and ages of all the others, plus information on who has taken psychological testing. Since all had previously given permission, I give this information to Carol. Jennifer lets me know that there has been a fair amount of turmoil lately. Their boss at work, who has been very supportive of her problem and is aware of the multiple personality issue, has informed her that he needs more consistency at work and therefore wants only one of them (Jennifer) to function at work. The others are feeling quite rejected. The boss has also informed them that he can no longer be their friend but has to be their boss. It would appear that he is getting pressure from his superiors.

We have for the first time co-consciousness between Jennifer and Katharine, who are the respective leaders of each of their groups. Both are tickled with their budding friendship.

August 27, 1984

I now have co-consciousness with Katharine. It wasn't something that was planned; she requested it during last week

Thursday's session. Dr. Ted talked it over with me first and I must admit that I did have some hesitation. I asked Dr. Ted to inform Katharine that I would agree to it, but that I did have some conditions. First, I do not want to become involved in a power struggle with her. Second, she remains advisor/mother figure to the first group; I, to the second group. We do each have "problem children" in our groups so we will be able to seek advice from each other, which will be nice for both of us. It will seem good to have someone to talk to; in fact, it does seem good since we've already talked and have done some sharing. Third, I want us to have some time together so we can talk and have a chance to get to know each other. We've done some of that off and on since Thursday and it has been wonderful.

Jennifer

Carol is angry because she's worked hard this week but has had little therapy time to process her attempt to retrieve memories. I use hypnosis and age regression with her and we go back to age six. In order to understand her, I use Jennifer as my interpreter. Carol is in a room at her grandfather's in the Netherlands: it is dark, cold, and she is forced to stay there. Then he beats her, rapes her, tears off her dress; all because she had one shoelace untied. This is an emotional experience for her. It is very clear that some of the specific memories are returning. Many of the other personalities are cooperating with Carol by giving her time to process this significant information.

Mandy must now be seen because she wants to cut herself again. In her dreams she has been reliving the sexual abuse by her grandfather and great-uncle. This is experienced by her as a rape two days ago, even though it actually happened many years ago. We work at defusing her anger. I make a deal with her that I will agree to help her special friend Marie if she will agree to read *When Bad Things Happen to Good People*, by Harold S. Kushner, a small book given to her by her minister, Pastor Aaron, who has been an active therapeutic ally for some time now.

Jennifer and Katharine are together co-consciously in session. They speak of the resentment others are feeling toward them because of their new relationship. Both feel they do not share the same body. Both fear going out of existence.

Moiria feels that she is really sticking her neck out by talking with me and that she feels as if she's a "traitor" to others in Group 2 who are as private as she has always been. She describes Group 1 as past-oriented and Group 2 as future-oriented. She wants more spontaneity, and more of an adult life for herself now, including more outside relationships. Histori-

116

cally, she has only been comfortable when relating to children. She badly wants "another outside adult to understand me," and is in therapy with me to test out an adult relationship in a fairly safe way.

Jennifer is approaching crisis. She is having more and more difficulty denying the reality of all sharing one body and therefore all being one person. She reports that many of the others are very angry with her for beginning to believe this information. There are many tears and much anger in session,but she does stay for the entire two hours.

Moiria is now talking a lot in therapy but there is basically a social quality to her involvement. She finally acknowledges her anger: "My contempt for others is my contempt for me," she says, but refuses several opportunities to deal with this. She is still testing me out, but she also asks if she can make contact with the children of Group 1. "I can help you if I can deal with them," she states. Moiria is now becoming an ally to me. Carol also gives up therapy time, specifically for Jennifer. She shares a letter of support she'd written for Jennifer, who is struggling with anger at Carol because "Carol first creates me and now she wants to destroy me." She is also quite angry with me because "You made me realize what deep down I've really known anyway." It's a productive session.

I'm informed that Mandy, along with Miranda and Caroline, began experiencing their rage at the grandparents, left the apartment for awhile and came back with blood on themselves (from self-mutilation). This is, in fact, less serious physically than it is psychologically. Jennifer continues to worry about becoming too dependent on this worker. She does agree to co-consciousness with Carol if I will agree to help her deal with her fear of dependency.

Moiria wants co-consciousness with five-year-old Callie. This is done and goes well. Callie makes a card for "good grandfather," (Rev. Jansma). Katharine will continue to function as "Mom" for Callie; Moiria will be the "big sister" for Callie. Moiria and Callie have been getting to know each other through their co-consciousness. Moiria brings in a stuffed animal (a pig) for Callie who wants me to name it for her during session today. We finally decide on "Priscilla Pig." There is much giggling and laughter, innocence, vulnerability, naivete and trust. Moiria is now beginning to believe that she and Jennifer share the same body. This is very troubling to her.

Carol has a session to specifically deal with memories. Even though much rage shows, it is still very controlled. It is directed at parents and grandparents. Jessica comes in asking for psychological testing feedback. She says it confirms her suspicions about her problems and she asks for therapy time. Katharine has not been seen in several months but finally-comes in at the urging of Francesca. Her agendas include: an awareness

117

of sharing one body, regression, which appears to occur when she's remembering the abuse, feeling old, arthritis, and anger at missing a sense of future. Francesca has now become the advocate for her and is nudging Katharine to get therapy for the above.

Moiria is also pushing Katharine into therapy. Katharine's self-hatred has been bubbling to the surface. She is quite depressed and feels: "I've been phoney with you because I've not really opened up to you." She hates her body, her appearance, Carol's parents, the grandparents, and her siblings. She has become suicidal and is now in crisis. Others have had to step in to prevent a suicide attempt. While in her car she had the urge to drive it into a concrete abutment; she also had the urge to run in front of a train. She feels like a failure to the others and is especially depressed over this. She will not consider hospitalization. We attempt to defuse this in session. If she were hospitalized she would probably regress into a deeper sense of failure, so I am hoping to prevent hospitalization. I contract with Francesca to make sure Katharine is here for a special session.

I see Katharine again in two days and she is very surprised to be here in session and is quite angry about it. Her aunt has died the previous evening and Katharine has much hatred for her. There turns out to be a very large emotional outpouring with much sobbing.

It is now October 11, 1984. Carol was informed yesterday that she is being laid-off permanently from her job. In addition, the parents informed her yesterday of the one session I had with them at my request last spring. This appears to be a clear attempt to have her end therapy: "You have made up all those bad things that supposedly happened to you when you were growing up," they say. Those involved in this session include Carol, Marie, Elizabeth, Jennifer, Andrea, and Alexandra. There is much anger at the employer and much anger at me for having that session with the parents without asking them, or even informing them. They are willing to talk about it, however, and we have a difficult session, especially with Jennifer who has been the primary one to cope with outside people. Actually they handle it quite well considering the amount of anger going on.

Jennifer has activated herself and is dealing with the job issue. A number of them are still quite angry with me for having the session with Carol's parents. Her mother is still calling and working hard to get them to terminate therapy.

Rachel, age twenty-eight, and "been around a few years," introduces herself for the first time. She tells me there are a number of others. She is requesting that therapy time be given to Moiria. Katharine is again regressing.

Elizabeth brings in Lisa, age thirteen, who wants psychological testing feedback, which is given. Mostly she wants to have a family, and to call Jennifer "Mom." I inform Jennifer of this, who agrees to co-consciousness. All are pleased.

At this time, Carol's mother has surgery. Father is quite frightened at this and cannot take her to the hospital, which Carol does. Many of the personalities during this period of time do see a lot of sensitivity and loving behavior between mother and father, and some affection even begins to get directed toward Carol. They become very confused at this and quite angry because of their resentment at not having received these kinds of behaviors earlier in their lives.

One morning, after she has returned from the hospital, mother calls and there is a big fight. Jennifer views Mom as very manipulative and self-centered, especially after all that she did for the mother during her surgery. There are many tears and she pounds her fist on the floor. She feels it's time for "divorce." I let her know that Carol would have to be involved in this decision. With prodding she agrees to her first co-consciousness with Carol. Although Carol is caught off-guard with this, she also agrees and they spend time together. Jennifer fills Carol in on the latest events involving the parents. Over a number of days the two of them share more information of specific behaviors regarding the parents over the years. Carol is trying very hard to act "grown up," and not be dependent on Jennifer (therefore, no tears during this period of time).

Carol comes in to give me her thoughts on the events of the past few days. Good insight is showing. She is sticking to her guns regarding separation from her parents and the use of therapy time for herself. She is aware of a number of ways she runs away from her problems. A sense of confidence appears to be emerging. I recommend that she also do some writing about her relationship with her parents, just as Jennifer is doing.

This is Jennifer Christenson. The date is October 28, 1984, and I want to put down my thoughts about Carol's parents and my feelings concerning them, especially this past week, when her mother had surgery and I spent a great deal of time at the hospital. It's been a confusing week for me as far as my emotions, my feelings, my thoughts. It brought back a lot of memories. I remember them as being very cold people and there never was any open display of affection. There was no caring that was shown, especially not to Carol. In fact, she was very often physically and emotionally abused. That's continued all her life. I remember times when she would be totally, totally left out of things, as if she didn't exist. Her parents have never known how to project warmth or caring towards her. They didn't know how or maybe they didn't want to nurture her. Anything that was ever important to her or to any of us for that matter, was brushed off as not being important. It was not important to come and listen to my concert, to listen to Christina do a solo. And that's a shame because I feel that they missed something very, very important- they missed their daughter growing up and when I say daughter, I guess I have to include all of us in that, because we all took turns being the daughter. I understand the concept of one body, but the others don't.

There was never any time for her. It was if she was an outcast. As if they blamed her for something. I believe that this has to do with Carol being an unwanted child. Her parents had to get married. Her mother was pregnant with Carol and so they got married. I don't think they were ready to get married. I don't think they were secure in their love for each other. I remember the father being very, very dependant on the mother. He is prone to depression and sometimes these depressions would last for weeks and he wanted absolutly nothing to do with anyone.

Sometimes he would work during these times, and sometimes he would stay home. I remember him being a very proud man. (My memories are more those taking place in the United States than in the Netherlands.) He did work hard to provide for his family. But he was a manipulator in that if he didn't get his own way, he became depressed. Carol wanted him to love her so very much. She wanted him to pay attention to her.

They had things in common. They shared a love for music, they shared a love of art. It would have been such a wonderful experience for Carol if he would have taken the time to spend with her,if only he would have sat down with her and looked at her art work and said, "You could probably change this a little bit, or, why don't you try this." But it didn't even get a glance from him. And that's rejection. When she starred in the choir in school, in a musical perhaps, which she did do, the parents didn't attend and I think, in many cases, it was more important to have her father attend than her mother.

She was often beaten, not just physically, but emotionally as well. She was too fat, she was too ugly, she would never amount to anything, she was a whore. All her life she was denied and rejected by them. And that's not going to change now. Their eyesaren't suddenly going to open wide and say:"Oh my, we've been wrong." That's not going to happen, because what is important to them is that they maintain their position. What will the neighbors think, what will the rest of the family think? And so it had been much easier to tell her, "No, you don't remember that. That didn't happen. That's your imagination." That's what she does to herself now.

Her father's love was something she wanted so desperately. And Carol herself--so insecure, someone with no self esteem--thought of herself as the most unwanted child in the world. She wasn't far from wrong. She wanted love and approval from another human being who had a little self-esteem himself. A man that couldn't let emotions show, unless it was anger. Unless it was to lash out at her, to whip her with a belt, to humiliate her which he did quite often. I remember a time when, I guess she was maybe fourteen, maybe fifteen, and a girl at that age has her purse and her private things. And girls at that age tend to carry quite a bit in their purses. And she did. One day, because he did not approve of purses that were full, he turned upside down in front of her, and in front of some guests that the parents had for the evening. She was on her way to a friend's house and he wouldn't let her out of the door until she had given him her purse. In front of all those people he dumped it upside down then ridiculed her. It's these kinds of events we haven't been able to forget. To us they're as bad as the beatings we received from him.

And the mother wasn't much better. The mother would tell her that she wasn't intelligent, that she took after the father's family. They had this thing in their blood that they needed sex, and she was one of those kinds of people. This woman had to have known that Carol had been sexually abused. She physically abused her. She physically abused her from the time she was a child until a year ago, when she slapped her in the face.

Once she threw her down the basement stairs. She would always tell her she wasn't good enough, she was evil, that she came from the devil. This same woman would turn around and spout out a bible verse at us. Carol, for a long time, had a strong faith, and that faith began to waver more and more. And even now she vacillates: Do I or don't I believe?

One evening, when she wanted to go out with some friends, the mother, not believing her when Carol told them where she was going and what she was going to do, called her a whore. One that was no better than the gals walking the streets. And she may as well join them, because that's what she was. Anything that she ever wanted to do--spend time at the beach with classmates from high school, go on the senior trip; go bowling with kids from school; go to the circus, the Shrine Circus; go to outings at the park with the school class, all of these things were vetoed. "You can't go." "The circus is sinful." "If you go bowling, you're going to drink."

It continued through her marriage. One of her mother's favorite lines was, "I know something about you that you don't think I know, but I do. Someone told me. But I'm not going to tell you who and I'm not going to tell you what I know." I call that mental cruelty. I'd call that mental cruelty when a daughter is called a whore in front of her siblings.

Trust was broken a long time ago. I don't have faith in what they say anymore. And yet, a part of me wants to see some real parenting, some real nurturing. And so there's a little bit of hope. And inside of my head there's this little warning bell that keeps going off. It says, be careful, you're going to get hurt again.

17

A Christmas Story

She heard the old man laughing as she ran out of the room--heard the door slam hard behind her. She ran as fast as she could, coat unbuttoned, scarf and mittens forgotten in her hurry to get away. Her feet slipped on the ice and she fell, badly scraping her knees and the palms of her hands. Scrambling to her feet, she continued running, up the hill, past the school and the church, not feeling the cold, not seeing the blood that ran down her legs, not feeling the pain in her arms. She ran until she couldn't run any more, and wanting to hide, she collapsed in a heap behind a large snowbank. She curled up in a ball and covered her head with her arms. It was hard to breathe and her chest hurt. She stayed there for a long time, not moving, not daring to make any noise in case the old man had followed her. When she was sure she was safe, she sat up and noticed for the first time that she was shivering and that she was terribly cold. Her head was pounding and her whole body hurt, especially between her legs. She remembered why she hurt there and bit her lips very hard because she did not want to cry. But she could not understand why the old man had hurt her again. He had made her take off her clothes again, just like the other times and he had pinched her breasts very hard. She looked down at herself and wished she didn't have breasts and that she didn't have to be a girl. She knew she was going to have scratches and black and blue marks all over her body again. She always did after the old man beat her up. She wondered why she was so bad because she was very sure he wouldn't do that to her if she was good. When he finished beating her up, the old man made her get on her hands and knees in front of him, and then he put his thing in her, first in one place and then in the other place. He always laughed when he did it and this time he had said it was his present to her. She had not even tried to fight back like she usually did because she could not get away anyway. She had been gone a long time and she had better get home now. She stood up slowly and almost cried when she looked down because she had ruined her kneesocks and that meant she was going to get another whipping when she got home. It was hard to button her coat because her fingers were very cold and stiff. Her wrists hurt very badly too, probably from the fall on the ice. She tried to think of something to say when she got home to explain why she was such a mess. If she told them what really happened they wouldn't believe her anyway, so she had better think of something else. All she wanted to do was wash herself, as many times as she could, just like she always did after the old man hurt her. But even if she washed herself over and over again, she still could not feel clean. She never did. She remembered why she had gone over to the old man's house and she almost cried again. She had gone over there to bring cookies, homemade Christmas cookies. Tomorrow was Christmas Day.

Lisa
December 24, 1961

On Christmas Day the girl was very sick and everyone was very angry at her because she had spoiled everything. Merry Christmas.

Lisa
December 28, 1961

This was given to me in November, 1984 by eleven-year-old Lisa, who wrote it "a few weeks ago" in 1961. Carol is unaware of writings such as this, because the other personalities kept them well-hidden. If she did come close to discovering a hiding place, one of the others would take over.

Carol reports in session that her parents have told sister Dena, who then told Carol, that the parents were coming to my office to demand that I tell them what Carol told me about them. Carol made it very clear I am not to talk with them, and I agree to this.

Much of the rage accumulated over the years is beginning to spill out, and much of this is directed into suicidal thinking. We have a crisis intervention session, arranged by phone by Maggie.

Jennifer has been in semi-crisis for about two weeks now. She adamantly refuses hospitalization and promises that she will sabotage all kinds of things if this issue is forced. Hospitalization in fact has not been pushed and others are stepping in to prevent anything serious from happening to Jennifer. She does reveal on this date that she really emerged when Carol was age two and that she and Katharine have functioned as a leadership team for years but Katharine did not know this. It is interesting to note that even within the personalities, amnesia for some experiences is possible. She reports that she has often had rages over the years but has always been able to privately control them. She adds that:"This scares me." Some of this is tied in with another phone confrontation with mother who has again tried to talk her out of therapy. Mom reached the patient by phone by having her son Nathan call, then mom took over the phone and the two of them "had it out." Jennifer's crisis is gradually defused over a period of a number of sessions. She now knows that her chronic anger is a major problem and has been repressed for most of her life.

We also now have an opportunity to see and meet the real "Eve" of the *Three Faces of Eve*, Chris Sizemore. Jennifer now volunteers to have co-consciousness with other interested parties (Katharine, Francesca, and Carol). All three of them are seen and all are willing and excited about this possibility. Carol has had total amnesia for the last two weeks and she is briefed. Because of Jennifer's crisis, she has had the great majority of time.

Nichole and Erika both are seen for the first time. Nichole wants to sabotage people going to see Chris Sizemore, ("Eve"). This is because going to see Chris represents a verification that multiple personality is something real, and if it happened to Chris, it could conceivably happen to them.Erika helps talk her out of it. Carol, Katharine, Jennifer, and Francesca--we have group co-consciousness in preparation for all going to see her the next day.

We go to Holland, Michigan to see and hear Chris Sizemore. I talk with Chris about thirty minutes privately and introduce her to Carol after her talk and they spontaneously hug. Chris has told me privately that she has often been able to identify another multiple personality in her audience merely by eye contact. It is fascinating and very touching to see her and Carol "connect." Jennifer's journal reports:

November 28, 1984

It has been an exciting, emotional day for us, including Carol. In fact, she is exhausted. She had a great deal to absorb today, first while watching the Phil Donahue Show, (who happened to have Chris Sizemore as a guest on his show), and then this afternoon while listening to Chris Sizemore speak in person. Chris is a former multiple personality. The book *The Three Faces of Eve* is about her and she also wrote *I'm Eve*. There was for all of us, some fear involved in meeting her. There were so many things going on in our minds, so many questions, that I'm sure it will be a few days before we recall everything that was said. Dr. Ted and a friend took us to hear Chris. I'm so very glad Dr. Ted is our therapist and it was so nice of him to arrange for us to meet Chris. I know that he must be exhausted after some of our therapy sessions, but he never gives up.

Jennifer

Meeting and speaking with "Eve" yesterday is a very powerful experience. Jennifer has many tears, and there is much emotion from both her and Katharine. Both find themselves particularly worried about "going out of existence."

I have been having contact with Erika who has been talking to Group 3. I have been hoping to get some feel for who and how many are involved in Group 3. Erika now informs me she has made a list and promises to bring it in to me. It appears she is functioning in the leadership role of this group, which is now our third major group of personalities. Carol has been regressing to age eight or ten and she wants evidence that I am not going to disappear like all of her other friends. Angelina from Group 3 comes in and introduces herself for the first time. She brings in a list of forty-two members of Group 3, which now brings the total number of personalities to seventy-three. Marie sends a note that a number of Group 1 personalities want to see me in the next session. I am feeling overwhelmed.

18

It is now December, 1984. Carol is quite regressed. We agree to deal with her feeling of blackness and of concrete blocks all around. We do a hypnotic time regression "to a time when you were feeling the same way." She winds up being age four, in a hole in the ground. She is retrieving memories of grandfather putting her in there because she got her dress dirty, and of herself crying. He finally lets her out and then is laughing at her. We do an age progression to age thirty-five. Although she is initially quite closed down, with prodding there is a tearful catharsis and she then remains an adult.

Marie had wanted therapy time but changed her mind after Carol's work in session today because, "She won't need me anymore." Erika is very angry at "having to die," which is how she views integration.

Stephanie comes in for the first time. Stephanie, Nicole, and Moiria later function as a co-conscious triad.

Moiria brings in descriptions of all forty-two members of Group 3. Jennifer and Katharine advertise their resentment of lost therapy time. Lisa and Margaret want my opinion of a trick Karl and Paul played on them. (The boys told them that "The exhaust of jet planes is really men going to the bathroom in the plane.")

Erika now appears to be primarily functioning as the leader and organizer. The only family contact they have is with sister Dena. They are staying away from everyone else. There are still no good job prospects. They are also checking out grants-in-aid for returning to school. I have for a couple of months now been encouraging Carol to draw and this is beginning to bring positive fruition, especially the drawing project of a "special" city in the Netherlands. Jennifer gets drunk over the weekend.

Her walls are up and she resents my questions about her parents, both of whom she says, "died when I was young."

Moiria brings in a number of belongings and mementos of several of the others in an effort to "prove to me how each of them is an individual and functions autonomously." She greatly resents the idea of integration. Katharine and Francesca are wanting to resume contact with me, which they haven't wanted during the past few weeks. They are, on their own, consulting with Carol by using their own co-consciousness. They are especially interested in her drawing of the Dutch city. Carol is quite excited about this project and has made arrangements to give her gift directly to Rev. Jansma (this is the city of his birth in the Netherlands). We do some work with age progression and age regression to get a comfortable age for Carol to operate right now, and sense that she is now comfortable only at an age in her early teens.

On December 24 approximately twenty personalities come in and take turns in this session to share Christmas greetings and Christmas cards. This is happy contact for most of them. Nicole, a bitter, depressed Group 3 representative who wants to sabotage others, especially Maggie's therapy, accepts a Christmas hug for the first time in her life. And there are many tears. They are going to spend Christmas Eve with their friend Sandy and her family. They receive a number of gifts and attend their first Christmas Eve Mass. It's a very special evening for all of them, especially since Sandy and her family are aware of the multiple personality issue, which means the personalities can be themselves.

Lisa and Margaret bring in separate Christmas cards. Lisa asks for co-consciousness with seven-year-old Alice, "because she has no playmates," and this is done.

Carol now asks for session time to retrieve more memories. With age regression she goes back to age thirteen when she brought cookies to grandfather and he made her disrobe and have sex with him and then beat her. She then ran away, hid in the snow, lost her mittens and tore her stocking. When she finally got home her parents were extremely angry with her. She became sick at that point and had to spend all of Christmas day in bed. There are many tears. Clinically, this is a good experience for her. Many in Groups 1 and 2 are now resenting Group 3 taking as much time as they have.

In the next session Carol refuses to talk. All the others refuse to take over for her. These others are clearly trying to help her to get better. Her anger builds into a crescendo of tears, especially relating to anger at grandfather. Many are now wanting therapy and there are many agendas. It is difficult to keep track.

Alice has refused to speak with anyone for a number of days now. I discover that she is quite frightened as a result of her hatred of hearing anyone talking about the grandfather. I encourage contact among Lisa, Margaret and Alice and they all agree to have some play time together. I spend time with Francesca who has always had a strong artistic bent. Because of the physical abuse of tying her hands down, she has not been able to free up her hands to return to her artwork. We use hypnosis today, trying a technique with a dial, and her pinkie toe for storage of pain in order to free up that hand. She will now try returning to her artwork, which she gave up as a young child.

Jenny gets very interested in woodcarving as a sideline. We work out a deal where she will give up her excessive exercising at the spa if others will give up time so she can pursue her interest in woodcarving.

A number of personalities are now involved in various art projects. Francesca finishes her first watercolor, a panda, which is quite lovely, but she wishes to destroy it. I talk her out of it. Marie complains of not having anyone to talk to and asks if it's okay to write Chris Sizemore ("Eve"), which I encourage.

Katharine is now asked to do three watercolors for a fee, but she doesn't think she's good enough. I bring Francesca into the discussion and encourage them to do it together. Katharine and Erika spontaneously have co-consciousness for the first time over the weekend. I have a group meeting with all seventy-three personalities at this point. I tell them that clinically I want all of them to have co-consciousness with each other, especially with Carol, but that I would not be pushing for integration.

January 17, 1985. I give Alice a little drawn image on a piece of wood of "E.T." from the movie of the same name. She decides she would like to spend some play time with Lisa and Margaret. Jennifer is feeling quite rejected by me because of the amount of time spent with many of the others. I work at getting her out of her consciously imposed shell. Erika has begun dating someone and he appears to be quite interested in her. She finds feelings of rage inside of her whenever anything of a romantic nature emerges.

A new friendship develops in church. She's brought in for part of one session. Many introductions are given because many want to meet this friend, who appears to be functioning as a very positive outside resource. The friend is fearful of doing damage and asks advice from me. I have contacts with Jacqueline, age nineteen, who is extremely preoccupied with physical appearance. She frequently uses diet pills, has used laxatives for over five years now, and also has been abusing Carol's Dyazide. She has suddenly asked for help, is acknowledging the problems listed above, and has on her own stopped all of the above. She is extremely hyperactive and shaking. We attempt hypnosis which has only a minimal

129

effect. She asks to enter an eating disorder program and we each share the load of checking out various programs. After some searching, we find one that seems to meet her needs.

Carol comes in one day very angry at the others for not giving her time in session. We have a session of group co-consciousness including Carol, Katharine, Jennifer, Marika, and Francesca. A very emotional Carol makes her feelings known. They inform her as to why they have taken the time and not encouraged her to have time; (Carol's anger has kept them away; they have, in fact, wanted to cooperate with her). This is a revelation for Carol. Suddenly her voice changes from the frightened thirteen-year-old voice to an adult voice. She is taken aback at this. It appears that the regression of the past six months is beginning to reverse. After this, Erika agrees to co-consciousness with Carol. As Carol describes it:

I must talk a little bit about the session that I had on January 24. I am still getting used to my voice. I can hear the change in it. In a way it amuses me, but it's also a bit scary. I have so many feelings about this session. I was so angry when I realized that I only had a half hour left, and the others had taken an hour and a half of the time. It was good to get their input on how they've been feeling towards me. They feel that I am angry with them much of the time, that I dislike them, and that I resent them. I think that's true. I do resent them because they have taken so much away from me. At least, that's how I feel. I know that much of it was done to protect me, but now I'm tired of losing time, I'm tired of not doing things for myself, of finding things in the apartment that aren't mine. I often feel as if I don't belong in this apartment, as if it's not my home. I said a lot of things in that session that have just been building up, that I have kept to myself. I resented Dr. Ted for awhile because I thought he cared more about the others than he did about me. I felt as if I just had the body but they inhabit it, they take over, they do everything. I don't want it to be like that anymore. I said that I wasn't helpless, and I'm not. I know that there are a lot of things that I have to learn, and a lot of things that I have to discover for myself. And I know that, at times, I'm going to fall flat on my face, but I also know that there are going to be times when I win. That feels good. I feel so left out. As if I don't matter. It is my life and I am not going to like remembering some of the things that have happened to me, but I came away from that session feeling better about myself than I have in a long time. Someone is taking a woodcarving class right now. It really surprises me that someone inside of me is taking a class, carving wood and that I know nothing about it. I'm glad for her because it's something she wants to do, but I feel as if something has been denied me. Maybe that's not really fair to them. I know that in the last few months I've been more like a child and my voice has been more childish. And there are things about the past few months I don't remember. Dr. Ted said that I sound like Kate now. I like Kate. We have some plans for the weekend. Even that makes me feel like a little girl, because we're going to have a baking and cooking lesson. Here I am, almost thirty-six years old, and I don't know how to do these things. I know I have to learn patience. Although I don't think that's ever been one of my virtues.

My therapy is important to me. It has been for a long time,but I have been afraid to find me. But I am also curious. I have more of a feeling that I'll get there, that there is going to be an end for all of this. Maybe I believe in myself a little more. I do trust Dr. Ted. I need to trust him because I can't do this alone. I feel that the rest of them have better relationships with him than I do. He knows Kate and Jennifer and he knows Francesca, but he doesn't know me. That makes me   sad,because I feel as if I haven't been given a chance.

From now on I want to be the one that tells Dr. Ted what is bothering me. They often tell him how I feel, so there's really no need for me to. And, from now on, when I remember things, I want to keep that memory and not put it away. I need to work through the feelings that I have associated with that particular memory, whether it's sadness or anger. I don't think there are going to be many happy memories.

I feel left out. Maybe that's where the sadness comes from. Not too long ago, when I was feeling so bad inside, I tried to get Dr. Ted to have someone else take over and he wouldn't. That was one of our better sessions because I cried and I ended up being angry and I needed to be that way. I needed to do those things. I want to be able to be angry without being afraid of it. And I want to be able to know what day it is, every day, what month. At least I know it's 1985 now. I don't want birthdays taken away from me anymore. I want to have my own friends.

I need to discover what was my past. I don't get enough time, but I hope after Thursday's session that will change and that theywill be more willing to share time with me. Just give me a little time on my own, give me a little time to spend with them. I don't need to hear everything that they say. I don't have to be part of every conversation. I don't have to know all their feelings. But just give me a little time, that's all I want. It's the only way I'm going to get anywhere.

Jacqueline has been off all chemicals for a week now and the physical problems, (headache, shaking, jumpiness), are now beginning to diminish.

Carol has her 36th birthday. Katharine, Francesca, Jennifer and Erika attempt to make it a good day for her. The parents are now trying to draw Carol back into the family fold and she is allowing some minimal contact. Today we have co-consciousness for the first time between Carol and Erika. Erika, although having volunteered this a week previously, has in fact become rather hesitant and I talk her into allowing this to happen. She probably will hold back somewhat but the co-consciousness is successful.

Moiria shares that Carol, although invited by her family to celebrate her birthday, is essentially ignored by the entire family except by her father. Group 3 is doing some sabotaging of Erika's therapy and I confront them. They eventually agree to cooperate.

Jenny is now involved in her woodworking class and this is calming her down very effectively. She is already demonstrating some talent. Lisa feels very guilty because she feels it's her fault that the grandfather sexually abused all of them when they were very young. I see Margaret, who cannot tolerate sitting in any chairs. This is because grandfather would tie her in the chair, naked, and molest her when she was very young. She has since then always had the need to sit on the floor. We experiment with her gingerly sitting in a chair in my office. She gets extremely frightened, bolts out of my office, and runs into the large snowbank in the parking lot. She does come back a little later with Lisa's help (I do not go out to get her), and she voluntarily asks to try again. This time she handles it quite well. Katharine later describes the issue as follows:

Margaret is ten years of age. She's part of Group 1, my group. She's a very special child. She has many fears, many nightmares. She still lives in the past. Margaret has a fear of chairs, she doesn't like to sit on chairs or couches or anything. She has a fear of noises. She's afraid that the grandfather will come back and hurt her, because he often tied her to a chair. Sometimes he would beat her first. Sometimes he would force her to perform oral sex on him while she was tied to a chair. She would be naked, and subsequently, she's afraid of chairs. She'll sit on the floor. I've tried to show her that in our own apartment, where it's safe, it would be all right to sit on a chair, but she's still very uncomfortable. She has done it in Dr. Ted's office in the "circus" chair as it's called. He has her sit in his chair, and it swivels and she likes that. I remember one of the first times that she sat in a chair in his office. She could only tolerate it for a short period of time before she had to run outside, without her shoes, without her coat, through the snow. Then she becomes angry with herself because she can't sit on a chair.

Someone who has been quite helpful to her has been Lisa. Lisa spends a lot of time with her, and it doesn't matter to her if Margaret sits on a chair or not. Lisa's thirteen, and so they play their games sitting on the floor. Every once in a while Lisa will sit on a chair and she will say, "Okay Margaret, now you try it. See, it didn't hurt me." It may not hurt Lisa, but it may hurt Margaret. If she sits on that chair she's afraid that grandfather will suddenly appear and tie her to it again. She's still living in the past. She's a very sensitive child. She wants to learn how to play the piano more than anything in the world. She loves music. I would like to be able to give her piano lessons. I've wondered though, if she would sit on the piano bench.

✳ ✳ ✳

Still being unemployed and with the fear of eviction from her apartment coming up, we work out a referral for Social Security disability help. We enlist the help of a social worker affiliated with the church and at this point it looks good. Margaret sits in two different chairs in my office on this date. Maureen Brandon, Group 3, age twenty-nine, arrived when Carol was five, comes in and introduces herself to me. We also focus in on Moiria's "anesthesia" for anger, just as Jenny has anesthesia for pain.

We do more age regression work with Carol, again using an interpreter. I bring her back to age six: she's in a small room, her shoelace untied, grandfather is beating her with a belt, forcing oral and vaginal sex on her, she runs out of the room into the rain, coming back to the kitchen and grandfather and grandmother scolding her. This is a very powerful session. She is beginning to connect some of the feelings with memories. She is bothered by this session, however, because she experienced it partly by observing Amanda (red hair) and Callie having the abuse done to them; she is therefore remembering it as an observer rather than as a participant.

Carol has finally gotten a new job possibility and today is her first day of training. Erika is actually the one who got the position and will be the primary one to handle the actual workload. I later discover that many are resenting Erika for taking over all of the job specifics.

We pursue co-consciousness with many different combinations, Moiria of Group 2 having co-consciousness with Francesca, Marie,

Andrea, Alexandria of Group 1. Secondly, Marika of Group 2 having co-consciousness with Christina and Diana of Group 1. Gretchen comes in for the first time.

There is now much discontent because of competing needs for therapy time. The following letter from Katharine demonstrates this:

February 20, 1985

Dear Dr. Ted:

Due to the time limit on our session yesterday I left many things unsaid and as a result I am very frustrated. You already are aware that I am angry and sad and I'm going to try and explain why.

I don't like what is happening to us collectively. I am, of course, more aware of what is happening in Group 1 than in the other groups. There are many of us that are depressed, myself included. We are all experiencing anger in one form or another. For some of us, it's an underlying current, a feeling within us we're not quite sure of, but nonetheless something we are afraid of. Some don't even view it as anger. To them it's a feeling that they can't define. For the rest of us, it's rage instead of anger. We want to hurt someone, destroy something, and it's becoming more difficult to control.

Many people in my group also feel left out, as if they don't count, and I'm going to give some examples. Jennifer and I have wanted another session, two hours if possible, to finish discussing the feelings we had after hearing Chris Sizemore speak. We had one session the day after and felt it was very productive, but we never finished. I haven't even dealt with all the feelings that caused my regression last summer. I'm still angry. Jennifer still has an immense amount of anger and she is still drinking. Marie still wants to speak properly. She is so sad and her poetry has definitely reflected that as of late. She hasn't resolved the abortion issue and has an immense amount of guilt and anger concerning that issue. Amanda is angry, period! Diana is struggling with the "one body" issue. Christina and Caroline are both extremely angry. They have both dealt with some of that in session, but neither one of them feel as if they can accomplish anything. In their minds, things were left unfinished. Alexis had all of two sessions. She didn't come near to completing what she had started. Paul and Karl are causing a lot of problems. They are little boys, not men, and have known only rejection in their lifetime. They are frightened. Simone had one or two sessions with you and she is a teenager with many problems, one of which is wanting to go to school and learn.

Katharine

We hit upon the idea of organizing the leaders: Katharine, Jennifer, and Erika representing Groups 1, 2 and 3 to better manage all of the others. They will have group meetings and discuss this on their own.

Jacqueline takes twenty-four laxatives, after going two weeks without taking any. After many false starts she is again scheduled to begin a wellness program next week and agrees to go. She asks me when she will feel clean. I tell her, "When you stop blaming yourself for being raped," Erika, for the first time, is feeling the physical effects of someone else's use of laxatives. Some subtle merging appears to be occurring.

Marie is extremely self-conscious about her sense of immaturity, which especially shows in her speech patterns and in her high-pitched and very immature-sounding voice. We use hypnosis for the first time to alter her speaking style. She wants to sound more like Katharine. The actual hypnosis work is not successful but we follow this up by having her consciously and intentionally talking in a lower voice and a very large difference is noted.

The group leaders are more in touch with each other and most of them are feeling quite good. Moiria introduces Paul and Karl for the first time and they talk with me briefly. They have no friends and have never before met an outsider. Margaret has been practicing sitting in chairs and it is becoming easier for her. Jacqueline has now begun her wellness program. Carol's finding herself grieving over the death years ago of her brother, Phillip. She literally lost several years of time between then and her divorce from James. Carol is quite angry with the others for having taken so many years. She is now having some spontaneous memory of Phillip and how mom would ignore them, be sarcastic with them, calling them "king and queen." She allows herself to cry during session.

I have an emergency session with Erika. Tanya and Alexis have loaded down with laxatives two evenings in a row. She has had bad headaches, stomach cramps, and is in very poor shape to do the psychological testing, which is necessary for her new job. We do hypnosis, (for removal of her headaches), and it makes a significant difference.

Laura White comes in for the first time. She appears to be a fairly healthy and reality-oriented personality. She asks to join our leadership group in order to take the place of Jennifer, who is refusing to accept a leadership position any longer. Co-consciousness is therefore effected with the other leaders of our leadership group including Erika, Carol, Katharine, and Moiria, who has been temporarily covering for Jennifer. Jennifer is backing away from her leadership role because she is so fearful of fusing. She feels that the best way to prevent this is to sabotage it. She's doing so by not providing any leadership functions. Erika passes all of her tests and out of all the applicants, she is the one who is first offered a job. She volunteers this information to no one, so others pass that information on to me.

Carol comes in. About fifteen minutes into the session she asks if someone has been role-playing her. She first asks Laura, but Laura will not "squeal." Carol then asks it in general and I support her question, at which point Stephanie emerges. She confesses that she has often role-played and explains that: "I don't want Carol to get better." She knows that if Carol gets better, it means that she will be going out of existence. She does promise that she will not do so again in session, but is also very clear about her wish to continue to sabotage Carol's therapy. I confront the entire "community" heavily on the issue of role-playing Carol, and deal with the dishonesty inherent in that. Francine (age 37) introduces herself for the first time and admits that she also, as well as others, have done so.

We have a one-hour joint session with Pastor Aaron. He has been involved with several of those personalities who have a religious value system, and has expressed his wish to be a support if that would be helpful. Approximately twenty personalities come out to meet him in session. I request that the kids specifically do not go in front of church when he calls all the church children up front for their children's sermon. We also do co-consciousness between Elizabeth and Marie, the two poets.

The leadership group is actively engaged in handling the logistics of training for Erika's new job. Carol is being nudged by a number of them on becoming more of an insider with this leadership group. There is now support for her growth by most of them.

19

The Secret

For a time it would be hers alone,
a secret not yet to be shared,
not yet to be exposed to the world,
a world that would not understand,
that will not try to understand.

For so long it was just a dream,
a dream kept within the
recesses of her soul,
an aching void within her,
a void she could not fill.

And her dream was now realized,
the aching void no longer there,
her fulfillment now complete,
a cherished secret hers alone.

And she was no longer restless,
quietness, elusive for so long
was hers at last,
and she knew contentment,
secure in her knowledge of
the child within her.

Elizabeth Austin

My name is Elizabeth Austin. I am twenty-one years old, and the date today is July 19, 1984. I haven't talked with Dr. Ted very often. Once, I had a whole session and the second time I talked with him briefly. I gave him some of my poetry. It feels a little strange to be doing this.

Of course, I've told Dr. Ted that I'm pregnant--and that I'm going to have a baby. It's a wonderful feeling, because it's something I've wanted for a very, very long time. And right now, it's a secret. I haven't told anyone else and I don't intend to because I'm afraid that there will be problems if I do, just like there are for Marie, and I don't intend to let what happened to Marie happen to me. (*Note: they both believe Marie had a baby and it was stolen.*)

137

This baby is so special. I don't know yet if it's a boy or if it's a girl, but it's a very special child already. And when it's finally here, I am going to love that child and take care of it and nurture it. It's going to be mine alone and there isn't going to be anyone that can take this child away from me. Not anyone. I don't think that anyone really understands how much it means to me to have a child. I used to feel so terribly empty inside and that feeling is gone. I have all these hopes and all these dreams and right now, they're centered around my baby.

I even write poems about it. I don't really like to call the child an "it," but yet, I don't feel right about saying a boy or girl, he or she, because I don't know. I guess I could be called rather quiet--I've been called "dreamy" because I do tend to daydream and I like my books and my music and my writing best of all. I don't need many people around me, maybe just a few good friends, which I don't have, but that's okay, because right now, my time is filled. I like to be outside. I sit and daydream at the beach or at the park. Find a quiet little spot all to myself. All I need is a blanket and a book and a pad of paper and a pen and I'm set.

Favorite Season

Brilliant sunshine, blue skies,
a crispness in the morning air,
scarlet and rust, orange
and bright yellow,
all the magnificent hues of autumn
not yet captured on an artist's canvas.
A time of lazy afternoons and burning leaves,
bulky sweaters, faded jeans,
Long walks down country paths,
mulled cider and hayrides,
grilled hotdogs, crunchy apples,
bonfires and guitars,
toasted marshmallows, mine burned.
Small children, eyes full of mischief,
faces filled with glee,
jumping in piles of just raked leaves,
all the wonderful sights and smells of autumn.

Elizabeth Austin

I would just as soon avoid people. I suppose that really is not the proper attitude, but sometimes I'm just not very proper. I'm

138

casual and unconventional, but that suits me. I don't like con-forming. I suppose there is a bit of a rebel in me, but a quiet rebel. I don't make scenes. I prefer to quietly go about my own business and do what is right for me.

I don't particularly care for men, but of course, that was a necessity in order to get pregnant. The act of sex was not enjoyable. I would just as soon do without it. I'm sure people would think me very strange for feeling that way, but I have my reasons. They're not something I care to discuss with everyone. For the most part, I keep things to myself, especially things like that. I don't brood about them, or at least I try not to.

Sometimes I wish that I could change a little bit, but then I look at what I have and what I would have to give up in order to change. I like my private world. I really don't care to venture out, or maybe I should say, to let other people venture inside of me. I observe others. I observe their actions, behaviors, things that make them tick, things that are seemingly important to them. I find the majority of people to be very shallow. They're out for themselves and to hell with everyone else. They often forget about the people that were once important to them. Like the people I wrote about in one of my poems: the old people, people that don't really have a present--they tend to live in the past. And those people are so often forgotten.

Aging

The past is now their present,
long-forgotten memories, events,
once again a reality,
husbands, wives, long since gone,
once more their companions,
children no longer recognized,
people once strong in youth,
now so dependent.
Old people. Precious people.

Elizabeth Austin

I guess I can daydream and fantasize a lot more now that I'm pregnant. And I hope my child does the same thing when it grows. I want it to discover the world, to discover life. To discover that life isn't all material things, it isn't keeping up with the next-door neighbors. And my child is going to be cared for. It's going to be shown that it's loved. It's going to be shown that it belongs and that it has a home and that it has a mother who thinks that it's very important. I want my child to be able to make decisions on its' own. But, most of all, I want it to like itself. I want it to know that it has a mind and I want to challenge that mind. That's so important, to teach a child to think for itself. To express opinions, ideas, feelings. Oh, and to especially express feelings. Emotions. I wish I could see it now. I can feel it.

My child is going to be allowed to get dirty because he is going to play in mud puddles and of course, then you get dirty. But there is something to be learned, even from a mud puddle. There's something to be learned from making cookies and getting flour all over one's self. It's all in the process of discovering. And it's going to learn that it is okay to sit down in a strawberry patch and eat the strawberries, even though they do get on your clothes. And that fresh blueberries are the best thing around, especially picked right off the bush.

I love to just sit and watch ducks swim up and down a creek or in the pond by the Veteran's Hospital. That's always been a favorite spot. The old veterans, you know, they come and talk and some of them are shy and some of them are grumpy and some of them are drunk. But they're people and they're human. And they all have feelings. And sometimes they don't smell so nice. But they're my friends and they know me and they wouldn't hurt me. And one of the men, he waits for me to show up and he always saves a bagful of bread for me--old stale bread--and we sit and we throw it at the ducks and we talk about the day's events, although his days don't change much. Sometimes he forgets who's president and I have to remind him and he laughs and he'll say, "Oh, Elizabeth, I guess I forgot." And then he'll tell me how much he likes the name Elizabeth. And he'll look at me and every once in a while he'll say, "You're different." And then I have to wonder why I'm different. Then there's the nursing home on Mayfair St. I go there and visit sometimes. I never really tell anyone who I'm visiting because I'm really not visiting anyone in particular, just all of them. And each one thinks that I belong to someone else. And so we sit and we shoot the breeze, as they say. I have a fondness for old people; people in places like the retirement home for the blind people. Those people that can't see, and yet, they can. These places are special places.

In the nursing home, I remember one lady in particular. She was old, probably eighty-nine or ninety. Had the softest, whitest hair. She couldn't take care of herself anymore. She didn't remember people. She didn't talk anymore and a lot of the time all she did was lay in her bed on her side. One day, I put a pink ribbon in her hair and she smiled. Maybe because I was talking to her. I'm not even sure if she was aware I did that, but I told her I was going to and I combed her hair and I put a pink ribbon in it and she smiled. And then she cried. I never did get to see her again, because she died. That made me sad because her family didn't care. They lived in town. They didn't come to visit her. I resented them so much. Every once in awhile, they'd make their duty visit at holiday time. I always got the impression they were just waiting for her to die.

We worked at a school for mentally impaired children. But boy, were they special. No hesitations in hugging and showing affection and crying, showing that they were angry. Some of them had tempers and were stubborn as mules. But they tried so hard. I shall never forget our first day at that school. Actually it was Kate's day, but was that funny. When the teacher went for his lunch time, and Kate was left alone with the class, the kids were supposed to put their heads down and take a short rest period after lunch, and all the lights were out, and that one boy didn't want to take a rest time and he took all his clothes off. And she couldn't get him to put his clothes back on. That was her very first day. But she won him over. But people, so often, just look at what these children look like and the way they walk and talk. And they don't see what's so good about these children.

It is now April, 1985 and we video tape some of the personalities for the first time. We wind up with about twenty-five personalities on tape. Emma introduces herself for the first time to any outsider, and does so on tape. This turns out to be a very powerful experience and we agree in our next session that we will all view this tape together.

We have a three-hour session to process the many reactions to this experience. Of special note is Elizabeth, who by viewing the video tape begins to approach crisis over the realization that she is not in fact pregnant and that she does not have long hair. Katharine is much more troubled than she or anyone else expects. The leadership group keeps them all in the office since many of them want to leave. The leadership group is becoming more effective in comparing notes and in providing support, especially now. Elena comes in for the first time during this session. "I am not a nice person." She comes to voice her anger at me for putting Elizabeth through the realization that she is, in fact, not pregnant.

At least seven personalities, Corrie, Karen, Christina, Victoria, Ellen, Alexis, Diana, come in joining forces together because of their joint anger at me for "what you did to Elizabeth and Marie." They confront me very heavily. Corrie functions basically as the spokesperson and she promises to terminate therapy for everyone. After letting them ventilate, I take a strong stand. When Corrie starts to walk out the door I block the door. Once they realize I am going to take such a strong stand, their anger

141

subsides and they acknowledge that they had been testing me. Marie and Elizabeth are both approaching crisis and we agree to emphasize individual therapy time with them. It is interesting to note that Carol saw approximately five minutes of the tape and there were no over-reactions on her part.

Carol requests to see the full tape. She finds out for the first time about the abortion (Stephanie had role-played Carol before when I thought Carol had remembered on her own, and confesses to this during this session). Katharine is struggling with the reality of "seeing" herself for the first time.

The leadership group brings in Elizabeth because she has become suicidal. Elizabeth will not talk with any of them and she demands her right to leave my office. I block the door. When she realizes that she cannot change my mind, her tears begin to flow and she begins to have her needed catharsis. She has denied anger all of her life. She's beginning to accept that there is in fact no baby, that there was an abortion, that she is suicidal, and that there is an immense amount of rage. There is much pounding on the sofa. Powerful. I cry too.

Therapist/Patient

He holds out his hand,
offers his friendship,
his help during the long
road ahead;
a road of risks and
struggles,
a road of tears and smiles.

She takes his hand,
the gesture tentative,
wanting to trust,
hesitant from past experiences,
past rejection,
willing to try one more time,
accepting the offer of friendship,
beginning to trust.

Elizabeth Austin

20

Lisa and Margaret are both having very bad nightmares, especially about abuse done by grandfather and others. I begin seeing each of them alone. Much of the clinical input centers around the issue of "Am I a bad person?" I recommend that each write down their story. Katharine and Jennifer will help them.

Laura presents her usual update. Carol's mother has called out of the blue this weekend and told Carol that she was, "No good. All you've ever been to us is a problem; you were a whore, but we fixed that when you were pregnant." Carol is very upset, not understanding what her mother has been talking about. She is also quite worried about time away from home during the job training, and if something happens to her, what becomes of all of her belongings. She clearly does not want her family to have many of her belongings and arranges for a list of specifics that she requests several friends to safeguard for her.

Carol is regressing again. The phone call from her mother really upset her. With hypnosis we determine she has regressed to approximately age fifteen. We age progress her to the present. She is confused and has little memory of her regression.

Jenny comes in to report that a fellow student (male) in her woodworking class kept teasing her. Then he did some some work on her project, unasked, and ruined a piece of it so she hit him in the face with her fist. He was quite embarrassed and threatened to sue her. The teacher observed all of this and clearly supported Jenny.

Extra time is now scheduled because of so much turmoil over training for their new job, and the subsequent cut-down on therapy time. We meet with group leaders (without Carol who has regressed to about age ten,

143

primarily due to Mom's confrontations and Mom's pitting Dena against Carol). We plan strategy. I encourage Margaret and Lisa to write. They need assurance that I won't be going away. A number of personalities are seen over specific issues. Elizabeth is, for example, still heavily grieving, but is beginning to open the doors in session: "Maybe I'll paint a picture." Miranda is seen in therapy. She is finally beginning to confront the issue of abuse and incest and tears happen for the first time. Moiria provides direction.

Erika calls me early in the morning. The car keys are missing and she cannot go for her training. I "order" the responsible person to speak up, and Miranda does so. I promise her a session if she'll come in to see me, which means she has to reveal where the keys are in order for her to get to my office. It turns out she had put them in a full bottle of fabric softener. She promises not to do it again. Melinda Daniels comes in for the first time letting me know that she knew of what was going on. It is later revealed that they have become very safe at home while being unemployed, and with the new training they fear for what will happen to therapy, how to handle meeting new people, and having to go out of town for that training.

Elizabeth is still grieving, but she has now given away all the baby clothes to a home for unwed mothers. Miranda has agreed to write an autobiography over the long weekend coming up, as a means of beginning to give direction to her unsettled feelings. We attempt to deflect anger in Miranda by having her experiment with touching a stuffed dog (Mishka) to retrain herself to begin appreciating softness. She later brings in a three-page autobiography and I encourage her to continue. Miranda also begins guitar lessons with Mother Josephine Bernadette, a wonderful and non-threatening nun who presents us with an excellent opportunity to build on the "community's" network. She proves to be a wonderful resource.

Jenny is introduced to Miranda for the first time. Miranda brings in two additional pages of writing and is much calmer. Elizabeth continues to grieve and there is still suicidal thinking. I encourage her to work on drawing, as a safe outlet. Carol is still regressed at age fifteen.

Erika calls and reports that she has lost most of her work day and no one has any idea who has taken the time. By phone, I request whoever it is to identify themselves and Lisle makes herself known. She agrees to meet me in session. Lisle, along with Krista, spend an entire hour volunteering availability to answer questions and advise on who needs what. They advise that there are twenty-three more personalities, then acknowledge that there are even more than this. They promise to provide a list of new personalities in addition to a list of those of the first seventy-four personalities who most need help. I can't help but wonder if it will ever stop.

Lisle and Krista now begin providing direction for therapy. They advertise that there are thirty-three members to their group which we now call the "Sunshine Group" because all the personalities appear to be very up and quite content. This now puts us at a total of one-hundred and seven personalities. They are invested in the first seventy-four personalities getting better.

Miranda feels that the others are angry with her because she has allowed her guitar teacher, Mother Josephine Bernadette, in on the issue of multiple personalities. Elizabeth is beginning to smile a little. I encourage her to continue to work on her drawing. I recommend that she do a weekly project even if it is quite small. She is grieving less.

It is now June, 1985. We do a review of events and of our situation. It has become clear that Katharine has been the host personality for much of their lives. This is the personality that is primarily in control of the body for the most time, a role which can be filled by more than one personality. Since many have given direct comments to this fact, I check directly with Katharine about her reaction to others merging into her. I recommend a group meeting of their own to discuss the issue for those that are interested. This is significant because we are now getting to the point in therapy where we can more realistically think of beginning to merge some personalities, which later hopefully will become full integration of all personalities.

Rebecca from the Sunshine Group comes in for the first time. Co-consciousness is done between Lisle and Erika to help handle her first on-the-job customers. Lisle also recommends that Katharine remain the host personality. Katharine and I discuss this further, and she agrees to accept this role.

Erika has an important session. Although there is much resistance on her part we begin reaching her anger. She finally shows some of it, letting some tears go. She has spent a lifetime of being "the robot." During the talk-down afterwards she says she is very embarrassed and will never come back to therapy again.

Laura is now providing much of the advisory and summary work, especially of Erika and Katharine. Most are now looking to Katharine as the central figure in their entire community. I'm informed that Katharine has discovered a lump in her breast and her physician has strongly recommended a biopsy, and mammogram at least, all of which she has refused to do. I confront her strongly and I get her to promise that she will do this. Erika, who did came back to therapy, for the first time discusses some of the history of parents locking her up in a closet, and of the grandparents and great-uncle sexually abusing her.

Lisle is also over-seeing therapy time and advising me whom to see and why. Katharine's mammogram looks suspicious. She has been "ordered" now to get a biopsy. She is quite frightened. Her breast biopsy proves

145

positive, but she refuses a mastectomy. She begins radiation therapy and chemotherapy will follow. Much clinical time will be spent with Katharine now. Erika now acknowledges that if Katharine has cancer, so does she.

Carol calls by phone and is not regressed: "Are you still my doctor?" I promise to have time with her next session.

Many are feeling ill from the radiation treatments, which means that many are having more difficulty denying that they are one body. Carol does an about-face, resisting therapy and will not come in at this point.

Maria S. (age thirty-five) from our Sunshine Group introduces herself for the first time. She presents as a bright, articulate, somewhat controlled but very well-functioning woman, who is also a poet. She advises me to see Katharine, but we wind up spending most of the session talking. I use this as an opportunity to gather support for proceeding with beginning merging also called fusion. (I might explain here that what we've called merging, or the blending of two or more personalities or fragments into one, is now generally called fusion. Integration is the full process and end product of the fusing of all personalities and fragments.) It now appears that our new group, the Sunshine Group, is a supportive, healthy, facilitating group, which will play a key role in this important stage of therapy. Katharine is struggling. She frequently feels ill, very weak, and very alone. We do co-consciousness between Katharine and Maria S. This is the first co-consciousness outside of the Sunshine Group. (Note: the Sunshine Group has awareness of all of the others but has never had co-consciousness with any of them.) I continue to talk with Maria S. who presents as a storehouse of information, memories, and especially presents as an overseer right now of the entire community of personalities. She has perceptiveness into the reasons for existence of many of the personalities. She is filling me in. She reports there is an additional group, Group 5, which consists of eleven members for a total of 118 personalities to date. She reports that she and Katharine came first at age two. She tells me that she has hung onto the good memories, while Katharine has hung onto the bad. Katharine is fast forming a close relationship with Maria. I am not sure whether we are experiencing the creation of new personalities or fragments, or the discovery of another group of personalities who have been there for years.

Maria S. agrees, probably with help from Katharine, to do an autobiography. Lauren (age sixteen) and from the Sunshine Group, comes in to meet me for the first time. She volunteers on her own to have co-consciousness with sixteen-year-old Miranda. This is an encouraging development.

The kids now need reassurance that Katharine will be okay. Katharine has decided to emphasize therapy for herself. Previously she has spent most of her time mothering the others. The issue that touches this off is the cancer and in particular how the cancer intrudes on her sense of femaleness. She is beginning to come out of her shell more.

A very powerful two-hour session with Katharine ensues, with the primary focus on her cancer, on her many "whys," her anger and frustration. Finally, for the first time, tears really happen. Maria S. functions as a support and advisor throughout this entire session. Katharine now volunteers to join Maria S. in doing the autobiography.

Carol now shows up for a session and is extremely angry. She feels cheated out of therapy. I let her ventilate and it is powerful. She announces her wish to terminate therapy. Katharine takes over at this point. Carol takes the spot back. (This is the first time Carol has had this much power for anything.) I let her ventilate some more. The end result will be time promised to her after my one week vacation.

Maria S. is working on her autobiography. I broach the subject of merging (or fusion), but she is quite worried and wants to talk about it. Carol is still very angry. She is maintaining her anger outwardly, (Note: Carol has, through her entire existence, absolutely and always avoided all anger.) She is taking time on her own now and is clearly developing strength. Right now the primary targets of her anger are Katharine and this worker. Katharine has some testing the next day to see if radiation is helping. She is very worried.

Carol remains very angry. She believes no one accepts her as her and therefore feels abandoned and unimportant. "I fantasize about getting a phone call just for me, from a friend." She wants all the others simply eliminated. I tell her that this is impossible. Maria S. reports that she is very worried about fusion at this point in time and about it happening too quickly. She has a list of six or seven of the others who will in all probability sabotage fusion unless they have some therapy for themselves.

In September, Maria S. lets me know that Moiria is beginning to fuse spontaneously with Carol. I spend most of this session with this combination of Carol and Moiria. Both are very confused. Moiria is very angry and Carol's very scared. It appears that the beginning of a very first fusion, as opposed to merely co-consciousness, is in fact occurring. (Be reminded that co-consciousness involves mutual awareness, while fusion is the process of two personalities actually becoming one.) This is both exciting, but also frightening. Maria S. encourages me to spend some individual time with Moiria. Moiria has become suicidal and interestingly, Miranda (age sixteen, the one who is having guitar lessons) has the most powerful impact on keeping Moiria together. I compliment Miranda on her effective work. She has previously been only preoccupied with herself and this is a very encouraging development. Moiria refuses to talk with me, but I

147

insist. She tries to bolt from the room and I physically restrain her. We have a verbal confrontation that is extremely strong on both parts. After approximately fifteen minutes of standing face to face, she finally breaks and a flood of tears happens--a very necessary emotional outpouring takes place.

Moiria continues to go through much "flip-flop" between being Moiria, being Carol, and being a combination of Moiria and Carol. Moiria comes in session only because Maria S. has brought her here and everyone else has just simply left them alone.

Monique Germaine comes in for the first time. She informs me that Katharine has refused to go for a biopsy of a lump on her neck, which might indicate a spreading of the cancer. Katharine is obviously quite frightened. Jennifer and Erika are quite angry with me for having given therapy time to others. Moiria continues to struggle. It turns out that at this point she does accept that Carol is a multiple personality, but does not accept that she is. Katharine finally gets her biopsy, which shows pre-cancerous growth. She is advised to get heavier radiation, but she initially refuses any treatment.

Katharine eventually allows some verbal and emotional letting go over the issues of cancer, parents' insensitivity, parents' threats of killing themselves: "And it will all be your fault." She speaks of not wanting fusion to begin because "We at least accept each other."

Karina Kastille, age thirty-six, arrived when Carol was age four, the first group of five to introduce herself to me, comes out in session. She's an accomplished artist and interestingly has sold much of her work, which no one knows about. Katie, from this last group of eleven also comes in and introduces herself for the first time.

We include Mother Josephine Bernadette in on part of a session. Since she has become quite invested in Carol's progress, and because a number of the others have now met her, it is viewed as appropriate to further nudge her involvement in their growing support system. About ten or twelve of the personalities come out to introduce themselves to her in session. This is also the last week of job training. It has been a long grind.

I have therapy time with Erika now. For the first time I do specific hypnosis work with her over a long-lasting migraine problem. It gives us a chance to deal with the reasons she keeps her neck muscles so tight and why her fingers remain clenched so much of the time.

Karina encourages me, as did Jennifer, to talk with Erika regarding contact with Carol's father. The father has reached out to her. They apparently spent several hours together during which time he volunteered a lot of history that he's never revealed before. This has left Erika very confused.

148

Amelia (Group 5 or the Sunshine Group) comes in for the first time and expresses her wish to join forces in helping with the first fusions. Jennifer's angry core is beginning to show. She has now defined herself as the seeker of revenge. She has a very detailed memory of all the slights and insults done through the years to all of them. She has defined herself simplistically as having a job to do: "get revenge," and once this is done she can kill herself. She also seriously intends to sabotage all efforts at fusion.

Katharine and Maria S., Laura and Erika have all joined forces and inform me that Miranda and especially Jennifer are approaching a "blow-up." I intervene. I extract promises from both that there will be no interference at work, and that they will see me next session, to which only Miranda agrees. I see Miranda first because in all probability the way to reach Jennifer is to go through Miranda anyway. Miranda's turmoil appears to primarily relate around the issue of sexuality and the total absence of sex education. She wants to know about sex, especially as it would apply to a sixteen-year-old. I recommend some reading material, and we spend some time discussing the issue. I see Jennifer who tries to sabotage by putting Carol in her place. I take a stand. She tries to physically leave and I physically restrain her. It appears that she is testing me out. She then begins to spill out her hatred, especially toward the parents.

Marissa (seen for the first time), Karina, and Katie (of Group 5) come in to brief me on Jennifer, who has had the urge to hurt herself and others. Jennifer's years of anger are now bubbling to the surface and I am working hard to re-establish her therapy relationship and re-developing trust. Jennifer's turmoil has prevented Erika from going to work and has resulted in three missed radiation appointments for Katharine. Jennifer is suicidal and fights with me. Her rage is ruling her now. She is basically in crisis, but she does agree to continue to talk with me. The next day Marissa brings Jennifer in again. I advertise that this time I will not stop her if she wants to leave. This appears to have defused her. She gives me a lengthy list of individuals who have hurt them and begins to unfold some of the specific details. At the times when she is about to cry, she gets up to leave (and actually did leave once) but comes back. She never does cry until the last twenty minutes of this two hour session, and then the tears flow. She asks if we can continue and I agree to the next session being hers. I see her again and we continue to defuse the rage. This gets tied into her lost sense of God, and her lack of religious belonging. This becomes a time of healing for her. She now even shows her willingness to go along with fusion.

21

Maria S. comes in for a session in early October 1985 and reports that someone has destroyed her first thirteen chapters and all of her notes. I confront "Miss X" very strongly and she appears but does not give her name. She acknowledges that she wants anonymity.

"Miss X" turns out to be Shana. She introduces herself to me by phone call that evening. She is eighteen years old and still hopes to have a family. She is fearful that the writing will sabotage that chance, believing that if their story were ever identified as Carol's, the entire outside family would disown them. She admits to having destroyed Maria's first thirteen chapters. We now need to defuse the turmoil Maria S. experiences over all of her lost work. She re-starts:

Maria

My name is Maria S. It's October 6, 1985. This is the second time I've attempted to write our autobiography; the first attempt was destroyed by someone who doesn't think that this is necessary; I think she's afraid.

I should start with Katharine St. Claire. She is thirty-eight years old. She is a member of Group 1 and she arrived when Carol was age two. Katharine is very special. She arrived very shortly after Carol's second birthday and has taken much physical and emotional abuse. She arrived when Carol was first badly beaten and she took over at that time. I am really quite at a loss as to how to explain how that happened; all I know from what Dr. Ted has said is that it's a defense mechanism and when one person cannot

151

tolerate that kind of punishment or abuse they unconsciously create another person to take over for them. Now, I don't feel as if I was created by Carol because I feel totally separate, and Katharine does also. We recognize that we are one body, but still we have our own looks, and our own dislikes and our own likes, and we've lived our own lives.

Katharine was one of those children that didn't try to make a fuss about anything. There was a very large hole dug and Carol's grandfather was very angry at her and he started to beat her. Only he wasn't beating Carol anymore, he was beating Kate. And it was rather cold outside, it was raining. He put her in this hole and he left her there. The first time he didn't leave her there very long. I imagine it was because she was so very little. It was so cold. She cried for a few minutes, and then she stopped. I don't know what made her stop, maybe a realization that the tears made him angrier and it was just defiance. But she stopped. And when he eventually came to get her he hauled her out of there, not gently. And she was muddy, and she was dirty. She was so cold. He took her inside and he stripped her, this little child. And he had a tub of water ready, only it wasn't nice, warm water. It was ice cold water. And he put her whole body in it. She came out of that water a hundred times colder than she was. Almost blue, but not one tear. She didn't make a sound the rest of the day.

In the Netherlands, where we grew up, there were a lot of poppies and she loved the color, the bright orange. She'd sit for hours and make daisy chains, and she'd pick buttercups. I remember seeing Kate with a fistful of flowers she'd picked. She'd bring them home and hold them out, and then I'd watch her smile just go away when it was ignored or pitched out. I have such a clear mental image of her, this tiny little thing, dressed in white like we usually were, and a little smile and all these flowers, and being ignored. I think that's where she got the idea that she was never good enough, because maybe in a way those flowers were peace offerings. "I'm giving you something so please love me and please forgive me." Only, there was nothing to forgive because she never did anything wrong. She was the perfect child. It still makes me want to cry.

I'd take her to places that I discovered. We had a secret place. It was close to the school. There were trees and bushes there and flowers, but it was way back in a corner of the field. We'd run away to our little secret place. One time I talked her into staying there for a long time. Kate wasn't able to sit down for, I think, two or three days after that one, she was beaten so badly. I knew it was my fault.

She was raped by her grandfather many, many, many times. He started out just by fondling her. I think that before she was age three he had penetrated her and he had sex with her. She bled.

She can be stubborn. I think she learned that when she was a child. "I'm not going to let them see me cry." So many of them have that attitude. "I'm not going to let them know how much they hurt me." Over and over again. I can remember her having welts and bruises on her body. They'd make her kneel in front of them and her back would be whipped--and not a sound. I could see that look come on her face--I will not cry; I will not show any emotion. A totally defiant look on her face.

No one should have to go through anything like that. She still has nightmares; she doesn't tell Dr. Ted about that. She shakes when there's thunder outside, and she wants to hide.

She dislikes being wet; I'm not saying that she doesn't take her baths or showers, she definitely does that, but she hates the feeling of being wet. Especially her face. I'm sure that's partly from being out in the hole when it was raining and cold outside. Kate was always cold. At home she's most often wearing a sweatsuit and a couple pair of socks and wrapped in a blanket, especially in the wintertime, but sometimes even in the summertime. It's been such an unfair life that she's had, yet she's still willing to see good in people.

That has started to change a bit now. We have some friends that care about us. We have Reverend Jansma. He has become one of Kate's most favorite people. They can talk together. Watching the boat races with Reverend Jansma was one of the most special moments she's ever had. With their lawn chairs and binoculars, they walk down to the lake together, talked together and watched. It was as if they had known each other for years.

Through all of it she's remained what Dr. Ted calls a "classy lady." One of the most special people I've ever known. She's had many disappointments. She has a lot to deal with.

Marissa suggests on October 31, 1985 a second video tape of the last two groups. This can then be shown to the first three groups who are on the first video tape. Dana (Group 5, age thirty-six) introduces herself for the first time. She agrees (with some prodding from Marissa) to give me another perspective on the other groups. She accepts the need for full integration. Jennifer is still doing some struggling but is no longer in crisis and wants to talk further about spiritual issues. Katharine now resents too much time being spent with Groups 4 and 5. Her chemotherapy starts next week. This will replace the radiation treatments. At this point in time, the cancer is not spreading any further. Her chemotherapy is expected to be necessary for some time.

It is November 1985, and Katharine has many questions regarding fusion and eventual integration. Even though she has daily treatments with chemotherapy, she is continuing her psychotherapy. She wants to fuse with Maria S. first. The effects of chemotherapy are showing more with Katharine--much need for sleep, skin drying out, food tastes like metal--she feels very alone.

22

I have an emergency session with Carol who had a bad dream the previous evening and is in crisis. She dreamt that she is about four years old, and some male begins to touch her, to fondle her, then proceeding into a full rape, both vaginally and anally. She then bit his finger. He then gagged her, and forced her to sleep with him. Throughout it all, she could not place his face. We do brief hypnosis and connect the face to the grandfather. There are many tears and there's a beginning acceptance of the reality of these events.

I have a session with Marie, who is now accepting the issue of one body and her own identity as a multiple personality. There is spontaneous fusion with Shana.

We do our second video taping with Groups 4 and 5. This is done so that Groups 1 through 3 can observe them. Katharine and Marissa are now beginning to spontaneously have co-consciousness.

Diana (Group 4) who has observed for some time, comes in on her own for the first time. She is twenty-seven years old and arrived to take over after Jacqueline's gang rape, which occurred when Carol was nineteen. Diana is quite aware of the issue of integration and fears it. We have several sessions to deal with this. Carol is just beginning to deal with the issue of not being able to have a baby.

Dana comes in to give Groups 4 and 5's perspective on the parents. There is much information on their suffering, especially at the hands of his parents who were physically and psychologically abusive to both of them too. Marissa's writings also deal with this. It is clear that there is a large group of personalities who have a much more positive picture of historical

155

events, and who view the parents in a much better light.

December 16, 1985. Lisa (age thirteen) and Margaret (age ten), who regularly appear together when I talk to them, are very fearful at this point of going out of existence. I put my emphasis on helping them to grow up. They are very conscious of their youth. I spend time talking with Carol. There is much spontaneous co-consciousness with Katharine. There is even spontaneous remembering of childhood in Holland, including some pleasant memories. This is quite unusual. We are now planning on the last week in January, 1986 for a major emphasis on fusion for many personalities.

I return to helping Lisa and Margaret grow up. It now turns out that Katharine is very reluctant for this because she then loses her "kids." I spend a good part of a two-hour session with Lisa and Margaret. Margaret gets in my "circus chair" (my swivel desk chair), laughing and having fun spinning around. Lisa takes over and is taken aback. She suddenly recognizes "one body" and goes to the sofa. Margaret finds herself on the sofa. She has a sudden realization of "one body" also. They now have spontaneous co-consciousness with Dana.

Many stop in to wish me a Merry Christmas. Of special note is that Lisa and Margaret spend much of the weekend talking with their "moms," Jennifer and Kate, the result being that both of them volunteer their wish to "grow up."

Katharine is informed that cancer has spread to the other breast. Double mastectomy is recommended, and she will get a second opinion. Much time is spent with Katharine who has been severely regressed for most of the weekend. This has occurred because of "losing her kids," the mastectomy issue, and fears of fusion. I talk with some Group 5 members who report that they, in fact, want full integration and know that it is necessary.

Carol now pushes hard to have her own session. The fighter in her is visible; this is a rare occurrence. We spend our time working on retrieval of memories. Lisa emerges spontaneously, but Carol then resumes control. Carol remembers age thirteen (there's probably some subtle fusion going on with Lisa) and she also remembers running away from her grandfather's house after he forced intercourse on her. Emotionally she acts as a thirteen year old. During the talk-down, she returns to age thirtysix.

Katie (Group 5), advises me to see Katharine because of her expectation that Katharine will sabotage our first age progression with her "kids." I talk with Katharine and she refuses to help, is very angry, is scared, has a migraine headache, and there is much sabotage: "You are not taking my kids away." I talk this through with her and finally Katharine allows me to work with Lisa (age thirteen) and Margaret (age ten). We do an age progression to age thirty-six. It is successful. Both are very uncertain and very aware of "something missing," (years of memory). Katharine advises me on her own that now she wants to help them. These are very emotional and difficult sessions.

I spend further time with Lisa and Margaret to help them to adjust to being grown up. We are now escalating the process and beginning to think in terms of age progression for Alice and the two Callies. We are also talking age progression for Paul and Karl all the way to age thirty-six, while Alice and the two Callies we will progress more slowly.

Dani (this is the one-hundred and nineteenth personality or fragment) spontaneously arrives. No one else knows about her. She had introduced herself over the phone to me two days previously. She has been an overseer and observer for the entire lives of all of the others. She has observed all five years of therapy. She volunteers a list of those personalities who should not be fused first. I keep Carol informed of the events occurring, especially the age progression. I talk with Alice who is looking for Margaret and cannot find her because Margaret has been age progressed to age thirty-six. We decide to celebrate a birthday for Alice, since she's never had a birthday, and we age progress her one year to age eight. I make her a "wind" birthday cake, an imaginary birthday cake made out of air. In an intentionally exaggerated manner: we bake it, put frosting on it, add candles and have a party. Alice can't sit still, she is so excited.

We are going very slow with this age progression because Katharine is so resistant. Katharine emerges spontaneously and asks to be age regressed to age thirty-six, (the age of Carol; Katharine has always been "two years older" than Carol). This presents as a peace offering. Dani informs me at the end of this session that she has one semester to go at a private local college for a degree in journalism! None of the other personalities is aware of this very surprising fact. It turns out that Dani has been secretly attending college for years, but doing so under her own name. I get her to recognize the importance of using Carol's name so all can get the benefits of this degree, and she does change it in the registrar's office.

Dani introduces me to two "friends." All three, including Dani, are unknown to any of the others. They have not newly emerged, but in fact have been around for a long time. These three include Dani, age thirty-six (arrived when Carol was two); Elsa, age thirty-six (arrived when Carol was four); and Sheila, age thirty-six (Elsa's twin, arrived when Carol was age four). This makes a total of 121 personalities and they advise me that this is now the complete total. I also talk with Paul and Karl who are working at breaking the ice with me. They have never had a relationship outside of their own community. Together we plan a practical joke on

Katharine. They also agree to age progress two years next session, from ages thirteen and twelve to ages fifteen and fourteen. Margaret is struggling with her new age. Dani is wanting to help in educating Margaret about her adult body, including information about sex. Lisa is too preoccupied with herself to be of help. Co-consciousness is done between Dani and Margaret.

I talk with Paul and Karl and we do our practical joke on Katharine: (they put a fake "antenna" on my head and call her out unaware). We attempt to age progress them two years but they resist. Dani informs me that they are afraid of losing their new "big brother," in me. I talk with the boys about this and we decide to proceed in the next session. We do so, with each of them being progressed four years to ages seventeen and sixteen. With Alice we do an age progression of two years, from age eight to age ten. With the two Callies we do age progression from ages four and five to ages six and seven. Both are now at an age where they can understand and speak some English and it will be much easier to communicate with them.

Dani and Katharine now have co-consciousness for the first time. Katharine lets me know she has been invited by her pastor to speak about multiple personality in a high school. Margaret is now writing, while she used to only print. She wants to show me how she's progressing. We keep Carol updated. Paul and Karl are progressed further from ages sixteen and seventeen to age twenty-one for both. It is hoped at this point that we can now begin to change their male identities. We do an age progression with Alice from age ten to age fifteen, so she can be the same age as Simone. The two Callies we equal in age so that both are seven years old and then age progress both to age eight. They can now fully speak English. Francesca now wants to go from age thirty to age thirty-six. There appears to be a bandwagon forming.

Katharine and Dani are very actively involved in planning strategy and keeping me updated. Both Callies, who are now age eight, are in session and we use hypnosis and do our first fusing of them together. We add to that an age progression to age fifteen. This frightens many of the others, especially Christina and Alice. I work with Simone and attempt to restructure her perceptual distortion of her body with hypnosis (long black hair and very obese) to an approximation of Katharine's appearance. Following this we do an age progression from age fifteen to age thirty-six. All goes quite well. I see Paul and Karl and begin to lay groundwork for accepting their femaleness. Katharine is doing much grieving over her "lost children."

Dani, Elsa, and Sheila take this time to introduce me to the two very protected "ideal" Carols who are: four-year-old Becky (Group 5) and seventeen-year-old Katie, who we call Dani's Katie. This is a big risk, especially for Dani, who has mothered and protected them all these years. They have never experienced a negative feeling. They are her "children."

I also see Paul and Karl (both are twenty-one now). I foster contact with Margaret to encourage further talking. Margaret and Lisa previously were their only real friends when they were much younger, and now that Margaret has been age progressed, it is hoped this will foster some good mutual modeling for the boys' acceptance of being adults.

I do co-consciousness between Amy and Elizabeth, (Group 5 who was the "voice" of Priscilla Pig). Maria S. has finished much of her autobiography. We work out an arrangement whereby she makes her work available to other personalities to read in case any of them are offended by anything in her reporting. Maria S. is feeling very awkward about her work and in particular is concerned about giving away confidences. Katharine has agreed to do the high school presentation on the subject of multiple personality, as long as there is agreement to keep her true identity hidden.

23

Lisa and Laura are quite distraught because of a phone call received from Carol's mother. For some reason her mother tells her that she, in fact, is not her daughter. Marie's anger and guilt over the abortion issue is now snowballing. She was contemplating suicide this weekend (by driving a car into an abutment), but Dani took over the driving once she became aware that this was Marie's plan.

It comes again--
the sound piercing,
shattering the silence,
filling the night.

There--on the fence,
Do you see it?
a rabbit--defenseless,
so helpless,
struggling to be free,
so desperate to flee.

Barbs tear its' flesh--
again the scream,
eyes, pain-filled eyes
look straight at me.
A spark of hope,

a silent plea
ever, ever so quietly
touches me.

Slowly I turn,
slowly walk away,
and the night is
silent,
quiet once more.

Marie
January 31, 1986

We have an emotional session regarding the abortion, being a bad person, guilt, anger at the doctor who did the procedure and parental anger. There is much pounding on the sofa and much weeping. Katharine is used at the end of the session as a support. It does appear that Marie's suicidal thinking is now diminishing. I see them again the next day.

Katharine, and especially Marie, need an extra session to process the previous day's dramatic session. Katharine now has a premonition of fusion that's going to happen spontaneously between her and Marie. Marie brings in a poem that was written yesterday. The sense of crisis appears to be over.

Past/Present

Memories, pain--
together they've torn,
ripped through my very soul,
making my past my present,
robbing me of life,
of living amongst the living.

And I weep for those years,
those seventeen lost years,
and the pain becomes a knife,
stabbing, twisting--
living within me,
deep within my soul.

But my mind, once so locked,
so imprisoned, now screams,
demanding it's freedom,
cries out for life,
searches for release,
release from the memories,
relief from the pain.

Marie
February 13, 1986

Karina volunteers to be introduced to Francesca. She brings in a newly started painting of Francesca's and with some prodding Francesca agrees to do co-consciousness. I see Shana, who destroyed the first set of

162

writings, and she updates me on her wish to read Maria's new efforts. Marie brings in the poem, "Solitude." Maria's prose and Marie's poetry are very moving. The last three poems so vividly portray Marie's clinical progression from the preoccupation with her abortion, to confrontation, and finally to acceptance. Katharine's sense of a spontaneous fusion with Marie is growing stronger, as we all sense that Marie's clinical work is finished. Because her reason to exist as a separate personality, that is to handle the pain of the abortion no longer exists, fusion should occur.

Solitude

Softly it envelopes me,
this gentle feeling of peace,
it whispers quietly,
ever so quietly.
bringing fantasies of
a mother's gentle voice,
freeing me from shadows,
the darkness in my mind,
and I welcome it,
embrace its comforting warmth,
and solitude becomes mine.

Marie
February 14, 1986

Jennifer is now re-accepting her role as "Mom" for Miranda who is struggling, primarily because of all the changes she is seeing in all of the Group 1 personalities. Karina brings in her unfinished painting. Marissa, who sells them for her, wants to make sure that I know that my opinion is very important to Karina. Her painting is close to being finished and there is fascinating symbolism in regard to the events happening with all of them. It appears that Katharine and Marie have now fused on their own. Katharine does complain of some difficulty with concentration--some confusion: "I don't know who I am anymore."

I see Maria A. of Group 3 for the first time. She has been observing for several years. "I'm the one who is responsible for everyone's weight problem." She is a compulsive eater, but there apparently is no purging or laxative use. She wants her own private, separate therapy: "Please don't let any of the others know," she says. I am now finally having some productive talks with Jennifer instead of the fighting and sabotage that has been going on for some time. She does tell me that she exists as the one to cope with rejection and as the one to get revenge. She is now sharing more in session than she has ever done before.

Katharine now reports positive results regarding her chemotherapy for the first time. Dani, Katharine, Jennifer, and Erika report spontaneous co-consciousness among themselves. There is also co-consciousness with Maggie, who is very shy and who is taking an art class. Jennifer advises me to see Carol who is not doing well. I see Carol and she presents as having regressed to age seven. I age progress her with hypnosis to her true age and we work on her retrieval of memories. She is progressing very slowly, but is remembering more.

I spend a session with Karina. Someone brought her painting to the parents. Her father ignored it and Mom said, "You're not bright enough to have thought of that on your own." Karina now refuses to finish the painting. (Note: Karina has always insulated herself from all negative feelings.)

Maria A. from Group 3 continues to have a one-hour session per week just for herself. She brings in a journal. She is troubled about weight and she knows the two main variables behind it: nurturing and fear of men. For the first time, she cries in front of someone else.

Aleisha presents herself for the first time. She has never before met any outsider. She is worried about Katharine and wants to urge me to spend time with her. Katharine is struggling physically. There is much nausea. Aleisha has now apparently taken on a role as Katharine's "secret ally" and "hidden supporter." She is very adamant that no one will get better unless Katharine gets better (and she is absolutely right). Katharine informs me that she has now been told that she has a degenerative bone disease.

Aleisha continues to update me regarding Katharine and makes her plea for therapy time for Katharine. There is a growing and common consensus that Katharine is central to everyone's mental health. I relay to her this message regarding her need for therapy, and its impact on everyone. Although she initially resents the responsibility inherent in this, she does use her time in session. She reveals that the most painful hurt of all is the abortion and sterilization. She leaves the room at this point saying, "So I wouldn't fall apart." This she later acknowledges to me. She then comes back and proceeds to open up regarding sexual abuse and the abortion. Although there are some clear limits on letting go emotionally, it is also clear she is no longer fighting the need to proceed in therapy. It appears that many, if not most, of the others will vicariously benefit by Katharine working through the groups' traumas.

Dani reports to me that Katharine has had almost all of the time outside my office. She is experiencing her anger much more and even took a stand with her father and with her sister-in-law.

It is April 4, 1986. Katharine is asking for extra time. She says she

164

realizes that something important has begun and she wants to pursue it. I arrange a two-hour session. Katharine spends her time talking about some of the memories, and feelings begin to spill out. As we proceed, a sense of change is showing, and "others" begin to be included. At the end of the session she has fused with the following, the specifics being later provided by Dani: Marie (her feelings), Moiria (her anger), Carol (her vulnerability), Margaret (the ability to accept being an adult), Maria S. (her ability with Callie to tell what has happened to them). What is showing in this session is that these other previously separate personalities have now begun fusing in some ways with Katharine. What is significant is that this process is now in Katharine's awareness. This is a very important session. I suggest that she write down her experiences:

This is Katharine St. Clair and it's April 4, 1986. The last few weeks have been really confusing for me. It's been a time when my childhood has come back to me very vividly. The abortion and the sterilization, and my inability to have children as a result of that has really been very foremost on my mind. I've been very, very restless. I've had most of the time, as if I couldn't stop myself. It made me uneasy, as if something was going to happen, only I didn't know what. Some of the others made sure that I got to Dr. Ted and talked with him. That was really difficult because I don't talk about myself very often. It's been much simpler to talk to him about all the others. But then I started to have this dream--in fact, it wasn't even always a dream because it didn't matter what I was doing. It would appear, all of a sudden, in a magazine, or on a page of a book I was reading. It was a face. A face that was put together by all these little pieces. Kind of like a puzzle. And in the middle of the face there was a piece missing. And somehow, that face would reach out to me, as if it wanted me to touch it. Only it scared me. I'd run away from it. I'd turn my back on it. Then when I'd look back at it, it had fallen into all these little pieces. Made me really uneasy. In fact, it made me even more restless. It was as if it was trying to tell me something and I didn't know what.

Yesterday, which was Thursday, April 3rd, I started to talk to Dr. Ted about my childhood. I felt kind of as if I'd been set up,' because I'm sure some of the others made sure I got to Dr. Ted--to the office. And I knew he wasn't going to let me get away without talking. And part of me really wanted to talk, so I told him about being locked up in a trunk, about having to kneel in it and having to put my head on the bottom of that trunk. I'd have to stay in that position for hours with that trunk locked. And I told him about Marie and how Marie is me too. I recognize her as part of me--or accept it. Maybe that's a better way of putting it. I'd always said that the parents made Marie have the abortion, they had Marie sterilized. They also did that to me, because we're a part of each other.

All of a sudden I had an urgent need to talk to Dr. Ted, to tell him about these things that were inside of me just waiting to come out. And I cried. I felt like this little kid that just needed to be held. I needed to be safe, and I knew I was in his office. When I left--I left because it was time to go--my session was up--it didn't seem right to leave because I wasn't finished. And at the same time, I was kind of aware of some other people in the background, only I wasn't quite sure who. I went through a lot of different feelings. I was angry. And as the night went on I just got more and more restless. And I haven't been able to sleep. So I called Dr. Ted and I asked him if he had any time available today, and he did, so I drove up to the Muskegon office. And I started to talk some more. Only I'd changed. And I was aware of the change. And then I knew that they were there. Marie was there, and Margaret and Maria S. and Lisa and Moiria, Callie, Alice, even Carol was there. And I was telling Dr. Ted about some of these things, about what the grandfather did, what I felt, telling him how he would put his penis inside of me, make me hurt. Strangle my throat too, so I couldn't yell. And I could hear my voice and it was different; sometimes it was my own voice and sometimes it was different again. And as I was talking to him, the others were gone. But they weren't because I could feel them. Only they were me. And Dr. Ted knew it too, because he asked me who was there and I could tell him. I could tell him about the things that had happened, because I was finally accepting them as being myself, and accepting the things that had happened to them as having happened to me. I wasn't sure about that at first, but Dani explained it to me. I'm a little mixed up because I don't understand why Carol became a part of me, because I thought I was to become a part of her. And I still find it a little confusing. (*Note: Katharine is verifying this totally unexpected and major development. We had always assumed that when integration was complete, the birth person, Carol, would be the remaining individual, that all the other personalities would fuse with her. It now appears that Katharine will become the final, integrated person.*) I wanted to talk to Margaret today, only she wasn't there, but I can feel her. She's not little anymore, she's grown up. I can't see them anymore. That really kind of bothers me, but at the same time that's all right, because I can still feel them. I don't know if that makes any sense at all. Dr. Ted told me that they've become part of me, that they've fused with me.

Even when I was growing up, you know, I'd feel terribly lonely, as if there's no one else. I used to feel that way when I was locked in that trunk or when I was in the hole. And what I would think about was that someday I was going to be a mother. I was going to have my very own baby. I know it isn't going to happen now. I always thought of myself as being a rather bad person, because I could never do anything to please anyone. They took my baby away. But I'm not a bad person. I didn't want them to do those things to me, and I didn't ask for it, and I didn't do anything so terrible to deserve it either. I was just a little kid. Later on, when I was older, I still didn't do anything bad. They didn't have to sterilize me. They didn't have to whip me, and they didn't have to put me in a trunk. They didn't have to do any of that. They were bad, not me. They're still bad. And I told Dr. Ted that I don't need them, I don't need them at all. They don't love me. All those things that they did to me; they said they did because they cared for me and they wanted me to turn out right. But that was all a lie. If you love children you don't beat them and you don't lock them up. And when I was little I knew I was never going to do that to my own baby.

I wish I could talk to Dr. Ted some more right now because I feel the same way I did before--I'm kind of lonely. I need a hug really bad. And I know somebody else is there, only I don't know who. Today, even when I was just me and I was talking to him, I wanted to call him "Our Dr. Ted," but the little kids do that.They call him "Our Dr. Ted."

I don't feel real safe right now. I don't know why. I'm terribly tired, only I can't go to sleep. Every time I go to sleep, I dream.

I don't know how to explain everything that happened to me today. I know I'm still me, but I still feel different. I kept trying to tell myself that I didn't feel quite right because of my cold and everything, but that didn't really explain why I was so restless. Nothing seemed right the last few weeks. I wasn't very patient and I normally am. I still feel as if something is about to happen. I was very glad I went to talk to him today. I almost didn't call him to ask if I could have extra time, but then I decided to, and I'm glad I did. For a little while, I wanted to leave, but I didn't. I wish getting better wasn't so confusing, because I am getting better.

I sure do wish I could talk to him right now. Monday seems a long ways off. It's just the strangest feeling. But I don't mind that I told him what I did at all. That probably means I'm getting better

167

too.

Sometimes, privately, I've called all the bad thoughts and feelings that I've had, and all the bad memories, I've called them my "feelings." Well, I want to get rid of them. I want to get rid of my feelings and get on with my life. And I want to learn how to accept just me; because if I can accept me, then I can like myself better. I do like myself a little bit better already, after I talking with Dr. Ted. And I feel better now that I've talked to him a few times about not having any children of my own. A lot of things I want to tell him. But it sure is strange with some of the others, with those others being gone, but I can still feel them. They're just not there where I can see and talk to them. I think that's all I want to say. I have to think about all of this some more.

I cried in Dr. Ted's office today, too. And I don't feel bad about that, either. And I have a poem on my mind that I want to work on. Usually Marie writes poems. I don't write poems. I don't ever write poems, but there's one in my head now. I would suppose that's because she's inside of me, she's part of me. I care about Marie a lot.

Dr. Ted seems to be very happy with all that's happened today. I trust him a lot.

It's really different how I could see them standing there and all of a sudden they weren't there anymore, but I could still feel them.

That's all I really want to say.

Dani continues to provide excellent overseeing and is a tremendous source of information to me. The fusions with Katharine have remained solid, although it has been a difficult weekend. There has been almost no sleep, almost no eating, and three phone calls to me.

In our next session, since Katharine is beginning to regress somewhat, I do an age progression with hypnosis at Dani's urging. Elizabeth Austin spontaneously emerges: Katharine's regression appears to be related to her lack of full acceptance of the abortion, which the formerly "pregnant" Elizabeth has finally accepted. We then proceed with a fusion between Elizabeth and Katharine, which occurs very quickly and easily once I say to Katharine: "There is no baby."

Katharine and I now do a two hour presentation together at a local high school on the subject of multiple personality, which was promised to Pastor Aaron who teaches there. Although the audience is large (about 300 students and faculty) Katharine and the others cope with this public situation extremely well. We do have a session afterwards and process the experience. I also talk with Jennifer who now acknowledges that her previous wish to "spontaneously" fuse with Katharine was in fact a "cop

168

out" on her part. It was really her way of avoiding her own therapy.She knows she has to deal with years of anger.

Katharine now reports that Carol's father has had serious heart attack. By-pass surgery is done and he is in critical condition. However, he does survive.

Katharine is still rather uncomfortable with being the central figure instead of Carol. She has a sense of an impending fusion. Laura spontaneously shows up to say that she is ready to fuse. I nudge the process with hypnosis. Katharine then becomes quite angry "...because there's a whole bunch more who want to fuse." I see her again later in the day for an emergency session. We spend our time processing her guilt over being the central person, and her anger over the other personalities wanting to fuse without asking her. The fusion of this morning with Laura appears to be holding well. We will hold off on any further fusions for the time being.

For the first time in a long time Jennifer really begins talking. She talks about her role, about her hate list, which especially includes the grandparents and father. She has more of a "soft catharsis" in session; and as she proceeds, it becomes quite clear that she has done a lot of work on her own. There is healing that has occurred. At the end of the session she says goodbye and spontaneously fuses with Katharine.

Jennifer has left a letter to "her daughter," Miranda, who is having much difficulty with Jennifer's departure: "She left me and went away." Miranda is reluctant to talk with me and walks out. Dani brings her back and there are many tears. At this point, Katharine now has experienced eleven fusions and it is all happening very fast for her. We will need to slow down.

According to Dani's updates, Katharine is doing some struggling. She has spent her entire life helping others to deal with their feelings and is still not used to knowing how to deal with her own feelings. She is quite scared, depressed, and is quite angry. Margaret has unfused briefly "...because Katharine has no one to talk with." Although it is difficult for us "outsiders" to comprehend, it is also possible for the fusion of a specific personality to be temporarily reversed. Although this is usually only possible shortly after the original fusion, when the psychological system hasn't had time to fully adjust itself, it is often an indication of "unfinished business" in which case this de-fusion might last for sometime.

Elsa and Sheila (Group 6) are introduced to Katharine. They have much laughter together and are clearly comfortable. I expect that they will fuse with Katharine shortly. Dani is struggling because of her awareness of how much she will miss Elsa and Sheila once they fuse.

Jenny and Amanda report in and show that they have been observing and that both now know that they all share the same body. Elsa and Sheila are somewhat hyperactive, but report that they are ready to fuse. They have said good-bye to Dani. We perform the fusion in session, which goes smoothly. Katharine immediately shows Elsa and Sheila's sense of humor once the fusion takes place. Dani has some grieving to do over their loss.

It is reported that Katharine had her picture taken by a friend and in the process, as several pictures were being taken, a number of the others "secretly posed," since none of them have ever had their pictures taken before. When the pictures are developed and they can see that all the pictures are of Katharine, it throws a number of them into turmoil.

Elsa has de-fused for one day to support Dani who has been grieving their departure. Katharine suggests that I talk with Dani who is struggling.

We have an open meeting with the entire group of personalities in attendance. There are a number of questions, especially regarding the pictures taken several days previously with "...all of us looking like Katharine." I take pictures of all personalities individually.

Dani reports that Katharine, for three weekends in a row,has washed all the walls in her apartment, and a number of the others are beginning to revolt against this. Erika is struggling with power issues. She is becoming troubled by her need to always be in power, and is apparently beginning to experiment with a more submissive style of operating. She is questioning her historical need to be so dominant.

Maria A. tries laxatives (twenty-four of them) as a means of weight control. She winds up vomiting. She has always been uncomfortable with her physical self and is also growing in her discomfort with her voice. She wants a more mature voice. She expresses the need to fuse, but only when "I'm all grown up." She needs to deal with the issue of incest by grandfather.

I show all of the personalities their pictures. There is much reaction. I see Elena who has regressed to about age seven. She could be suicidal at this time.

Katharine has regressed severely over the weekend. Her father had a second coronary and is in intensive care. This, combined with all of the others seeing their own picture, has produced much turmoil. I again see Elena, who Dani has been prodding to come in. Elena is from Group 3 and is quite aggressive, a perfectionist, forty-two years old. She apparently had to regress to age seven in order to come in for a session.

It is reported that Katharine is still washing walls every Saturday. There are daily struggles with whether to rearrange her bookshelves. I question her about this. For the first time Katharine shows anger instinctively.

Dani has become very central to the therapy process now, and in fact is getting rather saturated with all the responsibility that she is accepting, particularly with Katharine's frequent regressions now. I spend some therapy time with Dani.

24

Maria S. is continuing to write her community's autobiography. She is providing powerful and very helpful information about their very early years; and an especially fascinating "insiders" view of the inner-workings of the formation of multiple personality dynamics. She writes:

> I would often take Kate to our secret place, especially in the afternoon. She often would be very, very tired and I suspect that's because of the physical and emotional abuse she had gone through. It just seemed strange that a small child, aged four or five, should be that tired. Anyway, we'd go there and I highly doubt that anyone could have found us there, even have thought to look for us there. For two little girls, it was quite a hike from home. And we'd pick flowers, and pretty soon there'd be a little smile on her face, and I'd know that she was feeling happier. Pretty soon she'd lay down on the grass. I always made sure that we had a jacket or something with us because I needed to cover her up, I had to take care of her. And she'd sleep. Sometimes while she was sleeping she'd have bad dreams, and she'd start to cry, and then I'd have to wake her up and hug her. And I guess even at that age I felt more like a mother. I felt very protective of her. The problem was we would always have to go home. And so when we'd been gone for a couple of hours, I'd have to wake her up. Sometimes I think that we weren't even missed, and other times we definitely were because punishment would be waiting for us when we got back. Sometimes I took that punishment and

sometimes Kate did. It depended on which one of us happened to walk in the door first. When she was the one being punished, I felt so awful, because it would usually be my idea. As much as she loved our secret place, it wasn't her idea normally to go there. But we never told anyone about it. We never shared that very special place, and I sometimes wonder if it's still there. I'd like to go back. I'd like to see it and I'd like to lay in the grass and watch the clouds go by, just like we used to. Somehow, things were better there.

There was a difference between Kate and myself, and yet we were very close. But the more abuse she took, the further she started to draw away from me. And then, by the time we were six, I don't think I existed for her anymore. I would try to reach her and I would try to make her smile, but I couldn't break through whatever wall she had put up. I missed her. My little world, which sometimes wasn't that pleasant, kind of crumbled for awhile. I've missed her all these years because it became tougher without her. It became more difficult for me to trust, and to love. I still was very protective of the rest of them, but I couldn't see the outside world anymore. I couldn't see beauty anymore--flowers didn't mean the same thing, music didn't mean the same thing, art didn't mean the same thing. I had lost part of me. We had been so intertwined that when she pulled away I felt as if I had lost the other half of me.

I thought it would be easy to describe Kate, to talk about what she's gone through in her life, because I want so much for people to understand. But I find myself wanting to shut her off, wanting to shut off all the bad parts. I think I realized this ever since I was little. I feel guilty about the things that happened to Kate, and the others.

I'm playing her favorite music right now, Chopin. Her favorite album. And even though Kate is back, and we can talk to each other again and we can laugh, there's still a reserve. There's still a wall, and I try to break through that bit of reserve that she has left and she won't let me, I wonder if she's angry with me for not having taken better care of her. But I couldn't because those people were so nasty. Sometimes I would have run away, and I know I shouldn't have left her behind, but I did. I'd run away and I wouldn't come back for hours. When I came back she'd have these big bruises on her and I knew that they'd beat her again. Sometimes it was the parents and sometimes it was the grandparents, depending on who was around.

For us life became more and more complicated. Kate wasn't

part of me anymore, I couldn't help her at all because she wouldn't let me in anymore. One day when we were at home, Kate had been sleeping and she woke up. I tried to reach out for her, and all of a sudden, I felt myself knock over a glass of water. And just like that, father was there and he just whipped the living daylights out of Kate because I knocked the glass of water over. I guess Kate left, and suddenly another little girl was there. Her name was Callie, and when she wasn't there anymore, then there was another new person. I know what her name is now, but I didn't at that time. Her name is Lisle. I wasn't too excited about this, and I couldn't figure out where she came from. I could see her. I don't think Kate could. That seems pretty impossible, how you can't see someone, but I really don't think that Kate knew that she was there. I didn't say anything to this new girl, not for the longest time because I had no idea what she wanted, who she belonged to or where she came from. She was just there. It just kind of blew my mind. What a strange expression to use: "She blew my mind." That's a very American expression.

Well, this little girl didn't go away, she just hung around. And I'd kind of spy on her, I'd listen. But it seemed as if she didn't get many of the whippings, or much of anything. She seemed nice enough. I think she was kind of a tough cookie on the inside. She didn't take anything from anyone, not even then. She had to have been about the same age as Kate and I were. But I didn't let her get to Kate. She got there when we were all about two years old, but even when we were five and six years old, I didn't let her get close to us, because I wasn't too sure of her. She was fast. She kind of made me laugh because the parents and grandparents would have a dickens of a time catching up with her. If she did something wrong, or whatever they considered wrong, and they were going to give her a spanking--you can't really call them spankings, they were more like beatings--she would run. She was faster than I was, and I was pretty fast. She never seemed to learn that if she ran and hid, she'd have to come out sooner or later and they would get her then. What I didn't appreciate were the times she would do something, anything--not have her shoes tied, or something like that--and out would come the belt, and she would disappear and I'd get it. Now that used to make me mad. She didn't get spankings, she didn't get beatings, she didn't get locked in the trunk like Kate did. She never got put in the hole. The hole became kind of an every-other-day occurrence for Kate. The trunk was good for the weekend. It's kind of nutty--every Friday or Saturday evening being locked in that trunk. It seemed as if the family got together; what one had the other had, you know, kind of to keep us in line at all times. I don't think Kate really remembers, she just thinks of the trunk as being at the grandparents' house, and it wasn't. It wasn't just in the United States, it was in the Netherlands, too. Parents and grandparents had a trunk. I think she kind of blocked all that out.

You never could make any sense out of the mother. I think that to do the things she did to us something must have snapped inside her head. We were kept tied to a chair. We weren't your normal two-year-old. We couldn't run around. We couldn't play with toys--that was too messy. And we couldn't draw. I wish that I couldn't remember being a two-year-old, but I can. So can Kate, for the most part.

This other little girl, Lisle, she always seemed to be in her own little world. Nothing touched her. It didn't matter what happened around the house, she always seemed to be fairly happy. The few times that she did get a licking, she grinned. She actually smiled, and I'm sure it wasn't because it felt good. It think it was because she was not going to give them the satisfaction of crying. And I think, somehow, at a very early age, like ages three and four, we knew what would get to them and what wouldn't. It would get to them if we smiled and if we didn't cry. If we cried--which we just didn't do--then they'd just laugh that off and whip us a little harder anyway. And it would make them more angry when we didn't give in.

So for awhile, there were three of us. And like I said, Kate never really saw her, but I did. She didn't look like Kate and me. She was spunky. I'm not saying I wasn't spunky, I was. But she was just a little bit faster than I was. I kind of admired her for that. She could color and draw really well. She was always a bundle of energy, Lisle was. Couldn't sit still. And when she couldn't sit still, one of the parents would reach out to slap her and always missed her and got me or Kate instead. Don't know why. Now I understand it's because we're all one body, but it sure didn't seem like it then. I knew that somehow, something was different. This was as I was getting a little bit older. Didn't really understand what it was.

I remember one time when Lisle was being chased by the father, and she was running away from him--and she knew it wasn't a game, he was mad--she actually giggled. That's when I really started to realize we were different. What she did was she

174

slid underneath the dining room table, and he had to go underneath the table and drag her out; only he didn't drag her out, he
dragged me out. And that's when I really started to realize there
was something different about us. Something really strange, I
think. And Kate was just totally unaware of all this. And later on,
after I'd had my beating, and I'd had to go to bed and I hadn't had
my supper and they wouldn't let me have a drink of water, I saw
her again. And she was sitting in the corner of the dining room and
no one noticed her. But she was sitting over there and she was
sticking out her tongue at them. And she kept on doing it and I
wanted to say to her, "You're going to get caught," only I couldn't.
She didn't hear me. She was lucky she didn't get caught. I don't
know how they didn't notice her. She must have sat there about
an hour. I had to laugh a bit and I didn't want them to hear me, so
I put my face in my pillow--and all the time she's sitting there in
the corner making faces at them and sticking out her tongue and
all this kind of stuff, and she didn't get caught. Now, if I'd done
that, I'd have gotten caught. I still can't figure out how the father
managed to pull me out from under the table when Lisle was the
one who scooted under there. Of course, now that I'm older I
understand it's because we're all one body. But she could have at
least taken her own beating. At least, that's what I thought then.
I guess I understand more now and so I can accept that a little bit
better, but for awhile she really made me mad.

Sometimes I was glad that Kate couldn't see her or be aware of
her because she would have upset Kate. I talk to her now and she's
really neat, I really like Lisle. She's still the same way, she still
won't take anything from anyone. She's not going to let anyone
walk all over her. I guess that's the big difference between Kate
and Lisle.

Life, our life was so strange then. So many things didn't make
sense then and they don't make sense now. Kate and I, for the
short time that we spent together in the Netherlands, we were very
special friends. And often it was as if we were just a part of each
other--I'd know what she was thinking and she knew what I was
thinking. Sometimes, too, when I was sad she'd do something to
make me feel better; she'd come up and just hug me. I did the same
thing to her. And I wish I could draw a picture of the way we were.
She's always wanting to make everything better. She still has that
same quality. And I wanted to get even sometimes. I wanted to
run away a lot. And I wanted to stay gone. And sometimes I
thought I really could. And in that way I think Kate was more

realistic than I was. She knew that we had to come back home.

As we got older, Kate and Lisle and I kind of went our separate ways. I eventually did make contact with Lisle and I found her, as I said before, to be very nice. She was happy, and to me it always seemed as if nothing ever bothered her. She just took everything as it came, and she never changed. She had a gentleness about her too, but she saw the parents in a different light than what I did. She was very perceptive; she still is, and maybe understood more, or was willing to try to understand more about the parents. She is willing to help all the others with their problems, the ones in her group. She's very intelligent, I think, even though she still speaks with an accent. She's an excellent writer and we still have things that she wrote when she was little. She reminds me somewhat in her writings of Marie. She doesn't have to write six or seven verses to a poem to say what she has to say, she can do it very simply and be very effective. She's very sensitive and she's still very lively and sometimes that sensitivity doesn't always come through. Sometimes I wonder about her energy, if she doesn't have a bit too much. And then sometimes I wonder what is inside of her. She never spoke of things she saw going on, things that happened to us, and I wonder what's locked up inside of her.

For me, it's guilt. I think that's my biggest problem, and I have, until now, not been willing to show that, not been willing to let anyone else in on that. Guilt can be a terrible thing, and I think it has been eating away at me ever since I was little, because I couldn't do for the others what I wanted to do. And again, especially for Kate. Not that the others weren't dear to me, they were--and still are today. I hope that someday Kate and I can be as close as we once were.

25

It is now mid-June, 1986 and Elsa appears in session. She has temporarily de-fused because of worries regarding Dani. (The reader is reminded that in the healing process of a multiple personality, it often happens that a given personality's fusion might initially be only a "dry run," which can be reversed for a variety of reasons. Permanent fusion might not occur until a second, third, or maybe even fourth attempt.) She views Dani's role as very central one. She verifies that many of the others are getting their therapy vicariously and will fuse in spontaneously.

Later in this session Elena shows up. She is very distrusting and tends to be a manipulative personality. She feels she has always been used. She is able to share this and is able to ask me directly if I cared, and if I was going to use them. She has never before allowed anyone else in on her thoughts or feelings. I do ask her if she will continue to speak with me and she agrees to.

In our next session Alexis Germaine (Group 4, age twenty-eight) introduces herself for the first time. She feels she is close to fusing. She shares that she has been observing therapy for years, and that she wanted to meet me first and ask for some time to talk.

Dani, with prodding, continues to deal with her anger in subsequent sessions. Elsa especially encourages me to help Dani deal with her anger at God. Therapy time for awhile is rather varied, much of it focusing in on Dani. Mariana White (Group 3, age fifteen, very immature) comes in for the first time and tells me of her wish to fuse. I check with Kate who prefers not to do that, saying she is not ready. Kate has been struggling with her arthritis quite a bit lately.

Dani is now finally initiating her own talking. She does acknowledge

177

chronic problems with men, with sex, especially with her view of her own body. These are good sessions for her. Many tears are shed.

Elsa now shares that she is "...going to let the others deal with their own problems." My interpretation is she will blend into the background and eventually fully re-fuse when she feels the timing is right. Dani is now resuming her role as the overseer and interpreter, having dealt with much of her anger. I have spent a fair amount of therapy time with Dani, dealing with her memories of growing up traumas and with religious issues.

I again see Maria A.; she is finally beginning to open up. Parents and grandparents gave her ridiculed her about her weight and her intelligence. She has always viewed herself as dumb and ugly. I am encouraging her to bring in either writings or paintings as a means of reaching her and a means of building her self-image. She eventually brings in a powerful story symbolizing revenge on her grandmother:

> The old woman whimpered softly as she struggled to sit up. Why wasn't anyone there to help her? Everyone knew she couldn't sit up without some assistance. She fell back on the pillows, crying in frustration, and pulled the blankets closer around her. It was becoming darker in her room, cold too. She could hear the rain beating on the roof and the first rumble of thunder off in the distance. She shivered as she lay there, eyes closed, frightened and alone. The thunder was closer now, a loud crashing sound that seemed to shake the bed. She wished some-one would come and turn on the light. It would be very dark soon and she hated the darkness. Her eyes flew open with the next crash of thunder and she moaned out loud. Not again, they couldn't be back again. She quickly closed her eyes again. Maybe they wouldn't be there if she kept her eyes closed, but she knew they were still there. She could almost feel them staring at her. She opened her eyes slowly and saw that they were all still there, forming a circle around her bed, all those eyes looking at her. She moaned again as she looked at them, frightened by their silence. So many faces, most of them children, a few of them older, more adult, all of them expressionless except for their eyes--huge eyes, accusing eyes, eyes full of hate. As she lay there, too frightened to move, they all quietly moved back, all of them except one, and the old lady cried out in relief as she recognized the one that was left. The child moved across the room and turned on the light. She then came back and stood by the bed, unsmiling, still expression-less. The old lady struggled to sit up. Why wasn't the child helping her? Couldn't she see that the old lady needed help? And why was

she moving her pitcher of water across the room? The old woman was thirsty and she needed a drink of water. "Please give me some water," she said, but instead the child started taking the blankets off the bed. Why was she doing that? The old woman was so cold. She needed her blankets. "Please don't," the old lady said. She started to cry, but the child didn't care. Instead, she folded the blankets and put them in a corner, across the room where the old lady couldn't reach them. She felt the child's' hands pull at her nightgown, tugging at it until it came off and the old lady was lying naked in the cold. "Stop it!" She screamed the words at the child. "Stop it! I'm your grandmother! Don't you know me?" The little girl gave a small smile as she heard those words. "No," the child said. "No, you are not my grandmother. In fact, I don't know you at all." With that, she turned, switched off the light and left the room. The old lady screamed as she heard the door being locked from the other side, screamed again when she saw that all the other faces were still there, faces that were smiling now, no longer expressionless. The old lady lay there, cold, hearing the thunder, knowing that the faces would never leave, realizing that no one would come to help her--and she was afraid.

Maria A.
July 13, 1986

✳

Following a two-week vacation, I get an update from many of them.Interestingly, a new personality, Elizabeth S., introduces herself to me and gives her own personal data to me. It would appear that Elizabeth S. is going to serve as a key transition personality, helping the therapy process by revealing "inside information" and helping non-fused personalities to adjust to the "loss" of their friends.

Dani informs me that Elsa has fully fused again on her own, while Alexis also has fused on her own during my vacation. Paul and Karl are beginning to accept the concept of one body and are thinking of having individual talks with me.

Elizabeth S. continues to inform me of some of the others' reactions and private thoughts. As it so happens, she has functioned as the silent protector of Paul and Karl over the years. She reports that the two of them have purchased a train ticket for Sunday to go to Chicago: "They're looking for a place to belong." I make a deal: if they will stay, I will have specific session time with them.

Becky arrives to be assured that I have in fact returned from vacation. She wants to have some more birthdays (she knows about my "wind birthday cakes," and that age progressions have occurred with some of the others. She asks that Dani (her "mom") be involved.

Elizabeth S. fills me in on what is happening with Paul and Karl. Paul is extremely angry (actually he's really scared and lonely, but is using anger as a cover), because he is afraid I will break up his friendship with Karl. I have a two-hour session with Paul. With prodding, even after walking out of the room once, he does eventually open up with me. He wants to go to Chicago and is angry with Karl for not having gone.

Maria A. has lost four pounds at Weight Watchers. She finally begins to talk about some of the bad sexual experiences with the grandfather. She has experienced sex only with him and only after the age of thirteen. She has never experienced sex with anyone else and at the age of thirty-eight, she feels and acts as if she were a much younger age. Erika has been her protector over the years.

Paul and Karl are seen again. It turns out that Paul (the silent one) has been actively observing me for our entire five-and-a-half years of therapy. He has even researched multiple personality extensively and has kept his own "chart" on therapy! The relationship is growing and we will continue to talk.

I see Kate briefly and am reminded I need to see much more of her. She informs me that she has agreed with Elizabeth S. on a certain question. (Kate has never been informed by anyone of Elizabeth's existence. This means that spontaneous co-consciousness is occurring.)

Shana tells me that approximately fifteen personalities from various groups are now ready to fuse. (We have fused sixteen thus far.) Kate is not yet ready for this step and clearly expresses her wish not to, so Shana and the others have agreed to wait while I spend some individual time with Kate.

At the end of August, Katie (a four-year-old regressed form of Kate) shows up at Erika's workplace. Wanting to find a way to communicate, I agree to see them on an emergency basis. I decide to use her stuffed animal, an all white dog we've named Mishka. I tell stories, using Mishka as an example of good, and the grandfather as an example of bad. In our next session Kate, having regressed again, arrives for the session and is elated because Mishka, who had been lost the previous evening, has now been found. We do a hypnotic age-progression at this point to age thirty-seven so that Kate will re-emerge in her adult form. I get her to report last evening's dream (I had been cued in by Elizabeth). She begins to interpret her dream herself. A key part of this dream is the theme of alienation from God. She also agrees to write about key events of the last few days.

*

I'm Katharine St. Clair. Today's date is September 1, 1986. I need to think about integration today--what it's like to be fusing with other people.

I find I'm having a problem with that because I don't understand myself anymore. To be fused with sixteen other people tends to change a person. It's as if I can't sort things out, and that's so unlike me. It makes me want to go back to the time when I first met Dr. Ted, when I first started to talk with him. It was simpler then. I gave information about everyone else and kept him updated on what the others were doing, what they were feeling. I never talked much about myself. I realize now that was a mistake, but that's just the way I've always been.

I find myself doing things that I wouldn't normally do-things that aren't me. I write poems. I've never written poetry. I know that's the Marie part of me. But it doesn't feel right, it doesn't fit, because I knew Marie prior to all of this. Marie was in my group, she was my friend, and I took care of her, and now she's gone. And I'm not supposed to really say that she's gone, she's fused inside of me, but it doesn't feel right.

There was also another person that wrote poetry and her name was Elizabeth Austin. I didn't know her very well, she wasn't in my group, and I didn't meet her until much later. She wrote a very different kind of poetry from Marie, and so when I write poems I'm constantly trying to figure out which part of me wrote the poems--sometimes I can tell.

And there are other things. I'm much more indecisive. It's as if I can't get myself together. There are so many things crowding my mind. To find myself doing something that isn't really me just kind of upsets me. And I want the old Kate back.

I guess I also feel very guilty about being the person that people are being fused into, because it doesn't seem fair somehow. They all wanted their own lives, their own identity. They wanted to keep their own names. Who am I to take all these people into me? What right do I have to take their identity away from them?

I miss some of them. Certainly, I didn't know all of them well, but there were some I knew very, very well. I miss the little ones, but at times I can see parts of the little ones in me. When I regress (sometimes I can feel that coming on) I feel myself becoming younger. I want different things. I'll have an urge to color, or to read a children's storybook, or to just sit on someone's lap and be

181

hugged and cuddled--things I wouldn't normally feel if I were an adult. Of course, regression for me happens when my mind is just too tired to sort things out anymore. I'm saturated, I guess. Or something that I do as an adult may seem wrong, it may seem childish. So in order to make it acceptable I have to become a child.

Sometimes I just want to scream. This has been going on since January, since fusing first started. I just want to scream, "Don't do this to me, I can't handle it."

I dream about the others. How they didn't want to go away, how they're trying to come back to being themselves, and I keep shoving them away and telling them, "No, you can't come back." Dreams like that terrify me. It makes me feel very, very selfish--like I'm not being fair to the others.

Sometimes I find myself wanting to run away and hide for no reason at all. Nothing spectacular has really happened to upset me, but just all of a sudden everything seems to close in on me and I want to run and hide and just be by myself, where no one can find me.

I had a belief that integration would make things easier. I don't feel that way anymore. I think that makes things even more difficult. I'm one person that's split apart; that can't put all the feelings into one person. I don't know if that makes any sense. I never felt as if I was a multiple personality before. Now, when I'm fusing with people I feel as if I am, because those people are being fused into me, only I don't feel as if I'm a whole, complete person.

Things just aren't the same, and it makes me very, very, discouraged. It makes me doubt my ability to become well. I have to wonder if there will ever be a time when I am well. Dr. Ted is so optimistic. I just don't have that same feeling. I don't have that same belief, and I don't know where to get it from. He says he sees changes and he's encouraged and those changes are good and there's healing going on. I don't have that same feeling. I feel as if things are more messed up now than they ever were. It's really difficult to explain what's in my head. I can't even find the right words. And that makes me angry because I used to be able to explain things. Now it gives me trouble.

The only thing that I seem to be able to do, that I've been able to put my energy into, is yard work. Sometimes I'm really, really tired when I do that, but it's as if I have to. There's something inside of me--I need that hard work.

But that's not going to last much longer. I mean, fall's coming.

Nights are not as long as they were. And so then what do I do with all this stuff that's inside of me, all this--energy--is the only way I can describe it. Energy for something--for physical work. I suppose that somehow for the winter I'm going to have to direct my energy back to the guitar and drawing and things like that, but how? It's all become very, very confusing for me.

There's a part of me that says, "I can't take therapy any more. I've had it." It would be really foolish to quit now. Now is when I should be talking, when I should be asking advice, getting out feelings. I feel as if I'm a person that's been torn into little pieces, and I'm just barely being held together by a little band-aid.

I had a bad week last week. I remember that Mishka, our stuffed animal, was gone. Dr. Ted told me that I pounded on the couch and told him that I was bad, and that grandpa's bad, but I don't remember that. And then the night that Mishka disappeared I had that dream--that I was looking for her and I couldn't find her. I looked all over and all of a sudden she appeared, only she was all cut up. She'd been beat up and she was hurt. But I didn't do anything to help her. I abandoned her. And then, Dr. Ted was in the distance and I tried to get to him, and I'd walk toward him a little ways and then I'd stop and I'd draw a line on the ground, trying so hard not to step over that line to get to him. Then eventually I would, I'd give in and I'd step over that line, but then when I'd start to walk toward him he'd get further away again. And then I'd stop and I'd draw another line and I'd try really hard not to step over that line, and eventually again I'd step over it, and he'd get further away. I'd try to get to him but I never could. I was reaching for him, and I felt so lost. And all of a sudden this mist would swirl around me and I wouldn't be able to see anything-- nothing. Like a fog. And all of a sudden, it would clear away and I'd try to reach Dr. Ted again, and I'd reach out my hand and he'd reach back but he was so far away, and I'd quick draw that line on theground again and I would tell myself: "You can't step over that line," but I did. And I'd try to reach him again and it was as if he kept backing off. Every time I got closer he'd get more and more distant, but he'd still reach out his hand and I would try, I would reach out my hand, and I couldn't touch him.

And all the while, I was thinking about Mishka--how I'd left her. I had looked and looked and looked so hard for her and then she was there all of a sudden. I abandoned her and she was hurt, and to me that's unthinkable because I would never do that to anyone. I don't understand where that came from.

I've had that dream every night since last Wednesday. And even though I knew later on in the evening that they'd found Mishka; that she was back in Dr. Ted's office, I still have the dream.

I don't feel as if I'm being very much of an adult now, because if I were really an adult I would be explaining all of this better.

I don't know what everything means right now. I don't know where I'm going. I need something more concrete, something to hang on to, I guess, because sometimes I just want to hold my head and scream, just shake it, so that everything inside of it is gone. It's like having a tremendous noise in my head at times.

I need some help. That's how I feel right now. I need some help. I need someone to tell me that things are going to be all right, everything's going to be all right, because I don't have that feeling. I have a bad feeling right now.

26

In early September, 1986, Anita Sanderson (age twenty-one) comes in for the first time. She agrees to come in at my request, after I was informed about her by Elizabeth S. She is ready right now to accept a job in Kalamazoo (no one knows about this except Elizabeth S.). She must decide by Friday. I spend as much time as I can dealing with her, since this could turn into a major crisis. She does agree to come back if "You will prove to me I really am a multiple personality."

I spend several sessions talking with Anita. She still intends to move to Kalamazoo, but she is talking with me. Attempts to use logic on her do not work well. What does seem to offer promise is an effort to reach her on a feeling level. Maggie's painting of a clown appears to offer good entre' in that regard. I share with her my frustration, and Krista steps in to interpret to me what is going on in Anita. Krista informs me, unbeknownst to the others know, of what happened to Anita. She tells me that when Carol was age twenty-two, a resident physician who had become somewhat of a confidant with Carol, eventually had sex with her and he proceeded to tell everyone about it. This sabotaged her job in the hospital, and she eventually found out that this was all done, that is, having sex with her in the first place, based on a bet. It is a major step for Anita, who experienced this humiliation for Carol, to subsequently share this information with me. The net effect of all of this is that Anita finally cries. The facade appears to have broken. The emotional out-pouring subsequent to this tells me that this crisis has now passed, and she does not go to Kalamazoo.

Time is spent with Paul and Karl, particularly talking about the subject of multiple personality. Anita is processing her experiences in therapy.

185

She shares with me more about her bad experience with the resident physician and why she subsequently closed down in all her relationships. I encourage her to again try painting as a further means of expression. Over the previous weekend she had destroyed her last two paintings.

Maria A.'s voice is changing and becoming more mature. There is a sense of humor showing. She is reading on the subjects of self-image and body-image and is now able to talk much more about her own comfortableness with her physical self. She is clearly growing. Cancer tests results continue to be negative, but there is much fatigue.

Amanda from Group 1 is very angry, especially at Kate. She views Kate as changing a lot, but does still want to help me to help her. They have a car accident a half hour before one session. Kate handles it all--and handles it quite well. This is encouraging. I confront Kate at this point on her "time out" in regard to therapy. She agrees that it is time to resume therapy. She is, however, still not ready to allow the new fusings.

In late September, we have an "all group" meeting so that I can educate people regarding Kate's diabetes, which has now been diagnosed, and its implications for everyone. Once this is finished, I am informed that personalities are sitting in various places in my office. I move over and sit in a different chair. A number of them, especially Rachel, take great offense at this. I stand to block the door because they are ready to bolt. It turns out that Elizabeth S. "came out of the closet" over the weekend, and after introducing herself invited everyone to today's session. Many feel betrayed. I finally do arrange a talk down with Rachel, who is the spokesperson, and Angie, who was "sitting" in the chair that I sat in. They are angry because I have blatantly confronted the visual distortion that occurs in multiple personality, and several of them are not ready to accept that as yet. We have several sessions trying to allow some healing from the confrontation. Angie and Rachel in particular appear to be working it through, and a number of them are more accepting of the reality of them being a multiple personality. Anita lets me know that she now accepts it.

In mid-October, Shana (age eighteen) appears to request an update on the next sixteen fusions. I let her know that I have not forgotten. I spend most of the session with Kate, dealing with her resentment at Carol "for coping out," and her resentment at me for being less sensitive than she wished regarding her fear toward the fusions. This is an emotional session. We finally agree that on Thursday, October 30, 1986, we will do the fusions. I will commit myself to emphasizing time with Kate to deal with her reactions. Later on, Shana brings in a list of what is now fifteen personalities to be fused. Dani's Katie decides to join in, but I find out that Elizabeth S. is dissuading her decision to fuse. I have a session with

186

Dani's Katie and we decide to postpone her fusion once it becomes clear that she is primarily only copying the group of fifteen. Kate is accepting even better the concept of our plan, but Elizabeth S. begins to grieve the fifteen who will leave and time is spent with her.

On October 30 we perform the fusion of fifteen personalities at one time with Kate. I meet with each one ahead of time and perform age progressions so that all are the same age. I have them each identify a favorite color. I use these individual colors in imagery while they are floating on a cloud in a hypnotic trance. As the colors blend together, the fusions occur quite easily and quickly. Kate is extremely sad when all this is over (she had been very anxious prior to the actual fusions). Elizabeth S. lets me know that Kate's sadness is also Shana's sadness (she had led the fifteen into the group fusion; her sadness is a result of knowing she now has to give up many of her dreams).

We have an extra two-hour session the next day. This is primarily to deal with Kate and her adjustment to the events of the day before. She has done some regression back and forth and needs some time to rock and to have a safe and peaceful, quiet time in session. There is a soft catharsis, which is extremely effective. I note that her "eye flitting" of today, which previously indicated that she was involved in some sort of communication with another personality, now is a sign of trying to sort out her feelings, communicating with herself. "I can never remember being this happy, or this peaceful," she says at the end of the session.

Over the course of the next week, Kate processes all these multiple fusings by regressing off and on. She rocks in session. She is very cold and she remembers the hole in the ground.

I have an emotional session with Kate over the repression and never worked through the grief over the death of her brother a number of years ago. She does so in the regressed state (approximately age ten). She emerges at the end of the session again at the age of thirty-seven.

By the following week Kate has decided that she wants even more fusings. She presents this as a statement of wanting to be well, but I encourage her to recognize that we need to be quite careful at this point. There is now no further regression in session, although it is reported that this happens at home.

Francesca (Group 4) comes in session for the first time. She brings in a list of ten personalities who wish to be fused next (she had called me by phone prior to the session to inform me of this). We have a group session with everyone. This is to deal with anger, scare, confusion, etc. associated with their wish. Many want evidence that those who are fused are in reality continuing to be a part of Kate. They meet on their own that evening and tape-record their discussion:

❊ ❊ ❊

It's November 18, 1986 and I'm Elizabeth S. We're going to have a group meeting to discuss some of our thoughts about fusing; how people are feeling about our situation in the last couple of weeks. People have been very up-tight, very unhappy. Anyone that wants to speak may do so. Each person will say their name before they speak, and hopefully, this will help us to understand each other a little bit better, so maybe they can be of help to each other. I guess that's all I need to say right now. So if anyone wants to talk, go right ahead. Everyone seems to be a little shy because of this tape recorder. That really shouldn't bother any of you, okay? It's going to help us in the long run. So, come on, you guys. We're all here to help each other, and I think that we all love each other, so the best thing to do is to get everything out in the open, okay?

Well, I'll start off. I'm Janice and I'm a little sad because I'm going to be one of the ones that's fusing, and I'm starting to have second thoughts because if everyone is going to be this unhappy about something that I want to do, then I'm not sure I want to do it, because I've never wanted to make anyone unhappy. It's just important to me to do this right now because it has to be done sometime, and I'm really rather tired of the way our life is going. I think that we've all been just kind of coasting along, and just because we didn't know that we were multiples up until a few years ago, doesn't mean that our lives were going all that smoothly. There were a lot of problems. We just didn't stop to really look at them, we just glossed over them. We can't do that anymore. We are different from other people, and being different can be pretty difficult. I think we've all learned about that. Dr. Ted says that we have an opportunity to have a better life, so I think we need to grab onto that. I mean, after all, we've had some terrible problems and people haven't been very kind to us. I would like to see that change. I don't think I'm going to die. I think it was Erika yesterday that said she saw a part of Jennifer in Kate, and she's right. I see different parts of people in Kate. We'll be just a little bit different than what we are now. But we've been different all of our life, so what the hell?

I think that's the first time I've ever heard you say "hell." I'm Rachel.

188

Well maybe. I'm Janice again. Maybe it's the first time I've said hell, but I feel pretty strongly about what I'm going to do and I'm sorry if it offends some of you. But I'm going to do what's right for me, and for me fusing is what's right. I don't think it's so bad. It didn't hurt any of the ones that fused last time. And sure, sometimes Kate has some setbacks after the fusions, but she didn't have as much trouble this last time. And I don't think I'll have any trouble either. I don't feel very peaceful a lot of the time, and I'm going to. I know I will.

Well, is anyone else going to say anything or are we going to have a stop and start every time? You know you guys can just talk as you always have, okay? This tape recorder means nothing. Try not to even look at it. Excuse me, I'm Elizabeth S. It's not so bad.

I'm Diana and I'm from Group 1, and I don't know...I don't know everyone very well, okay, but I think that, especially in the last few weeks everyone from all the groups have started to get to know each other better, and I hate to see that break up. I miss some of the people that have fused from my group terribly and I think that we need to stick together. I think it will be very lonely if some of you people go again, and it's going to get worse. I mean, we've always been there for each other. I don't want to see anyone else go.

Diana, you know me. I'm Esther and I'm going to be fusing too, next week. But that doesn't mean that I'm going to go away from you, or from anyone else here. I think if you really tried, and this goes for everyone, not just Diana, I think if you really tried you could see us inside of Kate. Some of us have reached a point where we can't keep on denying that we're a multiple and that we're separate people inside of one body, and I think that's the key. We have looked at each other as totally separate people for so long that it's very hard for us to reverse our way of thinking and say "but we have one body." One body! You know, that makes us so different from other people. I mean, you have to look at other people outside of us, okay? Look at Dr. Ted. He's pretty special to all of us, and we're getting to know him pretty well, but when he talks to us, and say he's unhappy, he doesn't change to a different person, or if he's being in a silly mood, he doesn't change to a different person. But that's what other people that

189

know about us as a multiple see in us. They see that one body, but it changes into a different person on the inside, not on the outside. And I think that's what you're not looking at. Other people on the outside don't have to become a different person to deal with an emotion, a feeling, or a thought. We do. That makes it very, very difficult, and all of us here would like to be able to use our own names twenty-four hours a day. When have you ever done that? Oh, sure, there have been instances where someone's had all of the time for a whole day or more, but it's not on an on-going basis.

Well, I don't like it. I'm Jill. I don't like this at all. I feel as if something is trying to destroy us. And I think that it's wrong. It's as if this whole family is breaking up. And I think that's what we were all about. We were a family and we still are a family and we're very, very close, and now there's this big wedge. And I don't especially like the idea of dying, because that's really what it is. I don't want to die. It scares me. I don't mind saying that.

Well, I don't understand. What scares you? Seriously, what scares you about it? Oh, I'm sorry, I'm Angie. Do you know what I mean? I know dying's not something that's really pleasant to think about.

This is Jill again. I guess what scares me is the thought of never being free. And I think that's what I've always wanted. I think that's what everyone else has always wanted. Our own freedom to be what we want to be. And so now if we have to merge, blend, or fuse, or whatever you want to call it. If you want us to do that, then we lose our freedom. I mean, we've always been so re-stricted.

I'm Victoria S. I know that you don't know me very well, but I'd like to say something anyway. I'm not in the group to fuse next week, but I have been thinking about it. We've had so many ups and downs, and I guess probably my group--I'm from Group 5--has had the least ups and downs. But, on the whole, we're only a small part, there's only eleven of us in that group, and two of those people have fused, and sure, I miss them. But it's not the end of the world, because they were my friends and I have pretty neat memories of them the way they were. I can look at Kate and sometimes I can see glimpses of them. As far as the ups and downs go, on the whole, our whole community has experienced terrible things. And I think that to become well again, fusing is going to have to happen sometime. I don't think that it's something that can be avoided.

Well then, I guess I have a question for you. I'm Allison Saunders. If this is something you say we can't avoid, then why aren't you fusing, or are you going to next week?

(Victoria) No, I'm not.

(Allison) Okay then, why not? If you're so sure about all of this, and if you're saying it has to be done anyway, why don't you go ahead and do it? You don't seem willing to. You must be afraid of something.

I'm Victoria again. Yes, I'm scared about it, and so I don't want to necessarily bring my fears into a fusion. If I am afraid (because of the unknown) then I don't think that I should fuse yet. But I know that I will. Maybe I will the next time. I just can't do it right now. I suppose to some of you that means that I'm chicken, but I don't see any of you doing it, except for the people that have decided they're going to do it next week. I think the rest of you are being pretty unfair to the ones that have decided to fuse. We've always had a respect for each other and we've always taken care of each other. We've listened and we've laughed and we've cried together, and so now if they want to fuse, you give them the cold shoulder, and I think that you're being pretty selfish because that's what they want to do.

I don't think that's being selfish. Excuse me, I'm Corrie. I'm on the list to fuse next week and I will go ahead with it. There isn't anyone in this group that will stop me. But I do think that your word "selfish" was a bit strong, Victoria. I don't think that anyone is intentionally being selfish. I think that people are looking at this as, well, we've always been so close and we've always needed each other and some of these people are fusing, they're leaving. I think that's what the people are looking at. They're losing a friend, a companion. Let's face it, we've been part of each others' lives for a very long time now. Some of us may have arrived later than others, but we've still been a part of each others' lives. Just look at us, look at how many of us there

191

are. Everyone wants time to do their own thing, everyone wants a different career--to be a homemaker, or whatever--and not one of us has ever had a full year to even start to accomplish what we want. Say one of you wants to be a nurse. That's great. A wonderful profession. But it sure is hard to go to nursing school full-time, isn't it? Can't do it. Because you've started, been there a couple of days, and guess what? Someone else wants to do something else. We've never stopped to think about that. How come we can't finish anything? You know, our goals are always out of our reach. Oh sure, maybe someone gets to finish a painting, but isn't it strange that if someone wants to take a trip somewhere and they get ready to go, they end up not going because someone else has gone someplace else? Know what I'm saying?

I think you're playing with our minds. (unidentified)

(Corrie) No, I wouldn't do that. We're not the kind of people to play with each other's minds.

I think I would like to say something. I'm Kate. I think some of you resent me, and I feel very badly that you do.You know, it was not an easy decision for me to have everyone else fused into me. I had a lot of guilt about that, and I still do, because I want everyone to have a chance at doing what they want. But we can't go on like this. I don't always enjoy having the problems I do after I fuse with someone, because I take on their problems if they haven't been worked through. And I have to experience new emotions and thoughts. Sometimes those emotions and thoughts are pretty difficult to deal with. I don't enjoy it when I regress. But I'll tell you one thing, I'm getting better. And I'm going to keep on getting better. And you know why? Because I want it. And you can all sit here and feel sorry for yourselves, or you can get off your butts and do something about it. Some of you tonight have said that we've all been very close. Well, I don't see too much of that lately. You fight more than you get along with each other. I realize that some of you are scared, but you have to talk--and you have to talk to Dr.Ted.

Kate, I'm Francesca S. I have a little problem with what you're saying. You see, because each one of us watches you when you fuse, and the first time you fused with someone it took you ten months. All you did for ten months was back away from people. You didn't talk to Dr. Ted.

(Kate) Yes, but I did talk this last time.

That's fine. You may have this last time--I'm sorry, I'm Francesca S.--and I have to admit you've done very well this last time, but I don't think anyone, especially the people in your group, could take another ten months of what they went through the first time. And if that's what it does to you, we don't want any part of it. I think that's what I'm hearing from a lot of people, also. Not today especially, but off and on. You probably have spent more time talking with Dr. Ted the last month than you ever have, but it took you a number of years to say anything to him outside of the usual things that you talk about. It took you a long time to allow him to get to know you. So I think you're coming down a bit hard on everyone as far as our getting off our butts and doing something about it. I know I'm a multiple, but maybe I don't fully accept that emotionally yet. I don't think that I do. But it's something that needs a lot of thought, and it's not something that you can just say "Okay, I'm a multiple, I fully accept it, and I'll fuse." That's not the way it goes.

I'm Kate. I know that's not the way it goes. And believe me, I'm sorry now that I didn't spend more time talking with Dr. Ted in the last ten months. I should have, but I couldn't.

Why? I'm Francesca S. again. I thought that he was your very good friend. See, it wasn't just him. You backed away from everyone. And you have to admit, that causes confusion.

I'm Dani. I think that I can understand Kate's backing away. And sometimes it takes a long time to come back. Kate needed that time to be able to realize "This is what I have to do. This is what I am, this is who I am, and now I'm ready to get some help." Asking for help isn't that easy for any of us because we've been so dependent on only each other. We haven't been dependent on anyone else, not outside of our family. And I know that's how we had to grow up, we had to depend on each other only, because there was no one else.

Well, maybe that has to change. I'm Erika, and I think that Dani has a point. For some of us it is extremely difficult to trust and to really talk, but I think that we really need to talk with Dr. Ted. I don't think that any of us are ever going to get better without his help. I don't think that we will ever stand a chance of having a normal, happy life without his help.

Why not? I'm Jill. Why not?

(Erika) Because, Jill, he really understands us. Sure, he can't fully understand what a multiple is because he can't be in our position, but he sure tries.

I think that...I'm Jill again, let me get my name in here...we did okay before.

I'm Elizabeth S. That's the problem, Jill. We only did okay. We didn't have a very full life, now did we? We never have had.

(Unidentified) I think this is just a bunch of bull. All I see is that our family is breaking up. And I think that's pretty sad. I don't want to be locked up forever, and that's what's going to happen. We're going to be locked up forever.

What do you mean, locked up?'

Well, it's like...

Go ahead, go ahead.

It's like when we were little, you know? We'd be locked up. So you put us inside of Kate and we're locked up again. I don't think that's fair to any of us. I don't want it. There's still a lot of things I want to do and I'm gonna do them. It's like putting us in a box again and just throwing away the key--or sticking us in that hole. You can't get out. Just stuck inside. And it's dark.

See, that's what I was talking about too. I'm Diana. That's what I was talking about, too.

Yeah, I feel the same way. I'm Caroline. We want a chance too. We got a chance at being something. No one ever gave us a chance. There's a whole big world out there, and I want to see

194

it too. I mean, if you're going to put us inside someone else, that's like killing us. I don't want to be locked inside someone else. I only want to stay just the way I am. I got punished enough. I don't need anyone to punish me any more. It makes me feel as if I did something really bad, and so I'm going to be sentenced for life by being put inside somebody else. It's like being all caged up. And I won't have it.

Oh, Caroline. This is Kate. You know me so well. You're part of my group. And you too, Diana. I don't know what to do to make you understand that you're not going to be all locked up. I got locked up too, but I don't know what's worse, being locked in a box or being put in a hole you can't get out, or being so locked up inside that you can't feel. I can feel some things. I can't feel anger. I know I'm supposed to be very angry at some of the things that happened to me. I think Dr. Ted's trying to help us by putting us all together so we'd never have to be apart again. We'd never have to be separated. It's not being locked up. We'll be able to do whatever we want--and when we want. There isn't going to be all this waiting time.

Sure, that's easy for you to say. I'm Caroline. I don't like us having to say our names. It's easy for you to say, Kate, because you're going to be here.

(Kate) I don't know what to say. I don't know how to make you understand. You have to try and look at it the way that Dr. Ted's tried to explain it to us. We're part of a puzzle. See, I feel as if I'm starting to come together. Sure, I have lots of things I don't understand yet inside of me, but I'm starting to sort them out. And sometimes they threaten me, and then I don't want to talk about them. But I think that Dr. Ted has a wonderful view of things that we can't even really see, you know. He sees us in the future as one person that's happy. Now we have all of us that are unhappy.

I'm Marissa. I'm from Group 5, and I think that what Kate is saying is very, very true. We can't, you can't go on being this unhappy. That's all you've ever wanted, happiness and freedom, and someone to love you. Don't you see that it will be much easier?

I'm Maureen and I'm from Group 4. I tend to be very cautious, so I'm not sure about all this integration stuff yet. I need to hear more. I guess I need to see another fusion happen. But maybe, and I'm not saying I agree with it yet, but maybe it is the way for us to go. We need some peace, and we're not getting it. Erika has not gone to work in a couple of days, and she can't stay home forever because we're upset. There has to be a different way. Some of you have had bad headaches lately. So many people have been so terribly unhappy that I think we need to consider change.

What kind of change?

I think that, for one thing, we need to spend more time with Dr. Ted in group discussion. I think that we need to make some changes within ourselves, because I see us as people that fight more than they care about each other lately. At least that's the way they act. I think that--this is still Marissa--I think that we really need to talk with Dr. Ted more. And I don't think that the same people should do all the talking. I guess I was happy to see that so many different people spoke up today.

I'm Elizabeth S. I think that we should wind this down now. I know that we could probably go on talking for many hours, but I don't think that we should make this tape any longer than what it is. Since I didn't hear any before, I'm going to assume that there are no objections to my giving this tape to Dr. Ted and will do so. So, I don't know what to say, except that I'm unhappy because of the way all of you feel. I don't want you to be this unhappy. Maybe we should do this more often. I know that we're planning some big things. Maybe I said that wrong, everyone's not planning to do a big thing here. The people that are planning on fusing next week are going to do so. So, I guess that's it for today. I'm glad that you agreed to this, and I hope it helps. I know that a lot of you are upset right now, and maybe it would be good to continue talking with each other for awhile.

27

At the end of November, with Elizabeth S. setting the stage, I have a two-hour session with Dani, dealing with the sexual abuse issues. For the last month or so the reality of her life has been setting in on Dani, and she is having increasingly more difficulty. She had always viewed the traumas as having occurred to the other personalities but not to her. She has lately been thinking more and more about these early childhood traumas, which has much to do with her withdrawal from everyone else. This appears to be the culmination of a fair amount of private processing on her part. Our time involves the verbal recall of many of the specific sexual abuse memories. She appears to now accept that these, in fact, are her own memories and these are not events that only occurred to the others. This is important time for Dani. She gives me a short story she wrote in 1983:

Freedom

Julie was having trouble staying awake. The room was so warm and stuffy and she could hear Mrs. Beacham's voice droning on and on in the background. Mrs. Beacham was Julie's third grade teacher. Julie likes her a lot, especially when Mrs. Beacham hugs her. Julie tries sitting up straighter and concentrating on the story Mrs. Beacham is reading to the class, but it just isn't working. Suddenly, she feels a hand grasping her shoulder and her eyes fly open to see Mrs. Beacham standing next to her and the whole class looking at her. She hears Mrs. Beacham ask her if she had any sleep last evening, but she is too embarrassed

197

to answer. Then she hears Mrs.Beacham say something that makes Julie feel icy cold. Mrs. Beacham is saying that she's going to have to contact Julie's grandparents about her constantly falling asleep in class. The hand on her shoulder feels heavier and heavier. In fact, the hand is resting right on one of the bruises from last night's beating and it hurts. She wrenches herself free from the hand and runs from the classroom, hears Mrs. Beacham's voice calling her from a distance. Faster and faster she runs out the school doors, through the school yard--faster and faster, feeling the wind blow through her hair. I can't go back, she's thinking--can't go back to that house, can't go back to the beatings, the fear, the hate. Why didn't you see, Mrs. Beacham? Why didn't you see that they hurt me? I thought you loved me? Why doesn't anyone see? Why doesn't anyone help me? She wants to cry, but she can't because if she does then she won't be able to think and right now she needs to think. Pretty soon the grandfather will be looking for her and she can't let him find her. Julie stops running and looks around her. She realizes she's in the woods about a mile from town. She sits down beneath a large tree to rest, but all she can think about is the grandfather and how much he hurts her. Julie knows that what she wants is her freedom, freedom from a life in which no one loves her. She gets up and slowly walks until she reaches the small shack she sometimes goes to when she needs time alone--time away from people. She knows she shouldn't stay too long, because he'll look here too. But she's so tired of running, so very tired of living. She sits down and looks around her, taking in everything, absorbing it--the dust, the dirt, the filth in the shack and she feels a sense of belonging. The shack reminds her of herself and in that instant--in that moment of recognition--she knows what she must do: she grabs the huge coil of rope lying in the corner. Her small fingers struggle with the rope, but finally she succeeds in making the knot.

It is already dark when they find her--a small, frail child hanging from the end of a rope in a shack in the middle of the woods. The grandfather laughes.

Dani
September 21, 1983

We identify our next group fusion day, and I speak individually with all ten volunteers. Two of the Group 2 members (Cynthia and Corrie) have changed their minds and no longer wish to fuse. It turns out that Corrie was only "copying" Cynthia. I spend approximately an hour with Cynthia (the storehouse of much anger). She eventually volunteers to change her mind, (probably only because she's never particularly liked Kate: "She's too perfect," and wants Kate to be stuck with her anger. We do age progressions as indicated and then the actual fusions.

That evening, Cynthia de-fuses. I later have an individual session with her. She is an angry woman who arrived when Carol was age thirty-one to handle the rejection and guilt over the divorce and her relationships with men. This is the first time she has ever talked of any of these things, and we have an excellent session.

I spend the second hour with Kate. There are no age regressions on her part, but there is some sense of confusion, being alone, being uncertain, wanting to rock. It would appear at this point that she is handling the second group fusions quite well. Elizabeth S. continues to be my confidant, advisor and interpreter. Over the weekend Kate does some regressing, although it is rather minor. It appears to be related to the anger of Corrie who has just fused with her. Cynthia has wanted Corrie back. We focus on Cynthia's anger and we make a deal: I will try to de-fuse Corrie if Cynthia promises to talk at length with Corrie. It would appear at this time we need to take some backward steps in order to be able to take some forward steps later. Maria A. agrees to give up her individual session (and is very unhappy about it) to Cynthia and Corrie. Most of the session is spent with Corrie. We focus on her anger, mostly at the grandparents. She is one of the very few who will, in fact, fight back. She does agree to come back and she is now more reconnected to Cynthia, who is mellowing and maturing. Her anger is dissipating. She's been doing some good work on her own and I am supportive of this.

Corrie is testing me. She has functioned in a very emotionless manner. "I am the repository of anger." Cynthia describes her as different from everyone else. She has never felt cared by or cared for anyone else (or loved or trusted). Behind her hard exterior, however, it is coming through that she likes me.

Maria A. now lets me know that she is furious. She's filled with self-hate. She feels no progress. She wants her entire community back. She is quite manipulative, testing me in session. Elizabeth S. is reinforcing the message that Maria A. is very serious about her feelings. What she is really reacting to is her lack of therapy time. She wants to focus in on body image issues, and time is given.

Kate continues to vacillate in age. We focus on the connection between her need to sleep in the fetal position, and her being in the fetal position in the trunk in the grandfather's house when she was a very little girl.

In mid-December, there are three spontaneous fusions: Karin, Group 2; Jacqueline of Group 3; and Stephanie of Group 3. The three of them fuse and de-fuse all weekend. Kate continues to remain regressed as a seven-year-old. She tells me: "If you don't put me in the box, I'll do anything you want." (Elizabeth interprets that she is referring to sexual expectations from the grandfather.)

By the following week Kate has regressed to a four-year-old. I get on the floor with her and I physically force her to be "in the box" with me. I become the grandfather for her at this moment and she becomes extremely emotional and a strong outburst occurs. Following approximately fifteen minutes of this, I become (by my choice this time) "Dr. Ted." I call her "Kate" and thirty-seven- year-old Kate emerges. There's more emotion at this point, some back and forth regression, and then we are able to talk. This is an extremely powerful experience for me.

Over Christmas, Kate remains to five or seven-year-old. By December 30, she has had four straight days as adult Kate, but then three as a seven-year-old. I tell Kate a parable about a dog named Napoleon, a Samoyed who has fallen in a hole. A kindly man finds him, takes him home, and he grows up in one day. After identifying with the dog in this story, she unregresses and has a soft, verbal catharsis. I see her again two days later, and we have a quiet session with much reminiscing and opening up. She needs no encouragement from me during this session. This is very effective and quite moving. She has remained thirty-seven-year-old Kate, even though there's been the urge to regress since our last session.

By January 5, 1987, Kate finally outlines one of the regular rituals, which was required of her: she must sit in the grandfather's lap, having to kiss, undress, and have sex with him. The grandmother must watch, then she undresses and forces her to perform cunniligus on her. There are many tears, and about five minutes of regression to a seven-year-old. The others now want Kate to have much therapy time, and Maria A. gives up her next session to Kate. Kate volunteers more information: the grandfather makes her (as Amy) wash floors naked with a dog collar and a leash around her neck. Objects are forced up her vagina while she is doing this. As Kate volunteers these traumatic memories, part of the time she regresses to approximately age seven.

I spend the next two hour session with Kate regressed at age seven. (Kate had phoned me the evening before to add information regarding the grandparents taking pictures of her naked, and also selling her to strange men.) The seven-year-old uses a stuffed animal, Priscilla Pig, as a conduit through which information is translated to me. (Priscilla Pig is telling her to tell me secrets. But Priscilla is really Elizabeth S. who is helping in therapy this way.) We have quite a long and powerful session, which has clearly been nudged by "Priscilla Pig." The seven-year-old finally cries. Kate then resumes her own identity as Kate with the help of hypnosis. While in trance, I suggest that she not lose these important memories, as has so often done in the past. This is extremely fruitful clinical time.

The following day we have an extra two hour session. It had been a very rough evening for all of them. Jenny Montgomery(Group 1) is especially struggling, as are many who are vicariously reliving events experienced by Kate. Jenny is seen first (she did not run out of the apartment last evening; she had promised me she would not do that after Elizabeth S. called me to warn me that Jenny was about to panic). I spend more time with the seven-year-old who gives more details: "The grandparents shoved something inside me, and made me stay that way all day while we visited (a customary Sunday visit) and threatened to cut Mom all up if I said anything." She suddenly becomes the adult Kate and she retains the memory and is able to talk about the healing that is happening in feelings toward her parents, who were apparently also physically and emotionally abused by the grandparents.

In subsequent sessions, we continue to use "Priscilla Pig" as a means to communicate direction. Priscilla urges the seven-year-old to open up and a large amount of previously unshared information emerges. Some of the specific information given through the prodding of Priscilla: the grandparents tell her she is ugly and stupid and of the devil. She is hung up by her arms and is whipped and twisted around. She is given enemas with cold water, because "They said I was dirty inside." She is tied up, put into a potato bag, "...to throw in the river or to be laid across the railroad tracks." There's a large group of all males who have anal, oral, and vaginal sex with her. One of the men urinates on her. She had to eat dinner with her plate on the floor, her hands tied behind her back, and eat like a dog. She was tied up and forced to eat rice on the floor. (These last behaviors were experienced by Callie, Cesca, Jenny, and Mandy.) She is also tied up on her knees in their bedroom, is forced to have oral sex with the grandfather, then with the grandmother, and they then go to sleep. She would have to stay in that position and when she fell over, he would awake and yank her back into that position. They also "...made me all dirty by spraying something all over me," (speaking of all the strange males ejaculating on her).

At this point the adult Kate suddenly emerges and says, "I always thought my father would save me." The seven-year-old takes over and reports that they would not allow her to go to the bathroom, and when she had to go and could not hold it back any further, they would make her smear it all over herself, both urine as well as feces.

201

The next day we arrange an extra two-hour session because of the dramatic content of the previous one. The seven-year-old is seen again for further opening up of memories. Elizabeth S. shares that she and Dani are functioning as "Priscilla." The seven-year-old now shares her two absolute worst events: one being anal sex, and the other being hot liquid poured into her vagina with admonitions to never be able to have, "Bad girls like you." She reports that once she was ordered to smear her feces all over herself and she then had to stay in an outside shed, naked, cold, and alone for an entire evening. She also shared that once when she was untied the grandfather brought her to the shed, had anal sex with her and said, "You have to be a good girl, and not tell anyone or I will kill Mom and Dad and your brother, Phillip." "He put his thing in my butt." And, "They hurt me right here,"(patting her uterus).

Priscilla continues to nudge her to keep sharing throughout this period of time. The seven-year-old regularly takes Priscilla's mouth to her ear to hear what Priscilla is encouraging her to share. "It was hot. They said that when girls grow up, they have to have babies, but you can't because you're a devil, so they put something hot in there." The grandfather would put clothespins on her nipples and her vaginal lips: "It hurt, and then I had to lick her." (They apparently first put a tube in her vagina, then something liquid and quite hot was poured through the tube, then they took the tube out and put in a hard rubber object to keep the liquid in. "When you grow up, you'll never have bad girls like you.") I note that Elizabeth S. continues to interpret using "Priscilla Pig" as the medium to reveal this. I see thirty-seven-year-old Kate at the end of this session for her talk-down. She is quite afraid.

Kate remains quite regressed for much of the weekend. We continue in session, with Elizabeth S. and Dani enabling me to talk to her through Priscilla Pig. The seven-year-old shares that they tied her spread-eagle on the bed, and the grandfather, the grandmother, and a series of strangers, would take sexual liberties with her. Sometimes another female would be brought in and she would be forced to lick her. She observes money being exchanged for these favors.

The primary underlying personal question throughout all of this reported trauma is: "Am I a good person or a bad person." It is clear that we are going through a major clinical event and although it is very draining and frightening, it is also exhilarating because of my sense of the significant progress being made. I continue to talk with the seven-year-old Kate. Two major memories are shared. Number one, a stick is rammed up her vagina making her bleed, while another object is inserted in her anus. She is also gagged. She is ordered not to tell the doctor, who the grandparents know she will have to see because of the bleeding, and she is ordered to report that she had fallen. Secondly, the grandfather picks her up by her feet and drops her to the floor intentionally. He did this often, beginning at age two!

The seven-year-old Kate agrees to share her last major traumatic memory at our next two-hour session. I wonder if this will finally turn out to be "the bottom line" for her. She reveals a party of especial degradation. At age five, her mom and dad have to go far away to the hospital to see Phillip (he has a crushed leg). The grandparents have planned a party: twenty-three men and women ("..in a large room with one light in the middle of it.") Kate must take a bath, is bound by her hands in a hanging position, and the guests begin to arrive. She has no clothes on. The guests take off their clothes and swing Kate back and forth, and "things" are put in her (..my butt, my mouth, and between my legs.") She is then let down on the floor, and she must lick the women between their legs. The men put their "...things in my butt and between my legs." She has by now vomited for a second time, the grandfather is angry at this and he makes her clean it up. A strange man "...beats me up for being bad." A dog collar is now put around her neck and she must serve them all food. Water is then poured into her vagina and up her anus and she has to keep it in her body and walk around under these circumstances, "They all laugh." The grandfather and the grandmother tell her that they can continue to keep doing these same things as much as they want. "The grandfather then put his thing in me and they all watched. Then they put the collar on me again, they put a chain on the collar, bringing me outside and chaining me to something." She is given no food or water during any of this time. She reports the most difficult part: "They all laughed." The degradation, abasement, and humiliation was the worst part for her.

She pounds on the sofa. There is much crying after reporting these terrible events. The crying and wailing continue unabated for some time. The pain is so profound I can only allow it to expend itself. Kate finally takes over spontaneously. The information is now out and Kate may not need to function much longer in a regressed form. This is obviously an extremely significant session. We are both drained.

I agree to an extra two hours on Saturday because of the emotionally traumatic material that is being dealt with at this time. We spend much of this time attempting to process the terrible material revealed previously. Kate finally takes over on her own. She is, however, not consistently stable at this point.

Added information is given: before any of the strangers did anything, Kate had to tell them that she liked it; that they could do anything that they wanted because she was dirty. She had to put makeup on her lips, eyes, and cheeks. They took pictures of all of these bizzare situations.

203

Elizabeth informs me privately that the primary struggle right now in session is whether she should be an adult now and talk, or whether she should stay safe as a child. Kate says:

>"Being a child is the only way I could force myself to tell you these things."

Is Anyone There?

Hello! Is anyone there?
Is there a soul within
that shell?
Is there warmth, and joy,
and peace?
Is there life that flows within?

Hello! Is anyone there?
A response does not come,
it cannot come,
because, you see, the shell
is vacant, void of feeling,
void of life.

Marie

They Say

They say that I am evil,
that I cause destruction,
they say that I am without feelings,
that I am without a conscience.
They must be right, for are
they ever wrong?

Marie

28

With such powerful events bursting out in therapy, it is sometimes easy to forget that there are many "stories" going on simultaneously. Although Kate's dramatic regressions and revelations are extremely gripping, and are obviously affecting many of the other personalities, I find myself remembering Maria S. At this point in time, Maria has fused. Even though I didn't know it at the time, it has turned out that her "purpose" became that of the chronicler. As she put it earlier:

This writing gets to me. I am, at times, so tired of it, and it hurts. It really hurts. But I'm just starting to realize that I should be proud of us because we made it! There were tremendous odds against us, but we made it! We're alive! If I could only make the others see that we should now take another step and become one person, a whole person. It really doesn't work this way. I don't like to admit that because I feel as if I'm almost being unfaithful, not only to the others, but to myself because I am a separate person, but I'm not whole. And I wish I could make some of the others see that. I think that as one person we would probably enjoy life to its fullest. There would be all kinds of things out there for us to experience, things that we can't do now, things that we don't have time for now. We don't have time for them because none of us have 24 hours a day.

I look around me and I see one person that does watercolors, another that uses pencil, another one that does oils, and a woodcarver, and someone that plays guitar--a couple of people that play guitar--someone that plays piano without ever having any lessons, people that do needlepoint, that knit, crochet, that write poetry. I find that someone went to college, almost has a degree in Journalism. And then collectively we don't think we have anything to offer. That we don't have any talent. And we're wrong. But we do have talent.

We've spent years thinking we're nothing. We allowed them to determine the way we felt about ourselves. Dr. Ted told us that "We give them power." No one should have that much power over another person.

I think it's time I started my own therapy. So when you read this, Dr. Ted, I'm ready whenever you are, because we have a lot of things to talk about.

I haven't revealed much of myself to you since the time that I introduced myself to you, because that's the part of me that doesn't want to dredge up the hurts and the pain from the past. But the more writing I've done, the more I hurt. And I dream about the past. I feel so guilty for not having been able to take care of the others the way I should have been able to. I've been watching therapy a very long time now, and I want to live, I want to be like anyone else in the world. I want stability. I want us to be able to stop therapy some day, not because I don't want to see you anymore, but because I want to be normal. I want the nightmares to go away. All their pain has become my pain, all their hurts, my hurts, and I can't sleep. I feel used and dirty sometimes, like Jacqueline. And sometimes the feeling that no one cares is so strong. I need to learn how to let go of the past and grab onto the future. Because you're very right, you know. The past has ruled us; it's overpowered us. Maybe when we were much younger we didn't have a choice, it was our way to cope, because I don't doubt that we'd be dead. Carol certainly would have killed herself. And I worry about that with some of the others, but I don't want to die.

I had to be ready for therapy, Dr. Ted, and I wasn't before. For awhile, after you first suggested that I write, I thought, "Oh well, that's not too tough, it's really easy. It won't take me that long." But at times I've hated it. I can't look on this as a fun project, and I resented you at times for suggesting it to me. But I understand what you mean when you say that this book will also be a catharsis for us, because I'm feeling it--I'm living it. To know so much about everyone else's lives within the group can be emotionally wearing, and for awhile I really didn't care about anything.

Sometimes I feel guilty, as if I'm squealing. As if I'm telling you things I have no business saying. And yet I know that somehow it's going to help you understand us a little more.

I'm very glad you've become our friend. At times we fight with you, we argue with you, and you know that we're stubborn. But you're still our friend. And that means a lot. It means more than we can even tell you, because you're the first person who really cared. We've spent five years in therapy now--just about five years--and any one of those times you could have just said, "I've had it," but you never did. And that in itself has been as important to us as therapy has been. You didn't walk away from us. So I figure if we can do it with you, if we can hold onto a friendship with you, then we have to be able to do it with other people. We've learned to trust a little. Some of us more easily than others, but we've learned.

I guess I now want to sit a minute because I feel some quietness inside. Maybe it's hope. Everything's been in such a turmoil lately. At times my insides felt as if they were being torn apart. But now I'm ready. Actually, I think it's called "getting well." I'm ready to get well. It's taken me a very long time to reach this point.

29

As we enter February of 1987, Kate continues to sporadically regress. I explain it to the others that analogously to the "ripple effect" of a pebble thrown into a pond, that Kate's system needs time to process these tremendous jolts, and her way of doing this is to have time as a child, which after all, was basically deprived her when she was a child. Therapy time is spent in helping Kate deal with this important time of transition. It also becomes evident that many others have been struggling, but also healing vicariously. It is reported that Jenny and Amanda are getting close to spontaneous fusion. Erika and Dani continue to observe and advise, which is very helpful to me.

At one point, the seven-year-old regressed Kate spontaneously becomes the adult Kate: "They had no right!" In other sessions I require Kate to repeat after me, "I, Kate, am a good person." She is now experiencing many nightmares. She brings in more childlike but very graphic drawings of the sexual abuse. Kate is getting stronger. I continue to spend time with her. She remembers also that the grandparents threatened to sell her if she didn't do exactly as they said.

By mid-February, Kate brings in an impressionistic outdoor painting, which is absolutely magnificent. She has been taking weekly painting classes, off and on, for about a year. Her instructor, Paula, has become a friend by now. Paula has therefore become another key element in Kate's growing network of healthy relationships. We review the recent events together, and she remains as an adult Kate. She is beginning to talk about her dreams and can remain as an adult as we process this.

Elizabeth S. reveals that seven fusions have occurred spontaneously in the last two days. These include five members of Group 4: Jessica Allen,

209

age twenty-seven, arrived at age four; Rachel, age twenty-eight, arrived at age eleven; Katrina, age eighteen, arrived age fifteen; Allison Andrews, age twenty-one, arrived at age fifteen; and Sonya, age nineteen, arrived at age fifteen. In addition, two members of Group 4: Allison Saunders, age twenty-four, arrived at age ten; and Maureen, age twenty-six, arrived at age thirteen, have all fused spontaneously in these last two days. These events speak clearly to the power of these last few weeks and the vicarious clinical significance and productivity that is occurring. This is very encouraging but is very trying on all parties.

I spend additional time with Kate who has further memories, like the time that an electrical prod is used on her (and she produced a drawing to depict this specific instance of abuse). The electrical prod consisted of two electrical wires, each with two forked ends on it. One would be inserted in her vagina, and the other would be inserted in her anus. They would first wet her down and then shock her, watching her body twitch--then all laugh. They would also hold her head under water in a tub. In addition, the grandfather made a board with arm and footholds, put her in it, turned it upside down and spun it until she got sick. They did this until she became unconscious. They would then throw cold water on her to rouse her. In addition, the grandfather made another board with a wooden penis attached to it and would force her to have that between her legs for hours at a time. Throughout these detailed abuses above, there would be constant derision and laughter by whatever adults were around.

I might add at this point, that even though the abuse sounds bizzare and cruel, there is every reason to believe that these reports are totally accurate. Research into the history of multiple personalities inevitably reveals abuse that is far beyond most people's comprehension. For the next several sessions we have fairly quiet talks. Many of the people are extremely affected by the powerful revelations that have been revealed.

By March, some of the other personalities are now beginning to resent the amount of time being given to Kate. Jenny and Amanda, in particular, are getting fed up and no longer wish to fuse. It is noted that they experienced much of the abuse also. Elizabeth S. is feeling very saturated. I'm not certain how much more of this she can take.

The seven-year-old regressed Kate brings in the rest of her drawings. There is much discussion with her. She reveals a traumatic event for her. She says that she "...stabbed a bad, fat man, and I think I killed him." Elizabeth S. interprets this sentence: at around age five, after much abuse (shoving hard objects in her vagina, intercourse) this man would often take her home with a dog collar around her neck and would abuse her. She did stab him with a knife that she saw lying on the floor. The grandfather and grandmother told her he was dead. But Elizabeth S. never did believe

210

that he had actually been murdered. Elizabeth prods me to encourage the seven-year-old to look at her own drawings, which I do. Emotion comes spilling out of her.

Kate continues to allow the seven-year-old to be her "insecurities." It is time to confront Kate on allowing her "child" to do all the work for her. I give to her an article on child abuse to read and Kate refuses. I confront her strongly. She tries to leave the room and I have to physically restrain her. She handles her anger by tuning me out. She finally does read the article and, with further confrontation on my part, also looks at the seven-year-old's powerful childlike drawings of the sexual abuse. She eventually does agree to a special session to confront the pain. In this session, Katelin (age thirty-eight, arrived when Carol was four, she is the "twin" of Elizabeth S. and also an "overseer") introduces herself to me in person for the first time. I spend the entire two hours in confrontation with Kate regarding her anger. This is essentially a psychodrama session and a very powerful one. It is very physical in terms of pounding and screaming, but is a very constructive two hours.

Elizabeth S. and Katelin are now publicly functioning together. The seven-year-old wants to go home whenever I raise my voice in role playing. Her need to deny anger is showing. Elizabeth and Katelin are prodding me to do more confrontation of Kate. The seven-year-old later announces that she will not be back because, "You cannot fix me." It turns out that someone has told her she is being a nuisance. We decide to have another confrontation session with Kate. She brings in some drawings. We try to activate Kate's anger. I take a somewhat less active role this time, until she finally decides on her own to sit in a corner: "I can't experience anger unless it's physical." I point out to her that she is having a conversation and that there's nothing particularly physical about it. Elizabeth S. and Katelin interrupt at this point. I tell them to go under my desk (the trunk), and I then call Kate out who immediately takes over. Finding herself "in the trunk" produces a very physical and emotional reaction. At times I must physically restrain her. Kate immediately regresses to the seven-year-old under my desk, but vacillates through various ages. She really stands up to me with hitting and kicking and verbally telling me off. She spontaneously again becomes age thirty-eight Kate after about thirty minutes of major confrontation. We then proceed to our talk-down.

By the next day, the seven-year-old is assuming an identity of her own, instead of being merely a regressed form of Kate. She has taken on the name of Amy, Kate's nickname in the Netherlands. Amy is beginning to sense that she is "...going to go away." We have a quiet and sharing time.

Kate is now beginning to grow in confidence. We are able to talk and she

211

voluntarily now brings up specifics about the abuse. This is a fairly quiet but appropriate session and the need for her to regress to the seven-year-old Kate appears to be waning. It therefore appears that Amy is also correct in her assessment that she will be leaving soon.

For two weeks we have a quieter time. We spend our time on many day-by-day living kinds of issues.

On April 6, Erika collapses at work. She is under doctor's orders to rest because of exhaustion. She missed three days of work and she cancels the first session that has ever been cancelled in therapy with me. Soon after, Kate is informed that she may have Hodgkin's disease. She will have to go to the university hospital for tests. This is especially discouraging since she had previously been given a clean bill of health regarding her breast cancer.

A bit later, Kate reveals for the first time to brother Patrick over the phone that the grandparents sexually abused her. She also tells him that her parents forced sterilization on her. This is very powerful for both of them. This is the first instance of sharing with someone in her immediate family, anything to do with any of the traumatic things that have happened to her. He appears to believe her, but barely comments on it.

Sessions continue to go from personality to personality; many practical, logistical issues are being dealt with. Most now say that they want therapy to be finished this year, although a few still don't. Kate is getting very invested in the concept of writing the book.

I see Amy: a very "raw nerve" regarding Phillip has been touched and she suddenly understands that he is dead. (Be reminded that Amy is so young she could not have awareness of events that will occur later in "her" life.) It would appear from the above that Amy's "adult life" is beginning to activate.

Kate now senses on her own that a new personality has emerged. I question the group and Kristina S. emerges (age thirty-eight, has been here for about six months, is lively and social). Kate also has let it slip that all of the group barriers have dropped completely on their own. There is total co-consciousness among all personalities except for this new one. It would appear that Dani was the instigator for this work and the community is becoming one group. Dani has now signed up for her last college course and it appears that she will graduate in August. I have known about this for some time, and have been quietly prodding her to continue this.

Kate has her siblings over for dinner in her apartment over the weekend. This is highly unusual and it goes quite well. Kristina is now becoming the new "reporter." Dani says that therapy is ending. She encourages me to especially see Erika, Francesca, the boys, and a few others. She reports that Amy and Kate are emerging even more now as separate personalities;

that Amy is no longer functioning as a regressed form of Kate, but is taking on her own separate identity. This should then allow for a fusion to happen, which will be permanent. It also means that the dam of repressed trauma, which broke wide open has been processed and true healing has been happening.

Kristina now introduces two others who have come into being with her (Juliet and Julia). This now makes a total of one-hundred twenty-nine personalities and fragments. Kate is quite nervous because she must go to the hospital tomorrow to be evaluated for Hodgkin's disease. I spend some time also with Dani focusing on her upcoming fusion and the things she will not be able to do any more because of this.

At the end of April, the university hospital confirms the diagnosis of Hodgkin's disease. Kate will now go to there every other weekend for treatments. She is struggling but coping. Dani is also struggling but coping. On the advice of Elizabeth S., I spend time with these two to process this next traumatic turn of events. I wonder if their pain will ever stop.

Most of the next session is spent with Amy. She is very quiet. I use Mishka to prod her. She finally reveals the new "ultimate" degradation. She shares that she had become pregnant at age twelve by either the grandfather or one of the strangers that forced sex on her. The grandparents arranged for an abortion, which was done sparing anesthesia. There was much pain, and she fainted several times. Then, two or three days after the abortion, the grandfather forces sex on her again. This time, according to Kristina, she could not stop screaming, "...which only added to his enjoyment."

Katelin now advertises to me that she resents having "...all our secrets out in the open and in your awareness at this time." I see Amy for most of this session. She would not talk at first. I let her rock and she begins to cry. She reveals that they made her clean herself up after the abortion. The house was very dirty. She finally begins to let go and there are many tears. She is able to say now that she is very angry. For several sessions I continue processing with Amy. She is opening up and growing. She is even writing about the subject of abortion. She now knows she is a multiple personality and she now knows that somebody else talked for "Priscilla." There is much insight developing. Her writing is a further extension of her healing.

By May 12, Amy is growing fast. She has realized on her own that her purpose was to reveal "...the last big piece of the puzzle" (the abortion at age twelve). These are dramatic developments. Elizabeth S. has agreed to write about her overview of all of this. She says this really represents the ending of therapy! Elizabeth S. reports that Amy needs to build some childhood memories: for example, kite flying, which is something she's always wanted to do, but was never able to. Amy is beginning to get childhood memories. She has bought a kite. She writes me the following letter:

May 14, 1987

Dear Dr. Ted:

Tomorrow is going to be a very special day for me, my day to fuse with Kate. Part of me wants to stay, to remain separate. I have learned in therapy, however, that it is necessary for me to fuse and that it is part of growth.

I'm excited, yet frightened when I think about tomorrow. I hope it's a beautiful day. I'm looking forward to spending some time talking with you tomorrow. I have my kite ready to go. It's so pretty with it's bright colors. Thank you for helping me select it. I've wanted to go kite flying since I was a small child and now I'll get my wish.

But most of all, Dr. Ted, thank you for your encouragement and for always being there for me. It's wonderful to hear you say that you care, not just for me, but for all of us. It's even more wonderful to actually be able to believe it. I'm sure it has played a major role in helping me reach this point in my life.

I do have two requests that I would like to make of you. The first is to ask that you take good care of Elizabeth S. She has spent a great deal of time with me and I don't want her to be sad or lonely. Secondly, I would like you to really work hard at helping everyone to stop smoking. They really should, you know. I hope you won't forget me. I know you will say you won't, but I just needed to say it. I've learned a great deal from you and I appreciate it more than I can say.

Thank you, Dr. Ted, for your gift of friendship. It's the best gift I have ever received. I will always remember the stories, the hugs, the chance to have things other children have. I needed those things to become an adult and with your help, I made it.

Love,
Amy

Amy and I arrange a special time on a Friday to implement fusion, which has been planned ahead of time. I am aware of the significance of this day. The entire community of personalities is aware, and has been preparing for this day. Amy seems to symbolize so much of what is good to everyone. And now is the time for everyone to say good-bye to her. Now is the time to allow her to fuse in her own way.

I can sense powerful feelings being experienced (mostly very privately) by the whole community. I have very ambivalent feelings myself. Even though we've all been through many fusions already, this one is so special, and so difficult. It is clear many will have to and have begun to, grieve.

Many have helped prepare for this day: time and space have been given to Amy, both in session and on their own time. A number of us have gone shopping with her and have helped her buy a kite.

We drive to a public beach on Lake Michigan. Since this is springtime, and a week-day, there are only two other people there. It is breezy and beautiful. The sky is blue, two-foot waves produce a continuing background of rhythmic and relaxing lapping sounds on the beach. The entire scenario speaks of freedom and peace, and of feelings--and being oneself.

All the others are staying away, respecting Amy's space, but I know they're watching.

We take our shoes and socks off and get our toes wet. Amy is all smiles and very excited. Although no words concerning fusing are spoken, she knows that today is important. There is a naturalness and a freedom and a serenity that is very visible, along with her very excited energy. It's as if all she symbolizes can be experienced now. And how very sad to also know that her purpose has now been completed at last. And how exciting to know that the wonderful wide-eyed child's qualities will become a part of Kate.

She asks me how to fly her kite. We assemble it and slowly let the wind pull it away from us. She is so excited. She takes the spool of string on her own and begins running up the beach. "Dr. Ted, Dr. Ted, this is so much fun!"

I join her and for about fifteen minutes we stand together and watch her fly her kite. She talks and talks about nature, and freedom and friendships and God. About my family and others in her community. She wants to know what I'm feeling and if I ever flew a kite when I was little. Her catharsis, her need for freedom and completeness is a wonder to watch and I say very little.

And now she becoms very quiet. Little Amy, who is already a merged product of many of the little ones, who symbolizes so much good, is getting ready to fuse. I take the kite string and look into her eyes: there is no fear, there is now tension, and her smile is natural. I give her a hug and whisper: "Amy, it's now okay for you to be really you." In about thirty seconds I can see a slight rigidity in her posture and I look at Amy's face and see Kate is here. Tears begin to slowy roll down my cheeks--she's gone.

30

Elizabeth S. now demonstrates her need to grieve for Amy much more than anyone has expected. Kristina, Julia, Juliet, Dani and Elizabeth S. are all encouraging me to push Kate into maintaining her memory of the abortion at age twelve. All are very invested in her progress and offer excellent lead-ins into what she is capable of handling. These personalities appear to be transition personalities. I have a low-key session with Kate. She now knows that she not only fears the actual memory of the abortion, but also (and especially) what her life will be like once all of the fusing is complete and therapy is over.

Concurrently, Kate must undergo her hospital treatments and is very sick each time she goes. Although some regression does occur, much of Amy shows through her. There is even some silliness. She is needing a sense of things becoming okay. She does have some pieces of memory: e.g., "Fat Simon with greasy hair."

By June 1, three new transitional personalities introduce themselves, Elizabeth Katherine, Debra, and Lisa. Kate has emotionally regressed, has lost the memory of the abortion, but has not age regressed. These three new personalities know that they came into being shortly after Amy's fusion and that they are here to help with full integration. Kate needs session time in our next session, Kate, with some prodding, does retrieve the memory of "Fat Simon" and the abortion at age twelve. This is done, however, without expressing any real feeling. Many of the others are staying in touch with me and all are bothered by her lack of affect. We have an extra session at Elizabeth's strong request. Kate is having nightmares and is struggling. I prod and she gives in detail the events of her abortion

217

at age twelve. This, however, is again done with almost no emotion, but at least all of the memory has returned.

With hypnosis and age regression, I bring Kate back to age twelve, and we proceed through the abortion. There is a very large emotional outpouring with anger, tears, much desperation: "I want my mommy, I want my mommy." This is a very powerful and necessary session. She submits the following:

It is June 3, 1987, and I'm Katharine St. Clair. I am writing about something that happened to me when I was twelve years old.

My grandparents asked that I be allowed to spend a week with them, and I didn't want to go; I never wanted to go to their house, they always hurt me, and I was hoping my parents would say no, but they didn't. And so I had to go.

I went to my grandparents' house late in the afternoon. It was in the summer. And as usual, I dreaded going there because any other time there was a lot of sexual abuse and physical abuse, not just from my grandparents but from other people that they invited.

This first day, this first evening, was different. Usually my grandfather or my grandmother would have sex with me; this time they didn't. The only thing that happened was that my grandmother gave me an enema. That in itself was not unusual, I was accustomed to that from her, but it was definitely unusual that neither one of my grandparents had sex with me. And I was kind of half awake all night waiting for one of them to come up to the room where I was, but neither one of them did.

The next morning I was told to get dressed, get ready, and I was not given any breakfast, and again, that was not unusual. Often when I was staying with them I wasn't given anything to eat. Early in the morning we left their house, and I assumed that they were taking me to someone else's house to be used for whatever purpose, be it sexual, be it housecleaning, be it to be abused physically. The house where we went was really not far from my grandparents house. I could still find it even now. There was a man there, and I automatically named him "Fat Simon," because that's how he appeared to me. He was maybe five-feet eight inches or five-feet nine inches, he was extremely heavy, and he had very dirty, greasy, dark hair, parted on the left, and a mustache. He had very bad skin, it was pockmarked. He had very small, beady eyes. His hands were dirty. He had on a tee-shirt that

218

I suppose at one time had been white, but had all kinds of stains on the front of it. I had no idea why I was there. We were led through the house, to the back of the house, to a room back there, and I was told to undress. There was a table in the room, and something that resembled a sink.

At this point I became frightened. Any other time when I was taken to strange people's houses I more or less had an attitude of resignation. Because I knew what was going to happen, there was going to be some kind of sexual abuse or sexual games that they played, and I was always the object. This time I had a different feeling, as if something very, very bad was going to happen. And there was a small table there that held what looked like knives to me. Anyway, I was told to get on the table, on my back. By this time I had taken my clothes off--I always did what I was told, I was a good little girl in that respect, I always did what I was told--and Fat Simon tied my legs down, my knees were bent. My grandfather did not come into the room. My grandmother was there and she stood behind me, and she ran her fingers and her hands all over my body, like she usually did. She loved to feel me. Sometimes I'd have to do that to her, too. Many, many times. Anyway, she was standing behind me and I remember being terribly, terribly frightened because I had no idea what was going to happen. Usually, in this type of a situation, the other people involved--another man, my grandmother, or my grandfather-- would have their clothes off by now. But no one took their clothes off. I was the only one with my clothes off. And I remember "Fat Simon" looking at me, and now I can put a word on it; he had a look of contempt. And he said to me, "So you got yourself into trouble." I had no idea what he meant by that. With my grandparents, I was always in trouble. I was always being punished for something. But I had no idea what he was talking about. And as he said that, he pushed down on my stomach very, very hard, and at the same time he inserted something into my vagina that was very, very cold. Then he inserted something else that was sharp, and I felt this tremendous pressure. And then I started to feel pain, because whatever he had inside me whatever he was using, was very sharp, and it seemed to go inside me for a very long way-- way up. And I could hear a noise starting. I can't really describe the noise, I suppose it was something like a scraping sound and I can't describe the pain, either. But it intensified--it got worse and worse. And I must have screamed or something because of the pain, and when I did, my grandmother slapped me and "Fat

Simon" told me to shut up. And I think that I tried to concentrate on something else and not the pain, but I couldn't because the pain blocked out everything else. There wasn't anything else but the pain. And I could feel something wet between my legs. I didn't know what that was. I remember now that I tried to struggle with my grandmother, but she was much stronger than I was, much bigger. She was holding me down by my shoulders and my upper arms. And then the pain got so bad, and the noise became so terrifying, that I must have passed out, because I don't remember anything after that except waking up and my face was all wet, and I think that they threw some water on my face.

And now that I look back at that time, I remember waking up and my grandparents and "Fat Simon" were standing in a corner of the room, and I could see them, and my grandfather was giving him something, and I assumed that he was paying him. Because now that I'm old enough and remember, I know that "Fat Simon" performed an abortion on me. I didn't even know I was pregnant, but I guess my grandparents must have kept track because they abused me frequently and they knew that I had started menstruating a few years prior to that. I don't know who got me pregnant. I don't know if it was my grandfather, or if it was one of the other men that had sex with me. Anyway, "Fat Simon" came over and he untied my legs and told me to get up. And I remember trying to sit up, and the pain was so bad at first I couldn't do that, so I had to roll over on my side and I kind of slid myself down off the table. And when I tried to stand up, I felt as if I were being ripped in half. I looked down and there was all this blood all over me. Then I got sick. And they became very, very angry. I had to clean myself all up. And all the time I was having these pains and these terrible cramps in my stomach. I had to walk out of there, and as I was walking out of there. "Fat Simon" looked at me and he said, "You'll be back." I assume he meant that I would need his services at another time in my life. I'm sure he thought I was a whore. I didn't even know what pregnant meant, where babies came from.

And my grandparents took me back to their house and pretty much left me alone that day, but the next day they started all over again. My grandfather forced himself on me, had sex with me. My grandmother watched. And then I had to make her feel good. I had to make her have a climax, and then my grandfather had sex with me again. And each time I'd bleed and then I'd have to clean myself up again, and I was so nauseated; I was so sick to my stomach and wanted to go home so badly, but they told me I had to stay there. And they told me if I ever told anyone they would kill my parents. The strange thing is, my parents only lived a few streets away from my grandparents and my parents never knew. My mom would call my grandparents during the week, the week that I was there, and I would be allowed to talk with her on the phone, and she would ask if I was having a good time and how things were going. I would say, "Yeah, I'm having a good time, I'm having a real good time, Mom." And I could never, ever, and I still don't ever, understand why my grandparents did all those things to me, and why my mom and dad never knew, and why my mom and dad never came to help me. I wanted to tell my mom so badly when she called that they'd hurt me, but I was sure that they would kill them and I did love my parents. But I could never understand why. I could never, ever understand why they did all that to me. I wonder if "Fat Simon" is still alive.

Sometimes I guess I get down on myself for not having realized what was going on, for not knowing that what they were doing to me was not their right, that it was wrong. How I could believe then and at times still now, that I'm a bad person. I was so afraid that they'd hurt my family, my mom and my dad and my brothers and my sisters. I was so afraid. I never knew "Fat Simon's" real name. "Fat Simon" is the name I gave him. He was younger than my grandparents. And I have to wonder about how many women he did this to, how many girls? I do think about that.

I think I should have known that I was pregnant. I wonder what my mom and dad would have done if they'd found out. I think about confronting my grandparents with all of this, but I think that they'd still laugh at me. They used to laugh at me all the time, like when I would beg to go home. I knew it wouldn't do me any good anyway. I learned that a long time ago, when I was a very little girl. They really, really had no use for me other than to be used for their own sexual satisfaction. And the enemas, their purpose was to cleanse me. Something my grandmother always said. They were supposed to make me clean inside.

And I wonder if it was a cleansing routine for her, for herself. I can't believe that they were just sick. I wish I could. Maybe it would make everything a little easier to understand, to accept and understand, but I don't believe that they were just sick. I think they fully knew what they were doing and I think that they enjoyed it. I think they were evil. It's one thing if a person is sick and their mind is not what it should be, but my grandparents were devious,

221

they were manipulative. They manipulated my mind to a point where sometimes I still feel as if I'm that little girl and their disapproval is still there, and I'm still afraid that they're going to hurt me. Part of me wants to confront all of them.

A lot of the time, though, I'm just sad. Sometimes the sadness just comes without any warning, it's just all of a sudden there. And I think that's when my past is working on my subconscious mind and I may not have an actual memory there, but something way down deep inside me knows it's there.

That's all I want to write tonight. I don't want to write any more.

Kristina now tells me that Lauren (age sixteen) progressed totally on her own to age thirty-eight. She arrived at age four and is in Group 4 and has written a paper on multiple personality. She winds up fusing completely on her own.

31

By mid June, Kate reports that she's now begun to think about bringing a lawsuit against the grandparents. Juliet reports that her motive is purely for revenge. They decide on their own to have an "open" meeting and all voice the wish to at least see if there is a possibility of a legal case.

Kate also reports that she has gone to her grandparents' house for their sixty-ninth wedding anniversary; she had been prodded, as she always is, by her parents to show respect for them. She tells the grandparents: "I remember it all, all you did to me." She finds out for the first time that she is no longer so intimidated by them. The others are worried that she thinks too much about getting revenge. She is now volunteering a lot of information. "I don't think I'll let you rescue me any more."

Carol calls me at home! I see her the next day in session. It turns out that she had de-fused, according to Christina, because of Kate's fear of confronting the grandparents. She (Carol) was furious because she believed she had been intentionally shoved underground by all the others--and by me. She does not accept the explanation that she fused. She does have all the memories that have been retrieved since she fused, including the abortion at age twelve. We have a talk-down. I give her a hug and she again fuses on her own. Christina is "directing" these events and providing the analysis, which is extremely helpful.

An unidentified personality informs me that someone is now hiding the journals, the writings, the diaries of all of the others. They have a group meeting on their own. They are able to get a few minutes alone with this unidentified personality and it turns out that seventeen-year-old Dani's Katie is the only one who knows her. "The others know her only as a feeling." I ask if she'd be willing to see me next week, and she agrees.

223

I meet Cassandra Brandon: age twenty-nine, arrived when Carol was age five. No one has known her, except for Dani's Katie. Katie tells me that "She represents the need for privacy; her presence therefore keeps the others from opening up." Cassandra has been obsessively reading all the others' written materials that she's hidden. She's aware of the idea of the book and is concerned. She decides that if she can type some of the book, she can then censor, if necessary, and I agree. Christina tells me privately that when Cassandra heals, many of the others will heal with her. Cassandra is personality number 134.

Dani is taking therapy time and is changing. Her memories of the sexual abuse are now "her" experiences. She writes me a letter:

July 3, 1987

Dear Dr. Ted:

I have spent a great deal of my time lately thinking about fusing with Kate. The very thought of fusing causes a great deal of turmoil for me. I have so many thoughts and feelings that are unresolved, so much that remains unfinished. Some of these you been very important to me. I wanted a close friend all my life and you finally gave that to me. I remember our sessions, the good times and some bad ones too. I learned much from you. Please remember, I always trusted you. I still do.

Take care and thanks,
Love, Dani

Together with Dani, we plan a strategy of therapy for her "three kids." Over a series of sessions we will age progress Becky to age seventeen, then age progress all three of them to age thirty-eight, and then fuse all three with Kate. While we are proceeding with this, Paul and Karl report that they've been observing and now know that: "Dr. Ted, we're going away too, aren't we."

Cassandra has been typing the book. In so doing, she (and many of the others who are observing) is having to confront many memories. There is much turmoil. Erika, who holds down their job, is struggling badly and can no longer cope with her work. We must arrange for a medical leave of absence, which eventually lasts for six months. However, this does now give us the luxury of being able to go through the many impending fusions, with all its associated grief, without having to worry about its effect on job performance.

By mid-August, clinical events are occurring quickly. Paul and Karl fuse. Soon after Dani, Victoria, Dani's Katie, and Becky follow suit. Then Jenny and Amanda fuse, followed by Angela and Angelina, then Karina

224

and Marissa, and finally Elizabeth W. Angela (age twenty-one), and Angelina (age twenty-four) of Group 3 explained to me how the process of becoming acclimated to fusion occurred with them. They explained that they had been vicariously experiencing therapy for several years, and were able to accept that: "We all have one body. We realized that we had to grow up. It's just a matter of thinking this out." Both of them age progressed to age thirty-eight on their own, and then they fused with Kate. Rachel must now grieve.

Within a few more weeks, all of Group 3 fuses, except for Erika who must continue to hold down their job. We now have a total of thirty-eight personality fusions.

In early autumn, the grandfather finally dies. I am expecting many reactions, but all is calm. By October 19, I go over the entire list of 135 personalities with Rachel W., and discover that there are only thirty-eight personalities left! It's obvious that much is occurring privately, and I'm not sure what to make of all this.

Kate continues her chemotherapy treatments throughout all this, which is gruelling. Her physicians keep telling her to cut out the stress in her life, but she does not want any of them to know of her multiple personality dynamics and they therefore have no idea what is really going on. I do hypnosis with Kate to help her quit smoking, which is successful and at least have eliminated this hazard. It is interesting to note that this is vicariously also experienced by the others who smoke and all have now quit. This is further evidence of the more subtle process of integration that is occurring in many ways throughout their entire system.

By early November, Kate feels strong enough to confront the grandmother, which she does. The grandmother totally denies all accusations, screams virulent accusations back, and threatens to call the police. Although shaken, Kate is actually handling it quite well.

We are now down to only thirteen active personalities, excluding Kate. Rachel W. lets me know that, since Kate's confrontations with the grandmother, more than twenty fusions have occurred. They appear to now know that: they have a right to their anger, they need not feel guilty about their anger, and they can respect themselves for having the courage to confront the grandmother.

On December 2, Kate is informed there are no further signs of any cancer!

Debra and Lisa now spontaneously fuse. This leaves ten personalities, plus Kate. To me, this means we finally have the luxury of being able to spend more concerted time with Erika, who's still on her medical leave from work. It appears that Erika has become the repository for phobic reactions, which probably stems from her time in the trunk as a child. We deal with her anger at me. She had been a very helpful informant very early

225

in therapy, and has felt left out because of so much therapy time with so many of the others. And finally, we can deal with her childhood. She arrived when Carol was four--her purpose was to "deflect" the bad feelings during the abuse into some diversion. By February 1, 1988, Erika returns to work. We are now down to eight personalities: Kate, Erika, Rachel W., Cassandra, Tanya, Jacqueline, Karina, and Francesca.

We are now entering our eighth year in therapy. We have written a lengthy account of our experiences and even dare to think of the possibility of publishing it together. I am getting very excited about the prospect of total healing and Kate truly having a life of her own. Kate is now even beginning to think of her first trip back to the Netherlands since she immigrated at age seven. I am aware that this will entail a visual confrontation with the actual places where her terrible abuse took place, and am hopeful that this will allow for the final confrontation necessary for full integration. I also know there still remains much potential for regression.

Rachel lets me know that Erika cannot fuse until she confronts her past in the Netherlands. Cassandra brings in a number of poems that Erika has written. It is interesting to see Marie's personality in Erika's work:

In her mind she
screams,
tearing him to
shreds with words,
bastard!!
bastard! bastard!
bastard! a litany
that never ends--
one he never hears.

Erika
February 9, 1988

please!
oh please don't
take it away.
I beg you, please
let it stay.
a sardonic grin,
the slam of a
lid.
Darkness.

Erika
February 9, 1988

Whisperings in my head
tell me stay me,
and relief floods me.
yet never ceasing demons,
louder, still louder,
tugging, pounding
in my brain,
let go, let go.
Can't.

Erika
February 9, 1988

32

Rachel W. arrived in May, 1985, when therapy was becoming difficult, especially over the issue of fusing. Others were quite frightened and Kate was still regressing. She now is the main transition personality, and is beginning to feel like she has been left "holding the bag." The others have learned to count on her steadiness, but by April, 1988, she's no longer feeling steady. She's getting tired of feeling responsible for the others. Depression is visible. She has her first migraine headache. However, she's still helping in session, especially with Francesca Germaine. 'Cesca has begun to reveal new degradations forced on them by "Heer Doktor," a man who apparently regularly bought time with Kate for various sexual reasons. And now, beginning information about beastiality is emerging. Everyone is now feeling especially dirty, shameful, heartbroken. 'Cesca has regressed. The child now lets it slip that she has a secret friend who no one else knows about. Her name is Callie J. (#135) and she represents everyone's shame.

The focus now switches to Rachel W., who is especially experiencing the feelings of dirtiness, taking on Callie's function. Although she arrived in 1985, she has all the memories and feelings of the abuse. She is suicidal. The "straw that broke the camel's back" is the memory of sex with animals: "We were no better than animals." Others describe her as: "...a whistling tea kettle, who never explodes. We can't begin to describe how she feels about the dogs. Rachel needs a catharsis session bigger than any of the others. She is all of us and therefore has no one else! She needs to find out if she can survive alone." I encourage her to begin writing down her reactions.

*

"Talk to me," says my therapist, but I don't know how to say the words anymore. I don't know where to start, how to explain, describe what is in my mind. Part of this is caused, I'm sure, by my own confusion and part by my anger at him. I don't quite understand all of the anger, but it's there. As far as talking to him about the past, I just don't know. Where should I start? Sometimes my head feels as if its in a vise, as if an outside force is pushing on my head. I feel so incompetent and so very evil. Maybe I should tell him about the bruise, the large, ugly black one on the left side of my chest. Grandpa hit me there, with his fist, hard. Knocked me over. I wanted to pretend it wasn't there, to push it away, but I couldn't. I touched it, pushed on it every day it was there, pushed on it so that I felt the pain all over again, so that I wouldn't ever forget how badly it hurt.

"I'll be good, please, I'll be good." It's disgusting to beg, but I do it anyway, can't stop myself this time. The hard thing slides all over me, all over my face, my whole body, between my legs and I'm flipped over, a pillow put over my head and I can't breathe. I can't move and the hard thing slides between my legs, again strong hands hold me down, other hands pulling at my butt, exposing me and then pain. Please stop the pain, I can't take it, please stop it. But the hard thing is inside me now, pumping in and out, almost lifting me off the bed, and then I feel the stickiness and he's done, please let him be done. When he stops, he hits me hard across my butt, hitting harder and harder, calling me a whore, then hitting me with a belt, screaming at me, telling me I'm filthy, dirty and I feel something sticky squirting all over my back and he tells me again that I'm a dirty whore and I know I am.

I can't tell my therapist these things. I can't say the words. It's all a secret, my secret. I've become a hostage of my past, I know that. There are so many "should have's" in my past and my present. I should have stopped them; I should not have begged that one time; I should have killed them. There are a lot of people I still want to kill, to hurt.

I don't think my therapist understands my not being able to say the words, and so I feel worse. I think about dying all the time. I would like to die, to be released from all the filth, from my dirty, ugly life. I don't like looking in the mirror. He used to make me do that, stand in front of the mirror and tell the image in the mirror, me, that it was dirty, ugly, a worthless whore. For hours I'd stand there, repeating the same words over and over, until he allowed me to stop, to crawl away on my hands and knees, the proper procedure for a slave to follow.

I'm still a slave--even now, so many years later. All those thoughts and I can't sort out the feelings to go with them. I should cry, but that would be false, it wouldn't go along with whatever feelings are there. Often times I think I'm crazy, totally off the wall. That used to frighten me, but it doesn't anymore. Maybe it would be better to be crazy. I read about child abuse and it doesn't seem to apply to me. No one ever pretended they were playing a game with me, tried to tell me it would feel good. Would it have made a difference? They didn't tell me I was special. They didn't play with me, didn't play with my hair, didn't caress me. I'm dirty, dirt, dirty and ugly, ugly, ugly. I despise me, I spit on me, I'm nothing, zero, worthless. I don't deserve to be here.

They went to church, grandpa and grandma, Heer Doktor too. They had their "outside" manners and their "inside" manners. To the outside world they presented themselves as Christians and me as their darling granddaughter who loved to stay with them. So polite, they were. So respected, always ready to help anyone. Such bullshit--such hypocrisy.

Please! I don't want to go home with Heer Doktor--please don't make me! But I go anyway, silently, dragged along behind him, then stripped when we get to his house. Next, a belt whipping, all over my body red welts appear. He wants me to cry. I know this and refuse. Next comes the dog collar, brown leather, attached to a harness of chains and leather, then my hands tied behind my back, the rope then pulled down to be tied around my ankles. He's perspiring by now and I can see his thing is hard, even though he still has his clothes on. Pretty soon the other people will be here. Show time! He drags me along the floor, pulling me along by the leash attached to the collar. I choke. He slaps me. He doesn't talk to me at all. The other people (all men) arrive and they all strip and touch each other. Then they sit down, waiting, in anticipation for the big event. Heer Doktor then brings in the dog, a large German Shepherd. I'm already nauseated. It happens every time and I talk to myself so I won't puke. I'm on my knees and my muscles are tense, waiting for the first lick, the initial shock as the dog's nose goes between my legs, and it's tongue starts to lick me. I try not to be sick, but I puke anyway and I hear them laugh, louder and louder. They push my face in it, rub it in and I feel the dog as it climbs on my back, paws on my shoulders, feel it rubbing against me faster and faster and then I'm all sticky. Please let me die, I think. My hands are untied, water thrown in my face and another dog is brought in. This one I have to masturbate and I get sick again. When I'm all done, they tie me, spread-eagle on the floor, and they spray their sticky cum all over me. I puke again and I hate me--I'm so weak! It's all my fault they're laughing at me, my own fault.

How do I say all these things and so much more? How do I actually say the words? I can't even explain the anger directed at my therapist to my therapist. I can't explain all the confusion, the fact that to me love is not a reality, but rejection and hurt are. I no longer believe that it will change for me either, that I will change. I hate. I think that's the only remaining emotion. And I want to die. That's real. No one has ever wanted me, just the sexual part of me.

I would have been an abusive parent. I know this. "Anthony, get off the coffee table right now. Anthony, please get down." He's two years old and walking all over my new coffee table. I've been married six months and I'm baby-sitting my husband's nephew. I'm running out of patience; with Aaron, with James and his sister. "Anthony, get down--now! Grab him by the arm--crack--hard, very hard, across the upper back part of his legs. It leaves the imprint of my hand. He screams and sobs and I cry. I rock him and rock him. "I'm sorry, Anthony, oh, little baby, I'm sorry." Rock him, sing to him. Cry and cry. I'm so sorry. I'm evil, bad, dirty, ugly, evil. I would have been a bad parent. No wonder they made sure I couldn't have children.

There is too much in my mind. I don't want to think anymore. Can't sort it out. No one can help me. I'm so lost--all by myself. Pressure is building inside and I can't explain it.

Did I mention the physical pain? I think I experienced what is called exquisite pain. I need to describe it, I need the words. Please, help me find them.

I feel as if my world is a windowless prison. No way out.

Rachel W.
June 24, 1988

After two more sessions:

I must protect myself, keep myself safe. I shouldn't have let today happen. I need to keep in control. I'm giving too much of myself away. I feel as if all my defenses are systematically being broken down against my will and I won't allow it. I allowed it when I was little, when I begged them not to hurt me, and I won't ever do that again. I promised myself a long time ago that I wouldn't. I can't be that weak again.

"Please don't hurt me, please, please. I'll be good. I'll do what you want. See? I'm taking my clothes off, I'm spreading my legs open, wide open, just the way you want me to, the way you like it." He hits me anyway, hard, the palm of his hand cracking across the side of my face; makes me bite my tongue and I taste the blood in my mouth, feel my lips swell. I know he likes it when I humiliate myself by begging and I vow never, ever to do it again.

He smells, grandpa always smells; so does Heer Doktor. They smell of sweat and urine, filthy. I throw up. They laugh, loudly--it hurts my ears. "Stop laughing, damn you." I yell it silently, not enough guts to scream it so they can hear. They shove their hard things in me, Heer Doktor in my mouth; he smells so vile. I try to think of something else but can't. I'm feeling the pain too much. I gag and choke. Slap, slap! That's not allowed. Grandpa inside me, deep, and I feel myself tearing inside, bleeding. When he comes he shoves a broom handle in my butt. Pain, hot, searing pain in my whole body and I pass out.

"Daddy, Daddy, please save me, please come get me." Of course, he never does, never hears me. He's scared too. Afraid of grandpa, even though he's grown up. Sometimes I hate him.

I often think of ways to die, as an adult just as I did when I was a child. It would solve everything--how wonderful to not feel anything. No pain--total release. Something, an unknown source always stops me. That too makes me angry.

Please don't touch me. Leave me alone. You don't mean those soft words, those soft touches. I know what you're going to do--you're going to hurt me again--just like always. No change, nothing's ever different.

Let me be, please let me die. Do you want my soul too? You've taken everything else; all my dignity, my pride, my self-worth. I'm totally empty, an absolute nothing. Can't you see that I'm not here anymore? Damn you! look at my eyes. Look at them--you've taken my life, you and your bitch, now let me be.

Dear Therapist:

I think I'm insane. Maybe you should lock me up, throw away the key; why don't you just forget about me. I'm going to blow up in a million, zillion pieces if you don't let go of me--I'm not going to cry, cry, cry anymore, ever again. You don't know, you can't feel the pain. You said it hurt you, how does it hurt you, why does it hurt you? Do I trust you? I don't know anymore. My mind isn't quiet--it hurts inside my head--a loud roar is always there. Please let me sleep.

Time for the dogs again. New position, new show. On my back, dog collar around my neck--all animals wear collars. Legs up in the air, spread widely apart, held there by chains hanging from the ceiling. Two dogs at the same time. One positioned over my face so I can stroke him, masturbate him. People stroking each other while they watch. The other dog between my legs, licking me, down on its haunches rubbing against me, trying to get in me. Their paws are scratching me. They cum, all over me and then they pee on me. I puke and choke, puke and choke. I have to lick it off the floor. A woman lies down and I have to service her, lick her with my tongue while someone butt fucks me with a dildo. I'm lower than an animal, dirtier that dirt--filthy, filthy, filthy. Guilty of terrible things. Off with my head! I'm not fit to lick their feet, they say, but they make me do it anyway. My stomach feels sick again. I have to learn to control that and eventually I do.

I used to have dreams when I was little, important dreams. I was important to someone, someone's special little girl, special granddaughter, maybe even a fairy princess who wasn't evil or dirty. I don't have such dreams anymore. It's no longer important. Life never changes.

Do I love anything or anyone? Do I even know how? Is it worth it? I used to try to love but it never worked. I tried to be a very good, good little girl but I was never good enough.

I'm thirsty, haven't had anything to drink or eat in two days. Please, I need some water. Little whore bitch, you must walk across the room on your knees fifty times first. So I did. Then, little girl, little nothing, you must stand naked without moving for one hour. Ah ha--I saw your finger twitch, that's an extra hour. Finally a sip of water, one sip. Please let me die. I can't give any more, I can't take any more.

The humiliation never goes away, even now. It inhibits me, makes me cringe inside. I don't want anyone else to know. I need to hide myself, keep my bad self hidden forever, locked away.

Sometimes I hate children. I don't mean to but I do. I watch them play, eat, sleep, with my mommy and daddy loving them and I get so angry inside that I now avoid all children. I hate them when they cry, or whine or throw temper tantrums. Please, please forgive me. I know I'm bad.

I'm standing in front of the mirror again, naked. I'm cold. I have to look at myself and repeat after him--my body is ugly--it's the worst body in the world--it's only purpose is to receive cocks and punishments. That's all that bad girls get. I hate looking at me. Please let me cover up.

I still smell all the men and women, smell the sweat and urine and sex, feel the dogs and it never stops. I'm not meant to forget!

Rachel W.
June 30, 1988

After two more sessions:

Sometimes, when I'm alone at night, in bed, I feel as if I'm a little girl again. I'm lying there, waiting for the footsteps, the hands on my body. Many times I stay awake all night, listening, waiting. Sometimes, in my head, it still happens.

Please let me sleep tonight. I'm so tired and I can't stay awake another night. But they pull me out of bed anyway. Time for fun and games. Bring the little tramp over here, someone yells. Mustn't let them see me cry, please, please, don't let me cry. I don't. I'm learning. Dildo time. Bend over, touch my toes--that's the rule. And the pain is there again as the dildo goes in my butt. No one to rescue me--no one ever rescues me. The dildo goes in and out, in and out and pretty soon I can see the blood hitting the floor. Please, please stop the pain. I'm going to die from it--please let me die from it.

Dear Therapist:

Try to imagine the pain, damn it, try! All that pain--I can't get rid of it. You can't begin to comprehend it, to feel what I'm feeling inside every minute that I think about it.

Please don't do that, it hurts. But I don't have a chance. On my knees, chain around my waist, legs pulled up behind me and fastened to the chain--I know it will be hours before they undo them--my arms are tied overhead--and then it starts--one by one the men stand in front of me, some with cocks already hard, others still flaccid, but eventually they all cum, every one of them, in my mouth. I don't gag because I've taught myself not to, but the nausea is almost too much to handle. I'm almost beyond accept-

233

ing the pain, my mind won't take it anymore and when they finally unchain me the pain is so intense that I almost scream.

You, dear therapist, can't understand that kind of pain--you don't even understand the emotional pain. I am sorry if I behave in a grandiose manner, that my therapy isn't progressing any faster than what it is, but damn it, I'm lost and I don't know where I'm at anymore. You tell me I'm not going crazy, but I don't believe you!

Rachel W.
July 6, 1988

We have two months before their trip to Europe. Rachel is now refusing to go. It seems clear that she represents the transition between being able to "fall back on" each other, having to handle things on her own. She greatly fears having to confront visually the scenes of her childhood, especially because she has to do it alone. During these two months of sessions, she does eventually accept that she really does need to go and even agrees to keep a journal, as do several of the others. I will be eager to see Rachel's journal.

33

Friday

I'm here--in this airplane and I don't want to be here. I wish I had never agreed to go after Dr. Ted asked me to--I want no part of the Netherlands, it's people and our supposed family. The sunset is magnificent, though. Makes me want to open the window so I can jump out and touch it. I wonder how many minutes it would take me to fall to the ground and hit it with a big splat! I'm glad the plane's not full--I can't stand having a lot of people around me. Kate is very nervous--I can tell. At least there is not someone sitting next to her, except for the young man that keeps coming over and flirting with her. If he doesn't go away soon I'm going to have to rearrange some of his body parts--he's obnoxious. I wish I had said no to Dr. Ted--I feel very confined. I can't get out of this plane and I can't breathe right--no fresh air. I want to go home, to my own apartment, my own bed and I want Priscilla. Please help me.

Sunday

Went to church today. Understood very little of it. Doesn't matter. I seem to have lost part of the day. I felt so cold and all the people looked the same, only I couldn't place the face that they all seem to represent. I almost got hit by a car, but Kate pulled me out of the way. It made me physically sick to be there.

Went to Lyonne next (the home of her grandparents). Somehow I lost time there. I found River Drive--that's where they lived. When I walked down the street my legs felt as if they were lead--I felt the same dread I felt as a child when we would visit there or when I had to stay. The closer I got to the street, to the house, the more nauseated I became and there was this tremendous pounding in my head. I felt as if everything was in slow motion--not just while I was at River Drive, but immediately on entering

235

the city. I felt something evil and it was going to grab me and pull me apart, consume me. I looked at all the faces and thought I recognized people--maybe I did. And I found the houses--or one house and one empty lot--I felt as if everything was still the same, just like it was when I was little. My legs felt as if they were going to give way and I wanted to scream, to lift my head up and howl, but I couldn't move. And then I saw the lady, the neighbor across the street, Mrs. Veenstra, and she wanted to get her husband because she said he would enjoy talking with me too, that she remembered me and that she was sure her husband would too. But I never met him, that's all I remember. Now there's rage and I don't know what to do with it. I need some help. I feel as if I'm going to suffocate. I feel abandoned. Dr. Ted should be here but he left me too. Someone has to help me but there is no one here. I'm a stranger, even to myself.

Monday

Today wasn't too bad--I'm jumpy and so very cold but at least I'm away from Lyonne. I didn't go with the others to the little towns they wanted to visit--I really didn't want to see churches and museums. I wanted to be by myself and I would really like to call Dr. Ted. I need to talk with someone about yesterday, about the flashbacks. They are so vivid today--me as a little girl, all those people standing around--no clothes, belts, chains--but they are all fragments--nothing fits together. I'm glad I'm alone today. I can't handle the noise, the constant chattering. It makes me so angry. I want to smash everything in sight. Most of the time being alone is best for me. I can't deny that this is the country of my birth -- wish I could!

Tuesday

Went to Port Seine today. It made me want to cry. The "good" grandparents lived here. House is still standing. I remember sleeping upstairs and waking up to the sound of a train whizzing by. I loved it. Went to the cemetery and looked at their graves. Why couldn't they have been here? They should have kept me with them. I feel so sad and so terribly alone. I remember walks with my grandfather and the monkeys and french fries on the street corners; or maybe fish, but mostly french fries. Ate them with mayonnaise. Still like them that way best, but I never do that in public because everyone eats them with ketchup in the U.S. I remember the canals and the market. It's so beautiful. Kate is very moved by being here too. I see tears in her eyes every once in a while. I think we both wish we could show Dr. Ted. The train station is as beautiful as I remember it. I ache inside. It hurts bad.

Wednesday

I went to East Byron today (Kate's original hometown). It's the same. It hasn't changed. I wanted to destroy it. The same pounding in my head as in Lyonne and I wanted to scream. I wanted to destroy the quietness in the village, to bring fear to it's people, to let them see my rage, but rage at what? I still have only fragments, but I want to tell the people that they have to right the injustice done to me. It's becoming more and more difficult to concentrate on the immediate family I'm spending time with. I want to go on a rampage, to destroy. It hurts to be here and it crushes me inside so that it's hard to breathe again. Just like in Lyonne. I need fresh air, only I'm outside. What am I going to do? I need someone to help me.

Monday
Back to East Byron. It was so hard to control the anger, to remain civil. I wanted to use all the obscenities I've ever learned, to smash glass, create havoc. I don't want to see the inside of the school, the inside of the church where I was baptized. It means nothing. Some of these people know me. Where were they when I needed help? Aren't children supposed to be precious, magic? I was abused, damn it. I was beaten and physically and emotionally degraded. Where the fuck were all these friendly do-gooders then. I spit on them. Maggots, all of them--no better than those who did this to me. I wonder if my mind will ever be right again. Everyone ignored me, ignored my pain. They must have seen it, must have known I needed help. But no one cares and there's no one I can truly trust. Trust doesn't last--it backfires.

Kate went to her hiding place for the second time. I'd like to stay there for awhile. It's so very peaceful. But it's not my hiding place. I haven't ever had anything that was just mine. I really hate it here. I want my own home--it's safer. Here I feel violent inside and so very confused. I need to be held and rocked. I want to be a normal little girl first and then an adult. I'm homesick.

Tuesday
Went to "our" childhood girlfriends' home this morning for coffee. I didn't recognize her but Kate did. Maybe because I was so angry with her. She should have helped me when we were young. I tried to tell her, tried to tell her mother but they wouldn't listen, wouldn't believe me anyway. I shouldn't have gone there. I wanted to tell her off. I keep visualizing myself as a very little girl, one who was very alone and still is, even as an adult. I hate it.

Wednesday
The others did more sightseeing today. I didn't go. I needed time to myself. I'm so angry and frustrated. I need to put everything together, to

find the thing that ties all the fragments together. I don't know what that thing is, though. I keep searching and my mind is blank. Maybe part of me doesn't want to know. I know I'm afraid and I'm so sad. I was sad before I came to this country but it's different, it's heavier and combined with the anger it's almost too much to handle. I want to let it all out but now is not the appropriate time and I know I'll bury it again if I wait too long. I'm still cold. It would be wonderful to feel warm, even for a little while. I could use some good sleep too. I have nightmares or else I can't sleep at all. I know I'm looking for something and that I'm supposed to find something here but I can't find it.

Thursday

We stayed home this morning. I was glad we did; so were the others, I think. I'm so tired. I keep seeing a little girl in my mind, no clothes, all tied up, unable to make a sound. I'm so scared and I need a hug and someone to talk with. I shouldn't have come here. I should never have agreed to this. I feel so threatened. I look at all these strangers around me when I'm out and wonder if they know, if they recognize me from a long time ago. There are always faces I think I recognize and I want to stop and ask them if they remember. I feel bad inside, as if I've done something bad. And ever since I've been here, I feel so terribly dirty. I also have to try and not think of Lyonne. That only intensifies all the bad feelings--the anger and the sadness.

People

They mask their faces with
smilesand tell you that they care,
They tell you that they like
youand that they want to
be your friend,
but look out for what's
behind the mask, for in
truth, they really
don't give a damn.

Marie

The Child

She sits silently, not moving,
eyes vacant, staring blankly,
seeing nothing.
There are sounds around her,
but she does not hear,
Hands touch her, but she
does not feel.
She gives no response to
life around her, so people
turn away.
And two tears silently
roll down her cheeks.

Marie

34

It's Sunday, November 6, 1988 and we've arranged a special session for Rachel. The consensus of opinion is that she is ready to break through her memory blocks and confront the terror. We are doing it on the weekend so that we need not worry about repercussions on her job.

Cassandra continues as my primary therapeutic ally and interpreter. Rachel is aware of the process she's involved in even though she has no memory as an adult of the key childhood events. As is typical, she too will have to first experience the terror at the age it occurred before she can heal as an adult.

Rachel represents the very key stage in Kate's "evolution" back to wholeness--the stage of having to function alone and without the security of any of the other personalities to help her or to "cover" for her. Even though she arrived in 1985, and did not herself experience any of the childhood trauma, Kate's system has unconsciously created her to assimilate everyone's memories and feelings, and to confront them alone. I think I know ahead of time the two key people and issues that will provide the clinical means to break through the blocks that separate thirty-nine-year-old Rachel, and Rachel as a regressed five-year-old. The thirty-nine-year-old Rachel has no awareness of, or memory of herself as aged five.

Cassandra arrives for our session. She has brought some key pictures taken on their trip this summer, especially one of Mijn Heer Veenstra, the man who lived across the street from the grandparents. He is now a very old man who reluctantly allowed Kate to take his picture (be reminded that Rachel has taken over Kate's experience for her). He actually began walking away when Kate introduced herself, but his wife pulled him back and let Kate take the picture. I have been informed in previous sessions by Cassandra that at age five, Kate went to Mijn Heer Veenstra and told him, "Grandpa and grandma beat me, hurt me, put things inside of me; would you please help me." Mijn Heer Veenstra promised to help, gave her candy, soothed her. Approximately three days later, Kate is blindfolded and tied down. A large naked man is on top of her forcing intercourse. She can hear many others all around them laughing and breathing heavily. Someone takes off her blindfold so she can have her special surprise. The man on top of her is none other that Mijn Heer Veenstra! Kate's sense of betrayal was monstrous and all-encompassing.

Cassandra and I begin our session with time taken to decide on our strategy. I then ask for Rachel, who gives me the journal of her trip to the Netherlands. I do hypnosis for age regression and five-year-old Rachel arrives. She trusts me now as a result of a number of previous sessions of groundwork. I give her a stuffed animal ("Garfield" the cat) and she is wide-eyed with excitement as she sits on the floor at my feet.

I show Rachel a picture of Kate's secret hiding place, taken in September. It is basically the same as it was 35 years ago: much undergrowth next to a curving canal with many trees on the other side. "Dr. Ted, Kate could lie down there for hours and nobody could see her."

I show Rachel a picture of Mijn Heer Veenstra and she quickly puts her hands over her eyes. "Please, Dr. Ted, don't make me look at that." She then grabs Garfield and crawls to the far corner of my office. I tell her that Mijn Heer Veenstra is now sitting in the chair next to her and ask her to look at him--she won't. I move to his chair and order her to look at me, knowing that in her eyes I will become him. She looks at me in terror and tries to crawl away. I hold her in position. She struggles, and cries, and begins to yell. I pin her down on the floor (Cassandra's description of what was done to her) and Rachel now clearly is fully re-living the terror of her abuse. The feelings come pouring out: the hatred and terror, alternating with: "I'll be good, Mijn Heer Veenstra," followed by punching and kicking and screaming, and then passive acceptance.

I move to her other side, while still pinning her down on the floor, and tell her I'm Heer Doktor and say "here dog." She screams, she struggles, she cries all the more. Her catharsis is huge. I have all I can do to hold her down. It takes a good fifteen minutes for the outpouring to begin to spend itself and we are both drained.

I now sit behind her, let her assume the fetal position and let her rock, re-assuming my identity as Dr. Ted. I put Garfield in her arms and allow the softer sobbing to just happen. I sense her beginning to spontaneously age: going to her teen years, becoming age five again, then age seven, then in her twenties, then age five again. Cassandra again sporadically appears

240

to interpret for me her thoughts and her age at given times. We have clearly broken through the amnesia, and I am relieved. By the end of the two hours, Rachel has returned to age 39 and has full memory of the entire session. I know that for awhile, she will vacillate between age 39 and various other young ages. I encourage her to write down her thoughts about today's experience and give her a big hug. I see her first faint smile.

November 6, 1988

I had an extra session today, one that has given me something to think about. For months now, there has been something missing, something I couldn't put my finger on. I wasn't sure if it was a thought, a feeling or maybe a specific memory. What I have been unable to accept is my fear of being afraid, if that makes any sense. Today, I was afraid and I wasn't able to hide it. Heer Doktor was back and he was going to hurt me again; I remember that. The problem is, I can't put it all together. I know there was more than Heer Doktor and the dogs but I can't remember. I know there was anger, but mostly there was fear. I was terrified of the hands touching me, all over, spreading my legs, of being used one more time. And there was rage. I wanted to hurt back, but I felt so weak, and it was so hard to breathe at times. I felt so exposed, so vulnerable and sad. I'm still sad. I wish there weren't so many gaps in my memory, yet at the same time I think it would be easier to not have a memory at all.

I remember Garfield. He's mine to hold whenever I want to. I feel as if I'm a child again and I want to be hugged and rocked again. My head is still pounding and I'm still sick to my stomach. I'm all alone again. I tried to be good and do what I was supposed to, but I think I was bad. It wasn't enough, so I'm still all alone.

I see Rachel the next day. She has vacillated between ages 5 and 39, and at age 39 has kept her memories of "Heer Doktor," but has lost all memory of Mijn Heer Veenstra. It is clear that there is either more feeling, or more memory that needs to be uncovered and we have another difficult and very emotional age regression and confrontation with me as Mijn Heer Veenstra. The most important new variable this time is the memory that she was blindfolded. This time I am hopeful that all her memory will remain intact.

November 7, 1988

Had a big session today. I'm not sure how I feel about it right

241

now, although Dr. Ted seemed pleased. I'm remembering things I don't want to remember, things I don't ever want to think about again. I know I've been afraid to remember for a very long time now. But now Mijn Heer Veenstra is there once again. I know he remembered me when I saw him during our trip. How could he forget? He abused me, raped me, but most of all he betrayed me. I really believed in my heart that he was going to help me. He was kind to me, listened to me and he promised me that everything would be all right. Instead, he lied and did the exact opposite. He didn't even wait very long after I told him. Just a couple of days. I remember, vividly, what that felt like. I was tied to a bed, blindfolded, and I could hear the people standing around talking and laughing. Hands were all over, touching me, pulling my legs wide apart, and then just when I felt someone's penis go inside of me they pulled the blindfold off me and there was Mijn Heer Veenstra, grinning at me, telling me I was going to be good to him. My grandparents were there too and when I yelled at Mijn Heer Veenstra that he had promised to help me--they all laughed. The physical pain that evening was terrible, something I'm not going to think about. The emotional pain of knowing I had been betrayed was overwhelming. I can't begin to describe how I felt, knowing that someone I had counted on for help, someone I had trusted had just thrown that trust away. The physical pain was even more intense than other times, a punishment and a warning to not talk to anyone else again. I never talked to anyone else about it again and I never trusted again.

Rachel

Three days later, when I see Rachel next, it is clear that we are not yet finished. Cassandra lets me know that there is much continuing vacilla-tion in age, that memory has basically remained intact, but that Rachel's rage is huge. We have our third extremely emotional, this time even more physically demanding psychodramatic catharsis in a row. Although it now appears that these memories will remain, I am growing in my expectation that there must be something further to uncover.

November 10, 1988

I'm all alone again, just like always. I had a session today. We talked about Mijn Heer Veenstra and I got very angry. I don't want to think about or talk about that time in my life. It makes me

242

feel bad and it makes me want to hurt people.

Rachel

Doubt

Are you really my friend?
Do you share like I share,
care like I care,
need me like I need you?
Are you really my friend?
Am I yours?

Marie
April 4, 1984

Solution?

It screams for relief,
for release from the pain;
don't want to remember,
don't want to feel,
don't want to think anymore.
Fifty altogether
different colors, different shapes,
cold water, swallow quickly.
An hour later a quiet mind.
Permanently.

Marie
July 17, 1984

The following night I get a call from Marika. Marika had been a Group 2 personality (age nineteen, arrived when Carol was sixteen) who did not endure any of the trauma. She had been fused for some time and her re-emergence tells me that there is additional trauma to be confronted. She had interpreted for me previously. She now tells me that these critical events in Rachel's life occurred at age four not five. She adds the comment that Rachel was told that her parents and her brother Phillip were dead, and

243

that Rachel believed this for one week. She also tells me that Rachel's anger continues to build; that none of them can control her; that they are having difficulty at work and they are fearful of the consequences. I must consider her potential for harm to herself, or to others and we arrange another emergency session. I will also use this time to evaluate for hospitalization. It does appear that the dam is bursting.

In session, Rachel arrives as the 39-year-old. Her anger is truly of rage proportions, but alternates now with moments of extreme fright. By now I have been able to "train" the five-year-old Rachel to arrive by signal (done with post-hypnotic suggestion), which I do. I am able to assume the role of Dr. Ted, or the grandfather, or Mijn Heer Veenstra depending on what chair I sit in, which allows me to have control over the direction of her experience. This means that if things are becoming too powerful for either of us--emotionally or physically--I can gain control quickly by assuming a new identity.

We proceed and the terrible details emerge. Rachel is told that she has to be punished for telling on the grandparents. The first part of her punishment is that her mother and father have been killed, and to make sure she "really learns her lesson," her brother Phillip, whom she loves more than anyone, also has to be killed. She is kept for the week out in a back yard shed and is regularly terrorized. Her head is held under water. A bag is pulled over her head with a rope tied around her neck. A pistol is put in front of her eyes and the trigger is pulled, which is when she finds out that "this time it isn't loaded." This goes on for one week, after which she is sent home alone on a bus, with her total belief by now that she is alone and will have to take care of herself. (In these small Dutch villages, a child alone on a bus is not uncommon, since everyone knows almost everyone else, and it is understood that the child will be safely transported to the correct location.)

As the information emerges, the emotion is immense. Her rage alternates with her terror. She tries to punch, she moans, she screams, she whines, she kicks, and swings, she calls out every foul word in their collective vocabulary, she cries, she promises to be good, she swears she'll kill me. I am amazed that after three lengthy catharses already, there remains this much emotion. I must work hard to prevent her from physically hurting me or herself. After about forty minutes of this outpouring, the energy is spent, and I am with a quietly weeping, thirty-nine-year-old Rachel.

When she is able to talk, her voice is firm and very adult. She is able to recount the entire experience and has retained complete recall. Her greatest surprise seems to be encapsulated in her comment: "It's hard for me to believe that I'm not dead."

I know that this has been a wonderful step for them all, clinically. I am elated that we seem to have finally broken down this huge wall, but know that both Rachel and Kate need time to process all this--as do I.

November 11, 1988

Had a special session this evening. I didn't ask for it; the others did. I didn't want to go back to age five or four either, but somehow I did. I remember I got very, very angry. I was little again, and so very frightened. I was told my parents and my brother were dead and then they "pretended" to kill me. I wish they had. I really believed my family was gone and when they sent me home alone I was convinced I was going to find pieces of their bodies all over my home, because that's what they told me they had done--cut them into little pieces and scattered the parts inside the house. For a week I alternated between fear and wanting to die, and an immense rage, a rage big enough to want to kill them. I know I fought and cried this evening, but I still don't feel as if I accomplished much. I still feel as if I lost. It is so difficult to retain the memories of those times because they are so very painful. I prefer to forget. When they are in the front of my mind, it becomes difficult to sleep and when I do sleep I have nightmares. Sometimes it's even difficult to breathe. It was this evening and that makes me panic. Even though I let a lot of anger show this evening, there is also a lot left inside of me and I think that it needs to come out. It scares me, though.

Head bowed,
knees bent,
softly saying the words,
Yes Sir, No sir--
I promise to
be good, Sir--
won't ever happen
again, Sir!
once again feeling
demeaned.

Erika
February 16, 1988

Chop! Chop!
tiny little pieces,
Chop! Chop!
smaller yet--
Dig! Dig!
Bury the little pieces,
Dig! Dig!
Deeper yet--

Jump! Jump!
everybody's jumping--
Jump! Jump!
the Bastard's dead!

Erika
February 16, 1988

During the following week, we have our regularly scheduled two sessions. After the events described above, we both badly need some quiet time to process and try to put into perspective what we've experienced. What troubles me is that Rachel continues to vacillate greatly between ages four and thirty-nine. I had hoped that her work was basically completed by now. Marika continues to advise and I realize how dependent I have become on her insights to guide our direction. She lets me know that the efforts to quietly talk through these horrible terrors at this point will not suffice. She lets me know of the clinical importance of one further male abuser named Hans, and the need to expel further rage at Mijn Heer Veenstra, Hans, and the grandfather for the week of terrorizing her into believing her parents and baby brother had been killed and cut up into little pieces.

I agree to another weekend session. Marika and I plan our strategy for exposing the two agendas noted above. Our plan basically involves my accepting the identity of Hans and forcing the four-year-old Rachel to relive her terror. The thirty-nine-year-old Rachel agrees to cooperate, knowing another difficult time is about to occur, without fully knowing why.

After arranging the four-year-old's presence, I show her the picture of Mijn Heer Veenstra. I quickly am able to assume the identity of Hans and what has now become our fifth confrontation in twelve days ensues.

Marika has let me know that Rachel badly needs some sense of having

246

won something from her terrorizers, and I intentionally allow her to physically dominate much of this catharsis. (She literally hits with a four-year-old's power, so I am able to prevent either her or myself from getting hurt.) Her rage is tremendous. When I describe how I've cut up her mommy and daddy and baby brother, her eyes are so big the terror is all-encompassing. We continue for a good forty minutes. Her screams are the loudest I've ever heard from her. The intensity of her need to kill me is scary. But the dam has burst. What a terrible way to have to heal.

As I sense the momentum of her outpouring is about expended, I assume my identity as Dr. Ted, and I encourage her to cry. And she cries. And she cries. All I can do is let it happen. And she cries. And she cries. I then softly begin to describe a swirling maelstrom of thirty-nine distinct colors, slowly and beautifully moving together in a magnificent scenario of blending and movement, which eventually integrates into one absolutely lovely, independent and strikingly bright and distinct color. This imagery is powerfully symbolic for her and suddenly a subdued 39-year-old Rachel opens her eyes. She's quiet for several minutes. She looks rather confused and uncertain. And she finally says: "I'm all alone." After a few minutes of silence she almost shouts: "But you don't understand. I'm all alone!"

It has finally happened. After almost forty years of being able to depend on other personalities to help her survive, one personality is having to now function as a whole person. This does not mean that full integration has occurred; it means that for the first time, one personality who has full memory of the horrible truth, no longer can communicate with or can even have any awareness of another personality.

How wonderful to realize that this little girl, abused and terrorized, humiliated and debased beyond what any of us could ever truly appreciate, is going to win the war. How inspiring it is to be reminded that the human spirit can be this strong. How grateful I am for my profession. How grateful I am for my own special two children. And I say a prayer for a frightened and now really alone Rachel, knowing that she has paved the way for Kate.

November 19, 1988

Last evening Rachel had a very emotional, important session. Even though I knew this session was necessary for her emotional health, it was difficult for me, as well as the others, to watch. Our first instinct was to protect her, but we also realized that doing that would not be helping her. It was Rachel's time to confront her fears, all the old demons from the past and she had to do it alone.

She had to relive the experience of betrayal, of being beaten and abused sexually and emotionally. The degradation, the shame she has felt for years has been immense. All of us have had these feelings to one extent or another, but none of us have had to experience them totally alone. We always had each other there to back us up, to take over, but Rachel has none of that. Her anger was, and still is, all hers. It doesn't belong to the rest of us. It is no longer shared. She has reached a point in her life where she has to depend on herself. All of us, at some point, will do the same, although I think it will be easier on all of us except Kate, who will be last. I found Rachel's rage frightening and it was painful to realize that it was strong enough to want to kill. I shouldn't be surprised, though. She was told her parents and brother had been killed, cut into pieces, and she believed them. That's a horrible thing to tell a child. They used mind games constantly to gain control of her. And Rachel couldn't show her fear or anger or sadness. As an adult, she has a very difficult time accepting and admitting to fear. I know that she is going to have a difficult time now that we can't cover for her anymore. She's going to feel betrayal, not just by us but by Dr. Ted also because he won't bring us back to her. She doesn't realize that he can't do that and it's going to be some time before she will be able to accept that. She feels a very deep hurt right now and it is going to instinctively make her back away because she's had so much hurt already. It is, of course, the worst thing she can do. Now is the time when she needs to be close, to talk and share her feelings. It will be overwhelming for her--so many details to remember and attend to. None of us have ever had to do this before. We've all had our own functions. In some ways, I'm afraid for her because she is confused right now and I'm afraid she will reject everything and everyone around her. It's a tremendous change for her mind to be "quiet." I can understand her panic. She's going to need a lot of patience and understanding.

Marika

Epilogue

Since the dramatic events outlined in the last chapter, it was revealed that ninety-eight other personalities, combined in their own sub-group-ings, have existed since Kate was a child, unknown to any of the others. They have silently observed the others all these years, but have also silently and vicariously experienced therapy. They made themselves known when their therapy was almost complete, and they were ready to integrate. In addition, one more transition personality has emerged. This makes a total of 233 personalities who have been operative in Kate's troubled quest for survival.

Even though the horrors revealed thus far are truly terrible, it turns out that they were not the full antecedents of Carol's multiple personality, we now know that one layer of personalities (and the horrors that resulted in their formation) can cover a deeper layer of personalities and even more traumatic horrors. The basic rule-of-thumb appears to be: the more traumatic and the more on-going the abuse, the greater the number of personality and personality fragments.

Kate also needed to regress to age four and confront further terrible events protected by amnesia. As four-year-old Kate, she had to retrieve grotesque, almost unimaginable tortures involving what clearly was an elaborate child pornography organization run by the grandparents. Kate was displayed and sold to the highest bidder--who were on three occasions two especially sadistic men. These memories involved mostly a series of unbelievably disgusting, degrading, brutally violent and terrifying animal tortures with blatant sexual violence--the many specifics Kate did not outline in this book.

There is, however, one event we have decided to share. Four-year-old

Kate is taken home by the two aforementioned men. They put a dog collar on her and force her to crawl and beg as a dog would. She is subsequently examined and found not to be a virgin, so they tie her down and literally sew up her vagina. Now they have the thrill of "de-flowering" her again.

In the more than a year of therapy that has occured since our last chapter ended, it was revealed that Kate's ordeal began when she was two. And, that her family was involved in a rather extensive underground Satanic cult. This, of course is the roots of her multiple personality.

I expect that almost anyone else would have either died or become psychotic. It would appear that Kate's intelligence was the key that allowed for this brilliant survival technique to be activated.

The details of Kate's cult history, which is essentially the story of the child who was chosen to become the new high priestess, are so troubling that she has decided not to include this information. But it is worth knowing that it did indeed happen.

It is also good to be able to report on developments in Kate's relationship with her parents. Although they've never accepted Kate's need for therapy, and still do not know she is a multiple personality, clear healing has been occurring anyway. All are able to say: "I love you" to each other. The parents see more stability in Kate's life. And Kate has come to see that her parents were also victims of the grandparents brutality. Kate can now much better understand her father's lifelong struggle with depression. The result of this is that for the first time in her life, Kate has actually achieved some sense of belonging in her own family.

At the time of this writing, we are close to full integration with most personalities and fragments having thus far fused. We expect this process will have been completed by the publication of this manuscript. The need for therapy will continue in order to help Kate adjust to her new life as the one unified person she was born to be.